ANALYSIS OF STRUCTURES

George Washington Bridge, spanning the Hudson River between New Jersey and New York (courtesy The Port Authority of New York and New Jersey).

WILEY

JOHN WILEY & SONS

New York   Chichester   Brisbane   Toronto   Singapore

2ND EDITION

ANALYSIS

OF

STRUCTURES

AN INTEGRATION OF CLASSICAL AND MODERN METHODS

HARRY H. WEST

Professor of Civil Engineering
The Pennsylvania State University

Library of Congress Cataloging-in-Publication Data:

West, Harry H., 1936-
Analysis of structures : an integration of classical and modern
methods / Harry H. West. -- 2nd ed.

    p.   cm.

Includes bibliographies and index.
ISBN 0-471-82949-8
1. Structures, Theory of.   I. Title.
TA645.W43 1988
624.1'7--dc19   88-5897
                          CIP

Printed in the United States of America

10 9 8 7 6 5 4 3 2 1

To Laurie

Give her the reward she has earned,
    and let her works bring her praise at the city gate.

Proverbs 31:31

ABOUT THE AUTHOR

Harry H. West is professor of civil engineering at The Pennsylvania State University where he has been engaged in structural research and teaching for 25 of the last 30 years. His research and teaching have encompassed a broad range of topics in structural analysis and design. Dr. West's research on suspension bridges was cited when he received the 1970 Moissieff Award of the American Society of Civil Engineers. His teaching was also recognized when he received the 1975 Excellent Teaching Award of the Penn State Engineering Society, the 1977 Western Electric Fund Award for Excellence in Teaching, and the 1986 Premier Teaching Award of the College of Engineering. Dr. West is a registered professional engineer in Pennsylvania and is a member of the American Society of Civil Engineers and the American Society for Engineering Education. He earned his Ph.D. in civil engineering (structures) from the University of Illinois after completing both his B.S. and M.S. degrees at Penn State.

ing problem for statically indeterminate structures. In addition to the classical formulation, a matrix approach is developed and applied.

Part III contains the fully generalized matrix methods. Force–displacement relations are introduced in Chapter 16, and the compatibility and equilibrium methods are developed in the basic elemental form as the modern flexibility and stiffness methods, respectively, in Chapters 17 and 18. These techniques are more general than the system formulations of Part II, but it is felt that the student is thoroughly prepared for this extension based on the foundational material of the earlier parts of the text.

One beauty of the modern matrix approaches is that the distinction between statically determinate and indeterminate structures is not really necessary. The separation is retained here, however, in order to maintain the tie with classical methods, where the distinction is important. In addition, there are behavioral differences between the two classes of structures that the analyst must understand.

The decision was made not to include formal instruction in matrix algebra in the book. Most engineering curricula require a course in linear algebra during the first two years of a student's schooling, and this book would typically be used in junior- and senior-level courses. It is recommended that the instructor review some of the basic matrix operations as they are needed; however, it has been my experience that students are fairly well prepared to handle the elementary concepts that are needed in this area.

Formal instruction in computer programming has also been excluded. There are several suggested problem assignments that call for the development of computer programs; however, these are given in broad conceptual terms. In using these problems, an instructor should designate the particulars for the assignment: size of structure to be considered (number of joints and/or members), structural configuration (continuous beam, multiple bay/story frame), suggested subroutine for the solution of simultaneous equations, etc.

As the second edition is published, the status of SI units remains uncertain. In some disciplines the conversion from conventional units has moved ahead, but structural engineering is not on the leading edge of this change. Both systems are used in this book. The conventional system is used to keep the student in touch with current practice, and the SI units are employed to prepare students for the probable transition. The emphasis is on development of familiarity with each set of units and not on training for a mere conversion capability. It is strongly urged that the student develop a dual capability in this area because structural engineers are going to be forced to operate in both domains for many years. There is a brief treatment of the SI system in the appendix for students who are not familiar with its use.

Although this textbook has been prepared to integrate the classical and modern methods of analysis, an instructor is not limited to this use of it. The book is organized so that a course may readily be taught in the classical methods simply by selecting the appropriate sections of Chapters 1 through 15. Conversely, greater emphasis may be placed on the matrix methods by omitting the sections that emphasize the applications of the classical methods in Chapters 2 through 15 and stressing Chapters 16 through 18.

The text is ideally suited for a two-semester sequence in structural analysis, which is a luxury that fewer and fewer instructors have. However, it may be used effectively in a single-semester course by selecting the chapters that are consistent with the goals of the course.

In preparing the manuscript, I have tried to develop a strong teaching textbook. It is fundamental in nature, and no effort has been made to include all that can be written on structural analysis. My goal was to provide a tool for integrating and teaching fundamental concepts rather than to prepare a reference text for practicing engineers.

It is my hope that this text will be an asset in the hands of both those who desire to teach and those who will assemble to learn.

State College, Pa
                                         Harry H. West

ACKNOWLEDGMENTS

Any work is dependent on the contributions of many; such is the case here. Those acknowledged in the first edition deserve recognition again because that volume served as the foundation for this one. The contributions of those cited there—colleagues, students, former professors, and family—are acknowledged with thanks. In addition, I wish to recognize the suggestions of users of the first edition, many of which have been included in this edition, and the faithful efforts of Debby Pruger, who typed new portions of the manuscript. Also, I reiterate my appreciation for the loving encouragement and assistance of my wife, Laurie, who once again utilized her skills in journalism to fine-tune the manuscript. And, thanks be to God, The Structural Engineer, Who has provided so many examples of structural form and beauty in His Creation and Who reminds us in His Word that the "wise man . . . built his house on a rock."

CONTENTS

PART II: STATICALLY INDETERMINATE STRUCTURES   351

9. FUNDAMENTALS OF STATICALLY INDETERMINATE ANALYSIS   353

10. ANALYSIS OF PIN-CONNECTED FRAMEWORKS BY COMPATIBILITY METHODS   371

11. ANALYSIS OF BEAM AND FRAME STRUCTURES BY COMPATIBILITY METHODS   410

12. ANALYSIS OF PIN-CONNECTED FRAMEWORKS BY EQUILIBRIUM METHODS   445

13. ANALYSIS OF BEAM AND FRAME STRUCTURES BY EQUILIBRIUM METHODS—DIRECT SOLUTIONS   481

ANALYSIS OF STRUCTURES

PART 1

BASIC CONCEPTS AND STATICALLY DETERMINATE STRUCTURES

View of Midtown Manhattan, New York City, featuring the Empire State Building, the Pan Am Building, and the Chrysler Building (courtesy New York Convention and Visitors Bureau).

INTRODUCTION

1.1 Structure

The word *structure* describes much of what is seen in nature. Living plants, from the frailest of ferns to the most rugged of trees, possess a structural form consistent with their needs. In each case, as can be seen in Fig. 1.1, the plant is the recipient of a structure—a gift of Providence—that supports its life. Insects and animals play a more active role in building the structures that they need. The delicate web of the spider, the vastly complicated societal complex of the termite,

Fig. 1.1 *Plant structures.* (a) *Fern (photo by Lew Sheckler).* (b) *Shrub with ribbed branches (photo by Michael Houtz).* (c) *Oak tree (photo by Harry G. West).* (d) *American elm trees (photo by Harry G. West).*

the carefully articulated dam or lodge of the beaver—each structure is fabricated to support the creature's activities. However, despite the intricacies of these structures, as can be seen in Fig. 1.2, the evidence clearly indicates that these creatures build from instinct and not from design.

Humans, too, are builders of structures; but more than that, they are conceivers and designers. If we suppose that the first structure was a tree that conveniently fell across a chasm and was subsequently used as a bridge, then since that meager and accidental beginning, humans have indeed advanced in their ability to design and build structures. When the structures of humans began to reflect their ability to conceive and design them as well as to construct them, structural engineering was born, and it has grown in sophistication as it has endeavored to meet the demands of humanity.

These demands may be highly functional and related to the basic needs of society, or they may be cosmetic and, therefore, related to the aesthetic or emo-

(a)

Fig. 1.2 *Structures of animals and insects. (a) The structure of a spider (courtesy of Richard B. Mansfield). (b) The structure of the compass termite (Australian Information Service Photograph). (c) Beaver dam (foreground) and lodge (background) (Neg. No. 238208 (Photo by B.M. DeCon) Courtesy Department of Library Services, American Museum of Natural History).*

(b)

(c)

tional sensitivities of humans. These two extremes are illustrated in Figs. 1.3 and 1.4, in which the basic functionality of the material handling structure is contrasted against the symbolic arch monument. Of course, most structures fall between these two extremes in that they serve a specific function and yet are designed in harmony with the aesthetic sensitivities of humanity. Most bridges and

Fig. 1.3 *Stacker/reclaimer handling coal (courtesy Dravo Corporation).*

Fig. 1.4 *The Jefferson National Gateway Arch, St. Louis, Mo. (courtesy Pittsburgh-Des Moines Steel Company).*

Fig. 1.5 *Sunshine Skyway Bridge, Tampa Bay, Florida (courtesy Figg and Muller Engineers, Inc.).*

Fig. 1.6 *Dulles International Airport Terminal, Washington, D.C. (courtesy Bethlehem Steel Corporation).*

buildings are in this category, and Figs. 1.5 and 1.6 show two classic cases in which the structure is highly functional and yet aesthetically pleasing. Each of the structures cited above, designed to meet its own set of demands, is the product of structural engineering.

1.2 Structural Engineering

The area of structural design might casually be associated with either science or engineering; however, there are important differences between the roles played by these two disciplines. These differences are perhaps best summarized by noting that science involves the investigation of what exists, whereas engineering engages in synthesis, to form that which does not exist. Although engineering requires the intelligent application of scientific principles, the creative nature of the discipline makes it an art. The list of modifiers that prefix the term engineering has become long over the years, as the number of disciplines to which the engineering approach is applied has grown.

Structural engineering centers about the conception, design, and construction of the structural systems that are needed in support of human activities. Although structural engineering is most directly associated with civil engineering, it interfaces with any engineering discipline that requires a structural system or component in meeting its objectives. Specific projects that involve structural engineering include bridges, buildings, dams, transportation facilities, liquid or gas storage and transmission facilities, power generation and transmission units, water and sewage treatment plants, industrial factories and plants, vehicular frames, and machine components. Each of these projects requires structural systems or components that must be conceived to meet the needs for which they are being built, designed to safely and serviceably carry the loads that will impinge upon them, and constructed to provide a final product consistent with the conception and design.

1.3 History of Structural Engineering

The evolution of structural engineering to its present form involved the development of several individual areas of endeavor. These were: the development of the theories of mechanics of materials and structural analysis; the formulation of the computational techniques necessary to solve the governing equations of these theories; the introduction of new building materials; the application of the theories and materials to the creation of new structural forms; and the inventive development of construction techniques. Some of these areas required the analytic talents of the mathematician, scientist, or engineer, whereas others required the daring and artistic skills of the entrepreneur or builder. Although each of these areas has its own historical chronology, their juxtaposition shows how the individual areas are nested—how a development in one area sparked a need and, therefore, a subsequent development in another area. However, the presentation of a complete history of structural engineering is not the purpose of this section. Such treatments are referenced in Section 1.11. The intent, instead, is to provide a glimpse of the skeleton of such a history. A few of the people and events that serve as benchmarks will be noted—especially some of those that are related directly to the topics in this book.

A beginning point for structural engineering as we presently view it would be at about 500 B.C. From that point until the time of Christ, the Greeks primarily used stone to build post and lintel structures, that is, structures whose columns supported short beams. An example of this form is shown in Fig. 1.7a. Even though experience and empirical rules formed the basis of this structural activity, Aristotle (384–322 B.C.) and Archimedes (287–212 B.C.) were establishing the beginnings of the principles of statics. Although some metals and wood were

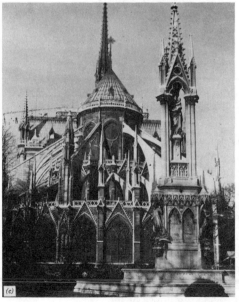

Fig. 1.7 *Notable historical structures. (a) Greek Parthenon (438 B.C.)—an example of post and lintel construction (courtesy of Greek National Tourist Organization). (b) Pont du Gard (13 B.C.)—a Roman aqueduct in southern France (courtesy of French Government Tourist Office). (c) The Cathedral of Notre Dame, Paris (13th century)—an example of Gothic buttress system (courtesy of French Government Tourist Office). (d) Coalbrookdale Iron Bridge over Severn at Shropshire, England (1776–1779)—first use of iron on a major scale (courtesy of British Tourist Authority).*

introduced, stone and masonry continued as the primary building material of the Romans until about A.D. 500. They introduced new structural forms such as the arch, vault, dome, and even the wooden truss. Some of these structures, such as the aqueduct shown in Fig. 1.7b, remain with us today as monuments of that era. However, the Romans were not analytic in their approach but rather were builders who concentrated on certain structural forms.

During the Middle Ages (500–1500), much of what the Greeks and Romans had developed was lost. The only major structural accomplishment during this time was achieved by the Gothic builders; witness their splendid cathedrals, characterized by pointed arches stabilized by "flying buttresses." Figure 1.7c provides an example of such a cathedral.

Following the inactivity of the Middle Ages, the Renaissance saw new impetus in many areas, including structural engineering. In the early part of this period, Leonardo da Vinci (1452–1519) formulated the beginning structural theory. However, Galileo (1564–1642), who published *Two New Sciences,* is generally credited with originating the mechanics of materials. He studied the failure of a cantilever beam, and even though his writings were not wholly correct, they did establish an important beginning. Europe was rife with the activity of rebirth, which spawned more than a few significant analytic accomplishments. The most important of these for our purposes were those of A. Pallidio (1518–1580), who introduced the modern truss; R. Hooke (1635–1703), who established the law governing the linear behavior of materials; Johann Bernoulli (1667–1748), who stated the principle of virtual displacements; Daniel Bernoulli (1700–1782), who contributed to the understanding of elastic curves and the strain energy of flexure; Leonard Euler (1707–1783), who examined column buckling and energy methods; and Louis Navier (1785–1836), who followed up the earlier work of C. A. de Coulomb (1736–1806) and published a book on strength of materials which dealt with the elastic analysis of beam flexure.

These accomplishments were paralleled by new developments in building and construction materials. Timber was used by German and Swiss engineers to construct bridges up to 300 feet long. And iron arrived with revolutionary impact. As a material, it exhibited elastic properties much better than those of wood or stone, and thus the new theories could be applied to enable more daring structural forms to be used with confidence. A whole host of "firsts" in both form and dimension followed—cast iron arch bridges (Fig. 1.7d), iron trusses, suspension bridges, etc.

However, the golden age of structural engineering is considered to be 1800–1900. During this period, most of the present-day theories of mechanics of materials and structural analysis were developed. A few relevant developments are as follows: S. Whipple (1804–1888), K. Culmann (1821–1881), and J. W. Schwedler (1823–1894) formulated the principles of statically determinate trusses; B. P. E. Clapyron (1799–1864) established the three-moment equation; J. C. Maxwell (1831–1879) developed the method of consistent displacements and the reciprocal theorem of deflections; O. Mohr (1835–1918) presented the method of elastic weights and worked on influence lines; A. Castigliano (1847–1884) stated the theorems that would carry his name; C. E. Greene (1842–1903) formulated the moment–area method; H. Müller-Breslau (1851–1925) published his principle for influence line construction; and A. Föppl (1854–1924) worked in the area of space frame analysis.

The golden age also saw new materials on the scene. Portland cement appeared early in the 1800s, and the first reinforced concrete bridge was constructed before the end of that century. Iron rolling mills made iron more usable, and quantity steel production was introduced by H. Bessemer. These developments led to new structural forms. In fact, the theoretical developments of the mid-1800s paved the way for the analysis of continuous beams and frames, and these forms grew in popularity by the early 1900s.

The twentieth century brought in some modest advancements in structural theory and some significant developments in solution techniques. A few are as follows: G. Maney (1888–1947) introduced the slope deflection method, which was the forerunner of modern displacement methods; H. Cross (1885–1959) contributed the moment distribution method, and R. Southwell (1888–1970) presented the more general relaxation methods (these two developments allowing systematic solution of statistically indeterminate structures and serving as the cornerstone of frame analysis for a quarter of a century); and several analysts contributed to the merging of matrix algebra and frame and continuum analysis to form the modern matrix and finite element methods of analysis. At the same time, the areas of inelastic analysis and strength methods were introduced.

The 1900s also saw a host of new materials, techniques, and structural forms introduced. Material developments brought forth aluminum, high-strength steels and concretes, special cements, plastics, laminated timber, and composites. Developments in technique include the introduction of experimental research, the use of electric welding and prestressed concrete, and the development of improved construction methods; however, the event that had the greatest impact was the introduction of electronic computation in the 1950s. New advances in structural form included the perfection of long-span bridges of many configurations, record-breaking heights in buildings, and newer forms such as shells, panels, and stress-skin structures.

1.4

The Engineering Design Process

The *engineering design process* encompasses much more than structural design. Although the primary role of the structural engineer is in structural design, he or she is necessarily enmeshed in the entire design process. This is illustrated by the following breakdown of the engineering process as it is related to a typical civil engineering project in which a structural engineer is involved.

Conceptual Stage. Any engineering project must be directed toward the satisfaction of a unique set of objectives. During the conceptual or planning stage, the specific needs are identified and the objectives are carefully articulated to meet these needs. These objectives must be consistent with the desires of the client and the interests of other involved parties.

This stage requires input from the client, architects, planners, the public as represented by elected officials, governmental regulatory agencies or civic organizations, and the engineer. During this stage, the engineer frequently serves as a resource person regarding the engineering feasibility and the economic soundness of the various alternatives under consideration. However, the engineer should not hesitate to state opinions on the aesthetic or environmental impact that the alternatives might have. Too frequently, engineers have remained silent on matters outside the purview of their narrow specialty, and they have come under sharp criticism for that silence.

The conceptual stage should bring forth a plan that maximizes the satisfaction of the stated objectives while minimizing any objectionable features of the project.

Preliminary Design Stage. The plan that emerges from the conceptual stage frequently includes several alternatives that are to be investigated through the preparation of individual preliminary designs.

Fig. 1.8 *Structures during fabrication. (a) New River Gorge Arch Bridge under construction, Ansted, W. Va. (courtesy of American Bridge Div. of U.S. Steel). (b) Greater New Orleans Mississippi River Bridge No. 2 during fabrication (courtesy Modjeski and Masters, Consulting Engineers).*

The preliminary designs are of vital importance, and the structural engineer plays a central role during this stage. Key decisions regarding the positioning of the structure, the structural form to be used, and the manner in which the structural components are to be connected will have a bearing on the final design. It is during this stage that the creative talent of the engineer is vitally important as he or she considers the options brought forward from the conceptual stage. Yet, the engineer must keep in mind that the structure he or she designs has to be built, and thus the construction and fabrication aspects must be carefully considered. In many cases, the most severe loading conditions occur during fabrication. Figures 1.8a and 1.8b show the erection of a large arch structure and a cantilever truss, respectively. The loading on these partially erected structures is vastly different from the loading that the final structures will support.

Each preliminary design involves a thorough consideration of the loads and actions that the structure will have to support, including the conditions that will occur during fabrication. For each case, a structural analysis is necessary—that is, the forces and deformations throughout the structure must be determined. It is this area, the structural analysis of the system, that is examined in detail in this textbook.

Frequently, the preliminary designs are based on approximate theories of structural analysis in order to minimize the time and effort invested in the preliminary phase. This phase must produce sufficient detail so that intelligent decisions can be made in the final selection of one of the alternatives that was proposed in the conceptual stage.

Selection Stage. Once the preliminary designs are completed, a selection must be made. At this point, the parties involved in the conceptual stage are reconvened so that they may participate in the selection process. The prime consideration centers on how each alternative satisfies the original objectives. Again, due consideration is given to any objectionable features of the alternatives.

The structural engineer is concerned at this stage with the relative economies of the alternatives, the impact that any unique features of each alternative might have on the structural behavior or practicality of construction, and any other areas related to the decision.

The result of this stage is usually a decision to proceed with one of the alternatives for which a preliminary design has been prepared. A growing trend is to carry two competing designs forward to the final design stage. Here, a final decision is delayed until after bid prices are established for the two alternatives. However, our discussion will proceed on the basis of the selection of a single design.

In some cases, all preliminary designs are rejected, and a return to the preliminary design stage is in order. However, before proceeding to the next stage, a selection among the alternatives is necessary.

Final Design Stage. The results of the preliminary design stage constitute a starting point for the final design stage; however, the structural engineer must proceed with greater care from this point. Here the loads are determined with greater accuracy than was necessary during preliminary design, and all plausible loading conditions and combinations must be considered. The structural analysis that is required for this stage must be carried out with great precision, and the approximations of the preliminary design stage must be eliminated. Again, it is this analysis process that is the central concern of this book. Each member is proportioned and the connections are detailed to ensure that the structure will behave in accordance with the assumptions made in the structural analysis.

The results of the final design stage are capsuled in a set of complete design drawings, which give a graphic portrayal of the details of the entire system. These are generally accompanied by written specifications that stipulate the materials to be used, the quality of workmanship, the pertinent codes to be employed, and many other items.

Construction Stage. The goal of this stage is to bring into existence that which was described in the final design stage. The completed documents of the final design stage serve as the basis for bidding by the prospective building contractors. The successful bidder frequently prepares additional drawings related to the fabrication of the structure. The bidder must have on staff competent structural engineers who can work closely with those who designed the structure. The role of the structural engineer is vital here, as is evidenced by the large number of structural failures that occur during construction. In fact, some of the most challenging problems in structural analysis are related to fabrication and erection.

The above breakdown is illustrated by the flow diagram in Fig. 1.9, in which the areas where structural analysis is employed are highlighted. This discussion is related specifically to the role of the structural engineer in a civil engineering project. The procedure is not followed in all engineering disciplines. Even in civil

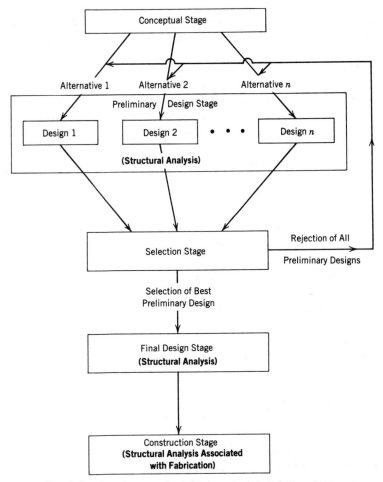

Fig. 1.9 *Flow diagram describing engineering design process.*

engineering projects, where the design and construction are done by a single organization, the sequential operations may be different. However, the key ingredients of conception, preliminary design, final design, and construction are present in any engineering design process.

1.5

Structural Analysis

Structural analysis is the process by which the structural engineer determines the response of a structure to specified loads or actions. This response is usually measured by establishing the forces and deformations throughout the structure. A given method of structural analysis is commonly expressed as a mathematical algorithm; however, it is based on information gained through the application of engineering mechanics theory, laboratory research, model and field experimentation, experience, and engineering judgment.

The earliest demands for sophisticated analysis, coupled with some serious limitations on computational capability, led to a host of special techniques for solving a corresponding set of special problems. These so-called *classical methods* incorporated some ingenious innovations and served the needs of the structural engineer very well for many years. However, the advent and subsequent development of the digital computer increased computational capabilities by several orders of magnitude and thus obviated the need for special techniques. The ingenious specializations of the classical methods were replaced by the sweeping generalities of the *modern matrix methods*.

The transition from the classical methods to the modern methods has triggered some revolutionary changes in structural engineering and in the education of structural engineers. Although the matrix methods have become the foundation of modern structural analysis as it is employed in the practice of structural engineering, it is not completely clear what the role of the classical methods will be—especially in the education process.

For reasons explained in the preface, the author has selected an integrated presentation for this textbook. The classical methods are used to introduce the fundamental concepts and are employed for the solution of small problems. These methods are then set in a matrix format to help the student adjust to the generalities of matrix formulations. Eventually, the generalized modern matrix methods are developed. This approach is intended to equip the student with the knowledge and understanding to proceed with hand calculations on small problems by the classical methods or to employ the modern computer methods of structural analysis on large problems.

By either classical or matrix methods, the analysis process can be a part of preliminary design, final design, or construction, as was described in the preceding section. However, it is important to note that structural analysis plays a limited role in the structural design process and an even smaller role in the overall design process. Furthermore, the role that it plays is entirely supportive of the design process. That is, structural analysis is not an end in itself. It is particularly important to understand the supportive nature of structural analysis as one studies the subject. It is easy for an engineering student to become enamored of this intriguing subject to the point of aspiring to become a structural analyst. However, such a goal is unrealistic. Good structural engineers are necessarily good structural analysts, and they will use their analytic ability intelligently as they fulfill their primary responsibility as structural engineers.

The form that a structure is to take is dependent on many considerations. Fre- **Structural**
quently, the functional requirements of a structure will narrow the possible forms **Form**
that can be considered. Other factors such as the aesthetic requirements, foundation conditions, availability of materials, and economic limitations may play important roles in establishing the structural form. The structural forms that are
available, along with their respective features, will be discussed only briefly here.
More complete treatments are given in the references in Section 1.11. The student
is encouraged to become sensitive to the various types of structures that appear in
nature or have been built by human beings; this discussion is intended to promote
the student's general awareness of structural form.

Tension and compression structures are composed of members that are subjected **1.6.1 Tension**
to pure tension or compression. Such an arrangement provides for highly efficient **and compression**
material usage because there is a constant stress level over the entire cross-sec- **structures**
tional area of each member. This is particularly true for tension elements, where
the stress level is limited only by the material strength coupled with an appropriate
factor of safety. Compression elements, however, are susceptible to buckling,
which can limit the allowable stress to a level lower than that dictated by material
considerations.

One of the simplest structural forms in which the elements are in pure tension
is the *cable-supported structure*. Structures of this type range from simple guyed
or stayed structures to large cable-supported bridge and roof systems. A few simple cable systems are shown in Fig. 1.10 along with their force-carrying mechanisms, and several examples of actual structures of this form are given in Fig.
1.11. Each of these structures contains cable elements that serve as tension mem-

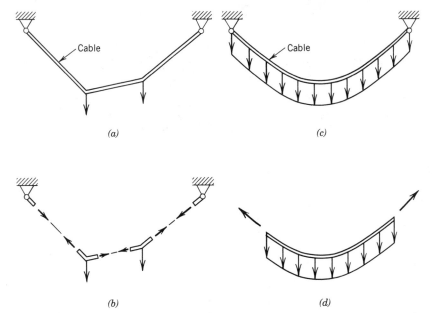

(a)   *(c)*

(b)   *(d)*

Fig. 1.10 *Force-carrying mechanisms for simple tension structures.*
(a *and* b) *Concentrated loads on cable.* (c *and* d) *Uniform load on cable.*

Fig. 1.11 *Typical tension structures.* (a) *Pipeline suspension bridge, Portsmouth, Ohio (courtesy Bethlehem Steel Corporation).* (b) *Radar–radio telescope, Arecibo, Puerto Rico (courtesy Bethlehem Steel Corporation).* (c) *Roof suspension system for Madison Square Garden, New York (courtesy Bethlehem Steel Corporation).* (d) *Golden Gate suspension bridge, San Francisco, Calif. (courtesy Bethlehem Steel Corporation).*

bers; some are primary members of the structure, whereas others are secondary hangers or bracing-type members.

The most common structure that carries pure compression in its primary element is the *arch*. Structures of this type, which have the configuration of an inverted cable, have a pure compressive thrust along the rib of the arch under the specific loading condition for which it was designed. Variations in this loading will introduce some bending in the arch, but compression will remain the dominant mode of action. Examples of arch action are shown in Fig. 1.12. Secondary members, which attach other portions of the structure to the arch, are compression members when they attach from above the arch or tension members when they attach from below the arch. Figure 1.13 gives several examples of structures in which the arch plays a major role. Note that the arch can be formed by a single rib or it can be built up as a trussed rib.

Of course, many structures have individual elements that carry compression. The most common of these elements is the *column,* which is a primary compressive member. These elements appear in nearly all structural forms. For instance, the towers in Fig. 1.11*a* and the vertical members between the roadway and arch in Fig. 1.13*a* are column elements.

A common structural form in which tension and compression elements are combined is the *pin-connected truss.* Here, each element carries pure tension or compression and acts in concert with other elements to form a stable structural system, as shown in Fig. 1.14. As will be seen later, true pin-connected frameworks are rare; however, under certain restrictions on member sizes and connections, and when the structure receives loads only at its joints, the tension–compression behavior is an appropriate assumption.

Truss-type structures are frequently composed of repetitive planar units that are analyzed as *planar trusses.* These are usually connected by beam systems, which transfer the loads from the slab or deck to the truss units. In other cases, the spatial relationships must be accounted for in the analysis; structures of this type are *space trusses.* Truss structures can take a variety of arrangements, as illustrated by the examples of Fig. 1.15.

1.6.2 Flexural beam and frame structures

A flexural element is one that is subjected to a bending action, as opposed to the pure tension or compression that was previously discussed. The reader will recall from mechanics that bending induces compression on one side of the element and tension on the other side and that a transverse shear may also be present in the member. The simplest structural member that displays this mode of response is

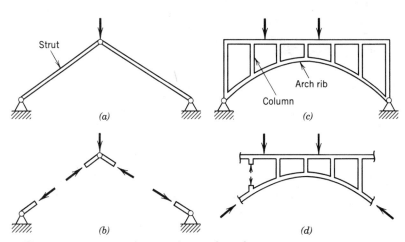

Fig. 1.12 *Force-carrying mechanisms for simple arch-type structures.*
(a *and* b) *Concentrated loads on simple strut structure.* (c *and* d) *Arch loaded by columns from above.*

Fig. 1.13 *Typical compression structures. (a) Concrete arch bridge, Butler, Pa.
(courtesy Portland Cement Association.). (b) New River Gorge steel arch bridge, Ansted,
W. Va. (courtesy American Bridge Div. of U.S. Steel). (c) Fort Pitt Bridge, Pittsburgh,
Pa. (courtesy Richardson, Gordon and Associates, Consulting Engineers). (d) Laminated
wood arches (courtesy American Institute of Timber Construction).*

the *beam* element, which is shown in Fig. 1.16. Flexural *frame structures* are
formed from a combination of beam and column elements, some of which are
subjected to pure flexure while others carry a combination of flexure and tension
or compression. A typical frame structure is shown in Fig. 1.17.

Structures that are composed of flexural elements can receive along their
member lengths loads that are directed normal to the orientation of the member.
They also are distinguished from pin-connected trusses by having flexural-resistant

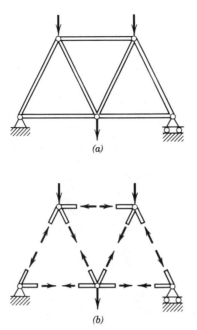

Fig. 1.14 (a) *Simple truss.* (b) *Load-carrying mechanism for a truss.*

connections that ensure continuity at the member ends. The concept of continuity is important when dealing with flexural beam and frame structures. This topic is discussed fully in Chapter 9.

Structures of this type can take a variety of configurations, a few of which are shown in Fig. 1.18. As was true with truss structures, typical building or bridge frames may be composed of repetitive planar units that are tied together by secondary systems. Figure 1.19 illustrates this arrangement for a highway bridge. The view here is from the underside of the structure. There are four main *girders* that span longitudinally between the *bents* that serve as the bridge piers. The transverse members that connect a pair of girders are called *floorbeams,* and the small longitudinal members that are placed atop the floorbeams are *stringers.* The floorbeam–stringer combination serves to transfer the applied loads from the deck to the girders, which are the primary planar units of the structure. The cross *bracing* between floorbeams is composed of tension and compression members to resist the lateral loadings caused by the wind. There are cases where repetitive planar units are not employed. In such cases, the full spatial assemblages must be taken into account.

1.6.3 Surfaces structures

The structures that have been considered thus far are composed of individual elements, each of which carries pure tension or compression, pure flexure, or a combination of flexure coupled with tension or compression. These elements are connected to form a skeletal structure in space that serves to support itself and everything that is attached to it.

Surface structures derive their spatial configuration through continuous three-dimensional surfaces, and the loads are resisted by the surfaces themselves. These structures carry tension, compression, and in-plane shear within the surface as

Fig. 1.15 *Typical truss-type structures.* (a) *Truss bridge across Ohio River, St. Marys, W. Va. (courtesy American Bridge Div. of U.S. Steel).* (b) *Space frame scaffold structure (courtesy of The Statue of Liberty and Ellis Island Foundation, Inc.).* (c) *Example of open-web steel joist trusses (courtesy Bethlehem Steel Corporation).* (d) *Electrical transmission tower under erection—both tower and crane boom are truss-type structures (courtesy Bethlehem Steel Corporation).*

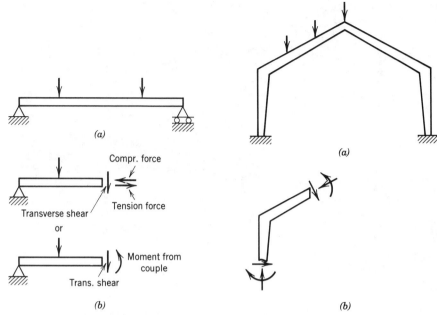

Fig. 1.16 (a) *Beam element.* (b) *Internal forces in a beam.*

Fig. 1.17 (a) *Frame structure.* (b) *Internal forces in a frame.*

membrane forces. Bending and transverse shear are carried either normal to or within the surfaces, depending on the loading and the surface orientation.

Structures that fall into the category of surface structures are *slabs, folded plates, shells, domes, skin-type structures,* and *inflatable membranes.* This class of structures provides some of the most efficient structural systems in terms of material usage, and this leads to inherent economies. Another major advantage of surface structures is that they give the designer tremendous flexibility to create an aesthetically pleasing structure. The manner in which a folded plate, a floor slab, and a thin shell support gravity loading is shown in Fig. 1.20, and Fig. 1.21 illustrates some typical examples of surface structures.

1.6.4 Closing notes on structural form

Most structures are composed of a combination of several structural forms in which each substructure serves in a unique way and the combination meets the functional objectives required of the structure. For instance, consider the Golden Gate Bridge, which is shown in Fig. 1.11*d*. The main cable and the hangers form a structural system of pure tension elements; the deck is a truss-type structure that stiffens the roadway and is referred to as the stiffening truss; the near approach span is supported by a trussed frame that contains a trussed arch in its low portion; the towers are frame structures that carry massive compressive forces coupled with some bending and torsion.

The structure of Fig. 1.15*c* is a second example of how different structural forms are used in a single structure. Here the roof surface is to be supported by the steel roof trusses; the roof trusses are supported by a system of beams and girders, which transfer their loads through flexural action to the columns; the columns act primarily as compression elements by carrying their loads to the foundation.

Fig. 1.18 *Typical beam and frame structures. (a) Slant-leg frame bridge, Charlottesville, Va. (courtesy Bethlehem Steel Corporation). (b) Continuous plate-girder bridge over Quinnipiac River, New Haven, Conn. (courtesy Steinman, Boynton, Gronquist and Birdsall, Consulting Engineers). (c) Typical rigid frame construction (courtesy Lincoln Electric Company, Cleveland, Ohio). (d) Moment-resistant frame structure, Dresser Tower, Houston, Texas (courtesy Bethlehem Steel Corporation).*

This textbook considers the analysis of the first two structural forms: tension and compression structures and flexural beam and frame structures. In these cases, the structure is composed of discrete structural elements. The third structural form is not treated in this textbook. It is much more difficult to analyze because the surface geometry and the three-dimensional material properties must be taken into account. Modern computer methods discretize these structures into a system of

Fig. 1.19 *Girder–floorbeam–stringer system for typical highway bridge (courtesy American Bridge Div. of U.S. Steel).*

individual elements, which can then be treated by an extension of the concepts that are used in frame analysis. This approach is referred to as *finite element analysis*.

1.7 Simplifications for Purposes of Analysis

A careful study of the photographs in the previous section reveals that the makeup of a structural system is rife with details concerning the individual members and their interconnection. In most cases, the overall structural analysis, which establishes the member forces and deformations, will ignore much of this detail. This

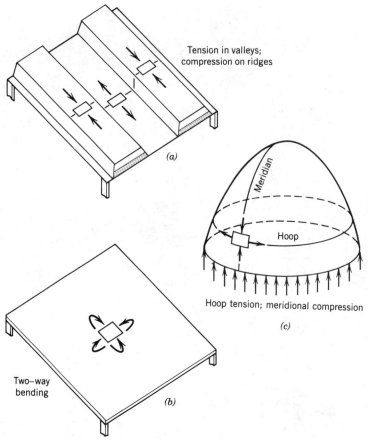

Fig. 1.20 *Surface structures under gravity loading.* (a) *Folded plate.* (b) *Floor slab.* (c) *Thin shell.*

is done to simplify the analysis; however, the process of simplification must be tempered with judgment to ensure that the results do not depart drastically from those in the real structure. There is no substitute for experience in establishing the limits in the simplification process.

For common truss- and frame-type structures, the member lengths are appreciably greater than their traverse dimensions. Also, the dimensions of the connections are usually small when compared with the member lengths. In such cases, the model that is adopted for purposes of analysis will be composed of line elements between the joints. This model is referred to as a *line diagram*. The lengths of the line elements are determined by the distances between the joint centers, which, in turn, are established by the intersections of the centroidal axes of the members that frame into a joint.

The member connections for truss-type structures are taken as frictionless pins with no moment-resisting capacity. For frame structures, rigid connections are assumed in which all member ends that frame into a joint have a common rotation. The support conditions are also idealized as frictionless pins or rollers, or points of complete fixity. More is said about the simplifications of member connections and supports in later chapters.

For cases where the members are nonprismatic—that is, their cross-sectional

Fig. 1.21 *Typical surface structures.* (a) *Folded plate roof structure (courtesy Portland Cement Association).* (b) *Hyperbolic paraboloid roof structure (courtesy Portland Cement Association).* (c) *Concrete dome, folded plate roof structure; University of Illinois Assembly Hall, Champaign, Ill. (courtesy Portland Cement Association).* (d) *Membrane roof supported by air pressure, Pontiac Stadium, Pontiac, Mich. (courtesy Bethlehem Steel Corporation).*

Fig. 1.21 (cont.) (e) *Floor slab system (courtesy of Portland Cement Association).* (f) *St. Louis Priory Chapel, system of parabolic shells (courtesy of Portland Cement Association).*

characteristics change along their lengths—these variations eventually must be taken into account in establishing the member characteristics. However, these refinements are frequently ignored in the early stages of analysis.

A casual review of real structures shows that they are three-dimensional assemblages of members. However, in many cases there are planar components of the structure that essentially function as independent units and can thus be isolated and analyzed as planar structures. There are, of course, cases where this is not possible or where the final analysis will require a full three-dimensional analysis, but there are numerous cases where the planar analysis of substructures is in order.

As an example of a simplification for analysis, consider the truss structure of Fig. 1.15a. The initial step in the simplification is to consider the structure to be composed of two longitudinal trusses that are connected by transverse-trussed frames. In representing the longitudinal trusses, the camber (slight curvature of the roadway) is ignored, the members are represented by straight-line elements that are located along their respective longitudinal axes, and pins are assumed to connect the members at their ends. The supports are taken as either pins or rollers. The effect of these assumptions leads to the representation shown in Fig. 1.22. This figure shows only part of the truss, but the effects of the approximation are clear.

As a second example of how a structure is simplified for analysis, consider the slant-leg frame structure of Fig 1.18a. This bridge is composed of five parallel

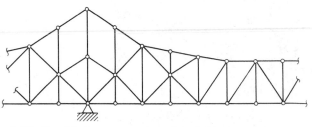

Fig. 1.22 *Simplified representation of the bridge truss of Fig. 1.15a.*

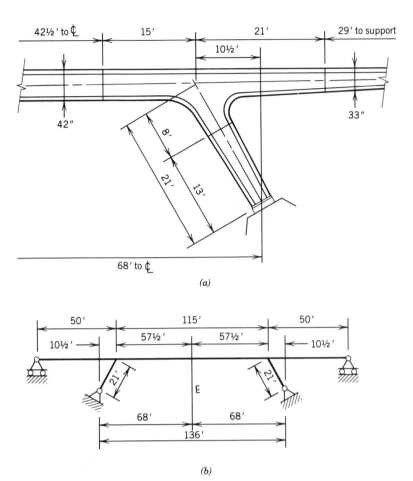

Fig. 1.23 *Simplified representation of the slant-leg frame bridge of Fig. 1.18a.*
(a) *Connection detail.* (b) *Simplified model for analysis.*

frames, and a separate planar analysis of each is appropriate. The details of the connection between the inclined column and the girders are shown in Fig. 1.23*a*. This structure can be modeled as a simplified frame, with the dimensions shown in Fig. 1.23*b*. The members are assumed to be rigidly connected, and the supports are simplified in the manner indicated.

The internal member forces that correspond to the points where the members interface with the haunched connections can be taken as the forces on the connection detail for an analysis of the connection itself. Also, it is possible to account for the member haunching (increase in depth) and the column taper if such refinements are desired.

1.8

Loading Conditions

Establishment of the loads that act on a structure is one of the most difficult and yet important steps in the overall process of design. The computer has made it possible to analyze structures that could not be analyzed a mere decade ago, but the accuracy of the results of the analysis is directly dependent on the accuracy of the loads used.

The loads that enter a system are of three different types. *Concentrated loads* are those that are applied over a relatively small area. A single vehicular wheel load and a load that one member transfers to another member are examples of concentrated loads. *Line loads* are distributed along a narrow strip of the structure. The weight of a member itself and the weight of a wall or partition are examples of this type of load. *Surface loads* are loads that are distributed over an area. The loads on a warehouse floor and the snow load on a roof are examples of surface loads.

The loads that act on a structure can be grouped according to three categories: *dead loads, live loads,* and *environmental loads.* These categories can be further divided according to the specific nature of the loading. Because the method of analysis is the same for each category of loading, all loads could be combined before the analysis is performed. However, separate analyses for the individual loading cases are usually carried out to facilitate the consideration of various combined loadings.

1.8.1 Dead loads

Dead loads are those that act on the structure as a result of the weight of the structure itself and of the components of the system that are permanent fixtures. As a result, dead loads are characterized as having fixed magnitudes and positions. Examples of dead loads are the weights of the structural members themselves, such as beams and columns, the weights of roof surfaces, floor slabs, ceilings, or permanent partitions, and the weights of fixed service equipment.

The dead loads associated with the structure can be determined if the materials and sizes of the various components are known. Standard material unit weights, such as those given in Table 1.1, are used for calculating these dead loads. A complete list of minimum design loads for various building components, such as typical flooring, ceiling, and roofing materials, is given in applicable codes and standards. Some of these references are listed in Section 1.11. The loads associated with service equipment are normally available from the manufacturers.

Whether these loads are treated as concentrated loads, line loads, or surface loads depends on the application of the loads and the arrangement of the structural members. Sometimes secondary structural analyses are necessary to determine the loads that are transferred from a substructure to a primary structure.

Accounting for the dead loads in design requires an iterative procedure be-

TABLE 1.1 Unit Weights of Typical Building Materials

Material	(lb/ft^3)	$(kN/m^3)^a$
Aluminum	165	25.9
Brick	120	18.9
Concrete		
Reinforced with stone aggregate	150	23.6
Block, 60 percent void	87	13.7
Steel, rolled	490	77.0
Wood		
Fir	32–44	5.0–6.9
Plywood	36	5.7

[a]SI units are described in Appendix A.1.

cause the load magnitudes depend on the member sizes, which are not known prior to the design. Approximate methods are available to estimate the dead weight of a structure, and these methods, or experience gained from previous designs, can be used as a guide for preliminary design. Each subsequent cycle of the iteration uses the latest information on member sizes to establish the dead loads, and, eventually, the design will reflect member sizes that are consistent with the assumed dead loads.

1.8.2 Live loads

In a general sense, live loads are considered to include all loads on the structure that are not classified as dead loads. However, it has become common to narrow the definition of live loads to include only loads that are produced through the construction, use, or occupancy of the structure and not to include environmental or dead loads.

These loads are dynamic in character in that they are fixed in neither magnitude nor position. Live loads where the dynamic nature has significance because of the rapidity with which change in position occurs are called *moving loads,* whereas live loads in which change occurs over an extended period of time, or where there is the potential for change whether exercised or not, are referred to as *movable loads.* Moving loads include vehicular loads on bridges or crane loads in industrial buildings. Examples of movable loads are stored material in a warehouse and movable partitions in an office building. Another type of live load is a variable load or a *time-dependent load*—that is, one whose magnitude changes with time, such as a load induced through the operation of machinery.

Recommended live load magnitudes are available through a carefully assembled record of experience in the form of building codes, design specifications, and research reports. For unusual cases, where suggested values are not available, the designer must embark on an independent study to establish the loading to be used.

1.8.2.1 Occupancy loads for buildings Occupancy live loads for buildings are usually specified in terms of the minimum values that must be used for design purposes. Some representative values are given in Table 1.2. Although this listing is incomplete, it is included to give the reader some appreciation of the range of

TABLE 1.2 Minimum Uniformly Distributed
Live Loads for Building Design[a]

Occupancy or Use	(lb/ft^2)	$(kPa = kN/m^2)$[b]
Residential dwellings, apartments, hotel rooms, school classrooms	40	1.91
Offices	50	2.39
Auditoriums (fixed seats)	60	2.87
Retail stores	75–100	3.59–4.79
Bleachers	100	4.79
Library stacks	150	7.18
Heavy manufacturing and warehouses	250	11.97

[a]Data are taken from ANSI A58.1–1982.
[b]SI units are described in Appendix A.1.

loadings that is used. These values are extracted from *American National Standard Minimum Design Loads for Buildings and Other Structures,* published by the American National Standards Institute (ANSI), and the reader is directed to that document or other references in Section 1.11.

1.8.2.2 Traffic loads for bridges   Bridges must be designed to support the vehicular loads associated with their functional use, and minimum loads are suggested for design purposes. In the case of highway bridges, these loads are enumerated in *Standard Specifications for Highway Bridges,* published by the American Association of State Highway and Transportation Officials (AASHTO) and cited at the end of this chapter. The approach is to specify standard loads for placement on the bridge, such as those shown in Fig. 1.24. These loadings provide for a set

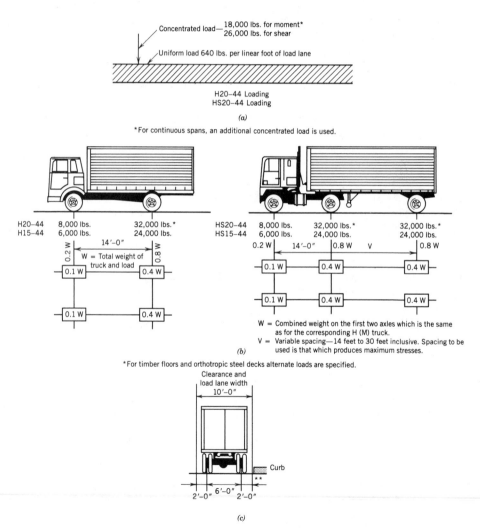

Fig. 1.24 *Typical highway bridge loading* (Standard Specifications for Highway Bridges, *Washington, D.C.: American Association of State Highway and Transportation Officials, copyright 1983. Used by permission.*) (a) *Lane loading.* (b) *Truck loading.* (c) *Lane arrangement.*

of concentrated loads, which represent either a truck or a tractor–trailer combination, or a uniform load, which simulates a line of vehicles. The appropriate set of concentrated loads or the uniform load is placed separately on the structure to produce the maximum value of a particular response quantity, and the larger of the two results is taken as the critical value of that quantity. Each response quantity of interest must be treated separately, and the critical loading arrangement must be established for each. Problems of this type are considered in detail in Chapters 7 and 15.

A similar approach is used in the design of railroad bridges. Here, the standard loading consists of a series of concentrated loads, separated by prescribed distances, followed by a uniform loading. This loading is described in *Specifications for Steel Railway Bridges,* which is cited in Section 1.11.

Bridges might also be subjected to longitudinal forces associated with the stopping action of a truck or train; or transverse lateral forces can be imparted by the wind acting on vehicles or by the centrifugal action of vehicles on a curved bridge.

1.8.2.3 Impact loads   Loads that are applied over a very short period of time have a greater effect on the structure than would occur if the same load were applied statically. The manner in which the load varies with time and the time over which the full load is placed on the structure will determine the factor by which the static response must be increased to obtain the dynamic response. This increase is normally expressed by use of an *impact factor*.

For building occupancy loads, the minimum design loads normally include adequate allowance for ordinary impact conditions. However, provisions must be made in the structural design for uses and loads that involve unusual vibration and impact forces.

One situation in which an impact factor is routinely applied is that of moving vehicular loads on a highway bridge. In this case, the movement of the vehicles is assumed to create an impact effect on the bridge. Although no effort is made to apply an analytic procedure, *Standard Specifications for Highway Bridges* requires the use of an impact factor, I, given by the relationship

$$I = \frac{50}{L + 125} \tag{1.1}$$

where L is defined as the loaded length of the span. In the form of the equation given, L is in feet, and I is not to exceed 0.3. According to this approach, the static response is multiplied by $(1.0 + I)$ to determine the combined static and dynamic response.

1.8.3 Environmental loads

There are numerous loading conditions that a structure experiences as a result of the environment in which it exists. Several of these conditions are described in the following sections.

1.8.3.1 Snow and ice loads   The procedure for establishing the static snow loads on a building is normally based on ground snow loads and an appropriate ground-to-roof conversion. A typical approach is that outlined by ANSI, which is cited in Section 1.11. Here, ground snow loads corresponding to a 50-year mean recurrence interval are specified by isolines on a map of the contiguous United States. These mapped data must be used with care—mountainous regions, high country, or other local conditions require adjustments. In some areas, the local variations are so extreme that no specific values are mapped. Table 1.3 gives some representative ground snow loads from the ANSI map to illustrate the range of loading.

The distribution of snow on a roof is complex, and many different approaches are used. According to the ANSI method, the snow load on an unobstructed flat roof, p_f, is given by

$$p_f = 0.7C_eC_tIp_g \tag{1.2}$$

where p_g is the site-specific ground snow load, C_e is a dimensionless exposure factor that ranges from 0.8 for a wind-exposed setting to 1.2 for a wind-protected area, C_t is a dimensionless thermal factor that ranges from 1.0 for a heated structure to 1.2 for an unheated structure, and I is a dimensionless importance factor that ranges from 0.8 for unimportant or temporary structures to 1.2 for essential structures.

The sloped roof load acting on a horizontal projection of the roof surface, p_s, is given by

$$p_s = C_sp_f \tag{1.3}$$

where p_f is given by Eq. 1.2 and C_s is a dimensionless slope factor that ranges from 0.0 to 1.0 and depends on the roof slope and whether the roof is warm or cold (C_t factor). For normal roof surfaces, $C_s = 1.0$ for warm ($C_t = 1.0$) slopes up to 30° or for cold ($C_t > 1.0$) slopes up to 45°.

Numerous other factors must be considered in calculating snow loads. These include, but are not limited to, the effects of unloaded portions of roof, unbalanced or nonuniform loads on various roof configurations, drifting, sliding snow, and extra loads induced by rain on snow.

Snow loads are not normally considered in bridge design because they are usually small when compared with other loadings on the structure. However, ice loads can be appreciable on bridge structures. The icing not only creates loads on the structure but also increases the member sizes, which in turn increases the magnitude of the wind-induced loads.

TABLE 1.3 Typical Ground Snow Loads, $p_g{}^a$

Location	(lb/ft^2)	$(kPa = kN/m^2)^b$
Portland, Maine	60	2.87
Minneapolis, Minnesota	50	2.39
Hartford, Connecticut	35	1.68
Chicago, Illinois	25	1.20
St. Louis, Missouri	20	0.96
Raleigh, North Carolina	10	0.48
Atlanta, Georgia	5	0.24

[a] Data are taken from ANSI A58.1–1982.
[b] SI units are described in Appendix A.1.

1.8.3.2 Rain loads Roof loads that result from the accumulation of rainwater on flat roofs can be a serious problem. This condition is produced by the ponding that occurs when the water accumulates faster than it runs off, either because of the intensity of the rainfall or because of the inadequacy or blockage of the drainage system. The real danger is that as ponding occurs the roof deflects into a dished configuration, which can accommodate more water, and thus greater loads result.

The best way to prevent the problem is to provide a modest slope to the roof (1/4 in. per foot or more) and to design an adequate drainage system. In addition to the primary drainage, there should be a secondary system to preclude the accumulation of standing water above a certain level. ANSI suggests that roofs be designed to sustain rainwater loads corresponding to the elevation of the secondary system plus 5 psf (0.24 kN/m^2).

1.8.3.3 Wind loads The wind loads that act on a structure result from movement of the air against the obstructing surfaces. As was the case with snow loads, there is no single code that is uniformly applied throughout the United States. However, the procedure suggested by ANSI will again be used to illustrate a typical approach. Wind effects induce forces, vibrations, and in some cases instabilities in the overall structure as well as its nonstructural components. These wind effects depend on the wind speed, mass density of the air, location and geometry of the structure, and vibrational characteristics of the system.

Wind speed data have been accumulated over the contiguous United States, corrected to a standard height of 10 m (32.8 ft), and the highest values for a mean recurrence interval of 50 years have been mapped by ANSI using interpolation. Table 1.4 gives a few representative wind velocities from the ANSI map. Of course, as was true for snow loading, extreme local variations can occur, and the engineer must be cognizant of these possibilities.

The *design wind pressure* that is used to establish the wind load on a structure is directly related to *velocity pressure,* which is given by

$$q = \frac{\rho V^2}{2} \tag{1.4}$$

where ρ is the mass density of air and V is the wind velocity. Taking the unit weight of air to be 0.07651 lb/ft^3, Eq. 1.4 can be expressed as

$$q = 0.00256V^2 \tag{1.5}$$

TABLE 1.4 Representative Wind Velocities and Resulting Dynamic Pressures[a]

Location	Wind Velocity (mph)	Velocity Pressure (lb/ft^2)	(kN/m^2)[b]
Miami	110	31.0	1.48
Houston	90	20.7	0.99
New York City	80	16.4	0.78
Chicago	75	14.4	0.69
San Francisco	70	12.5	0.60

[a]Data are taken from ANSI A58.1–1982.
[b]SI units are described in Appendix A.1.

where q is in pounds per square foot and the velocity is in miles per hour. ANSI modifies this equation to take the form

$$q_z = 0.00256K_z(IV)^2 \qquad (1.6)$$

where the z subscript identifies the height above the ground, K_z is a dimensionless exposure coefficient that ranges from 0.12 for a low building in a sheltered location to 2.41 for a tall building in an unobstructed location and is equal to 1.0 for open terrain without obstructions at $z = 10$ m, and I is a dimensionless importance factor that ranges from 0.95 to 1.11. Table 1.4 gives the velocity pressures for the corresponding wind velocities assuming values of unity for K_z and I.

The design wind pressure for structural design is given by

$$p = qG_hC_p \qquad (1.7)$$

where q is evaluated from Eq. 1.6 at a designated height, z, above the ground for each surface, G_h is a dimensionless gust factor that ranges from 2.36 for low buildings nestled among other structures to 1.00 for tall buildings in an unobstructed location, and C_p is a dimensionless pressure coefficient. The pressure coefficients may be positive for direct pressure or negative for suction. A typical set of pressure coefficients for a structure with a gabled frame is shown in Fig. 1.25. The resulting force on any surface is determined by multiplying p by the appropriate area.

For nonstructural components such as cladding. Eq. 1.7 is augmented by a negative term to provide the net effect between the external and internal pressures.

1.8.3.4 Earthquake loads   A common dynamic loading that structures must resist is that associated with earthquake motions. Here, loads are not applied to the structure in the normal fashion. Instead, the base of the structure is subjected to a sudden movement. Since the upper portion of the structure resists motion because of its inertia, a deformation is induced in the structure. This deformation, in turn, induces a horizontal vibration that causes horizontal shear forces throughout the structure.

The resulting earthquake loads are dependent on the nature of the ground movement and the inertia response characteristics of the structure. In the United States, the ground motion expectation is given by a seismic risk map, which as-

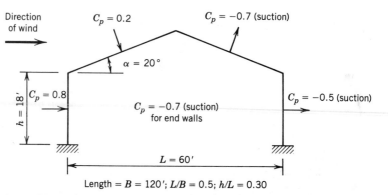

Fig. 1.25 *Pressure coefficients for gabled frame with wind normal to ridge.*

signs a risk factor according to a regional distribution. Each region is assigned a single number, ranging from 0 to 4, to reflect the severity of earthquake activity in that region. For example, a high-risk area such as southern California receives a risk factor of 4, whereas central Pennsylvania has a risk factor of 0.

Current trends are toward using the theories of structural dynamics to analyze a structure subjected to time-dependent earthquake motions. However, common practice continues to use a static approach. Here, the earthquake action is represented by a set of equivalent static forces, and a static analysis is used to establish the resulting member forces.

For example, ANSI requires that the building be designed for a minimum total lateral seismic force, V, given by the expression

$$V = ZIKCSW \tag{1.8}$$

where Z is a dimensionless seismic zone coefficient (1 for a seismic risk of 4 to 0.125 for a seismic risk of 0), I is a dimensionless occupancy importance factor (1.0 to 1.5), K is a dimensionless factor related to the lateral load-resisting characteristics of the structure (0.67 to 1.33 for buildings), S is a nondimensional soil-profile coefficient (1.0 to 1.5), W is the total dead load of the structure in the same units as V, and C is given by

$$C = \frac{1}{15\sqrt{T}} \le 0.12 \tag{1.9}$$

where T is the fundamental period of vibration of the structure. The total force V is then distributed over the height of the structure according to key building characteristics.

This approach is admittedly empirical, and it will certainly undergo improvements as a fuller understanding of the earthquake phenomenon is realized.

Consider the gabled frame shown in Fig. 1.25. Assume that it is located in a suburb of Chicago, Illinois.

1.8.4 Example problem

(a) If the structure is heated, is considered to be a nonessential structure, and is situated in a wind-protected, rural environment, determine the roof snow load, p_s, acting on a horizontal projection.

Determination of p_f: For a flat roof, Eq. 1.2 gives

$$p_f = 0.7C_eC_tIp_g$$

where

C_e = 1.2 for wind-protected, rural environment
C_t = 1.0 for heated structure
I = 1.0 for a nonessential structure
p_g = 25 lb/ft^2 for Chicago, Illinois (Table 1.3)

Therefore,

$$p_f = 0.7 \times 1.2 \times 1.0 \times 1.0 \times 25 = 21.0 \text{ lb/ft}^2$$

Determination of p_s: For a sloped roof, such as the one in Fig. 1.25, Eq. 1.3 states

$$p_s = C_s p_f$$

where

C_s = 1.0 for a warm, normal roof surface with a slope <30°

Thus,

$$p_s = 1.0 \times 21.0 = 21.0 \text{ lb/ft}^2$$

   (b) If the wind acts normal to the ridge line (as shown in Fig. 1.25), determine the design wind pressure on each roof surface. Again, assume a rural, protected environment and a nonessential structure.

Velocity Pressure: The velocity pressure is determined from Eq. 1.6, where z is taken as the vertical distance to the midheight of the roof. Thus, we have

$$q_z = 0.00256 K_z (IV)^2$$

where

z = 20 + 15 $\times$ tan 20° = 20 + 5.5 = 25.5 feet
K_z = 0.46 for given exposure at z = 25.5 feet (ANSI)
I = 1.0 for a nonessential structure
V = 75 mph for Chicago, Illinois (Table 1.4)

Therefore,

$$q_z = 0.00256 \times 0.46 \times (1 \times 75)^2 = 6.6 \text{ lb/ft}^2$$

Design Wind Pressure: The design pressure is obtained from Eq. 1.7, which in this case gives

$$p = qG_hC_p$$

where

q = 6.6 lb/ft for z = 25.5 feet
G_h = 1.54 for given exposure at z = 25.5 feet (ANSI)
C_p = 0.2 for windward side and
$\quad\;$ -0.7 for leeward side (Fig. 1.25)

Therefore,

$$p = 6.6 \times 1.54 \times 0.2 = 2.0 \text{ lb/ft}^2 \text{ on windward side}$$
$$p = 6.6 \times 1.54 \times (-0.7) = -7.1 \text{ lb/ft}^2 \text{ on leeward side (suction)}$$

1.8.5 Other loads

Numerous other loading conditions can impinge on a structure, and whether they are dead, live, or environmental would depend on the root cause associated with the load. A few examples are cited in the following sections.

1.8.5.1 Water and earth pressures Liquid pressures act normal to the surfaces of any structural element that is submerged in the liquid. The intensity of this normal pressure is given by

$$p = \gamma h \tag{1.10}$$

where γ is the unit weight of the liquid and h is the difference in elevation between the liquid surface and the point being considered. There is thus a triangular distribution in the pressure intensity, p, with increasing depth, which acts on the exterior of submerged structures or on the interior of vessels or tanks.

Structures that extend below ground level, such as retaining walls or foundations, are subjected to earth pressures. The vertical earth pressure is governed by Eq. 1.10, where γ is the unit weight of soil, which ranges from 90 to 120 lb/ft^3 (14.1 to 18.9 kN/m^3). The associated lateral soil pressure may be considerably lower than the vertical pressure because of the combined effects of cohesion and friction. If the structure extends below the water table, the combined effects of the water pressure and the earth pressure must be considered.

1.8.5.2 Induced disturbances Numerous disturbances can occur within a structural system and induce deformations that may cause forces to develop within the system. These disturbances result from support settlements or member length changes. The latter might result from temperature change in a member, forced fit of a member of improper length, or shrinkage. For certain classes of structures, the forces that are induced by these disturbances are of considerable magnitude.

Although these disturbances are not actually loads, the effect of each of them is equivalent to a loading on the structure. This is evident when the detailed equations for the analysis of the structure are considered. Problems of this type are treated in detail in later chapters.

The final design of a structure must be consistent with the most critical combination of loads that the structure is to support. However, some judgment is necessary in selecting loading conditions that can reasonably be combined. Obviously, the maximum effects of all loading conditions should not be combined because it is unlikely that they will all occur simultaneously. In fact, certain combinations are highly unlikely: full snow load is not likely to occur with full wind load because the wind would undoubtedly blow much of the snow off the structure; likewise, an earthquake is unlikely to occur at the instant that the structure is subjected to full wind load. The load combinations that must be considered are normally specified by the governing codes; in some cases, designers must exercise their judgment.

1.8.6 Load combinations

For example, ANSI permits two different design approaches, and each carries its own patterns for load combinations. The first is *allowable-stress design* or *service-load design,* where the computed elastic stress associated with any load combination must not exceed a designated value. Some typical load combinations to be considered are

1. Dead.
2. Dead + live + snow.
3. Dead + live + wind or earthquake.

A *load combination factor* is used for certain cases. For instance, to reflect the improbability that the full wind or earthquake effect will occur when the full live load is on the structure, a 0.75 factor is used for situations involving these combinations. This is equivalent to permitting an increase of $33\frac{1}{3}$ percent in the allowable stresses for these combinations.

The second design approach is *strength design* or *load-factor design*. Here, the strength of the structure must be adequate to support the most critical combination of factored loads, where a factored load results from multiplying a given load by a prescribed load factor. Some typical load combinations with the corresponding load factors are

1.   1.4 (dead).
2.   1.2 (dead) + 1.6 (live) + 0.5 (snow).
3.   1.2 (dead) + 0.5 (live) + [1.3 (wind) or 1.5 (earthquake)].

Cases 2 and 3 reflect the philosophy that the combination of dead, live, and snow loads is more probable than the combination of dead, live, and wind or earthquake loads.

1.8.7 Effects of loads on members

The loads described in the previous sections are imposed on the structure; they generally enter the structure by being applied on some surface, such as the roof, siding, or slab, and are then distributed to the supporting members as forces. These forces are, in turn, transmitted to other members, and this succession continues until the effects of the imposed loads have been delivered to the foundations. The response of the structure and the individual members is studied in detail in the remainder of the book. Our immediate concern is the manner in which the imposed loads filter through the structure.

The floor system shown in Fig. 1.26 is assumed to behave in the following fashion: The *surface load, q* (force per unit area), is applied to the slab; each beam receives this load from its *tributary area*—for instance, beam *cd* receives load from the shaded area of Fig. 1.26*b*, which extends over its full length *l* and a distance of $h/2$ on either side. This results in a *uniform line load* (force per unit length) of intensity $q(h/2)2 = qh$. Beam *cd* is supported at each end—end *c* is supported by girder *ae*, and end *d* is supported by girder *bf*. In providing this support, each girder receives a *concentrated force* of magnitude $(qh)l/2 = qhl/2$. Of course each beam is delivering like forces to its supporting girders; therefore, the full girder load at any point, such as point *c*, must include the effects of the beams framing in from both sides of that point. If the panel to the left of point *c* is of length *l*, then the total concentrated load at point *c* is qhl. Each girder then transfers all of the loads on it to the supporting columns. For instance, girder *ae* brings loads to the columns at *a* and *e*. The magnitude of each column load depends on the number of, and the configuration of, the girders supported at each column.

In this discussion, only the superimposed load, *q*, has been considered. One should also consider the effects of the dead load for each component: the slab load would be treated similarly to *q;* the beam dead load would add an additional component of uniform load along the beam; and the girder dead load would, likewise, enter as a uniform load on the girder.

Some floor systems are more complicated than the one described here. For

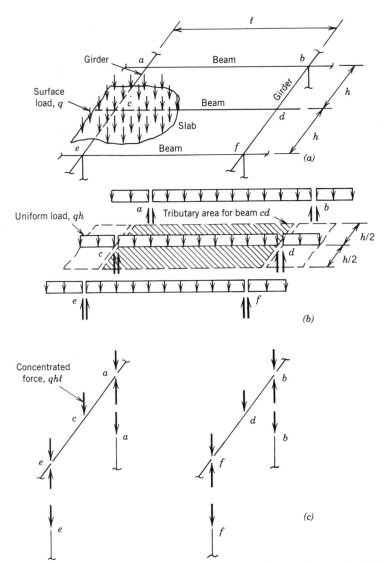

Fig. 1.26 *Typical floor system.* (a) *Surface load on slab.* (b) *Beam loads.* (c) *Girder and column loads.*

the bridge system shown in Fig. 1.19, the slab rests on stringers, the stringers are supported by transverse floor beams, and the floor beams frame into girders, which are mounted atop the bridge piers.

The building system of Fig. 1.27 provides yet another example. Snow or wind loads that act on the roof surface are transferred to the supporting frames by *purlins*. For wind loads on the sides of the building, the surface loads are picked up by the *girts* and carried to the frames. Figure 1.18c provides an illustration of this type of system.

More complicated structures have correspondingly more complicated mechanisms for supporting the applied loads. However, the basic concepts hold—there is a progression of transfers from member to member until the full effects of the loads have been carried to the foundations.

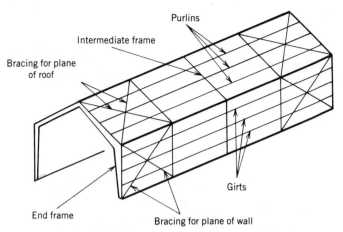

Fig. 1.27 *Typical skeletal system for frame building.*

1.8.8 Example problem

For the floor system shown in Fig. 1.26, assume that the surface load, q, is applied on panel *abef*. If $q = 100$ lb/ft^2, determine the resulting effects on beams *ab*, *cd*, *ef*, girder *ae*, and the columns at *a* and *e*. Assume $l = 20$ ft and $h = 10$ ft.

*Beam **ab** and Beam **ef**:*

$$\text{Tributary area} = (h/2)l = (10/2)20 = 100 \text{ ft}^2$$

$$\text{Uniform load} = w = q\,(h/2) = 100(10/2) = 500 \text{ lb/ft}$$

$$\text{End support load} = \frac{\text{total load}}{2} = \frac{wl}{2} = \frac{500 \times 20}{2} = 5{,}000 \text{ lb}$$

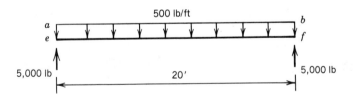

*Beam **cd**:*

$$\text{Tributary area} = 2(h/2)l = 2(10/2)20 = 200 \text{ ft}^2$$

$$\text{Uniform load} = w = qh = 100 \times 10 = 1{,}000 \text{ lb/ft}$$

$$\text{End support load} = \frac{\text{total load}}{2} = \frac{wl}{2} = \frac{1{,}000 \times 20}{2} = 10{,}000 \text{ lb}$$

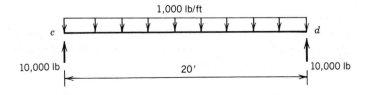

Girder **ae:** The end support loads from the beams at points *a, c,* and *e* must be applied to the girder.

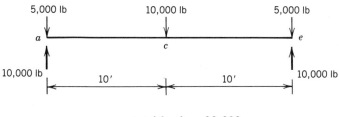

$$\text{End support load} = \frac{\text{total load}}{2} = \frac{20,000}{2} = 10,000 \text{ lb}$$

Column Loads at **a** *and* **e:** The girder end support loads for girder *ae* are applied to the column top. Column load at *a* and *e* = 10,000 lb.

Notes:

- The end support loads for both beams and girders are equal to half the total load because the members are symmetrically loaded. Procedures for unsymmetrical loading are covered in Chapter 2.
- If loading extended beyond beams *ab* and *ef*, the uniform loads on these beams would be increased.
- If loading extended beyond girders *ae* and *bf*, additional beam loads would enter the girders.

1.9 Building Materials

For a structure to function satisfactorily, it must have sufficient strength to support the applied forces without experiencing excessive deformations. The ability of the structure to satisfy these requirements depends largely on the material from which it is constructed. Although most structures experience complicated stress states, the important material properties can be determined through simple specimens loaded in tension or compression.

1.9.1 Steel

Steel is one of the most commonly used structural materials. It possesses essentially the same properties in both tension and compression, and a simple tension test on a small specimen is normally used to establish these properties. A simplified stress–strain curve for a mild carbon steel (A36 structural steel) is shown in Fig. 1.28, and this curve shows a yield stress of 36 ksi (248 MPa) and an ultimate stress of 60 ksi (414 MPa). Other steels exhibit different yield characteristics and ultimate stress values; however, they all exhibit the same initial slope of the stress–strain curve. This property is called the *modulus of elasticity, E,* which has the value 29×10^3 ksi (200 GPa).

One of the major advantages of steel is its *ductility,* which is reflected by the large deformation (approximately 25 percent) that occurs after yielding, before rupture. However, steel can exhibit a brittle mode of failure under certain conditions, such as repetitive loading or cold temperatures. Also, steel is susceptible to corrosion.

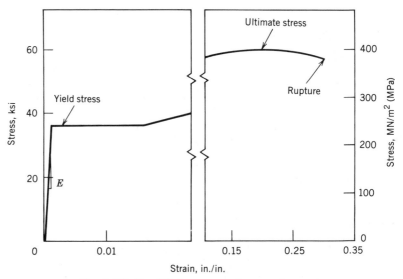

Fig. 1.28 *Simplified stress–strain curve for steel.*

1.9.2 Concrete   Structural concrete possesses considerable strength in compression, but it is very weak in tension. A typical stress–strain curve for a compression test of a standard cylindrical specimen is shown in Fig. 1.29. The compressive strength, f'_c, varies from 3 to 7 ksi (20 to 48 MPa), and the initial portion of the stress–strain curve reflects the near-elastic behavior. For this reason, concrete structures are normally analyzed by elastic methods. For concrete of normal weight, the modulus of elasticity is approximated by

$$E = 57,000 \sqrt{f'_c} \qquad (1.11)$$

where f'_c and E are in units of pounds per square inch. For $f'_c = 3,000$ psi, $E = 3,120,000$ psi (21.5 GPa), which is about 10 percent of the modulus for steel.

Figure 1.29 shows that there is no distinct yield point, nor is there any ductile behavior. However, the strains corresponding to the maximum strength, $\epsilon = 0.002$ in./in., and the point of rupture, $\epsilon = 0.004$ in./in., remain essentially the same for varying concrete strengths.

The tension strength of concrete is quite low—in the range of 0.5 ksi (3.4 MPa). For this reason, steel reinforcing bars are embedded in concrete members where tension exists, to form reinforced concrete.

1.9.3 Wood   Wood is a material whose properties depend on the grain orientation and the cell structure. Tests show that wood exhibits linear behavior over the useful stress–strain range. The modulus of elasticity varies depending on the species, but a typical value for Douglas fir is 1500 ksi (10.3 GPa).

Wood does not behave ductilely, but it has a high strength-to-weight ratio. Modern glue-lamination procedures enable the fabrication of members of widely varying shapes and sizes and have greatly broadened the range of usefulness of wood products.

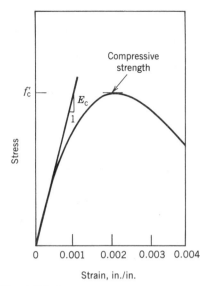

Fig. 1.29 *Stress–strain curve for concrete.*

Numerical Computations

Nearly every aspect of our lives has been touched by the computer. Tasks are routinely done today that were not even considered possible a mere 10 or 20 years ago. In structural engineering, the computer has enabled the development of methods of analysis that were virtually impossible by hand calculations. The matrix methods that are discussed in this textbook have been understood conceptually for years; however, their detailed development had to wait for the computational capability provided by the computer.

Even though modern methods of analysis require the computer for their practical application, much of the formal instruction centers on the formulation and solution of simple problems that are designed to help the student understand the methods. Thus, the student is still engaged in hand calculations for small problems where the computations are manageable. In this area, the modern hand-held calculator has had as profound an impact as the computer had on the practicing profession.

As it presently stands, the student has access to the hand-held calculator for the routine computational work associated with small problems, and highly generalized programs are available through computation centers for the solution of complex problems. In addition to these extremes, minicomputers and microcomputers (personal computers) are playing an ever increasing role in solving problems that range from simple to complex. In all these cases, the results are available in at least eight significant figures. Although this is desirable, and even necessary in certain situations, it must be remembered that the solution can be no more accurate than the data that are originally used. For instance, if the loads are known to only two significant figures, it is unreasonable to report the resulting displacements and member forces to eight significant figures. The final results should be rounded off to the number of significant figures that is consistent with the input data.

1.11

Additional Reading

American National Standard Minimum Design Loads for Buildings and Other Structures, ANSI A58.1-1982, American National Standards Institute, New York, 1982.

Kinney, J. S., *Indeterminate Structural Analysis,* Chapter 1, Addison–Wesley, Reading, Mass., 1959.

McGuire, William, *Steel Structures,* Chapters 2 and 3, Prentice–Hall, Englewood Cliffs, N.J., 1968.

Norris, C. H., Wilbur, J. B., and Utku, S., *Elementary Structural Analysis,* 3rd Ed., Foreword and Chapter 1, McGraw–Hill, New York, 1976.

Pannell, J. P. M., *Man the Builder,* Crescent Books, New York, 1964.

Sack, Ronald L., *Structural Analysis,* Chapter 1 and Appendix B, McGraw–Hill, New York, 1984.

Salvadori, M., and Heller, R., *Structure in Architecture,* Prentice–Hall, Englewood Cliffs, N.J., 1963.

Specifications for Steel Railway Bridges, American Railway Engineering Association, Chicago, Ill., 1965.

Standard Specifications for Highway Bridges, 13th Ed., American Association of State Highway and Transportation Officials, Washington, D.C., 1983.

Straub, H., *A History of Civil Engineering,* MIT Press, Cambridge, Mass., 1964.

Timoshenko, S., *History of Strength of Materials,* McGraw–Hill, New York, 1953.

Uniform Building Code, International Conference of Building Officials, Whittier, Calif., 1976.

von Frisch, Karl, *Animal Architecture,* Harcourt Brace Jovanovich, New York, 1974.

White, R. N., Gergley, P., and Sexsmith, R. G., *Structural Engineering,* Chapters 1 through 4, Wiley, New York, 1976.

"Wind Forces on Structures," *Trans. ASCE,* **126,** Part II, p. 1124, 1961.

1.12

Suggested Problems

1. For each of the structures listed below, construct a simplified sketch of the structure and indicate the kind of action transmitted by the major elements of the system. Use the following key on your sketch to show the *action on the adjacent joint* for the ends of each member.

Typical member between end joints

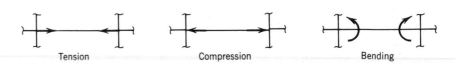

    Tension   Compression   Bending

  a. Pipeline bridge of Fig. 1.11*a.* Consider each of the following loading conditions separately:

    • Uniform dead load of pipeline.

    • Uniform wind load on pipeline.

b. Reinforced concrete arch of Fig. 1.13*a*. Assume uniform live load along the roadway.

c. Typical gabled frame of Fig. 1.18*c*. Assume that concentrated loads are transferred from longitudinal members in the plane of the roof (purlins) to the inclined girders of the frame.

2. Consider the gabled frame structure that is shown. The frames are spaced at 30-ft intervals along the building, and purlins and girts span between the frames as shown. It is assumed that each frame supports a section of roof that extends midway to the next frame in either direction. Likewise, purlins and girts receive loads from tributary areas that extend midway to the next purlin or girt, respectively.

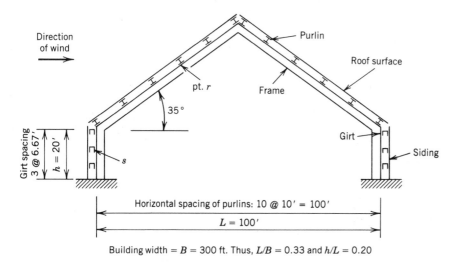

Building width = B = 300 ft. Thus, L/B = 0.33 and h/L = 0.20

a. Determine the snow load, p_s, acting on a horizontal projection of the roof surface based on the following information:
- Structure is located in a suburb of Minneapolis, Minnesota.
- Building has occupancy of over 300 people (I = 1.1).
- Building is heated (C_t = 1.0) and the roof is warm (C_s = 0.9).
- Structure is in a windy, exposed site (C_e = 0.8).

b. For the loading of part (a), determine the concentrated load transmitted to the frame at point r.

c. Determine the wind pressure, p, on the windward and leeward walls based on the following data:
- Structure is located in a suburb of Houston, Texas.
- Building has occupancy of over 300 people and is in hurricane area (I = 1.11).
- K_z = 0.37 and 0.55 for windward and leeward walls, respectively.

- G_h = 1.65 and 1.48 for windward and leeward walls, respectively.
- C_p = 0.8 and -0.50 for windward and leeward walls, respectively.

d. For the loading of part (c), determine the concentrated load transmitted to the frame at point s.

e. Determine the wind pressure, p, on both sides of the roof based on the following data:

- Structure is a utility structure in urban New York City, nestled among tall structures.
- Building is a nonessential storage building ($I = 0.95$).
- K_z = 0.22.
- G_h = 1.92.
- C_p = 0.35 and -0.7 for windward and leeward sides, respectively.

f. For the loading of part (e), determine the concentrated load transferred to the frame at point r.

3. Consider the building floor system shown in plan view.

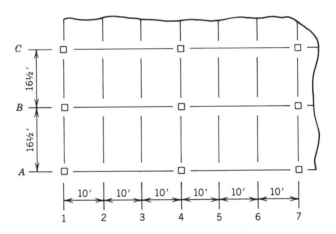

Reinforced concrete girders (12 in. wide and 20 in. deep) span along lines A, B, and C, and reinforced concrete beams (8 in. wide and 14 in. deep) run along lines 1 through 7. The reinforced concrete floor slab is 5 in. deep.

a. Determine the dead load carried by the following members:

- Beam along line 1 between A and B.
- Beam along line 3 between A and B.
- Girder along line A between 1 and 4.
- Girder along line B between 1 and 4.
- Columns at intersections of A and 1, A and 4, B and 1, and B and 4.

b. Determine the live load carried by the members enumerated in part (a) if the structure is to be used as an office building.

4. Consider the structure shown, which has a flat roof with an octagonal plan. If this building is to be constructed in Portland, Maine, determine the distribution and corresponding magnitudes of the line load caused by snow that must be supported by a typical radial girder ab. Assume that the roof surface rests directly on the radial girders.

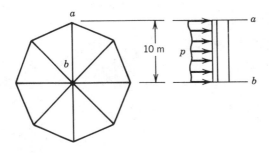

CHAPTER 2

Delta-frame steel bridge, Lexington, Va. (courtesy Bethlehem Steel Corporation).

EXTERNAL EQUILIBRIUM OF PLANAR STRUCTURES

2.1

Forces

Whatever the structure, it is subjected to a loading condition that results from its own dead weight, from the weight and functional input of the elements that it supports, and from the impact of the environment in which it exists. This loading may be very complicated and difficult to determine, as was explained in Chapter 1. However, with the assistance of building code information, research, and testing, a loading condition is finally assumed. Further assumptions might be necessary regarding how this loading condition is applied to the structure. But eventually, the result of these considerations is a set of *applied forces* of determined magnitudes, acting in specified directions at designated points on the structure. In the present context, these applied forces have a general meaning in that they may include moments; however, whatever the nature of these forces, they are count-

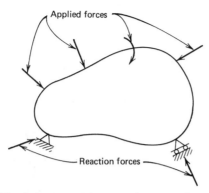

Fig. 2.1 *Applied forces and reaction forces.*

ered by *reaction forces* at the support points for the structure as illustrated in Fig. 2.1. The structure must have sufficient strength to support the total array of forces corresponding to the most severe loading conditions.

It should be noted that the forces that act on a structure may be either static or dynamic. Static forces are of fixed magnitudes and are independent of time, whereas dynamic forces have time-dependent magnitudes. This book will treat only static loading conditions or dynamic loading conditions that can be approximated by static loading.

2.2 Free-Body Diagrams

In structural analysis, one of the most useful tools at the analyst's disposal is the free-body diagram. A *free-body diagram* is simply a sketch of a structural component with all of the appropriate forces acting on it. It may be of an entire structure or it may be of a subsection of a larger structure.

For instance, consider the structural body shown in Fig. 2.2*a*. Figure 2.2*b* shows a free-body diagram of the entire structure along with the applied forces and the reaction forces. If the body is cut along the line *a–b* shown in Fig. 2.2*a*, a free-body diagram of either section must show the internal stresses (force intensities) that act on the cut face. These internal stresses act in an equal and opposite fashion on the two free-body diagrams of Fig. 2.2*c*. If the two separate free-body diagrams are brought together, the internal stresses cancel, and the total free-body diagram of Fig. 2.2*b* is restored.

A free-body diagram of an entire structure leads to relationships between the external forces on the structure, whereas free-body diagrams of subsections of the structure will lead to relationships between the external and internal forces. In any case, each free-body diagram represents a structural unit that is subject to all of the considerations of structural analysis. The judicious selection of free-body diagrams and the subsequent analysis of each structural component are fundamental to the field of structural analysis.

2.3 Equilibrium

A structure is said to be in equilibrium when it is initially at rest and remains at rest as it is acted on by a set of forces. When a structure satisfies this test for

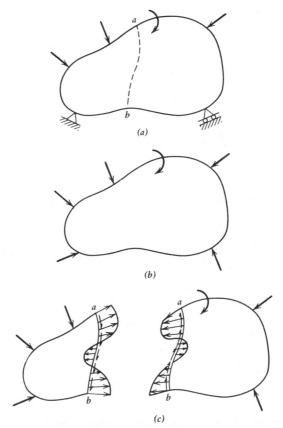

Fig. 2.2 *Free-body diagrams.*

equilibrium, any constituent part of the structure will also be found to be in equilibrium.

For the condition of rest to be satisfied, there must be a balance in the force
tendencies that would act to disturb the structure in any way. If we consider the
planar body shown in Fig. 2.3, with the cartesian coordinate axis system given,
this balance can be expressed by

$$\sum P_x = 0; \quad \sum P_y = 0; \quad \sum M_z = 0 \qquad (2.1)$$

where the first two equations refer to the summation of force components in the x
and y directions, respectively, and the third equation gives the summation of the
moments (forces times moment arms) about the z axis. These equations are referred to as the *equations of static equilibrium*. When these equations are satisfied,
the body is in balance, at rest, and thus in equilibrium.

Equations 2.1 represent the most common form of the equilibrium equations
for a planar structure; however, alternate forms are possible. For instance, we
could use

$$\sum P_y = 0; \quad \sum M_a = 0; \quad \sum M_b = 0 \qquad (2.2)$$

where a and b are two points in the xy plane as shown in Fig. 2.4b, and $\sum M_a$ and
$\sum M_b$ represent the summation of moments about a z axis through points a and b,
respectively. This procedure is valid as long as the line connecting points a and b

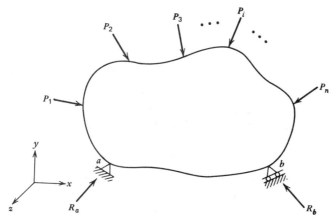

Fig. 2.3 *Equilibrium of a planar structure.*

is not perpendicular to the y axis. Or, in yet another form, we could have

$$\sum M_a = 0; \quad \sum M_b = 0; \quad \sum M_c = 0 \tag{2.3}$$

where point c also lies in the xy plane as depicted in Fig. 2.4b and ΣM_c is the summation of moments about a z axis through point c. These equations sufficiently express the equilibrium condition as long as points a, b, and c do not lie along a straight line.

Two special cases for planar bodies deserve mention. The first deals with a *concurrent force system* in which all forces pass through a single point as illustrated in Fig. 2.5a. Here, the condition that $\Sigma M_z = 0$ is automatically satisfied, and Eqs. 2.1 reduce to

$$\sum P_x = 0; \quad \sum P_y = 0 \tag{2.4}$$

A second special case deals with a *parallel force system* such as the one shown in Fig. 2.5b. In this case, the summation of forces normal to the direction of the forces is automatically zero, and Eqs. 2.1 reduce to

$$\sum P_p = 0; \quad \sum M_z = 0 \tag{2.5}$$

The term ΣP_p represents the summation of forces in the direction of the parallel forces.

In both of these special cases, alternate forms of the equations of equilibrium are possible, just as Eqs. 2.2 and 2.3 were alternate forms of Eqs. 2.1.

All of the cases considered here may be applied to a structure as a whole or to a portion of the structure. In each case, the portion under consideration is

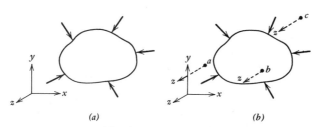

(a)   (b)

Fig. 2.4 *Planar body.*

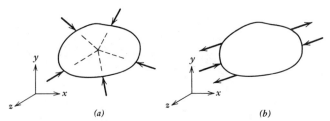

Fig. 2.5 *Special force systems.* (a) *Concurrent.* (b) *Parallel.*

isolated as a free-body diagram, and the appropriate equations will express the conditions that the forces must satisfy for equilibrium to exist.

2.4

Specification of a Force

The structure of Fig. 2.3 is acted upon by two different types of forces—some are action forces that have applied to the structure, whereas others are reaction forces that develop in order to provide equilibrium. Each of these forces is completely specified by its magnitude, direction, and line of action. Figure 2.6*a* shows a graphical representation of the force P_i applied to a structural element in which these three force characteristics (P_i, α, l) are shown. For the application of Eqs. 2.1, it is convenient for the force P_i to be replaced by its components in the x and y directions. These components are represented by P_{ix} and P_{iy}, respectively, as shown in Fig. 2.6*b*, and are given by

$$P_{ix} = P_i \cos \alpha$$
$$P_{iy} = P_i \sin \alpha$$

(2.6)

The force components collectively give the magnitude and direction of the force by the relationships

$$P_i = \sqrt{P_{ix}^2 + P_{iy}^2}$$
$$\alpha = \tan^{-1} \left(\frac{P_{iy}}{P_{ix}} \right)$$

(2.7)

However, the line of action, l, is still required for a complete specification of the force as shown in Fig. 2.6*b*. This fixes the point at which the force acts on the structure.

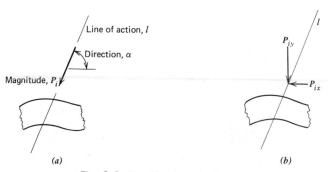

Fig. 2.6 *Specification of a force.*

Each of the applied forces is part of a prescribed loading condition, and all three of its force characteristics are specified. However, for each reaction force, certain force characteristics are specified by the nature of the support, whereas others will result from overall equilibrium requirements.

In considering the reaction force at a support point, it is convenient to translate from the three force characteristics mentioned earlier to the reaction force components of horizontal force, vertical force, and moment. Figure 2.7 shows the support point "a" isolated as a free body. In Fig. 2.7a the resultant reaction force is described in terms of the force characteristics (R_a, α, l), while Fig. 2.7b shows the resultant reaction force in terms of its force components (R_{ax}, R_{ay}, M_a). The two forms of specifying a force are completely equivalent and are related by the expressions

$$R_{ax} = R_a \cos \alpha$$
$$R_{ay} = R_a \sin \alpha \qquad (2.8)$$
$$M_a = R_a \cdot e$$

where e is the eccentricity of the line of action with respect to point a.

Table 2.1 summarizes the known and unknown elements of a reaction force by the two different approaches for a variety of support conditions. For each support condition considered, the symbolic representation is given along with the restraints that are provided by that condition. These restraints are then expressed in terms of the force characteristics of the resultant reaction as well as the reaction force components. It is important to note that whether force characteristics or force components are considered, the same number of known and unknown quantities results.

The support conditions given in Table 2.1 are idealized representations of the real conditions. In actuality, since friction-free pins and rollers are nonexistent, all supports for planar structures will have horizontal, vertical, and rotational restraints. However, special hardware is available that provides support conditions that come very close to those simulated by the idealized cases. Figure 2.8 shows

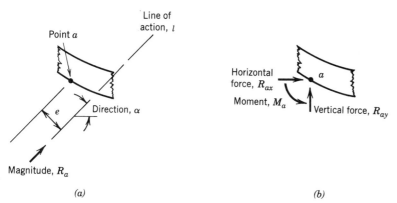

(a)   (b)

Fig. 2.7 *Elements of reaction force at point* a. (a) *Force characteristics.* (b) *Force components.*

TABLE 2.1 Reaction Forces

Type of Support	Symbolic Representation	Restraints	Identification of Known and Unknown Elements of Reaction Force (see Fig. 2.7)	
			In Terms of Force Characteristics	In Terms of Reaction Components
Hinge or pin		Horizontal Vertical	Unkn: R_a, α Known: l through pt. a	Unkn: R_{ax}, R_{ay} Known: $M_a = 0$
Roller or rocker		Vertical	Unkn: R_a Known: l through pt. a, $\alpha = 90°$	Unkn: R_{ay} Known: $R_{ax} = M_a = 0$
Roller on incline		Inclined	Unkn: R_a Known: l through pt. a, $\alpha = \tan^{-1}\frac{s}{r}$	Unkn: R_{ay} Known: $R_{ax} = R_{ay}\left(\frac{r}{s}\right)$ $M_a = 0$
Fixed or clamped		Horizontal Vertical Rotational	Unkn: R_a, α, l (l must have eccentricity such that $M_a = R_a e$) Known: —	Unkn: R_{ax}, R_{ay}, M_a Known: —
Link		Inclined	Unkn: R_a Known: l through pt. a, $\alpha = \tan^{-1}\frac{s}{r}$	Unkn: R_{ay} Known: $R_{ax} = R_{ay}\left(\frac{r}{s}\right)$ $M_a = 0$
Guide		Horizontal Rotational	Unkn: R_a, l (l must have eccentricity such that $M_a = R_a e$) Known: $\alpha = 0°$	Unkn: R_{ax}, M_a Known: $R_{ay} = 0$

Fig. 2.8 *Typical pintle or hinge bearings.*
(a) *(courtesy M. L. Sheridan).* (b) *(courtesy Stewart C. Watson, Watson–Bowman Associates, Inc., Buffalo, N.Y.).* (c) *(courtesy Stewart C. Watson, Watson–Bowman Associates, Inc., Buffalo, N.Y.).*

examples of pintle or hinge support bearings and Fig. 2.9 illustrates typical roller and rocker support bearings. For such cases, the ideal assumptions are in order.

Even though all supports provide certain components of restraint, it should also be noted that all supports exhibit deformations in the directions corresponding to these force restraints. This fact becomes clear if we note that the supporting mechanism must undergo strains in order to develop the stresses that are associated with the reaction forces. These strains then result in support deformations. The larger the effective stiffnesses of the supports are, the smaller the deforma-

Fig. 2.9 *Typical roller and rocker bearings (courtesy Stewart C. Watson, Watson–Bowman Associates, Inc., Buffalo, N.Y.) (a) Roller bearing. (b) Rocker bearing. (c) Rocker bearing. (d) Large-displacement roller bearing.*

tions will be; however, theoretically such support movements will always exist. Thus, there is no such thing as a fixed support, although in many cases the deformations are so small that the assumptions of fixity are valid.

2.7
Statical Determinacy and Stability

Based on the discussion of Section 2.5, each support point for a structure can be evaluated regarding the number of unknown reaction components. For the structure as a whole, the total number of unknown reaction components, r_a, is given by the summation of the unknowns for all support points. The *statical classification* of the structure is dependent on the quantity r_a and the arrangement of the reaction components.

To determine the unknown reaction components, it is necessary to apply the equations of static equilibrium that were presented in Section 2.3. Because three equations are available, it is possible to solve for three independent reaction components. Thus, if $r_a = 3$, the structure is said to be *statically determinate externally*. If $r_a > 3$, there are more unknowns than there are equations for their determination and the structure is classified as *statically indeterminate externally*. Here, the equations of static equilibrium remain a vital part of the solution, but they must be augmented by equations that relate the deformation characteristics of the structure. For cases where $r_a < 3$, there are fewer unknowns that equations, and thus there is no solution. Structures of this type are *statically unstable externally* because there is no solution that can simultaneously satisfy the requirements of equilibrium.

In summary, the following criteria hold:

$r_a < 3$; structure is statically unstable externally

$r_a = 3$; structure is statically determinate externally

$r_a > 3$; structure is statically indeterminate externally

The above criteria must be applied with some discretion. If $r_a < 3$, the structure is definitely unstable; however, in cases where $r_a \geq 3$, the structure is not necessarily stable. It is possible that the reaction components are not properly arranged to ensure external stability. Structures in this category are termed *geometrically unstable*.

Table 2.2 gives the statical classification of several structures. Cases a through f represent straightforward applications of the criteria. The structure under case g would appear to be determinate according to the criteria; however, the three reaction components are not arranged in a way that would produce stability under all loading conditions. Only if the applied loads were vertical would the structure be stable. For this special case, the equilibrium equation dealing with horizontal forces is automatically satisfied, and only two equations are available to solve for the three unknown reaction quantities. Thus, the structure is indeterminate even if its stability is ensured by the loading. Case h also appears initially to be statically determinate; however, further inspection shows that all reactions pass through point p. Any external load that does not pass through this point cannot be equilibrated by the reactions and thus the structure is unstable under general loading. If the applied loading passes through point p, the structure is stable. For this special case, the moment equilibrium equation is inherently satisfied; however, the two remaining equations are not sufficient to determine the three unknown

TABLE 2.2 External Statical Classification of Structures

Structure	Independent Reaction Components, r_a	Classification
(a)	3	$r_a = 3$; determinate, stable
(b)	3	$r_a = 3$; determinate, stable
(c)	4	$r_a > 3$; indeterminate, stable
(d)	2	$r_a < 3$; unstable
(e)	5	$r_a > 3$; indeterminate, stable
(f)	4	$r_a > 3$; indeterminate, stable
(g)	3	$r_a = 3$; geometrically unstable
(h)	3	$r_a = 3$; geometrically unstable

reaction components. Thus, the structure is either geometrically unstable or indeterminate.

Cases h and g fall within a broad category of structures whose reactions form a system of parallel or concurrent forces such cases were discussed in Section 2.2. Situations of this type are potentially unstable even if $r_a \geq 3$ and are statically indeterminate for the special loading cases that lead to stability.

The equations of static equilibrium play a vital role in determining the reactions **Computation** for any structure. If the structure is statically determinate externally, these **of Reactions** equations provide all that is needed for the solution, whereas for statically inde- **Using** terminate cases, the equations of equilibrium still apply, but they are not sufficient **Equations of** for a solution. This section considers only structures that are statically determi- **Equilibrium** nate.

In determining the reactions, a free-body diagram of the entire structure is used. All of the known forces are shown to act in their prescribed directions, and each unknown reaction component is assumed to act in a specified direction. The equations of equilibrium are written in consistency with the assumed directions, and these equations are simultaneously solved for the unknown reactions. If the solution yields a positive reaction component, then the assumed direction is correct, whereas a negative result indicates that the wrong direction was assumed.

In the latter case, care must be exercised in subsequent calculations. If the negative sign is retained, then the originally assumed direction must also be retained. However, the direction can be reversed on the free-body diagram and a positive sign can then be used in subsequent work.

Compute the reactions for the beam structure shown.

2.8.1 Example problem

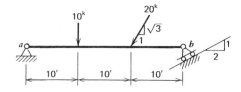

Free-Body Diagram: It is convenient to show all forces, known and unknown, in terms of their components. The reaction at point a has two unknown components, whereas the reaction at point b has only a single unknown component.

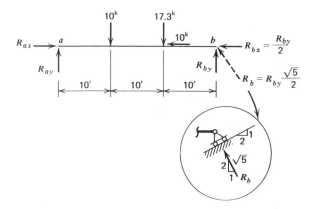

Determination of Reactions: Since $r_a = 3$, and the arrangement is stable, the structure is statically determinate. The reactions are determined by systematically applying the equations of equilibrium.

$$\sum M_a = 0 \; \circlearrowright \; +$$

$$(10 \times 10) + (17.3 \times 20) - (R_{by} \times 30) = 0$$

$$R_{by} = \underline{\underline{14.9^k}}$$

$$\therefore R_{bx} = \frac{R_{by}}{2} = \frac{14.9}{2} = \underline{\underline{7.5^k}}; \qquad R_b = R_{by}\frac{\sqrt{5}}{2} = 14.9\frac{\sqrt{5}}{2} = \underline{\underline{16.7^k}}$$

$$\sum P_y = 0 \; \uparrow +$$

$$R_{ay} - 10 - 17.3 + R_{by} = 0$$

$$R_{ay} = 27.3 - R_{by} = 27.3 - 14.9 = \underline{\underline{12.4^k}}$$

$$\sum P_x = 0 \; \xrightarrow{+}$$

$$R_{ax} - 10 - R_{bx} = 0$$

$$R_{ax} = 10 + R_{bx} = 10 + 7.5 = \underline{\underline{17.5^k}}$$

Since all reaction components are positive, each one acts in the direction assumed on the free-body diagram.

2.8.2 Example problem

Determine the reactions for the structure shown in the sketch below.

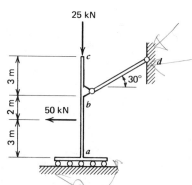

Free-Body Diagram: The known forces and the three unknown reaction components are shown on a free-body diagram.

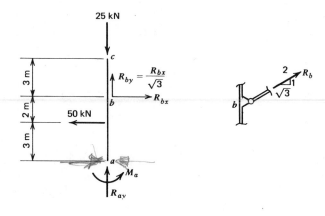

Determination of Reactions: The structure is stable, and $r_a = 3$; therefore, the structure is statically determinate. The equations of equilibrium are applied to determine the reactions.

$$\Sigma P_x = 0 \xrightarrow{+}$$
$$R_{bx} - 50 = 0; \quad R_{bx} = \underline{50 \ kN}$$
$$\therefore R_{by} = \frac{R_{bx}}{\sqrt{3}} = 28.9; \quad R_b = 2R_{by} = \underline{57.9 \ kN}$$

$$\Sigma P_y = 0 \uparrow +$$
$$R_{ay} + R_{by} - 25 = 0$$
$$R_{ay} = 25 - R_{by} = 25 - 28.9 = \underline{-3.9 \ kN}$$

The negative sign indicates that the vertical component of reaction at point a acts downward, not upward, as assumed on the free-body diagram.

$$\Sigma M_a = 0 \circlearrowright +$$
$$(R_{bx} \times 5) - (50 \times 3) - M_a = 0$$
$$M_a = -150 + (50 \times 5) = \underline{+100 \ kN \cdot m}$$

Calculate the reactions for the frame structure shown. $2 \times 15 \times (\frac{15}{2})$

2.8.3 Example problem

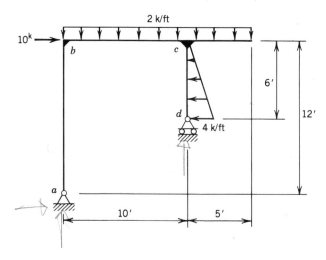

Free-Body Diagram: The distributed forces are replaced by resultant concentrated forces at their centroidal positions. The unknown reaction components are also shown.

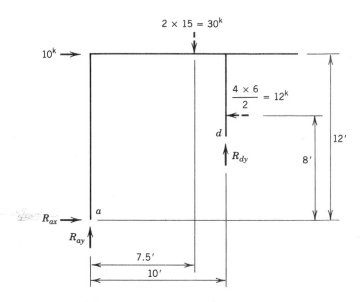

Determination of Reactions: The arrangement of the reaction components provides stability, and $r_a = 3$; thus, the structure is statically determinate. Application of the equations of equilibrium leads to the determination of the reactions.

$$\sum M_a = 0 \circlearrowright \; +$$

$$(10 \times 12) + (30 \times 7.5) - (12 \times 8) - (R_{dy} \times 10) = 0$$

$$\underline{R_{dy} = 24.9^k}$$

$$\sum P_x = 0 \xrightarrow{+}$$

$$R_{ax} + 10 - 12 = 0; \qquad \underline{R_{ax} = 2.0^k}$$

$$\sum P_y = 0 \; \uparrow +$$

$$R_{ay} - 30 + R_{dy} = 0$$

$$R_{ay} = 30 - R_{dy} = 30 - 24.9 = \underline{5.1^k}$$

2.9

Principle of Superposition for Force Quantities

Much of structural analysis is based on the *principle of superposition,* which states that the effects of individual actions on a structure can be superimposed to determine the total effects. We will limit our discussion here to force quantities; displacement quantities will be considered later.

Consider the structure shown in Fig. 2.10a, which is subjected to the concentrated force P and the uniformly distributed force w. The response of the structure includes the reaction forces, such as R_a, which is the vertical component of the reaction at point a. In Figs. 2.10b and c, the beam is subjected to the individual actions of the concentrated force and the uniform force, respectively, and the individual reaction components, R_{a1} and R_{a2}, result. According to the principle of superposition,

$$R_a = R_{a1} + R_{a2} \tag{2.9}$$

In the example problems of the previous sections, there are several instances where the use of the principle of superposition was implied. That is, in determin-

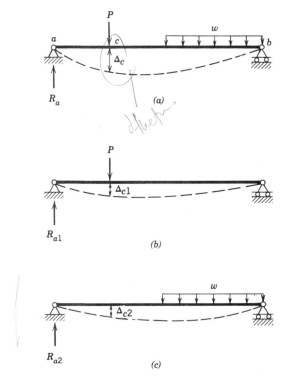

Fig. 2.10 *Principle of superposition.* (a) *Beam subjected to* P *and* w. (b) *Beam subjected to* P *only.* (c) *Beam subjected to* w *only.*

ing the total reaction components, the calculations reflect the superposition of the effects of the individual forces. For statically determinate stable structures, super-position can always be used for force quantities.

However, there are situations where the principle of superposition does not hold for force quantities. One such case corresponds to geometrically unstable structures that must undergo large displacements in order to achieve equilibrium. Case *h* of Table 2.2 serves as an example. An applied force that does not pass through point *p* causes large distortions of the structure before it is equilibrated. If a second force is considered, again not through point *p*, the structure undergoes a different large distortional pattern. In each case, the reactions are related to the adjusted geometry. The combination of the two forces would likewise be associated with yet another distorted structure; therefore, superposition is not valid. It should be noted that this group of geometrically unstable structures is statically indeterminate as explained in Section 2.7.

For statically indeterminate structures, such as cases *c*, *e*, and *f* of Table 2.2, force quantities can be superimposed if the constituent material possesses a linear stress–strain diagram. If such a material relationship does not exist, super-position cannot be used. Similarly, for a geometrically unstable structure, such as case *g*, which reverts to a statically indeterminate structure under special loading conditions, superposition can be applied only if linear elastic material is used.

The principle of superposition will be considered again in Section 5.12, where displacement quantities are discussed.

2.10

Condition Equations

For the structures that have been considered thus far, it has been possible to determine a maximum of three unknown reaction components for a statically determinate and stable structure. These reaction components have been determined by systematically applying the three equations of static equilibrium.

In certain instances, where special internal conditions exist, it is possible to

TABLE 2.3 External Statical Classification of Structures with Condition Equations

Structure	Indep. React. Comps. r_a	Numb. of Conditions n	Req'd. React. Comps. $r = 3 + n$	Classification
$M=0$ (a)	4	1	$3 + 1 = 4$	$r_a = r$; determinate, stable
$M=0$ (b)	5	2	$3 + 2 = 5$	$r_a = r$; determinate, stable
$M=0$ (c)	4	1	$3 + 1 = 4$	$r_a = r$; determinate, stable
$M=0$ (d)	5	1	$3 + 1 = 4$	$r_a > r$; indeterminate, stable
$M=0$ (e)	5	2	$3 + 2 = 5$	$r_a = r$; determinate, stable
$M=0$ (f)	4	2	$3 + 2 = 5$	$r_a < r$; unstable
$M = 0$ (g)	3	1	$3 + 1 = 4$	$r_a < r$; unstable
$M=0; H=0$ (h)	4	2	$3 + 2 = 5$	$r_a < r$; unstable

have more than three statically determinate reaction components. This occurs when a structure is composed of two or more substructural units connected in a manner that allows one or more statical equations to be written involving the reactions in addition to the overall equations of static equilibrium. These equations are referred to as *condition equations,* and each condition increases by one the number of statically determinate reaction components.

It was shown in Section 2.7 that for a structure without condition equations, a minimum of three reaction components are required for external stability. This number is increased by one for each condition equation that is available. Thus, if r is taken as the least number of reaction components required for external stability of a structure with n condition equations, then $r = 3 + n$. If r_a is the actual number of reaction components, then the same line of reasoning that was applied to structures without condition equations leads to the following criteria:

$r_a < r$; structure is statically unstable externally

$r_a = r$; structure is statically determinate externally

$r_a > r$; structure is statically indeterminate externally

As before, the criteria for $r_a \geq r$ are necessary but not sufficient conditions for statical classification. The reaction components must be arranged in a manner that ensures stability.

Table 2.3 shows the statical classification of several structures with condition equations. For example, consider structure a. In accordance with the notation introduced, $r_a = 4$, $n = 1$, and $r = 3 + n = 3 + 1 = 4$. Since $r_a = r$, and the reactions provide a stable arrangement, the structure is statically determinate. The procedure for determining the reactions is an orderly one. Using the equations of equilibrium for the entire structure, one can write three equations in terms of the four unknown reaction components. A fourth equation involving some of the unknowns can be written by taking a free-body diagram to either the left or right of the internal hinge and by applying the condition that the moment is zero at the hinge. Thus, there are four independent equations to solve for the four unknowns. It is important to note that only one independent condition equation results from the internal hinge. The equation for the section to the left of the hinge is not independent of the equation for the section to the right of the hinge and vice versa. Each one can be formed by combining the other one with the equations for overall equilibrium of the structure.

Compute the reactions for the cantilevered structure shown below.

2.10.1 Example problem

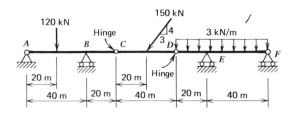

Two condition equations can be written: $M_C = M_D = 0$. Thus, $r = 3 + n = 3 + 2 = 5$. Since $r_a = 5$ also, and since these reactions are arranged so that the structure is stable, the structure is statically determinate.

It is possible to write the three equations of equilibrium for the entire structure and the two condition equations and solve the resulting five equations for the five unknown reaction components. However, it is more convenient to disassemble the structure into several free-body diagrams and to apply statics to each part. Here, the condition equations are used as each free body is analyzed.

Free-Body Diagrams: The structure is broken into three separate free bodies, and the unknown reactions and internal forces at the hinges are labeled.

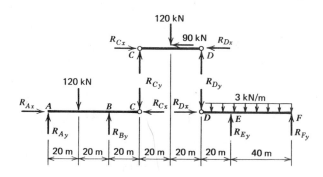

Determination of Reactions by Equilibrium:

Free Body CD: Here we treat CD as a simply supported beam and thus inherently use the conditions that $M_C = M_D = 0$. The reactions (internal forces transferred at hinges) are

$$R_{Cx} = +90\text{kN}; \quad R_{Cy} = +60\text{kN}; \quad R_{Dy} = +60\text{kN}$$

Free Body ABC: This segment is analyzed for the applied forces with R_{Cx} and R_{Cy} acting at point C. This leads to

$$R_{Ax} = +90\text{kN}; \quad R_{Ay} = +30\text{kN}; \quad R_{By} = +150\text{kN}$$

Free Body DEF: This segment is analyzed for the applied forces and R_{Dy} acting at point D. This leads to

$$R_{Ey} = +225\text{kN}; \quad R_{Fy} = +15\text{kN}$$

540↑

2.10.2 Example problem   Compute the reactions for the three-hinged arch shown.

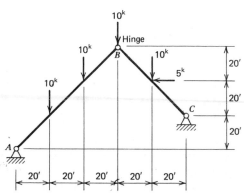

Classification:

$r_a = 4$; $r = 3 + n = 3 + 1 = 4$; since $r_a = r$, structure is statically determinate, and it is stable.

Free-Body Diagrams:

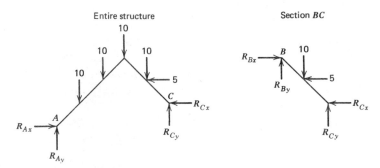

Determination of Reactions: For the entire structure:

$$\sum M_A = 0 \;\curvearrowright^+ \quad 10(20 + 40 + 60 + 80) - 5(40)$$
$$- R_{Cy}(100) - R_{Cx}(20) = 0$$
$$R_{Cx} + 5R_{Cy} = 90$$
$$\sum P_y = 0 \;\uparrow^+ \quad R_{Ay} + R_{Cy} - 4(10) = 0$$
$$R_{Ay} + R_{Cy} = 40$$
$$\sum P_x = 0 \;\xrightarrow{+} \quad R_{Ax} - R_{Cx} - 5 = 0$$
$$R_{Ax} - R_{Cx} = 5$$

For section *BC*:

$$\sum M_B = 0 \;\curvearrowright^+ \qquad 10(20) + 5(20) + R_{Cx}(40) - R_{Cy}(40) = 0$$
$$\text{(condition equation)}$$
$$4R_{Cy} - 4R_{Cx} = 30$$

In summary, the equations of equilibrium are

$$R_{Cx} + 5R_{Cy} = 90$$
$$R_{Ay} \qquad\quad + R_{Cy} = 40$$
$$R_{Ax} \qquad\quad -R_{Cx} \qquad = 5$$
$$-4R_{Cx} + 4R_{Cy} = 30$$

The solution is

$$R_{Ax} = 13.75^k; \quad R_{Ay} = 23.75^k; \quad R_{Cx} = 8.75^k; \quad R_{Cy} = 16.25^k$$

2.11 Matrix Formulation of Reaction Problems

Because of the generality afforded by matrix methods, and because computers can readily be programmed to perform the operations of matrix analysis, matrix formulation of structural problems is becoming more popular. Although there is no great advantage in applying matrix methods to simple reaction problems, it is done here to introduce some basic concepts of matrix methods.

Consider the beam problem of Fig. 2.11, in which there are three unknown reaction components. Applying the equations of equilibrium, we obtain

$$\sum M_a = 10P_{1y} + 20P_{2y} - 30R_{by} = 0$$
$$\sum P_y = R_{ay} + R_{by} - P_{1y} - P_{2y} = 0 \qquad (2.10)$$
$$\sum P_x = R_{ax} + P_{1x} + P_{2x} - \frac{R_{by}}{2} = 0$$

Once the applied forces are specified, these equations can readily be solved by elimination for the reactions. However, in this case a more general procedure will be used. If the reaction terms are retained on the left side of the equations and the forces are transferred to the right side, then these equations can be written in the following matrix form:

$$\begin{bmatrix} 0 & 0 & 30 \\ 0 & 1 & 1 \\ 1 & 0 & -\frac{1}{2} \end{bmatrix} \begin{Bmatrix} R_{ax} \\ R_{ay} \\ R_{by} \end{Bmatrix} = \begin{bmatrix} 0 & 10 & 0 & 20 \\ 0 & 1 & 0 & 1 \\ -1 & 0 & -1 & 0 \end{bmatrix} \begin{Bmatrix} P_{1x} \\ P_{1y} \\ P_{2x} \\ P_{2y} \end{Bmatrix} \qquad (2.11)$$

This equation can be abbreviated as

$$[q]\{R\} = [S]\{P\} \qquad (2.12)$$

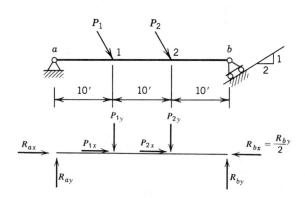

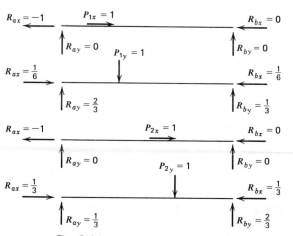

Fig. 2.11 *Beam structure and loading.*

where $\{R\}$ and $\{P\}$ are the reaction vector and force vector, respectively. Premultiplication by $[q]^{-1}$ results in

$$\{R\} = [q]^{-1}[S]\{P\}$$

or

$$\{R\} = [b]_{RP}\{P\} \tag{2.13}$$

where $[b]_{RP}$ is an equilibrium matrix that relates the reactions to the applied forces.

Evaluating the equilibrium matrix, we have

$$[b]_{RP} = [q]^{-1}[S] = \begin{bmatrix} \frac{1}{60} & 0 & 1 \\ -\frac{1}{30} & 1 & 0 \\ +\frac{1}{30} & 0 & 0 \end{bmatrix}\begin{bmatrix} 0 & 10 & 0 & 20 \\ 0 & 1 & 0 & 1 \\ -1 & 0 & -1 & 0 \end{bmatrix} = \begin{bmatrix} -1 & \frac{1}{6} & -1 & \frac{1}{3} \\ 0 & \frac{2}{3} & 0 & \frac{1}{3} \\ 0 & \frac{1}{3} & 0 & \frac{2}{3} \end{bmatrix} \tag{2.14}$$

and thus Eq. 2.13 can be written in an expanded form as

$$\begin{Bmatrix} R_{ax} \\ R_{ay} \\ R_{by} \end{Bmatrix} = \begin{bmatrix} -1 & \frac{1}{6} & -1 & \frac{1}{3} \\ 0 & \frac{2}{3} & 0 & \frac{1}{3} \\ 0 & \frac{1}{3} & 0 & \frac{2}{3} \end{bmatrix}\begin{Bmatrix} P_{1x} \\ P_{1y} \\ P_{2x} \\ P_{2y} \end{Bmatrix} \tag{2.15}$$

As shown in Fig. 2.11, the reaction component R_{bx} is dependent on R_{by}; however, the $\{R\}$ vector will now be expanded to include this reaction component. Thus, we have

$$\begin{Bmatrix} R_{ax} \\ R_{ay} \\ R_{bx} \\ R_{by} \end{Bmatrix} = \begin{bmatrix} -1 & \frac{1}{6} & -1 & \frac{1}{3} \\ 0 & \frac{2}{3} & 0 & \frac{1}{3} \\ 0 & \frac{1}{6} & 0 & \frac{1}{3} \\ 0 & \frac{1}{3} & 0 & \frac{2}{3} \end{bmatrix}\begin{Bmatrix} P_{1x} \\ P_{1y} \\ P_{2x} \\ P_{2y} \end{Bmatrix} \tag{2.16}$$

This is an augmented form of Eq. 2.13 in which $\{R\}$ is the expanded reaction vector and $[b]$ is now enlarged to a 4×4 array. This equation can be written as

$$\begin{Bmatrix} R_{ax} \\ R_{ay} \\ R_{bx} \\ R_{by} \end{Bmatrix} = \begin{Bmatrix} -1 \\ 0 \\ 0 \\ 0 \end{Bmatrix}P_{1x} + \begin{Bmatrix} \frac{1}{6} \\ \frac{2}{3} \\ \frac{1}{6} \\ \frac{1}{3} \end{Bmatrix}P_{1y} + \begin{Bmatrix} -1 \\ 0 \\ 0 \\ 0 \end{Bmatrix}P_{2x} + \begin{Bmatrix} \frac{1}{3} \\ \frac{1}{3} \\ \frac{1}{3} \\ \frac{2}{3} \end{Bmatrix}P_{2y} \tag{2.17}$$

In this form, it is clear that each column vector and associated load term on the right-hand side of the equation gives the reactions induced by that load. Therefore, the vector itself corresponds to the reactions induced by a unit value of the associated force. This is verified by the four unit load problems shown in Fig. 2.11. In general, any element of the equilibrium matrix, b_{ij}, is the ith reaction quantity associated with a unit value of the jth force. This is graphically shown as follows:

$$\begin{Bmatrix} \cdot \\ \cdot \\ \cdot \\ R_i \\ \cdot \end{Bmatrix} = \begin{bmatrix} \cdot & \cdot & \cdot & \cdots & \cdot \\ \cdot & & & & \\ \cdot & & & & \\ \cdot & \cdot & b_{ij} & \cdots & \\ \cdot & & & & \end{bmatrix}\begin{Bmatrix} \cdot \\ \cdot \\ P_j \\ \cdot \\ \cdot \end{Bmatrix} \tag{2.18}$$

$$R_i = \cdots + b_{ij}P_j + \cdots \tag{2.19}$$

Once a set of equations in the form of Eq. 2.16 has been developed, the reactions for any loading condition that is consistent with the loading arrangement assumed can be determined. For instance, consider the problem solved in Section 2.8.1. The structure and loading arrangement is consistent with that given in Fig. 2.11 and, thus, Eq. 2.16 is applicable. In this case,

$$\{P\} = \begin{Bmatrix} P_{1x} \\ P_{1y} \\ P_{2x} \\ P_{2y} \end{Bmatrix} = \begin{Bmatrix} 0 \\ 10 \\ -10 \\ +17.3 \end{Bmatrix} \text{kips}$$

Application of Eq. 2.16 gives

$$\begin{Bmatrix} R_{ax} \\ R_{ay} \\ R_{bx} \\ R_{by} \end{Bmatrix} = \begin{bmatrix} -1 & \frac{1}{6} & -1 & \frac{1}{3} \\ 0 & \frac{2}{3} & 0 & \frac{1}{3} \\ 0 & \frac{1}{6} & 0 & \frac{1}{3} \\ 0 & \frac{1}{3} & 0 & \frac{2}{3} \end{bmatrix} \begin{Bmatrix} 0 \\ 10 \\ -10 \\ +17.3 \end{Bmatrix} = \begin{Bmatrix} 17.4 \\ 12.4 \\ 7.4 \\ 14.9 \end{Bmatrix} \text{kips}$$

These results agree with those determined in Section 2.8.1.

2.11.1 Example problem

(a) Determine the matrix expression for the reactions for the beam structure shown. Permissible applied forces are identified.

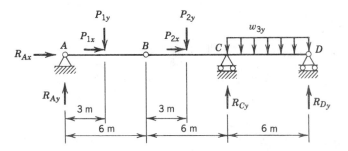

Classification:

$r_a = 4$; $r = 3 + 1 = 4$;   statically determinate, stable.

(b) Calculate the reactions for the following specific loading condition:

$$\{P\} = \begin{Bmatrix} P_{1x} \\ P_{1y} \\ P_{2x} \\ P_{2y} \\ w_{3y} \end{Bmatrix} = \begin{Bmatrix} 0 \\ 75 \text{ kN} \\ -75 \text{ kN} \\ 100 \text{ kN} \\ 20 \text{ kN/m} \end{Bmatrix}$$

Free-Body Diagrams:

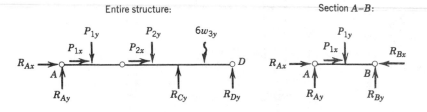

Entire structure: Section A–B:

Note:

The distributed force, w_{3y}, is replaced by an equivalent concentrated force of $6w_{3y}$.

Equations of Equilibrium: $\Sigma M_D = 0$, $\Sigma P_x = 0$, and $\Sigma P_y = 0$ for the entire structure and $\Sigma M_B = 0$ (condition equation) for section AB leads to

$$[q]\{R\} = [S]\{P\} \tag{2.12}$$

$$
\begin{bmatrix}
0 & 18 & 6 & 0 \\
1 & 0 & 0 & 0 \\
0 & 1 & 1 & 1 \\
0 & 6 & 0 & 0
\end{bmatrix}
\begin{bmatrix}
R_{Ax} \\
R_{Ay} \\
R_{Cy} \\
R_{Dy}
\end{bmatrix}
=
\begin{bmatrix}
0 & 15 & 0 & 9 & 18 \\
-1 & 0 & -1 & 0 & 0 \\
0 & 1 & 0 & 1 & 6 \\
0 & 3 & 0 & 0 & 0
\end{bmatrix}
\begin{Bmatrix}
P_{1x} \\
P_{1y} \\
P_{2x} \\
P_{2y} \\
w_{3y}
\end{Bmatrix}
$$

Solution for Reactions: Premultiplication of both sides of the foregoing equations by $[q]^{-1}$ and substitution of the given applied forces yield the final reactions.

$$\{R\} = [q]^{-1}[S]\{P\} = [b]_{RP}\{P\} \tag{2.13}$$

$$
\begin{Bmatrix}
R_{Ax} \\
R_{Ay} \\
R_{Cy} \\
R_{Dy}
\end{Bmatrix}
=
\begin{bmatrix}
-1 & 0 & -1 & 0 & 0 \\
0 & 0.5 & 0 & 0 & 0 \\
0 & 1 & 0 & 1.5 & 3 \\
0 & -0.5 & 0 & -0.5 & 3
\end{bmatrix}
\begin{Bmatrix}
0 \\
75 \\
-75 \\
100 \\
20
\end{Bmatrix}
=
\begin{Bmatrix}
75.0 \\
37.5 \\
285.0 \\
-27.5
\end{Bmatrix}
\text{kN}
$$

2.12

Displacements

In response to the forces that act on it, a structure undergoes *deformations*. At specific points throughout the structure, these deformations are manifested by a set of *displacements* as indicated in Fig. 2.12, in which each displacement is a translation or rotation at some particular point on the structure. Just as the structure must have adequate strength to withstand the forces that act on it, it must also have sufficient rigidity not to become unserviceable because of unacceptable deformations.

The dual considerations of forces and displacements are of major importance in structural analysis. In the foregoing sections of this chapter, we have seen how to calculate the reaction forces that result from designated applied forces on planar structures. In subsequent chapters, the methods for determining the forces carried by the individual members of a structure will be examined. Also, techniques for

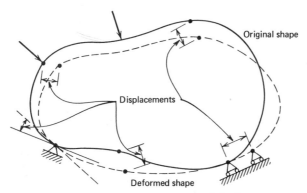

Fig. 2.12 *Structural deformations and displacements.*

determining the displacements that are associated with prescribed force conditions will be presented. At this point, it is sufficient to recognize that displacements do occur as part of the response of a structure to a loading condition.

2.13

Work

The concept of work is of prime importance in structural analysis. Consider the situation shown in Fig. 2.13a, where a force of magnitude P acts through the incremental displacement $d\Delta$. The *increment of work*, dW, done by the force is defined as

$$dW = P \cdot d\Delta \tag{2.20}$$

It is important to note that the displacement must be along the same line of action as the force in order for the force to do work.

If the magnitude of force is plotted against the displacement, then the increment of work at any load level is given by Eq. 2.20 and is represented by the crosshatched area of Fig. 2.13b. Thus, the *total work* W done over the full displacement Δ is given by

$$W = \int_0^\Delta P \cdot d\Delta \tag{2.21}$$

where P is a function of Δ, and integration is over the full range of displacement. Equation 2.21 represents the shaded area under the load–displacement curve between displacements of zero and Δ as shown in Fig. 2.13b.

One common case is when the load has a constant value over the full range of the displacement as shown in Fig. 2.13c. In this case, the work is given by

$$W = P \cdot \Delta \tag{2.22}$$

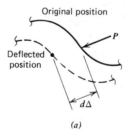

(a)

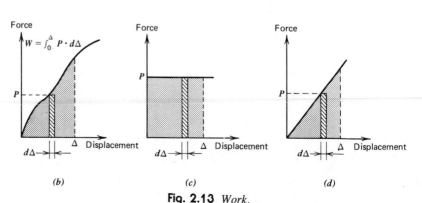

(b) (c) (d)

Fig. 2.13 *Work.*

This case corresponds to the situation in which the force is at its full magnitude on a structure and is subjected to a displacement caused by an action independent of the force. Another case of interest is when the force is proportional to the displacement. This case is represented by Fig. 2.13d and the work is

$$W = \tfrac{1}{2}P \cdot \Delta \tag{2.23}$$

This situation occurs when the force itself is causing the displacement that produces the work and the structure is linearly elastic.

Everything that has been stated thus far is related to a single force and a single displacement. For a system of n forces and n displacements, the integration indicated by Eq. 2.21 must be carried out for each of the n components. The total work is then given by

$$W = \sum_{i=1}^{n} \int_{0}^{\Delta_i} P_i \cdot d\Delta_i \tag{2.24}$$

For constant load values over the full range of the displacements, this becomes

$$W = P_1\Delta_1 + P_2\Delta_2 + \cdots + P_i\Delta_i + \cdots + P_n\Delta_n \tag{2.25}$$

or

$$W = \sum_{i=1}^{n} P_i\Delta_i = \{P\}^T\{\Delta\} \tag{2.26}$$

where $\{P\}^T$ is the transpose of the load vector and $\{\Delta\}$ is the corresponding displacement vector, each of which possesses n elements. For linear force–displacement relations, Eq. 2.24 leads to

$$W = \tfrac{1}{2}\sum_{i=1}^{n} P_i\Delta_i = \tfrac{1}{2}\{P\}^T \{\Delta\} \tag{2.27}$$

Equations 2.26 and 2.27 are multidimensional matrix versions of Eqs. 2.22 and 2.23, respectively.

2.14 Principle of Virtual Displacements

The principle of virtual displacements is credited by John Bernoulli in 1717. To develop this principle, consider the rigid body shown in Fig. 2.14a, which is in static equilibrium under the set of P forces shown. It is now imagined that the body is subjected to a translation, without rotation, as shown in Fig. 2.14b. This imagined translation is called a *virtual displacement*, and this is a fitting term because the word virtual means "being in essence or effect, but not in fact." The fact that this is a virtual displacement is indicated by a δ prefix on the displacement components δu and δv shown in Fig. 2.14b. As the body undergoes the virtual displacement, each of the P forces does virtual work by an amount equal to its magnitude times the virtual displacement along the line of action of the force. It is most convenient to work in terms of force components and the displacement components as shown in Fig. 2.14b. Letting δW_T represent the resulting virtual work of translation, we have

$$\delta W_T = P_{1x}\delta u + P_{1y}\delta v + \cdots + P_{ix}\delta u$$
$$+ P_{iy}\delta v + \cdots + P_{nx}\delta u + P_{ny}\delta v \tag{2.28}$$

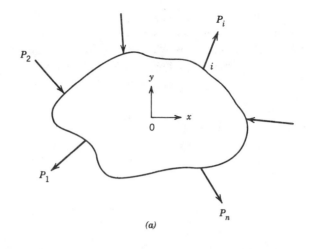

(a)

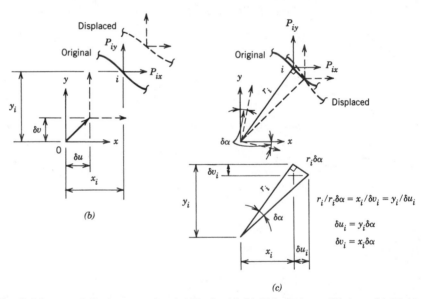

(b)

(c)

Fig. 2.14 *Virtual displacement for rigid body.* (a) *Rigid body in equilibrium.* (b) *Rigid body translation.* (c) *Rigid body rotation.*

In writing this expression, care must be exercised regarding the sign of each work term. Positive work results when the force and virtual displacement are in the same direction as shown in Fig. 2.14b, whereas negative work results when they are in opposing directions. Equation 2.28 can be rewritten as

$$\delta W_T = (P_{1x} + \cdots + P_{ix} + \cdots + P_{nx})\delta u + (P_{1y} + \cdots + P_{iy} + \cdots + P_{ny})\delta v$$

or

$$\delta W_T = \left[\sum_{i=1}^{n} P_{ix}\right]\delta u + \left[\sum_{i=1}^{n} P_{iy}\right]\delta v \qquad (2.29)$$

However, as the body is in equilibrium, each of the bracketed terms is zero as shown in Section 2.3. Thus,

$$\delta W_T = 0 \qquad (2.30)$$

It is next imagined that the body is subjected to a virtual rotation of $\delta\alpha$, which induces horizontal and vertical displacements of δu_i and δv_i, respectively, at point i, as shown in Fig. 2.14c. This set of virtual displacements causes the P force to do virtual work. Representing the virtual work of rotation by δW_R, we have

$$\delta W_R = (P_{1x} \cdot y_1\delta\alpha - P_{1y} \cdot x_1\delta\alpha) + \cdots + (P_{ix} \cdot y_i\delta\alpha$$
$$- P_{iy} \cdot x_i\delta\alpha) + \cdots + (P_{nx} \cdot y_n\delta\alpha - P_{ny} \cdot x_n\delta\alpha) \quad (2.31)$$

Again, care must be exercised in establishing the sign of each term. For the load P_i, Fig. 2.14c shows that $\delta\alpha$ induces positive work for P_{ix} and negative work for P_{iy}. Rewriting Eq. 2.31, we obtain

$$\delta W_R = \sum_{i=1}^{n} [P_{ix}y_i - P_{iy}x_i] \, \delta\alpha \quad (2.32)$$

Because the body is in equilibrium, the bracketed term, which represents the summation of moments about point O, is zero and we have

$$\delta W_R = 0 \quad (2.33)$$

Any rigid-body virtual displacement can be composed of a combination of translation and rotation. Thus, the total virtual work, δW, also must vanish. That is,

$$\delta W = \delta W_T + \delta W_R = 0 \quad (2.34)$$

Thus, the principle of virtual displacements can be summarized as follows: *If a rigid body is in equilibrium under a set of P forces and it is subjected to any virtual displacement, the virtual work done by the P forces is zero.*

2.15 Computation of Reactions by Virtual Methods

As shown in the previous section, a structure is in equilibrium if the virtual work is zero for any virtual pattern of rigid-body displacements. This fact can be used conveniently to determine the reactions for a statically determinate structure.

If the structure as a whole is subjected to a virtual rigid-body motion, an expression can be written for the associated virtual work. This expression will involve the virtual work done by the known forces as well as the unknown reaction components, and it must be set equal to zero as a condition for the structure to be in equilibrium. Each independent virtual motion will produce an independent virtual work expression.

In a simple statically determinate structure, it is convenient to select rigid-body motions of horizontal translation, vertical translation, and rotation. This forms an independent set of motion patterns, because any other motion can be formed as a combination of these three. The equations resulting from these three motions can be solved simultaneously for the unknown reaction components. In most cases, the solution of simultaneous equations can be avoided by carefully selecting the virtual motion patterns so that a single unknown reaction does virtual work for each pattern considered.

For statically determinate structures with condition equations, each condition entitles the analyst to form an additional independent virtual displacement pattern. Thus, there will always be a correspondence between the number of independent virtual motions and the number of unknown reaction components.

2.15.1 Example problem   Determine the reactions for the structure of Example 2.8.1.

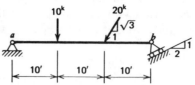

Free-Body Diagram: Again, it is convenient to show all force components on a free-body diagram.

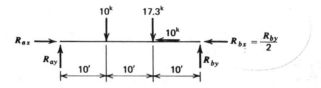

Virtual-Work Expressions: The total virtual work is the summation of the individual virtual work quantities done by each force acting on the structure. If the force and virtual displacement are in the same direction, positive virtual work results; however, negative virtual work results if the force and displacement are in opposing directions.

Vertical Motion:

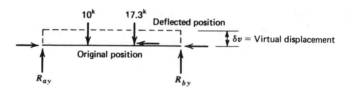

$$\text{Virtual work} = \delta W = (R_{ay} \cdot \delta v) + (R_{by} \cdot \delta v) - (10 \cdot \delta v) - (17.3 \cdot \delta v) = 0$$

$$R_{ay} + R_{by} = 27.3 \tag{a}$$

Horizontal Motion:

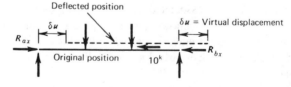

$$\text{Virtual work} = \delta W = (R_{ax} \cdot \delta u) - (R_{bx} \cdot \delta u) - (10 \cdot \delta u) = 0$$

$$R_{ax} - R_{bx} = 10 \tag{b}$$

Rotation:

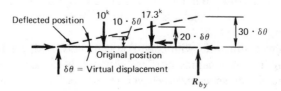

Virtual work $= \delta W = (R_{by} \cdot 30\delta\theta) - (10 \cdot 10\delta\theta) - (17.3 \cdot 20\delta\theta) = 0$

$$3R_{by} = 44.6 \qquad\qquad\qquad (c)$$

Solution:

Solution of Eq. c yields

$$R_{by} = +14.9^k$$

Then from Eq. a

$$R_{ay} = +12.4^k$$

And from Eq. b

$$R_{ax} = +17.5^k$$

These are the same as the values determined earlier.

Determine the reactions for the structure of Example 2.10.1 using the principle of **2.15.2 Example** virtual displacements.   **problem**

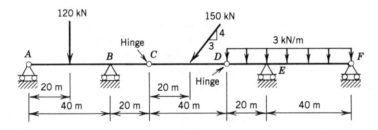

Free-Body Diagram:

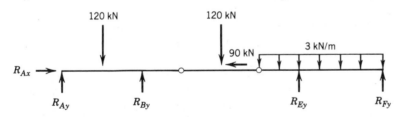

Determination of Reactions: A virtual displacement pattern is established to determine each reaction component, and the associated virtual work expression is written.

R_{Ay}:

Virtual work $= \delta W = (R_{Ay} \cdot \delta v) - \left(120 \cdot \dfrac{\delta v}{2}\right) + \left(120 \cdot \dfrac{\delta v}{4}\right) = 0$

$$R_{Ay} = 60 - 30 = +30 \text{ kN}$$

R_{Ax}, R_{By}, and R_{Ey}: Separate inducements of a virtual displacement at A, B, and E, respectively, each in the direction of the desired reaction, yield the following results:

$$R_{Ay} = +90 \text{ kN}; \quad R_{By} = +150 \text{ kN}; \quad R_{Ey} = +225 \text{ kN}$$

R_{Fy}: The virtual work of a uniform load is given by the product of the load intensity and the area swept by the virtual displacement.

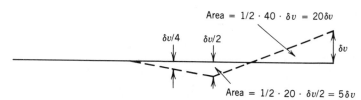

$$\text{Virtual work} = \delta W = \left(120 \cdot \frac{\delta v}{4}\right) + 3(5\delta v - 20\delta v) + (R_{Fy} \cdot \delta v) = 0$$

$$R_{Fy} = -30 + 45 = +15 \text{ kN}$$

2.16

Compatibility

As explained in Section 2.3, equilibrium requires that the forces on a structure satisfy certain relationships in order that the structure be at rest. On the other hand, *compatibility* places certain requirements on the displacements of a system to ensure that the constituent parts of the structure fit together without voids or overlaps.

To illustrate the concept of compatibility, two examples are considered. Figure 2.15a shows a two-span beam. If the spans are viewed separately, compatibility requires that they both have the same value for θ_b, the slope at the b end of each span. This ensures that when the two separate spans are rejoined, there is the proper continuity in slope at point b. It should also be noted that Δ_b, the vertical displacement at the b end of each span, is zero. This too is a requirement of compatibility.

Figure 2.15b shows two beams that are connected at point b, and these beams form a layered system that collectively resists the applied load P. As the system responds to the load, both beams deflect, and compatibility requires that Δ_b, the vertical deflection at point b, be the same for each beam.

It should be noted that the ensurance of compatibility requires the computation of displacement quantities. The techniques for computing the required displacements are covered in detail in Chapters 5 and 6.

2.17

Boundary Conditions

Structures are made stable through their interface with support points. The details concerning how a structure connects to its supports are enumerated by *boundary conditions*. Some boundary conditions are related to constraints that the supports provide on movement and are referred to as *displacement boundary conditions*. Other boundary conditions express conditions relative to the forces that can be developed at a support point. These are referred to as *force boundary conditions*.

For instance, consider the frictionless pin support shown in Fig. 2.16a. This support imposes a displacement boundary condition that constrains point a against

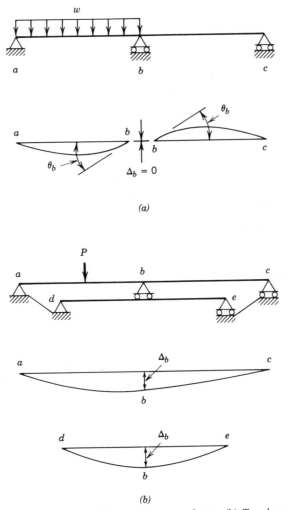

Fig. 2.15 *Examples of compatibility.* (a) *Two-span beam.* (b) *Two beams in parallel.*

translation. Thus, a reaction force must develop that will preclude translation at this point. However, the existence of a frictionless pin implies that there is no resistance against rotation and therefore no support moment. This force boundary condition requires that sufficient rotation occur to render the pin free of any moment.

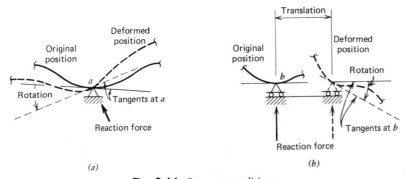

Fig. 2.16 *Support conditions.*

Similar reasoning can be applied to the roller support of Fig. 2.16*b*. This support imposes the displacement boundary condition that point *b* cannot translate normal to the surface on which the roller is mounted. However, there is no resistance to rotation or translation parallel to the support surface. These two force boundary conditions will require sufficient rotation and translation to free the support of moment or force parallel to the surface.

From these two cases, it is clear that a pattern exists that can be summarized as follows:

1. A displacement boundary condition places a constraint on a displacement quantity; however, the corresponding force quantity must be allowed to take on whatever value is required to ensure the specified displacement.

2. A force boundary condition places a constraint on a force quantity; however, the corresponding displacement quantity must be allowed to take on whatever value is required to ensure the specified force.

Care must be exercised not to overspecify boundary conditions. If a displacement condition is specified, the corresponding force condition cannot be specified, whereas if a force condition is specified, the corresponding displacement condition cannot be specified.

2.18

Additional Reading

Beaufait, F.W., *Basic Concepts in Structural Analysis*, Chapter 3, Prentice–Hall, Englewood Cliffs, N.J., 1977.

McCormac, Jack C., *Structural Analysis*, 4th Ed., Chapters 2 and 3, Harper & Row, New York, 1984.

Norris, C.H., Wilbur, J.B., and Utku, S., *Elementary Structural Analysis*, 3rd Ed., Chapter 2 and Art. 8.3, McGraw–Hill, New York, 1976.

Timoshenko, S.P., and Young, D.H., *Theory of Structures*, 2nd Ed. Chapter 1, McGraw–Hill, New York, 1965.

2.19

Suggested Problems

1. Classify each of the following structures with respect to statical determinacy and stability.

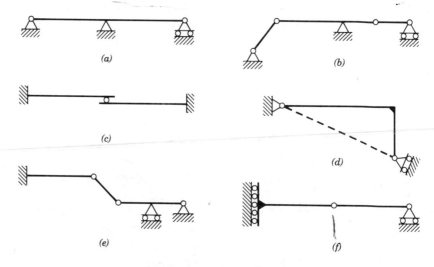

(a)

(b)

(c)

(d)

(e)

(f)

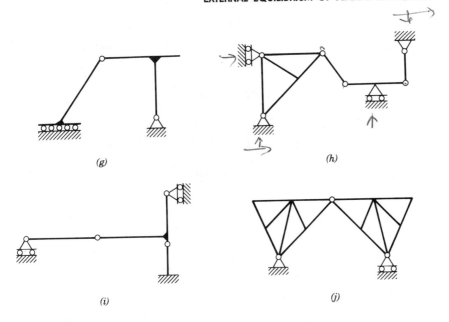

(g)

(h)

(i)

(j)

2 through 19. Use the conventional equilibrium approach to determine the support reactions for each structure under the prescribed loading.

2.

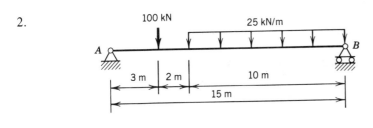

3.

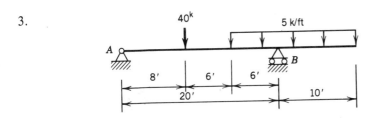

4.

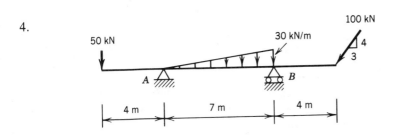

5.

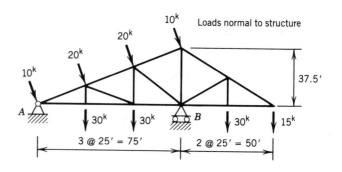

6.

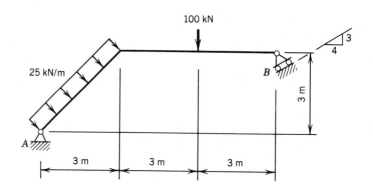

7.

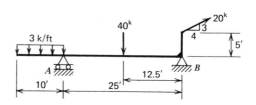

8.

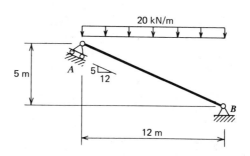

9.

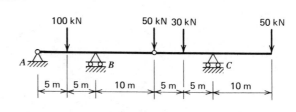

10.

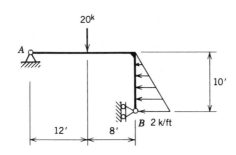

11.

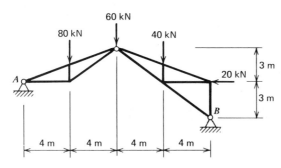

12.

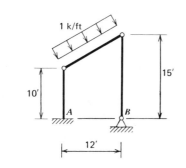

13.

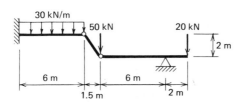

14.

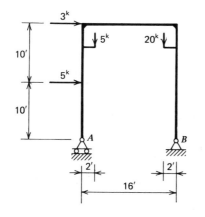

15.

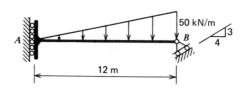

16.

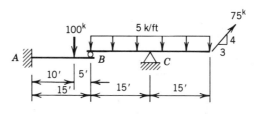

17.

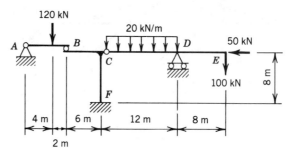

18.

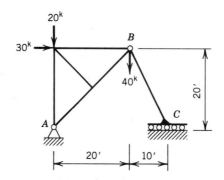

19.

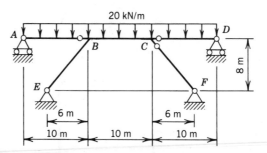

20 through 28. Use the matrix formulation method to determine the support reactions for the designated structure and loading. In each case, include only
sufficient applied force components for the prescribed loading.

20. Structure and loading of Problem 2.

21. Structure and loading of Problem 4.

22. Structure and loading of Problem 5.

23. Structure and loading of Problem 7.

24. Structure and loading of Problem 9.

25. Structure and loading of Problem 11.

26. Structure and loading of Problem 12.

27. Structure and loading of Problem 14.

28. Structure and loading of Problem 18.

29 through 36. Use the virtual work method to determine the reactions for the designated structure and loading.

29. Structure and loading of Problem 3.

30. Structure and loading of Problem 5.

31. Structure and loading of Problem 6.

32. Structure and loading of Problem 8.

33. Structure and loading of Problem 13.

34. Structure and loading of Problem 15.

35. Structure and loading of Problem 16.

36. Structure and loading of Problem 17.

CHAPTER 3

Truss bridge under construction over Ohio River, Vanport, Pa. (courtesy American Bridge Div. of U.S. Steel).

INTERNAL EQUILIBRIUM OF PLANAR PIN-CONNECTED STRUCTURES

3.1

Truss Structures

A structure that is composed of a number of members pin connected at their ends to form a stable framework is called a *truss*. If all the members lie in a plane, the structure is a planar truss. For such structures to be stable, there can be no relative movement between any two points in the structure beyond that which is caused by the deformation of the members of the framework. A common stable arrangement results from an assemblage of triangular units as shown in Fig. 3.1a.

In actuality, most truss structures are three-dimensional in nature, and a complete analysis would require consideration of the full spatial interconnection of the

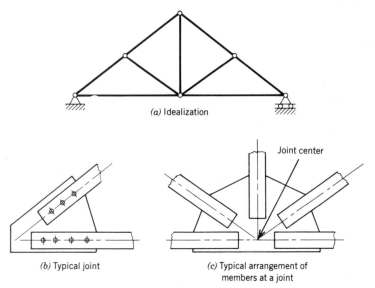

(a) Idealization

Joint center

(b) Typical joint

(c) Typical arrangement of members at a joint

Fig. 3.1 *Planar truss details.*

members. In tower structures, microwave dishes, complicated roof structures, and aerospace structures, such analyses are in order. However, in many cases, such as bridge structures and simple roof systems, the framework can be subdivided into planar components for analysis as planar trusses without seriously compromising the accuracy of the results.

Truss structures can be statically indeterminate in two different ways. If there are more reaction components than can be determined by consideration of statics, the structure is statically indeterminate externally. However, if there are more member forces than can be determined from statics, the structure is statically indeterminate internally. The topic of statical classification will be covered fully in Section 3.9.

This chapter will introduce the basic concepts involved in the analysis of pin-connected structures and will deal exclusively with planar trusses. In Chapter 8 some observations will be made concerning three-dimensional structures, and they are again discussed in Part III of the text.

3.2 Idealized versus Actual Trusses

In reality, those structures that are collectively called truss-type structures are not composed of pin-connected members. In fact, the assumption of pin-connected members is only one of several assumptions that are customarily made in the analysis of trussed frameworks.

The normal assumptions made in the analysis of planar trusses are enumerated below along with a comparison of how each of these idealizations relates to an actual structure.

Assumption 1

Members are connected at their ends with frictionless pins.

Most members are not pinned at their ends, but are welded or bolted as shown in Fig. 3.1*b*. Even if they were pinned, some friction would be present. However, this assumption gives reasonably good results for most truss-type connections.

Assumption 2
Loads and reactions are applied to the truss at the joints only.

This is realistic in terms of the applied loading because joints can be placed at points where loads are to be applied. Frequently, these loads come from other members that frame into the truss. For instance, a bridge might be composed of two parallel planar trusses as shown in Fig. 3.2. These trusses are connected by floorbeams that span between them and frame into each truss at a joint. The floor system of the deck, stringers, and floorbeams constitutes a complicated grid arrangement that is supported by the trusses at the floorbeam connection points.

There is, of course, the dead load of the truss members, which is distributed along their lengths. These loads are generally small and are lumped at the joints for purposes of analysis.

Assumption 3
The centroidal axis of each member is straight and coincides with the line connecting the joint centers at each end of the member.

To avoid eccentricity at a joint, the centroidal axes of all members framing into a joint should intersect at a point. This is called the joint center, as shown in Fig. 3.1*c*. To ensure that a member is not loaded eccentrically, the centroidal axis of each member should pass through the joint centers at its ends. This assumption can be fulfilled by careful detailing, and it is especially important for structures that are subjected to repetitive loads.

If all of the above assumptions are satisfied, then a free-body diagram of any truss member will show that it must carry axial load only, with no bending moment or transverse force present. Because of this simple loading condition, truss members are frequently referred to as *bars* and the corresponding member forces are called *bar forces*.

An analysis based on the assumptions of an ideal truss will yield the *primary*

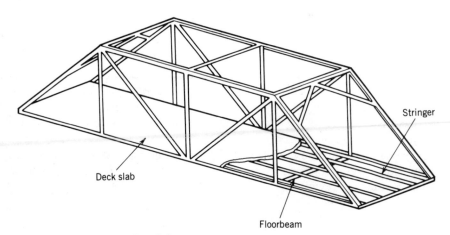

Fig. 3.2 *Typical truss bridge structure.*

bar forces. An actual truss, which will be in variance with the assumptions of an ideal truss, will have bar forces that differ from the primary bar forces. These differences are the *secondary bar forces*. In most cases, the secondary effects are negligible when compared with the primary effects; however, the analyst must be sensitive to the fact that they may be important in certain situations.

3.3
Variations in Truss Configuration

The simplest form of a stable truss is a triangular arrangement of three members as shown in Fig. 3.3*a*. Here, relative joint movements are precluded except for those that result from the member elongations that accompany the material strains. In the present context, these member deformations are ignored, and therefore the triangle forms a rigid unit. In contrast, the four-member arrangement of Fig. 3.3*b* admits relative joint movements through a realignment of the members as shown by the dashed lines.

If it is required to enlarge the structure of Fig. 3.3*a* to include an additional joint *d*, as shown in Fig. 3.3*c*, then two new members must be added to the truss. Further expansion of the truss follows the same pattern. The structure of Fig. 3.3*d* results from adding members *ed* and *ea* to establish joint *e* and members *fb* and *fe* to include joint *f*. A structure formed by this basic procedure is called a *simple truss*. Of course, in forming a simple truss, the original triangular unit must connect joints that are not located along a straight line, and each added joint must not lie along a line passing through the two joints to which it is being connected.

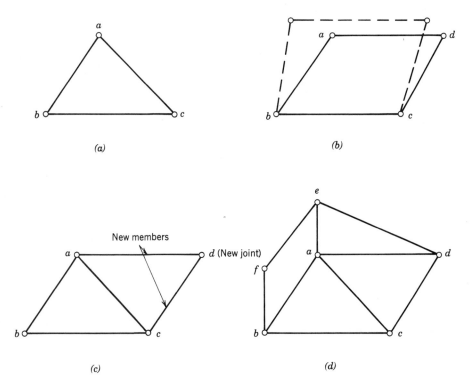

Fig. 3.3 *Arrangement of members for simple trusses.* (a) *Basic triangular arrangement.* (b) *Four-member arrangement.* (c) *Enlargement of truss to include joint at* d. (d) *Further expansion.*

Other stable truss configurations are possible that do not evolve from the basic procedure described above. For instance, a *compound truss* results when two or more single trusses are joined together to form a stable framework. Consider the two simple trusses shown in Fig. 3.4a. To join these separate trusses to form a stable compound truss, three conditions of connectivity are required to ensure that there will be no relative translation or rotation between the two sections. If the two trusses are connected at joint *a*, as shown in Fig. 3.4b, relative translations are precluded. The addition of member *bc* prevents relative rotation, and therefore the overall structure is stable.

An alternate connection scheme would be to connect the two simple trusses with three nonparallel members as shown in Fig. 3.4c. Here, members *ad* and *ac* eliminate relative translations, and member *bd* prohibits relative rotations.

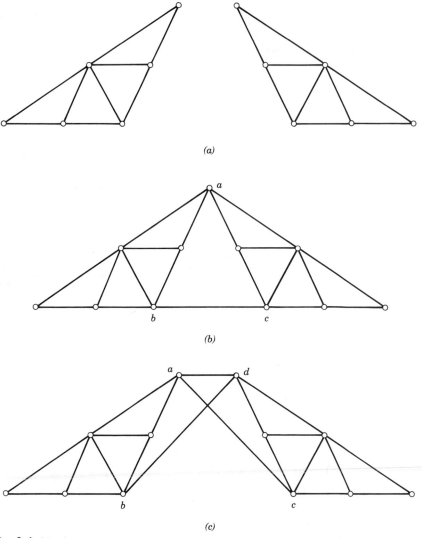

Fig. 3.4 *Member arrangement for compound trusses.* (a) *Two simple trusses.* (b) *Compound truss formation using common joint and one member.* (c) *Compound truss formation using three members.*

Trusses that cannot be classified as either simple or compound are called *complex trusses*. Structures of this type are discussed briefly in Section 3.12 and a complete treatment is offered by Timoshenko and Young, whose work is cited in Section 3.13.

3.4

There are numerous ways to identify the truss joints and to represent symbolically a given bar force. Any notation scheme is acceptable if it effectively communicates the meaning and if the analyst is consistent in its use. **Truss Joint Identification and Bar Force Notation**

In the development given here, each truss joint is labeled with a letter as shown in the portion of the truss in Fig. 3.5a. A bar force is represented by the letter F with a double subscript appended to designate the bar. For instance, F_{cb} represents the force in the member that connects joints c and b.

If a free-body diagram cuts through a member, then the force in that member must be shown to act on the cut face. Figure 3.5b shows a section through member cb and the force F_{cb} acting on the cut face of the member. Since the member must carry an axial force, the force must act in a direction that is coincident with the member direction. In the analysis process, it is frequently convenient to work with components of a bar force. Actually, a force can be resolved into any set of components that yield the given force as a resultant. However, the most commonly used components are those in the vertical and horizontal directions. In resolving a bar force into these components, the horizontal component is designated by X and the vertical component by Y. Again, a double subscript is appended to represent the bar. Figure 3.5b shows the bar force F_{cb} along with its components X_{cb} and Y_{cb}.

Because the bar force and member direction are coincident, similar triangles exist between the member and its geometric components and the bar force and its components. These similar triangles are lightly shaded in Fig. 3.5. From these similar triangles, the following relationships exist between the bar force and its components:

$$\frac{F_{cb}}{L_{cb}} = \frac{X_{cb}}{x_{cb}} = \frac{Y_{cb}}{y_{cb}} \tag{3.1}$$

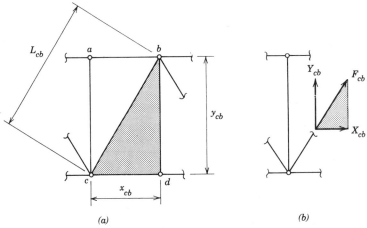

(a)   (b)

Fig. 3.5 (a) *Joint identification.* (b) *Bar force notation.*

From this, any element of the force triangle can be determined from any other. That is,

$$F_{cb} = X_{cb}\left(\frac{L}{x}\right)_{cb} = Y_{cb}\left(\frac{L}{y}\right)_{cb}$$

$$X_{cb} = F_{cb}\left(\frac{x}{L}\right)_{cb} = Y_{cb}\left(\frac{x}{y}\right)_{cb} \qquad (3.2)$$

$$Y_{cb} = F_{cb}\left(\frac{y}{L}\right)_{cb} = X_{cb}\left(\frac{y}{x}\right)_{cb}$$

3.5

Sign Convention and Member Force Representation

Along with a comprehensive and consistent notation, a sign convention must be adopted. In any analysis, there is no unique sign convention. However, whatever the analyst chooses must be used with consistency. In truss analysis, the sign convention that is almost universally employed designates a tensile force as a positive bar force and a compressive force as a negative bar force. This notation has a convenient physical interpretation in which a positive force is accompanied by a lengthening of the member, whereas a negative force is accompanied by a shortening of the member.

The process of determining an unknown bar force requires that the analyst isolate a free-body diagram in which the desired bar force is exposed as an unknown. This unknown force is assumed to act in the positive direction (tension) on the cut face as shown for member cb in Fig. 3.5b. Application of the equations of static equilibrium will yield the unknown bar force. If the solution produces a positive force, there is a dual significance in the plus sign. First, it means that the correct direction was assumed, and second, it means that the member is elongated or in tension. On the other hand, a negative force means that the assumed direction was incorrect and that the member is being shortened in compression.

Sometimes it is necessary to isolate a member and represent the forces acting on it. There are two ways in which it is convenient to represent such member forces. In the first, the forces that are exerted on the member by the joints are indicated, whereas in the second, the actions of the member on the joints are shown. These two approaches are shown in Fig. 3.6.

In the first method, a free-body diagram of the member is considered and the forces that the joints apply to its ends are shown. In the second approach, free-body diagrams of the joints are considered and the actions of the member on the joints are shown.

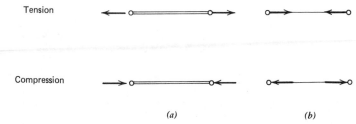

Fig. 3.6 *Bar force representation.* (a) *Action of joints on member.* (b) *Action of member on joints.*

The two bar force representations are equivalent—each has advantages in certain situations and both will be used.

The term *truss analysis* is used to describe the process whereby the bar forces in a truss are determined. For a statically determinate truss, these forces can be found by employing the laws of statics to ensure internal equilibrium of the structure. The process requires repeated use of free-body diagrams from which individual bar forces are determined.

In some cases, a single joint is isolated as shown in Fig. 3.7a. Here, the free body is acted on by a concurrent force system and moment equilibrium is automatically satisfied. Thus, two independent equilibrium equations are available, and only two unknown bar forces can be determined. The *method of joints* is a technique of truss analysis in which the bar forces are determined by the sequential isolation of joints—the unknown forces at one joint are determined and become known bar forces at subsequent joints. Simple trusses can always be analyzed by the method of joints. This is evident if it is noted that there are only two members framing into the joint that was positioned last in the formation processes. Therefore, these two bar forces can readily be determined by applying the equilibrium equations. Isolating the joints sequentially, in the reverse order from that used in forming the truss, ensures that the remaining bar forces can be determined.

In other cases, a larger portion of the structure is isolated by taking a section

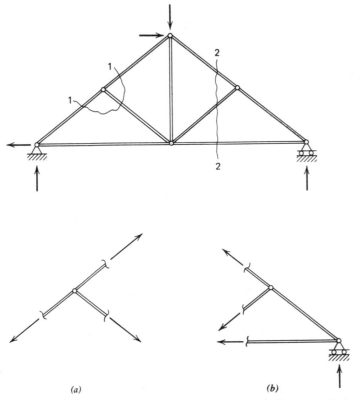

(a)   (b)

Fig. 3.7 *Typical free-body diagrams.* (a) *Section 1–1.* (b) *Section 2–2.*

through the structure as shown in Fig. 3.7b. Here, all three equations of static equilibrium are available and thus three bar forces can be determined. Analysis by this method is referred to as the *method of sections*. This technique is particularly useful when only certain bar forces are required. It is also necessary to use this approach when the sequential consideration of joints fails because there are more than two unknowns at a joint. This is frequently the case for compound trusses.

In many cases, it is found that a combination of the method of joints and the method of sections is employed. One should determine the desired bar forces by applying the equations of equilibrium to the most convenient free-body diagrams and should not be committed to some particular method.

3.7

Numerical Truss Analysis Problems

Several numerical problems will be considered in this section. Each problem has been selected because it demonstrates a specific concept regarding truss analysis. The first example illustrates the use of the method of joints for a complete truss analysis. The other examples involve both the method of joints and the method of sections.

Regardless of the method being employed, there are certain common points. For any free-body diagram, all known bar forces and the applied forces are shown to act in their proper directions. Unknown bar forces are assumed to act in tension—if the subsequent analysis yields a positive bar force, the bar is in tension, whereas a negative bar force corresponds to compression in the bar.

In most cases, there are opportunities to perform independent checks on certain bar forces. Such opportunities should always be used to ensure that the results are consistent with all of the requirements of equilibrium.

Although one would normally determine the statical classification of a truss before proceding with the analysis, a detailed discussion of this topic will be deferred to Section 3.9. All examples in this section are statically determinate and stable.

3.7.1 Example problem

Compute the bar forces in all members of the truss shown.

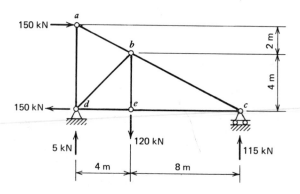

Determination of Reactions: The truss is statically determinate externally; and the reactions are determined and shown on the structure.

Determination of Bar Forces: Isolate joint a as *FBD:*

$$\sum P_x = 0 \xrightarrow{+}$$

$$150 + X_{ab} = 0; \quad X_{ab} = -150\text{kN}$$

$$Y_{ab} = X_{ab}\left(\frac{y}{x}\right)_{ab} = -150\left(\frac{2}{4}\right) = -75 \text{ kN}$$

$$F_{ab} = X_{ab}\left(\frac{L}{x}\right)_{ab} = -150\left(\frac{4.47}{4}\right) = 167.6 \text{ kN}$$

$$\sum P_y = 0 \downarrow+$$

$$F_{ad} + Y_{ab} = 0$$

$$F_{ad} = -(-75) = +75\text{kN}$$

Isolate joint d as *FBD:*

$$\sum P_y = 0 \uparrow+$$

$$75 + 5 + Y_{bd} = 0; \quad Y_{bd} = -80 \text{ kN}$$

$$X_{bd} = Y_{bd}\left(\frac{4}{4}\right) = -80 \text{ kN}$$

$$F_{bd} = Y_{bd}\left(\frac{5.66}{4}\right) = 113.2 \text{ kN}$$

$$\sum P_x = 0 \xrightarrow{+}$$

$$F_{de} + X_{bd} - 150 = 0$$

$$F_{de} = 150 - (-80) = 230 \text{ kN}$$

Isolate joint e as *FBD:*

$$\sum P_x = 0 \xrightarrow{+}$$

$$F_{ec} - 230 = 0; \quad F_{ec} = +230 \text{ kN}$$

$$\sum P_y = 0 \uparrow+$$

$$F_{eb} - 120 = 0; \quad F_{eb} = +120 \text{ kN}$$

Isolate joint c as *FBD:*

$$\sum P_x = 0 \xleftarrow{+}$$

$$X_{bc} + 230 = 0; \quad X_{bc} = -230 \text{ kN}$$

$$Y_{bc} = X_{bc}\left(\frac{4}{8}\right) = -115 \text{ kN}$$

$$F_{bc} = X_{bc}\left(\frac{8.94}{8}\right) = 257.0 \text{ kN}$$

$$\sum P_y = 0 \uparrow+$$

$$Y_{bc} + 115 = 0; \quad Y_{bc} = -115 \text{ kN} \quad \text{check}$$

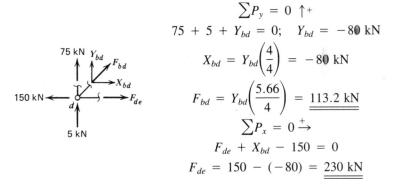

At this point, all bar forces have been determined, but joint b should be considered as a check.

Isolate joint b as *FBD:*

75 kN

150 kN →

b

80 kN →

← 230 kN

115 kN

80 kN   120 kN

$$\sum P_x \overset{+}{\rightarrow}$$

$150 + 80 - 230 = 0$ check

$$\sum P_y \downarrow +$$

$75 - 80 + 120 - 115 = 0$ check

The above treatment illustrates a complete analysis using the method of joints. Each joint has been isolated and the unknown bar forces have been determined by using the two equations of equilibrium. The unknown bar forces at one joint become the known bar forces at subsequent joints. This problem is so straightforward that it could actually be done by inspection.

The truss is lightly drawn in the sketch shown below with the applied loads and reactions. Starting at joint a, $\sum P_x = 0$ shows that bar ab must impart a horizontal component of 150 kN toward the joint. Since the bar force must act along the member, the horizontal component must be accompanied by a vertical component of 75 kN. Use of Eq. 3.2 yields the actual bar force of 167.6 kN. Figure 3.6b shows that this kind of member action at a joint corresponds to a compressive member force and there is a similar action on the joint at end b of the member. Application of $\sum P_y = 0$ at joint a shows that member ad must impart a force that pulls away from the joint with a vertical component of 75 kN. Since this member is vertical, the bar force itself is 75 kN, and Fig 3.6b shows that the member is in tension and imparts a force that pulls away from joint d.

By proceeding in the same fashion to joints d, e, and c, we are able to determine the remaining bar forces. Joint b is then used as a check to see that the bar forces determined are consistent with the requirements of equilibrium at that joint.

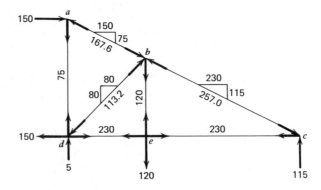

The above example was simplified by the fact that at each joint one of the two unknown bar forces acted either vertically or horizontally. Thus, the application of one equation of equilibrium led to one bar force and the second equation led to the second bar force. If both bars were inclined so that each had vertical and horizontal components, then the two bar forces would result from the simultaneous solution of the two equations of equilibrium. Frequently, the solution of simultaneous equations can be avoided by employing a moment equation, as illustrated in subsequent problems.

Determine the forces in members cd, Cd, CD, BC, and cC for the truss shown. **3.7.2 Example**
The loading and the corresponding reactions are given.   **problem**

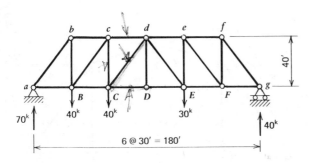

Determination of Required Bar Forces: Isolate portion left of panel c-d.

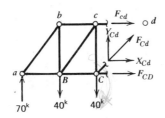

$$\sum M_C = 0 \; +\curvearrowright \quad F_{Cd} \; \& \; F_{CD} \text{ pass through } C$$
$$(70 \times 60) - (40 \times 30) + (F_{cd} \times 40) = 0$$
$$F_{cd} = \frac{-3000}{40} = \underline{\underline{-75^k}}$$
$$\sum P_y = 0 \; \uparrow+$$
$$Y_{Cd} + 70 - 40 - 40 = 0; \quad Y_{Cd} = +10^k$$
$$X_{Cd} = Y_{Cd}\left(\frac{3}{4}\right) = +7.5^k; \quad F_{Cd} = Y_{Cd}\left(\frac{5}{4}\right) = \underline{\underline{+12.5^k}}$$
$$\sum P_x = 0 \; \overset{+}{\rightarrow}$$
$$F_{CD} + X_{Cd} + F_{cd} = 0; \quad F_{CD} = -(7.5) - (-75) = \underline{\underline{+67.5^k}}$$

Isolate portion left of cut through cd, cC, and BC.

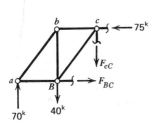

$$\sum P_x = 0 \; \overset{+}{\rightarrow}$$
$$F_{BC} - 75 = 0; \quad F_{BC} = \underline{\underline{+75^k}}$$
$$\sum P_y = 0 \; \downarrow+$$
$$F_{cC} + 40 - 70 = 0; \quad F_{cC} = \underline{\underline{+30^k}}$$

3.7.3 Example problem

Determine the forces in members U_1L_2 and U_1U_2 for the truss and loading shown. The reactions are given.

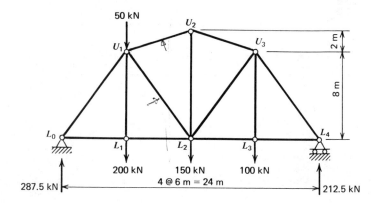

Determination of Required Bar Forces: Isolate portion left of section through U_1U_2, U_1L_2, L_1L_2.

Locate pt. o, at intersection of $F_{U_1U_2}$ & $F_{L_1L_2}$.

$$r = \frac{6 \times 8}{2} = 24 \text{ m}$$

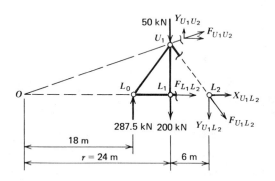

Resolve $F_{U_1L_2}$ into components at pt. L_2; then $X_{U_1L_2}$ passes through point o.

$$\sum M_o = 0 \; {}^+\!\curvearrowright \; (200 + 50)24 + (Y_{U_1L_2} \times 30) - (287.5 \times 18) = 0$$

$$Y_{U_1L_2} = \frac{-6000 + 5175}{30} = \frac{-825}{30} = -27.5 \text{ kN}$$

$$F_{U_1L_2} = -27.5 \left(\frac{10}{8}\right) = \underline{\underline{-34.4 \text{ kN}}}$$

$$\sum P_y = 0 \; \uparrow\!\!+ \; 287.5 - 250 + Y_{U_1U_2} - Y_{U_1L_2} = 0$$

$$Y_{U_1U_2} = 250 - 287.5 + (-27.5) = -65 \text{ kN}$$

$$F_{U_1U_2} = -65 \left(\frac{6.32}{2}\right) = \underline{\underline{-205.4 \text{ kN}}}$$

Determine the bar forces in members *ad* and *bd* for the tower shown.

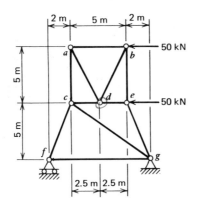

Determination of Required Bar Forces: Isolate joint *d*:

$$\sum P_y = 0 \uparrow+$$

$$Y_{ad} + Y_{bd} = 0; \quad Y_{ad} = -Y_{bd}$$

Inclinations of *ad* and *bd* are the same; therefore

$$X_{ad} = -X_{bd}; \quad F_{ad} = -F_{bd}$$

Take section below *a* and *b*. Isolate portion above section.

$$\sum P_x = 0 \xrightarrow{+}$$

$$X_{ad} - X_{bd} - 50 = 0; \quad \text{but} \quad X_{bd} = -X_{ad}$$

$$\therefore X_{ad} + X_{ad} = 50; \quad X_{ad} = +25 \text{ kN}$$

$$F_{ad} = +25 \left(\frac{5.59}{2.5} \right) = \underline{+55.9 \text{ kN}}$$

$$F_{bd} = -F_{ad} = \underline{-55.9 \text{ kN}}$$

Determine the bar forces for the truss shown below. The reactions are given.

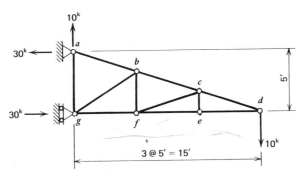

Determination of Bar Forces: The method of joints will be used, with the solution shown on the accompanying sketch of the truss.

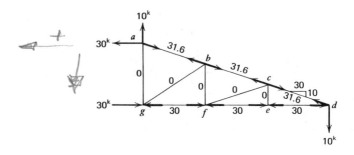

Joint d: $Y_{dc} = +10^k$; $X_{dc} = +30^k$; $F_{dc} = +31.6^k$
$X_{de} = -30^k$; $F_{de} = -30^k$

Joint e: $F_{ef} = -30^k$; $F_{ec} = 0^k$

Joint c: $\sum P_{y'} = 0$ $\therefore$ $F_{cf} = 0^k$
$\sum P_{x'} = 0$; $F_{cb} = F_{cd} = +31.6^k$

Joint f: $F_{fg} = -30^k$; $F_{fb} = 0^k$

Joint b: Similar to joint c: $F_{bg} = 0^k$; $F_{ba} = +31.6^k$

Joint g: $F_{ga} = 0^k$

3.8

Condition Equations

As discussed in Chapter 2, three independent reaction components can be determined through the systematic application of the equations of static equilibrium. Again, as explained in Section 2.10, additional reaction components can be determined if there are internal conditions that permit the expression of special condition equations.

In the case of truss structures, the conditions might take a number of forms, some of which are described in Fig. 3.8. The structure of Fig. 3.8a is merely a trussed version of a three-hinged arch. For a free-body diagram either left or right of the hinge, the condition at point b augments the equations of equilibrium applied to the entire structure to enable the solution of the four reaction components.

The structure of Fig. 3.8b is a statically determinate stable truss, and the three reaction components can be determined from statics. If this structure is modified to form the *cantilever truss* of Fig. 3.8c, the condition at point c allows for the determination of the additional reaction component at point b. Further modification yields the structure of Fig. 3.8d. Here, the link at point c provides two conditions ($M = 0$ at each end), and therefore one more reaction component is added to maintain stability—the horizontal reaction at point d.

Normally, the absence of a diagonal member in a truss panel, such as shown in Fig. 3.3b, would render the truss unstable. However, under proper support conditions, this configuration merely introduces a condition. In Fig. 3.8e, a section through panel bc reveals that no vertical force component can be transferred through this panel. This condition, coupled with the overall equations of equilibrium, enables one to determine the reaction components.

Other conditions are possible, and these are discussed in detail by Timoshenko and Young and by Norris, Wilbur, and Utku, as noted in Section 3.13. Special care must be exercised to make certain that trusses with conditions are stably arranged.

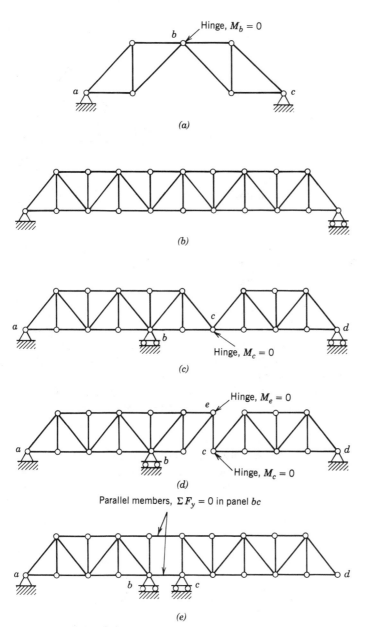

Fig. 3.8 *Condition equations for trusses.*

The most fundamental form of a stable truss is the simple triangular arrangement of three members shown in Fig. 3.9a. In this case, there are six unknowns (three bar forces and three independent reaction components) and six independent equations of equilibrium (two equations at each joint), and thus the system is statically determinate. Of course, the reactions could be determined first by applying the conditions of static equilibrium to the entire structure, and then the bar forces could be determined by joint considerations. However, in either case there are only six *independent* equations of equilibrium.

Statical Determinacy and Stability of Trusses

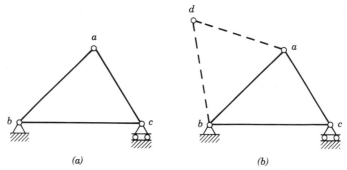

(a)   *(b)*

Fig. 3.9 *Formation of statically determinate truss.* (a) *Stable truss.* (b) *Enlargement of truss.*

If it is required to enlarge the truss to include an additional joint at d, as shown in Fig. 3.9b, then two members must be added to the truss. The new joint furnishes two new equations of equilibrium for the determination of the two new bar forces. If required, additional joints can be added; each additional joint requires two additional bars to include it as a stable portion of the truss, and it furnishes two additional equations of equilibrium. Expansion of the truss according to this pattern ensures that the truss remains stable and statically determinate, and the so-called simple truss results.

If a reaction component were added without including a condition that would enable its determination, the structure would be statically indeterminate externally. Or, if a member were added without increasing the number of joints, the structure would become statically indeterminate internally.

Based on the foregoing discussion, criteria can be developed concerning the statical determinacy of truss structures. First, the structure can be tested for external determinacy and stability according to the criteria developed in Chapter 2. That is, if r is taken as the least number of reaction components required for external stability and r_a is the actual number of reaction components, then the following criteria hold:

$r_a < r$; structure is statically unstable externally

$r_a = r$; structure is statically determinate externally

$r_a > r$; structure is statically indeterminate externally

As pointed out earlier, the conditions for $r_a \geq r$ are necessary but not sufficient conditions for statical classification. The reactions must be properly arranged to ensure stability.

For internal classification, in addition to the above definition for r, let m be the total number of bars and j the total number of joints. Then

$$2j = m + r \tag{3.3}$$

is the relationship that must be satisfied if the truss is to be statically determinate internally. The left-hand side of the equation represents the number of equilibrium equations that are available, and the right-hand side represents the total number of unknown bar forces and unknown required reaction components. Equation 3.3 can be rewritten as

$$m = 2j - r \tag{3.4}$$

In this form, m is the number of members required to form an internally statically determinate truss that connects j joints and has r reaction components required for external stability. If m_a is the actual number of bar forces in the truss, then the following conditions prevail:

$m_a < m$; truss is statically unstable internally

$m_a = m$; truss is statically determinate internally

$m_a > m$; truss is statically indeterminate internally

Care must be exercised in using the above conditions. If $m_a < m$, the truss is definitely unstable, but if $m_a \geq m$ it does not necessarily follow that the truss is stable. It is possible that the m bars are not properly arranged to ensure internal stability. Such trusses are said to have *critical form*. The subject of truss stability is discussed again in Section 3.10.

The foregoing criteria for the internal classification of a truss are clearly applicable for simple trusses because the development of Eq. 3.4 is based on principles that parallel the development of the simple truss. But what about compound trusses? The two simple trusses of Fig. 3.4a would each require three reaction components for external stability. Thus, according to the requirement of Eq. 3.4,

$$m_1 = 2j_1 - 3 \tag{3.5}$$

and

$$m_2 = 2j_2 - 3 \tag{3.6}$$

where the subscripts correspond to the two separate cases. Since $j_1 = j_2 = 6$, these equations produce $m_1 = m_2 = 9$, and since $m_a = 9$ for each structure, both are statically determinate. Addition of Eqs. 3.5 and 3.6 yields

$$m_1 + m_2 = 2j_1 + 2j_2 - 6 \tag{3.7}$$

and it is reasonable to expect this equation to reflect the requirement for statical determinacy for a compound truss that would be formed from the two simple trusses.

Rearranging the terms of Eq. 3.7, we have

$$(m_1 + m_2 + 1) = 2(j_1 + j_2 - 1) - 3 \tag{3.8}$$

If the two simple trusses are combined according to the arrangement of Fig. 3.4b, then $m = (m_1 + m_2 + 1)$ and $j = (j_1 + j_2 - 1)$ for the new structure, and therefore Eq. 3.8 expresses the requirement of Eq. 3.4 for the compound truss. Alternatively, Eq. 3.7 can be rearranged in the form

$$(m_1 + m_2 + 3) = 2(j_1 + j_2) - 3 \tag{3.9}$$

When the separate trusses are combined in the manner shown in Fig. 3.4c, then $m = (m_1 + m_2 + 3)$ and $j = (j_1 + j_2)$, and Eq. 3.9 again reflects the requirement of Eq. 3.4.

Therefore, it is clear that Eq. 3.4 is a valid requirement for the statical determinacy of compound trusses and the resulting criteria, already expressed for the internal classification of simple trusses, are valid for compound trusses.

Table 3.1 illustrates the statical classification of several truss structures. In each case, both external and internal classification is considered according to the criteria that have been developed. It should be recalled that r is equal to three (corresponding to the three equations of static equilibrium) plus the number of

TABLE 3.1 Statical Classification of Trusses

Structure	Structure Characteristics			r	External Classification Class.	$m = 2j - r$	Internal Classification Class.
	j	m	r_a				
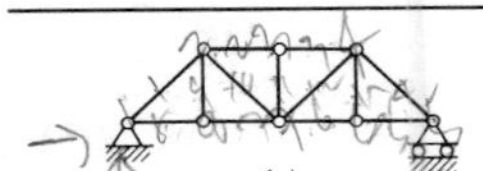 (a)	8	13	3	3	$r_a = r$, deter. stable	$13 = 16 - 3$	$m_a = m$, deter. stable
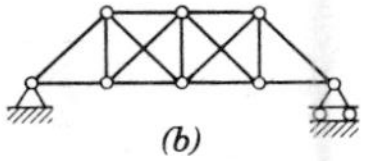 (b)	8	15	3	3	$r_a = r$, deter. stable	$13 = 16 - 3$	$m_a > m$, indet. stable
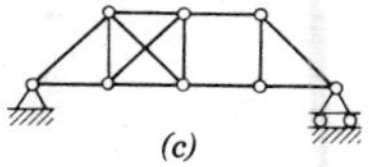 (c)	8	13	3	3	$r_a = r$, deter. stable	$13 = 16 - 3$	$m_a = m$, but unstable
 (d)	7	9	4	3	$r_a > r$, indet. stable	$11 = 14 - 3$	$m_a < m$, unstable
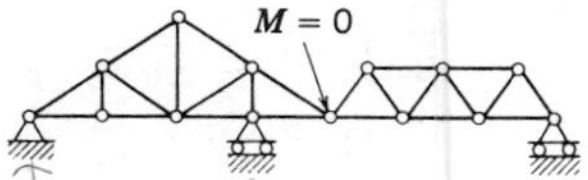 (e)	14	24	4	$3 + 1 = 4$	$r_a = r$, deter. stable	$24 = 28 - 4$	$m_a = m$, deter. stable
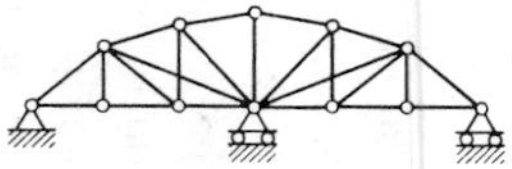 (f)	12	23	4	3	$r_a > r$, indet. stable	$21 = 24 - 3$	$m_a > m$, indet. stable
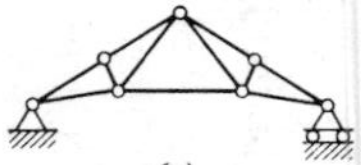 (g)	7	11	3	3	$r_a = r$, deter. stable	$11 = 14 - 3$	$m_a = m$, deter. stable
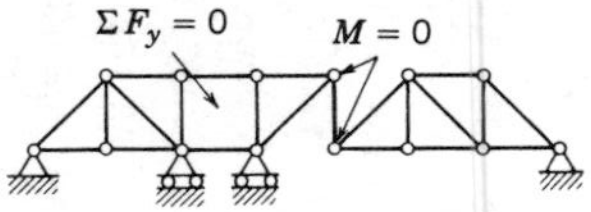 (h)	14	22	6	$3 + 3 = 6$	$r_a = r$, deter. stable	$22 = 28 - 6$	$m_a > m$, deter. stable

$M = 0$

$\Sigma F_y = 0$ $M = 0$

additional condition equations available for the determination of the reactions.

The procedures used in Table 3.1 separate the considerations for external and internal classification. These can be lumped together by replacing r in Eq. 3.3 by r_a. Equation 3.4 then takes the form

$$m = 2j - r_a \qquad (3.10)$$

Here, m is the number of members required to form a statically determinate truss that will connect j joints and the actual number of reaction components, r_a. In this case, the following conditions are valid:

$m_a < m$; structure is unstable

$m_a = m$; structure is statically determinate

$m_a > m$; structure is statically indeterminate

Again, $m_a \geq m$ are necessary but not sufficient conditions. Also, if the tests reveal that the structure is unstable or indeterminate, it is not clear whether the condition results from external or internal considerations. For instance, applying Eq. 3.10 to structure f in Table 3.1, $m = (2 \times 12) - 4 = 20$. Because $m_a = 23$, it is clear that there are three redundancies, but it is not immediately clear that one of these redundancies is external and two are internal.

3.10 Matrix Formulation of Truss Analysis

A matrix formulation for the analysis of statically determinate trusses is based on the method of joints. Each joint of an ideal truss constitutes a planar, concurrent force system for which two independent equations of equilibrium can be written. If the conditions of determinacy outlined in the previous section are satisfied, the method of joints will provide the requisite number of equations for the solution of the unknown bar forces and reactions.

One might argue that it would be more efficient to determine the reactions first and then the member forces. However, in order to maintain the unity of the method, the reactions and the bar forces will be treated as unknowns on an equal basis. This unity has advantages for computer applications. Also, for statically indeterminate structures, which will be treated later, the redundant forces might be either bar forces or reactions. Therefore, reactions should not be excluded a priori from the analysis formulation.

In order to generalize the method, consider the portion of a statically determinate truss shown in Fig. 3.10a. Joint i is a support point and the reaction R_i is shown in terms of its components R_{ix} and R_{iy}. Externally applied loads are shown to act at joint i and j; P_i has components P_{ix} and P_{iy}, and P_j has components P_{jx} and P_{jy}. The member ij is shown to act in tension with the member action on joints i and j shown in the figure. All forces are positive in the sense shown.

Figure 3.10b shows the isolation of member ij with the coordinates of the terminal points in the xy coordinate system. At end i, the force F_{ij} can be resolved into its components X_{ij} and Y_{ij} in the x and y directions, respectively, as follows:

$$X_{ij} = F_{ij} \cos \alpha = F_{ij} \frac{x_j - x_i}{L_{ij}} = F_{ij} l_{ij}$$

$$\qquad (3.11)$$

$$Y_{ij} = F_{ij} \cos \beta = F_{ij} \frac{y_j - y_i}{L_{ij}} = F_{ij} m_{ij}$$

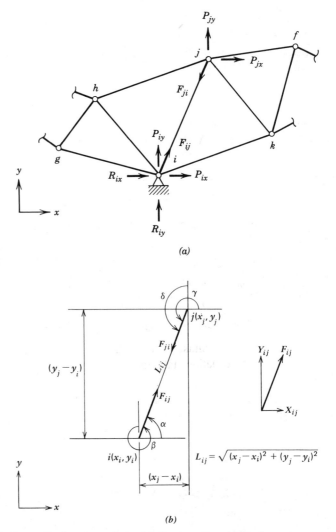

Fig. 3.10 *Generalized truss analysis.* (a) *Portion of statically determinate truss.* (b) *Member* ij.

In the above, the quantities l_{ij} and m_{ij} are called direction cosines because they give the cosines of the angles measured counterclockwise from the x and y axes, respectively, to the member ij. At end j, the force F_{ji} is resolved into its components X_{ji} and Y_{ji} in the x and y directions. These are

$$X_{ji} = F_{ji} \cos \gamma = F_{ji} \frac{x_i - x_j}{L_{ij}} = F_{ji}l_{ji}$$

$$Y_{ji} = F_{ji} \cos \delta = F_{ji} \frac{y_i - y_j}{L_{ij}} = F_{ji}m_{ji}$$

(3.12)

where l_{ji} and m_{ji} are the direction cosines for member ji.

However, because the magnitude of the bar force is the same at each end, $F_{ij} = F_{ji}$. Also, an examination of Eqs. 3.11 and 3.12 shows that $l_{ij} = -l_{ji}$ and $m_{ij} = -m_{ji}$. Thus, we have

$$X_{ij} = F_{ij}(l_{ij})$$
$$Y_{ij} = F_{ij}(m_{ij})$$
$$X_{ji} = F_{ij}(-l_{ij})$$
$$Y_{ji} = F_{ij}(-m_{ij})$$

(3.13)

Applying the equations of equilibrium to joints i and j, we get the following equations:

$$X_{ig} + X_{ih} + X_{ij} + X_{ik} + P_{ix} + R_{ix} = 0$$
$$Y_{ig} + Y_{ih} + Y_{ij} + Y_{ik} + P_{iy} + R_{iy} = 0$$
$$X_{jh} + X_{ji} + X_{jk} + X_{jf} + P_{jx} = 0$$
$$Y_{jh} + Y_{ji} + Y_{jk} + Y_{jf} + P_{jy} = 0$$

(3.14)

Or, using the relationships suggested by Eq. 3.13, these equations of equilibrium can be written as

$$F_{ig}l_{ig} + F_{ih}l_{ih} + F_{ij}l_{ij} + F_{ik}l_{ik} + P_{ix} + R_{ix} = 0$$
$$F_{ig}m_{ig} + F_{ih}m_{ih} + F_{ij}m_{ij} + F_{ik}m_{ik} + P_{iy} + R_{iy} = 0$$
$$F_{jh}l_{jh} - F_{ij}l_{ij} + F_{jk}l_{jk} + F_{jf}l_{jf} + P_{jx} = 0$$
$$F_{jh}m_{jh} - F_{ij}m_{ij} + F_{jk}m_{jk} + F_{jf}m_{jf} + P_{jy} = 0$$

(3.15)

If the structure has a total of n joints, then $2n$ equations of equilibrium can be written. In matrix form these equations take the form

$$
\begin{bmatrix}
\cdots & \cdots & \cdot & \cdots & \cdots & \cdots & \cdots & \cdots & \cdots & \cdots \\
& -l_{ig} & -l_{ih} & -l_{ik} & -l_{ij} & \cdot & \cdot & \cdot & \cdots -1 & \cdots \\
& -m_{ig} & -m_{ih} & -m_{ik} & -m_{ij} & & & & \cdots -1 & \cdots \\
& & & l_{ij} & -l_{jh} & -l_{jk} & -l_{jf} & \cdots & & \\
& & & m_{ij} & -m_{jh} & -m_{jk} & -m_{jf} & \cdots & &
\end{bmatrix}
\begin{Bmatrix}
\vdots \\ F_{ig} \\ F_{ih} \\ F_{ik} \\ F_{ij} \\ F_{jh} \\ F_{jk} \\ F_{jf} \\ \vdots \\ R_{ix} \\ R_{iy} \\ \vdots
\end{Bmatrix}
=
\begin{Bmatrix}
\vdots \\ P_{ix} \\ P_{iy} \\ P_{jx} \\ P_{jy} \\ \vdots \\ \vdots
\end{Bmatrix}
$$

(3.16)

$2n \times 2n \qquad 2n \times 1 \qquad 2n \times 1$

In this form, it is clear that there can be a total of only $2n$ unknown bar forces and reactions. If there were more than $2n$ unknowns, the system would be statically indeterminate, whereas if there were fewer than $2n$ unknowns, there would not be a unique solution. This latter case would indicate that the structure was unstable—there would not be enough members or reactions to satisfy the equilibrium requirements.

Equation 3.16 can be written in an abbreviated form as

$$[C]\begin{Bmatrix} \{F\} \\ \{R\} \end{Bmatrix} = \{P\}$$

(3.17)

where $[C]$ is the coefficient matrix of direction cosines or the overall statics matrix, $\{F\}$ and $\{R\}$ are the vectors of unknown bar forces and reaction components, respectively, and $\{P\}$ is the load vector, which includes the components of the applied forces. In a condensed form,

$$[C]\{Q\} = \{P\} \tag{3.18}$$

where $\{Q\}$ is the vector of unknown bar forces and reactions. The solution can be represented by

$$\{Q\} = [C]^{-1}\{P\} = [b]_{QP}\{P\} \tag{3.19}$$

where $[b]_{QP}$ relates the bar forces and reactions to the applied forces. We recall from Section 3.9 that having the same number of equations and unknowns does not necessarily mean that the structure is stable. If the structure were unstable because of a critical form, the matrix $[C]$ would not possess an inverse and the solution implied by Eq. 3.19 would not be possible.

Since the treatment given here is for statically determinate trusses, the equations used in the solution are based on statics. In Chapters 10 and 12, matrix formulations will be presented for the analysis of statically indeterminate trusses. These treatments take into account the deformation characteristics of the structure, and it happens that they are also applicable to statically determinate trusses. These methods have the advantage that the joint displacements are determined as a part of the solution for the bar forces. In addition, Chapters 17 and 18 present formal matrix methods of truss analysis.

3.11

Numerical Problems by Matrix Formulation

A matrix formulation does not lend itself to hand solutions except for the most simple structures. Thus, the procedure developed in Section 3.10 is of primary value when a computer is used for the solution. However, to illustrate the technique, a simple example problem will be treated according to the matrix formulation.

3.11.1 Example problem

For the truss problem of Example 3.7.1, set up the equilibrium equations in the form of Eq. 3.16 and solve for the unknown bar forces and reactions.

The accompanying figure shows the structure, the joint coordinates, and the assumed directions of the reactions and bar forces.

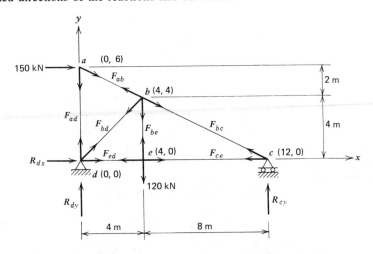

Directions Cosines:

$$l_{ij} = \frac{(x_j - x_i)}{L_{ij}}; \quad m_{ij} = \frac{(y_j - y_i)}{L_{ij}}; \quad L_{ij} = \sqrt{(x_j - x_i)^2 + (y_j - y_i)^2}$$

Member ij	$(x_j - x_i)$ m	$(y_j - y_i)$ m	L_{ij} m	l_{ij}	m_{ij}
ab	4 − 0 = 4	4 − 6 = −2	4.47	+0.895	−0.447
ba	0 − 4 = −4	6 − 4 = 2	,,	−0.895	+0.447
ad	0 − 0 = 0	0 − 6 = −6	6.00	0	−1
da	0 − 0 = 0	6 − 0 = 6	,,	0	+1
bd	0 − 4 = −4	0 − 4 = −4	5.66	−0.707	−0.707
db	4 − 0 = 4	4 − 0 = 4	,,	+0.707	+0.707
be	4 − 4 = 0	0 − 4 = −4	4.00	0	−1
eb	4 − 4 = 0	4 − 0 = 4	,,	0	+1
bc	12 − 4 = 8	0 − 4 = −4	8.94	+0.895	−0.447
cb	4 − 12 = −8	4 − 0 = 4	,,	−0.895	+0.447
ce	4 − 12 = −8	0 − 0 = 0	8.00	−1	0
ec	12 − 4 = 8	0 − 0 = 0	,,	+1	0
ed	0 − 4 = −4	0 − 0 = 0	4.00	−1	0
de	4 − 0 = 4	0 − 0 = 0	,,	+1	0

Equations of Equilibrium:

The equations of equilibrium are written for each joint and expressed in the form of Eq. 3.18:

$$[C]\{Q\} = \{P\}$$

$$
\begin{array}{l}
\Sigma P_{xa} \\
\Sigma P_{ya} \\
\Sigma P_{xb} \\
\Sigma P_{yb} \\
\Sigma P_{xc} \\
\Sigma P_{yc} \\
\Sigma P_{xe} \\
\Sigma P_{ye} \\
\Sigma P_{xd} \\
\Sigma P_{yd}
\end{array}
\begin{bmatrix}
-.895 & 0 & & & & & & \\
+.447 & +1 & & & & & & \\
+.895 & & +.707 & 0 & -.895 & & & \\
-.447 & & +.707 & +1 & +.447 & & & \\
 & & & & +.895 & +1 & & \\
 & & & & -.447 & 0 & & -1 \\
 & & & 0 & & -1 & +1 & \\
 & & & -1 & & 0 & 0 & \\
 & & 0 & -.707 & & -1 & -1 & \\
 & & -1 & -.707 & & 0 & & -1
\end{bmatrix}
\begin{Bmatrix}
F_{ab} \\
F_{ad} \\
F_{bd} \\
F_{be} \\
F_{bc} \\
F_{ce} \\
F_{ed} \\
R_{dx} \\
R_{dy} \\
R_{cy}
\end{Bmatrix}
=
\begin{Bmatrix}
+150 \\
0 \\
0 \\
0 \\
0 \\
0 \\
0 \\
-120 \\
0 \\
0
\end{Bmatrix}
$$

Solution for Bar Forces and Reactions:

$$\{Q\} = [C]^{-1}\{P\} = [b]_{QP}\{P\} =
\begin{Bmatrix}
F_{ab} \\
F_{ad} \\
F_{bd} \\
F_{be} \\
F_{bc} \\
F_{ce} \\
F_{ed} \\
R_{dx} \\
R_{dy} \\
R_{cy}
\end{Bmatrix}
=
\begin{Bmatrix}
-167.6 \\
+75.0 \\
-113.2 \\
+120.0 \\
-257.0 \\
+230.0 \\
+230.0 \\
-150.0 \\
+5.0 \\
+115.0
\end{Bmatrix}
$$

all in kN

These agree with the results of Example 3.7.1.

A solution can always be obtained by solving the complete set of simultaneous equations in the form of Eq. 3.18 as demonstrated in the preceding example problem. In this example, the coefficient matrix $[C]$ is quite sparsely populated. If the reactions R_{dx}, R_{dy}, and R_{cy} were determined by applying statics to the entire structure, then the simultaneous equations of the example problem can be solved by sequentially solving several sets of fewer simultaneous equations. Specifically, the first and second equations can be solved for F_{ab} and F_{ad}; F_{ad} is now known and F_{bd} and F_{ed} can be determined from the ninth and tenth equations; knowing F_{ed}, we now solve for F_{ce} and F_{be} from the seventh and eighth equations; and finally, the sixth equation produces F_{bc}. This is precisely what was done in the example problem of Section 3.7.1 when the truss was analyzed by the method of joints.

3.12

Complex Trusses

There are truss configurations that are neither simple nor compound. For instance, consider the simple truss shown in Fig. 3.11a. Beginning with the triangular unit *abc*, the truss is systematically assembled by adding two members to establish each new joint. A count of the members and joints verifies that Eq. 3.4 is satisfied, and therefore the structure is statically determinate. If members *cd* is removed and a member connecting joints *e* and *f* is added, as shown in Fig. 3.11b, the structure remains statically determinate; however, its form is no longer that of a simple truss. In addition, it is not a compound truss. Such a configuration is referred to as a *complex truss*. Complex trusses can be analyzed by a strategy involving a combination of the method of joints and the method of sections. Several special techniques have been developed for the analysis of complex trusses, and Timoshenko and Young have one of the most complete treatments. Their work is cited in Section 3.13. These procedures had greater importance in the precomputer age because of the need to avoid solving large numbers of simultaneous equations. However, given the present computer capabilities, the method given in Section 3.10 is viable for the analysis of complex trusses.

Complex trusses are particularly prone to possess critical forms. Although these forms are frequently difficult to detect by inspection, they always produce a singular $[C]$ matrix in Eq. 3.17 for which no inverse exists.

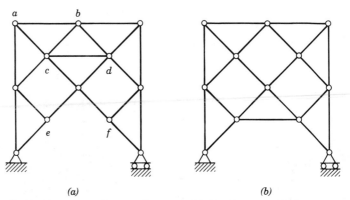

(a)   (b)

Fig. 3.11 *Formation of a complex truss.* (a) *Simple truss.* (b) *Complex truss.*

3.13

Additional Reading

Beaufait, F. W., *Basic Concepts of Structural Analysis*, Chapter 4, Prentice–Hall, Englewood Cliffs, N.J., 1977.

Hsieh, Yuan-Yu, *Elementary Theory of Structures*, Chapter 4, Prentice–Hall, Englewood Cliffs, N.J., 1970.

McCormac, J. C., *Structural Analysis*, 3rd Ed., Chapters 5 through 9, Intext Educational Publishers, New York, 1975.

Norris, C. H., Wilbur, J. B., and Utku, S., *Elementary Structural Analysis*, 3rd Ed., Chapter 4, McGraw–Hill, New York, 1976.

Sack, Ronald L., *Structural Analysis*, Chapter 3, McGraw–Hill, New York, 1984.

Timoshenko, S. P., and Young, D. H., *Theory of Structures*, 2nd Ed., Chapter 2, McGraw–Hill, New York, 1965.

3.14

Suggested Problems

1. Classify each of the following structures with respect to statical determinacy and stability. Consider both external and internal criteria. In addition, for each stable truss, note whether it is simple, compound, or complex.

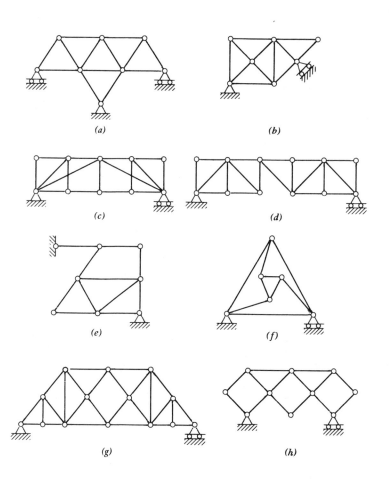

(a)   (b)

(c)   (d)

(e)   (f)

(g)   (h)

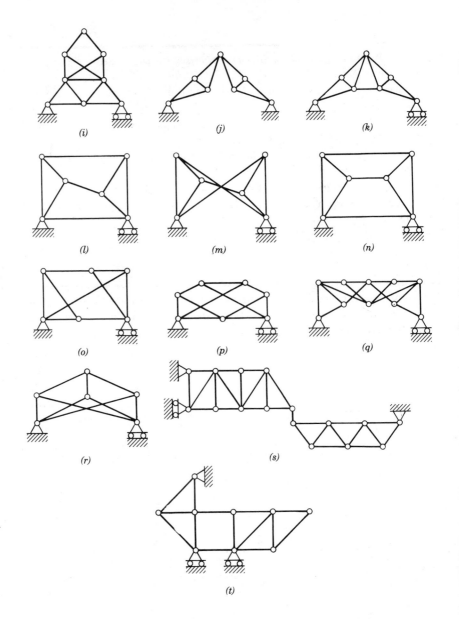

2 through 23. Determine the bar forces for all members in the given trusses.

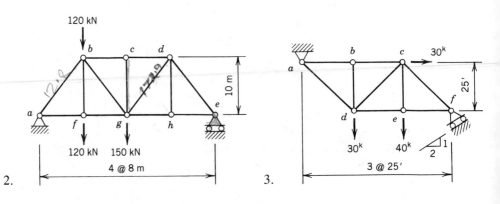

2.

3.

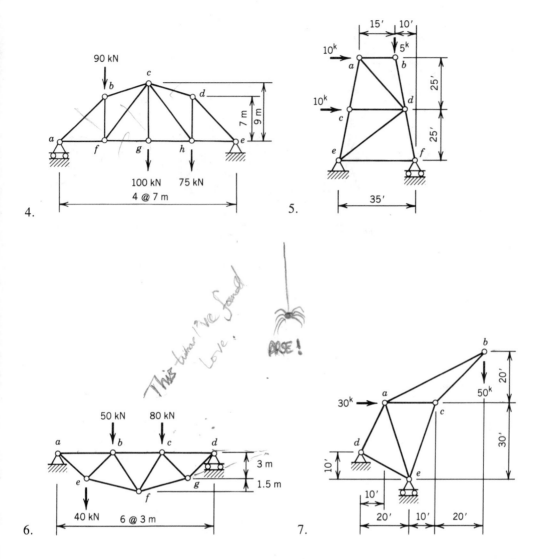

4.

5.

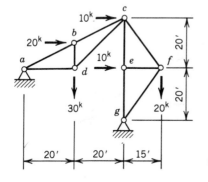

6.

7.

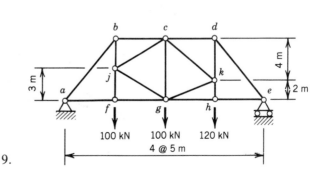

8.

9.

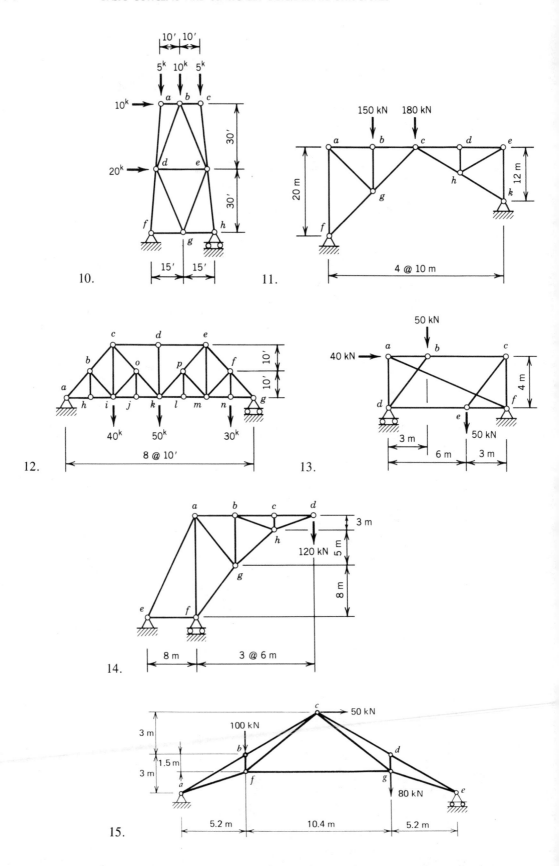

10.

11.

12.

13.

14.

15.

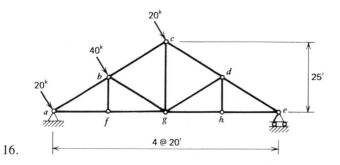

16.

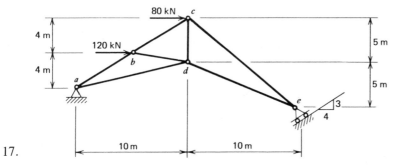

17.

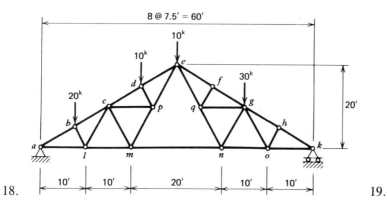

18.

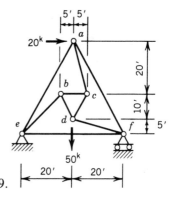

19.

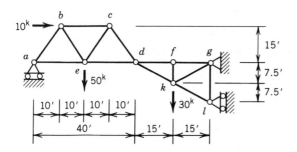

20.

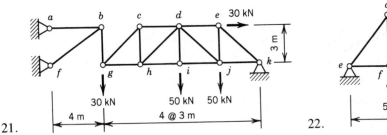

21.

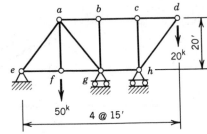

22.

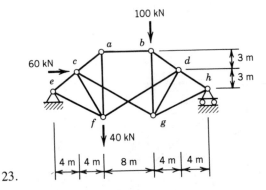

23.

24 through 29. Set up each of the following in the form of $[C]\{Q\} = \{P\}$.

24. Structure and loading of Problem 3.

25. Structure and loading of Problem 5.

26. Structure and loading of Problem 7.

27. Structure and loading of Problem 8.

28. Structure and loading of Problem 13.

29. Structure and loading of Problem 17.

30. Write a computer program for the method of truss analysis developed in Section 3.10. Use the program to solve Problems 3, 5, 7, 8, 18, 19, and 23.

31. **(a)** Set up the matrix solution for the structure shown in the form $[C]\{Q\} = \{P\}$. What can be concluded about the structure from these equations?

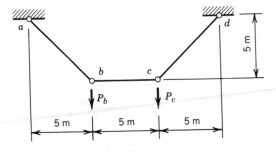

(b) If $P_b = P_c = 400$ kN, use conditions of symmetry to solve the equations of part (a) for the unknowns $\{Q\}$.

Whiskey Bay Pilot Channel Bridge, Iberville Parish, La. (courtesy Bethlehem Steel Corporation).

INTERNAL EQUILIBRIUM OF BEAM AND FRAME STRUCTURES

4.1

Beam and Frame Structures

In Chapter 3, we studied the internal equilibrium of pin-connected or truss structures in which all members were subjected to pure axial load. The assumptions that had to be satisfied for a structure to behave as a truss were enumerated in Section 3.2. If any of these restrictive assumptions are relaxed, the constituent members are subjected to flexure in addition to axial load, and the structure collectively supports the applied loads by a mechanism that is more complicated than that of a truss. Structures of this type are classified as beam and frame structures, and the members are flexural members. As was the case with trusses, we will limit our consideration to stable structures in which relative motion within the structure is limited to that which results from the deformation of members.

Most beam and frame structures are three-dimensional, and a complete analysis should recognize this fact. In many cases, however, it is appropriate to consider portions of the structural system as planar substructures. In this chapter, the emphasis is on the analysis of plane structures or planar substructures of more complicated systems. Nonplanar beam and frame structures are treated in Chapter 8 and in Part III of the text.

Planar beam and frame structures may be statically classified according to the usual practice. That is, if all the reaction and member-end forces can be determined by employing the laws of statics, the structure is statically determinate; however, if it is necessary to employ conditions of compatibility, the structure is statically indeterminate. As was true with trusses, an indeterminate structure can result from redundant reaction components, redundant internal force components, or both. Similarly, instabilities can result from an inadequate number of independent reaction components or internal force components. The topics of determinacy and stability are treated in Section 4.12.

This chapter focuses on the determination of the internal forces to which a member is subjected at any point along its length. In most cases, the structures treated here are statically determinate, but a few indeterminate structures are considered. In either case, once the member-end forces are determined, the technique for studying the internal forces is basically the same.

4.2

Internal Forces for Flexural Members

With flexural members, it is not sufficient to know the member-end forces because the member forces may vary along the length of the member. If the internal member forces are desired at some specified point, it is necessary to cut the structure at that point and determine the forces that must be applied at the cut surface to equilibrate the resulting free-body diagrams. Consider the beam ab in Fig. 4.1 and assume that it is desired to determine the internal forces at point c. If the member is cut at point c and the equations of static equilibrium are applied to portion ac,

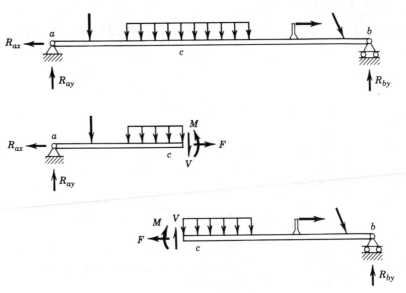

Fig. 4.1 *Internal member forces.*

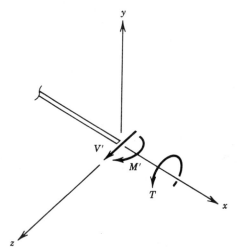

Fig. 4.2 *Torsion and out-of-plane bending.*

the internal forces at point c are found. That is, the summation of vertical forces reveals that a transverse force V is required, the summation of horizontal forces leads to the tensile force F, and the summation of moments shows that a bending moment M is required—all at point c. If portion bc is treated as a free body, equal and opposite forces are found to be required at point c. For segment ac, the equilibrating forces at point c are provided by portion cb, and vice versa. As the two portions are merged, the internal forces cancel and we are left with the original structure and loading.

Directing attention again to the internal forces, we define the force F as the *axial force,* the transverse force V as the *shear force,* and the moment M as the *bending moment.*

As has been noted, our present concern is with planar structures, and therefore the internal forces are limited to those shown in Fig. 4.1. For nonplanar structures, one must include a torsional moment about the longitudinal member axis and a bending action in the plane normal to the page. Figure 4.2 shows the torsion as T and the shear and bending moment as V' and M', respectively, for flexural action in the xz plane. These effects are considered more fully in subsequent chapters.

4.3

Notation and Sign Convention

The notation that will be used to represent the internal forces in flexural members was introduced in the previous section. That is, the axial force is represented by F, the transverse shear force by V, and the bending moment by M. In certain instances, these symbols carry a subscript to identify the point on the structure to which they correspond.

A sign convention must be adopted regarding these internal forces. Two different approaches are used; they are totally equivalent. In one case, the forces acting on each side of a cut section at a given point are considered, whereas in the other case, the forces acting on an infinitesimal element at the point in question are observed. The two methods of isolating point a are shown in Fig. 4.3a.

For axial force, tension is taken to be positive, as was the case with trusses. This condition is illustrated in Fig. 4.3b for both the section and the element

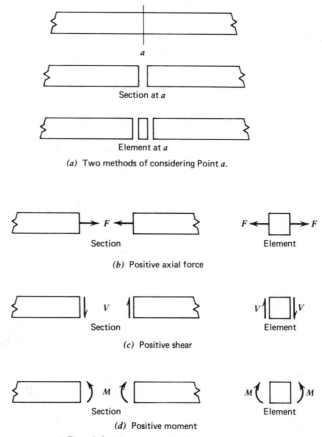

Fig. 4.3 *Notation and sign convention.*

representations. In either case, it is seen that positive axial force is represented by a vector directed away from the face on which it acts. For the element, positive axial force tends to elongate the element.

Positive shear force is shown in Fig. 4.3c by both representations. In either case, it is seen that positive shear acts downward on an exposed right-hand face, whereas it acts upward on an exposed left-hand face. Positive shear tends to deform an element by "shearing" the right side downward relative to the left side.

Positive moment is illustrated in Fig. 4.3d. As shown by either representation, positive moment places the top portion of a cut face in compression and the bottom portion in tension. Under positive moment, the element tends to curve so as to be concave upward.

4.4

Determination of Internal Forces

When the internal forces are to be determined at a given point in a beam or frame structure, the same general technique is employed that was used in trusses. That is, a free-body diagram is taken which cuts through the structure at the desired point, and each unknown internal force is assumed to act in the positive direction. The analysis proceeds, and if it produces a positive value for the unknown, then the assumed direction is correct and the internal force is indeed positive. If, however, the analysis yields a negative value for the unknown, then the assumed direction is incorrect and the internal force is negative.

In this approach, a separate free-body diagram must be considered for each point on the structure where the internal forces are desired. The analysis for each case yields the desired internal forces at that cut section.

Determine the internal forces (axial load, shear, and bending moment) at points c and d for the beam structure shown. The reactions are given.

4.4.1 Example problem

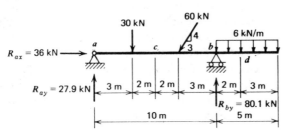

Point c:

$$\sum P_x = 0 \xrightarrow{+}$$
$$36 + F = 0; \quad F = -36 \text{ kN}$$

$$\sum P_y = 0 \uparrow +$$
$$27.9 - 30 - V = 0; \quad V = -2.1 \text{ kN}$$

$$\sum M_c = 0^+ \circlearrowright$$
$$M + (30 \times 2) - (27.9 \times 5) = 0; \quad M = 79.5 \text{ kN} \cdot \text{m}$$

The negative signs associated with F and V indicate that these forces are directed in the opposite senses from those assumed and are thus negative by our sign convention.

Point d:

$$\sum P_x = 0 \xrightarrow{+}; \quad F = 0 \text{ kN}$$

$$\sum P_y = 0 \uparrow +$$
$$V - (6 \times 3) = 0; \quad V = 18 \text{ kN}$$

$$\sum M_d = 0 \circlearrowleft +$$
$$M + (6 \times 3) \times 1.5 = 0; \quad M = -27 \text{ kN} \cdot \text{m}$$

Again, the negative sign for M indicates that the moment is opposite from that assumed and thus negative for our sign convention.

4.5

Relationships between Load, Shear, and Bending Moment

The procedure outlined in Section 4.4 can be used to determine the internal forces at as many points as are desired. However, there is a more efficient technique for determining the shears and bending moments associated with transverse loading.

To develop the necessary relationships, consider the beam shown in Fig. 4.4a, which is subjected to a distributed transverse load of varying intensity $p(x)$. We have already introduced a sign convention for shear and moment, and we now add to this the convention that upward load is positive and that x increases from left to right. An element of the beam, Δx in length, is isolated as a free-body diagram in Fig. 4.4b. On the left-hand face of the element, the load intensity is $p(x)$ and the corresponding shear and moment are $V(x)$ and $M(x)$, respectively. These functions are dependent on x as indicated; however, for convenience they are represented by p, V, and M as shown in Fig. 4.4b. Since these functions vary with x, there may be incremental changes over the distance Δx—hence, on the right-hand side of the element, the respective values of load, shear, and moment are $(p + \Delta p)$, $(V + \Delta V)$, and $(M + \Delta M)$. The load, which varies from p to $(p + \Delta p)$ over the length Δx, has an average value of p_a, and the resultant load of $(p_a \cdot \Delta x)$ acts at a distance of $(\lambda \cdot \Delta x)$ from the right-hand face of the element.

The desired relationships result from the application of the equations of static

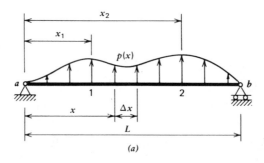

(a)

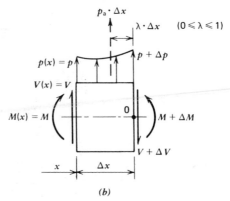

(b)

Fig. 4.4 *Statics of beam element.*

equilibrium to the element of Fig. 4.4b. Horizontal equilibrium is automatically satisfied, and the summation of forces in the vertical direction yields

$$V + (p_a \cdot \Delta x) - (V + \Delta V) = 0$$

from which

$$\frac{\Delta V}{\Delta x} = p_a$$

In the limiting process, as Δx approaches zero, the load intensity p_a approaches $p(x) = p$, and the resulting differential equation is

$$\frac{dV}{dx} = p \qquad (4.1)$$

The summation of moments about point O gives

$$M + (V \cdot \Delta x) + (p_a \cdot \Delta x)(\lambda \cdot \Delta x) - (M + \Delta M) = 0$$

or

$$\frac{\Delta M}{\Delta x} = V + (p_a \cdot \lambda \cdot \Delta x)$$

As Δx approaches zero, the term $(p_a \cdot \lambda \cdot \Delta x)$ vanishes, and the resulting differential equation is

$$\frac{dM}{dx} = V \qquad (4.2)$$

Equations 4.1 and 4.2 are differential equations of equilibrium that must be satisfied at any point along the beam length. Substitution of Eq. 4.2 into Eq. 4.1 gives

$$\frac{d^2M}{dx^2} = p \qquad (4.3)$$

which relates the bending moment at any section to the load intensity at that section.

Equation 4.1 can be written in the form

$$dV = p \, dx \qquad (4.4)$$

which upon integration gives

$$V = \int p \, dx + C_1 \qquad (4.5)$$

Equation 4.5 gives the expression for shear at any point along the beam. The constant of integration, C_1, that results from the indicated integration is evaluated by using the boundary conditions on shear. If Eq. 4.4 is integrated between sections 1 and 2, as shown in Fig. 4.4a, then we have

$$V_2 - V_1 = \Delta V_{1-2} = \int_{x_1}^{x_2} p \, dx \qquad (4.6)$$

where ΔV_{1-2} is the *change* in shear between sections 1 and 2.

If Eq. 4.2 is written in the form

$$dM = V\, dx \tag{4.7}$$

then integration will yield

$$M = \int V\, dx + C_2 \tag{4.8}$$

This gives an expression for bending moment throughout the beam, and the resulting constant of integration, C_2, is determined by using the boundary conditions on moment. Integrating Eq. 4.7 between the limits of sections 1 and 2 gives

$$M_2 - M_1 = \Delta M_{1-2} = \int_{x_1}^{x_2} V\, dx \tag{4.9}$$

where ΔM_{1-2} is the *change* in moment between sections 1 and 2.

4.6

Shear and Moment Diagrams

For any flexural member, once the load is specified, it is possible to use the relationships developed in Section 4.5 to construct graphs in which the shear and moment are plotted as ordinates against x as abscissa. The resulting graphs are called *shear and moment diagrams*. These diagrams are extremely useful because they simultaneously give the values for shear and moment at any point along the member, thus obviating the need to determine these functions point by point as was demonstrated in Section 4.4

Two methods are customarily used to construct shear and moment diagrams. Each approach is illustrated with a simple example problem in this section, and further examples are provided in greater detail in Section 4.9.

The first method is based on Eqs. 4.5 and 4.8, in which formal integration leads to mathematical expressions for the shear and moment diagrams. This method is most appropriate when the load intensity is represented by a continuous mathematical function that is readily integrated.

The second method is based on the systematic application of Eqs. 4.1, 4.2, 4.6, and 4.9. Each of these equations contributes to the construction of the shear and moment diagrams; the role of each is as follows:

Eq. 4.1: The slope of the shear diagram at any point is given by the load intensity at that point.

Eq. 4.2: The slope of the moment diagram at any point is given by the value of the shear at that point.

Eq. 4.6: The change in shear between two points is equal to the area under the load intensity diagram between these two points.

Eq. 4.9: The change in moment between two points is equal to the area under the shear diagram between these two points.

In this method, changes in shear and moment are determined by incrementally integrating the load and shear diagrams. These changes are accumulated to determine key points on the shear and moment diagrams.

Construct the shear and moment diagrams by the formal integration method for the simply supported beam under the triangular load shown. The reactions are given. **4.6.1 Example problem**

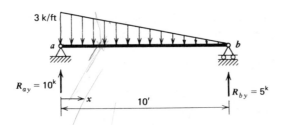

Load Diagram: The load intensity diagram is taken directly from the given problem.

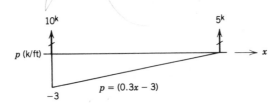

Shear Diagram: The shear diagram is determined from Eq. 4.5, where

$$V = \int p \, dx = \int (0.3x - 3) \, dx = 0.15x^2 - 3x + C_1$$

C_1 is a constant of integration, which is determined from the boundary conditions. In this case, for a section an infinitesimal distance right of point a, $V(x = 0) = + 10$ (see figure). Therefore,

$$V(x = 0) = 0.15(0)^2 - 3(0) + C_1 = +10$$
$$\therefore C_1 = +10$$

and thus,

$$V = 0.15x^2 - 3x + 10$$

The shear diagram is given by

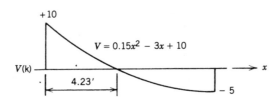

Substitution for $x = 10$ gives $V(x = 10) = -5$, which verifies the boundary condition on the shear at point b.

Moment Diagram: The moment diagram is determined from Eq. 4.8, where

$$M = \int V \, dx = \int (0.15x^2 - 3x + 10) \, dx$$
$$= 0.05x^3 - 1.5x^2 + 10x + C_2$$

C_2 is a constant of integration, which is determined from the boundary conditions. In this case, at point a, $M(x = 0) = 0$ (see figure), and thus

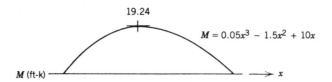

$$M(x = 0) = 0.05(0)^3 - 1.5(0)^2 + 10(0) + C_2 = 0$$
$$\therefore C_2 = 0$$

and therefore

$$M = 0.05x^3 - 1.5x^2 + 10x$$

The moment diagram is given by

19.24

$M = 0.05x^3 - 1.5x^2 + 10x$

M (ft-k) $\longrightarrow x$

Substitution for $x = 10$ gives $M(x = 10) = 0$, which verifies the boundary condition on the moment at point b.

The shear and moment can be determined at any x by substituting the value of x in the given formulae.

The point of zero shear is determined from

$$V = 0.15x^2 - 3x + 10 = 0$$

from which $x = 4.23$ ft.

Since $dM/dx = V$, the point of zero shear corresponds to the point of maximum moment. Thus, the maximum moment occurs at $x = 4.23$ ft and is given by

$$M_{max} = 0.05(4.23)^3 - 1.5(4.23)^2 + 10(4.23) = +19.24 \text{ ft-k}$$

4.6.2 Example problem

Construct the shear and moment diagrams for the beam shown under the given loading using the incremental change method. The reactions are given.

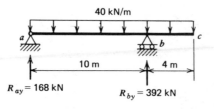

40 kN/m

10 m   4 m

$R_{ay} = 168$ kN $R_{by} = 392$ kN

Load Diagram: The load intensity diagram is taken directly from the given problem.

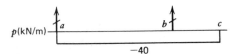

Shear Diagram: The shear diagram is determined by considering the boundary conditions on shear for each span and applying the changes in shear from Eq. 4.6. The slopes of the shear diagram are controlled by Eq. 4.1.

Point a:

$$V_a = +168 \text{ kN}$$

$$R_{ay} = 168 \text{ kN}$$

a to b:

$$\Delta V_{a-b} = \left\{ \begin{array}{c} \text{Area under load diagr.} \\ \text{between } a \ \& \ b \end{array} \right\} = \left\{ -40 \ \frac{\text{kN}}{\text{m}} \times 10 \text{ m} \right\} = -400 \text{ kN}$$

$$\frac{dV}{dx} = p = -40 \ \frac{\text{kN}}{\text{m}} \quad \text{throughout}$$

Just left of point b:

$$V_{b,\text{ left}} = V_a + \Delta V_{a-b} = 168 - 400 = -232 \text{ kN}$$

Just right of point b:

232 kN V_b, right

$$V_{b,\text{ right}} = 392 - 232 = 160 \text{ kN}$$

$R_{by} = 392$ kN

b to c:

$$\Delta V_{b-c} = \left\{ \begin{array}{c} \text{Area under load diagr.} \\ \text{between } b \ \& \ c \end{array} \right\} = \{-40 \times 4\} = -160 \text{ kN}$$

$$\frac{dV}{dx} = p = -40 \ \frac{\text{kN}}{\text{m}} \quad \text{throughout}$$

Point c: $V_c = V_{b,\text{ right}} + \Delta V_{b-c} = 160 - 160 = 0$

Thus, the shear diagram becomes

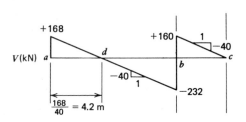

Moment Diagram: The moment diagram is determined by considering the boundary conditions on moment for each span and applying the changes in moment from Eq. 4.9. The slopes of the moment diagram are controlled by Eq. 4.2.

Point a: $M_a = 0$ (simple support)

a to d: (see shear diagram for point d)

$$\Delta M_{a-d} = \left\{ \begin{array}{c} \text{Area under shear diagr.} \\ \text{between } a \text{ \& } d \end{array} \right\} = \tfrac{1}{2}[168 \text{ kN} \times 4.2 \text{ m}]$$

$$= 352.8 \text{ kN} \cdot \text{m}$$

$$\frac{dM}{dx} = V \quad \therefore \quad \text{Slope varies from} + 168 \text{ kN to zero}$$

Point d: $M_d = M_a + \Delta M_{a-d} = 0 + 352.8 = +352.8 \text{ kN} \cdot \text{m}$

d to b: $\Delta M_{d-b} = \left\{ \begin{array}{c} \text{Area under shear diagr.} \\ \text{between } d \text{ \& } b \end{array} \right\} = \tfrac{1}{2}[-232 \times 5.8]$

$$= -672.8 \text{ kN} \cdot \text{m}$$

$$\frac{dM}{dx} = V \quad \therefore \quad \text{Slope varies from zero to} - 232 \text{ kN}$$

Just left of
    point b: $M_{b,\text{ left}} = M_d + \Delta M_{d-b} = +352.8 - 672.8$

$$= -320 \text{ kN} \cdot \text{m}$$

Just right of
    point b:

b to c:

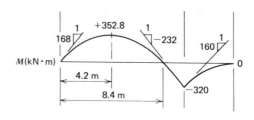

$-320 \text{ kN·m} \qquad M_{b,\text{ right}}$

$M_{b,\text{ right}} = -320 \text{ kN} \cdot \text{m}$

$$\Delta M_{b-c} = \left\{ \begin{array}{c} \text{Area under shear diagr.} \\ \text{between } b \text{ \& } c \end{array} \right\} = \tfrac{1}{2}[160 \times 4]$$

$$= +320 \text{ kN} \cdot \text{m}$$

$$\frac{dM}{dx} = V \quad \therefore \quad \text{Slope varies from} + 160 \text{ kN to zero}$$

Point c: $M_c = M_{b,\text{ right}} + \Delta M_{b-c} = -320 + 320 = 0 \text{ kN} \cdot \text{m}$

4.7

Qualitative Deflected Structures

The determination of deflections for flexural members is very important in structural analysis, and this subject will be developed quantitatively in Chapter 6. However, it is very useful to be able to sketch the general deflected shape of a structure in a qualitative fashion. This is easily accomplished by using the moment diagram in conjunction with the sign convention for moment.

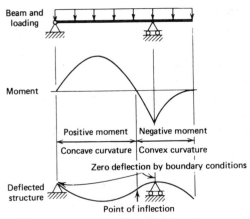

Fig. 4.5 *Qualitative deflected shape of structure.*

Recalling that positive moment is associated with member curvature that is concave from above whereas negative moment corresponds to convex curvature, we can sketch the deflected structure from the moment diagram. Points of zero moment are referred to as *points of inflection* and the curvature changes from concave to convex, or vice versa, at these points. Of course, in sketching the deflected structure, adherence to the boundary conditions must be observed. Figure 4.5 shows the qualitative deflected shape of the structure analyzed in Section 4.6.2. The same procedure can be followed for sketching the deflected shape of any structure once the moment diagram has been determined.

It is strongly suggested that the deflected shape of the structure be sketched at the conclusion of each shear and moment diagram problem. This will help the student to develop an appreciation for the way structures respond to loading. Errors can frequently be detected by alert analysts who have sharpened their intuition regarding structural behavior. Also, being able to sketch a qualitative deflected shape will be of great assistance when the quantitative deflection problem is studied in Chapter 6.

4.8

Detailed Construction of V and M Diagrams

The manner in which the key equations of Section 4.5 are used to construct shear and moment diagrams was described in Section 4.6. Two methods were described in that section—one used formal integration and the other accumulated changes in shear and moment. The second method has the greater practical value, and this section will focus attention on its application. The roles of Eqs. 4.1, 4.2, 4.6, and 4.9 were described briefly in Section 4.6. We will now expand on the use of these equations and offer some detailed guidance for the construction of shear and moment diagrams. The following summary points are given:

1. Construct a load diagram reflecting the loading on the structure in adherence with the sign convention for load. The diagram should have units of load intensity. Concentrated loads, which actually have an infinite intensity, are plotted as spikes on the load diagram with their magnitudes indicated.

2. Construct the shear diagram, noting the following:

    a. Based on Eq. 4.1, the slope of the shear diagram at any point is equal to the load intensity at that point. Specifically, this means that

- for uniform load, there is a constant slope of the shear diagram which is equal to the load intensity.
- for varying load intensity, the slope of the shear diagram will vary accordingly.
- at a point of concentrated load, the load intensity is infinite, and thus there is an abrupt change in the ordinate of the shear diagram.

    b. Based on Eq. 4.6, the change in shear between two points is equal to the area under the load intensity diagram between these two points. Thus,

- for a distributed load, the change in shear over a segment of the beam is given by the area under the load intensity diagram over the corresponding segment.
- for a concentrated load, the change in shear is given by the magnitude of the load. This is verified in Fig. 4.6.
- for general loading, the change in shear between two points is given by the total applied load between these two points.

3. Construct the moment diagram, noting the following:

    a. Based on Eq. 4.2, the slope of the moment diagram at any point is given by the value of the shear at that point. Specifically, this means that

- in regions of constant shear, there is a constant slope of the moment diagram which is equal to the shear.

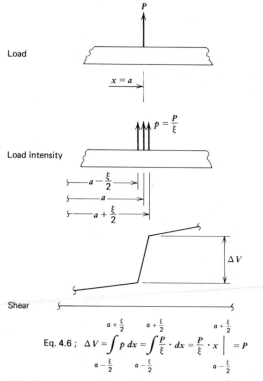

Eq. 4.6 ; $\Delta V = \int_{a-\frac{\xi}{2}}^{a+\frac{\xi}{2}} p\, dx = \int_{a-\frac{\xi}{2}}^{a+\frac{\xi}{2}} \frac{P}{\xi} \cdot dx = \frac{P}{\xi} \cdot x \Big|_{a-\frac{\xi}{2}}^{a+\frac{\xi}{2}} = P$

Fig. 4.6 *Change in shear associated with concentrated load.*

- for varying shear, the slope of the moment diagram will vary accordingly.
- at points of zero shear, there is a stationary point (zero slope) on the moment diagram.
- at a point of concentrated load, there is an abrupt change in the ordinate of the shear diagram. Thus, there is a corresponding change in the slope of the moment diagram—that is, there is a *cusp* in the moment diagram.

b. Based on Eq. 4.9, the change in moment between two points is equal to the area under the shear diagram between these two points. Thus,

- for a typical variation in shear, the change in moment over a segment of the beam is given by the area under the shear diagram over the corresponding segment.
- for a concentrated moment, there is an abrupt change in the ordinate of the moment diagram at the point where the moment is applied. This change is not evident from the transverse load diagram or the shear diagram.

In applying the techniques enumerated above, it is generally best to start at an end point of the structure. Based on the boundary conditions, the end values of shear and moment are determined, and the diagrams are then constructed by accumulating the incremental changes in shear and moment.

Some convenient area formulae for a variety of segments are given in Fig. 4.7. These are helpful in evaluating the incremental changes needed in the construction of shear and moment diagrams, and they are used in the example problems that follow. The centroidal positions that are shown are of no interest at this point, but they will be used in Chapter 6 when beam deflections are studied. The area formulae are presented for figures with a horizontal base. They are also valid for skewed-base situations when the base dimension, l, is taken as the horizontal projection of the base and the altitude is measured perpendicular to the horizontal. Two skewed-base situations are shown in Fig. 4.8. In Fig. 4.8a, the total area cannot be taken as $\frac{2}{3} lh$ because the tangent at B is not parallel to the base. However, by constructing AC parallel to this tangent, we can find the areas of the two subsections. The area of the upper section is $\frac{2}{3}lh_2$, where l is the horizontal projection of the skewed base and h_2 is the altitude of this subsection measured perpendicular to the horizontal. The area of the lower subsection is simply that of a triangle. Similarly, in Fig. 4.8b, if AC is constructed tangent to the curve AB at A, the area of the upper segment is $\frac{1}{3}lh_2$ and, again, the area of the lower subsection is that of a triangle.

4.9
Example Shear and Moment Diagram Problems

This section gives several example problems involving the construction of shear and moment diagrams. No commentary is given where a straightforward application of the principles of Section 4.8 is involved. However, commentary is provided where an extension of these principles is used. It is important to note that the locations of the points of zero shear and moment are shown on the shear and moment diagrams. Also, a qualitative sketch of the deflected structure is included for each problem.

It should be noted that all of the examples treated in this section are statically

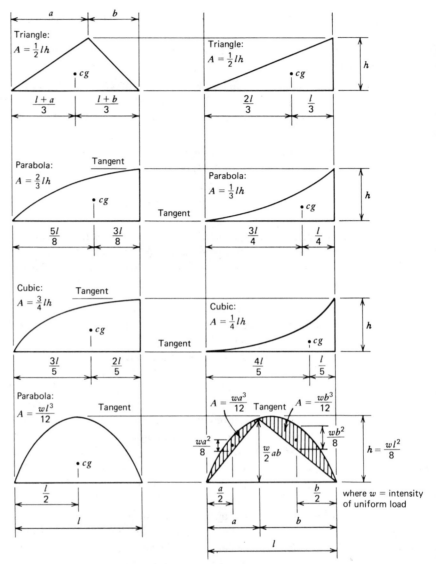

Fig. 4.7 *Frequently used area properties.*

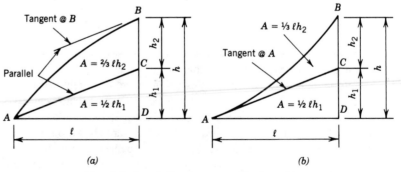

Fig. 4.8 *Parabolic segments with skewed base.*

determinate. One would normally ascertain, as a first step in the analysis procedure, the statical classification according to the criteria of Section 4.12.

Construct the complete shear and moment diagrams and sketch the deflected struc-   **4.9.1 Example**
ture for the beam and loading considered in Section 4.4.1   **problem**

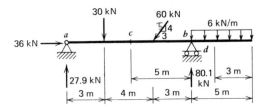

Transverse Loading Diagram: The transverse loading is taken directly from the
given problem.

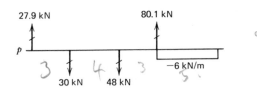

Shear diagram:

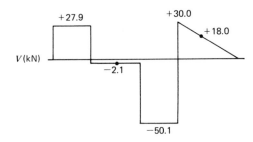

Moment Diagram:

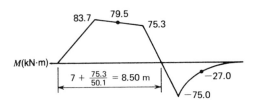

Deflected Structure:

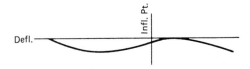

The values of shear and moment for points c and d, which were determined in
Section 4.4.1, are verified as shown on the shear and moment diagrams.

4.9.2 Example problem   Construct the complete shear and moment diagrams for the beam and loading shown.

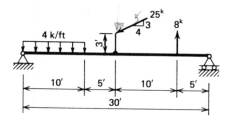

Loading Diagram: The reactions are determined and then the loading diagram takes the following form:

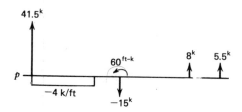

Shear Diagram:

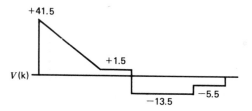

Moment Diagram:

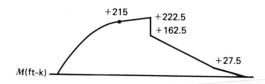

A problem arises regarding how the concentrated moment of 60 ft-k affects the moment diagram. This can be resolved by considering the moment as acting alone. It then becomes clear that there must be a 60 ft-k decrease in moment as one preceeds from left to right in order for the boundary conditions to be satisfied.

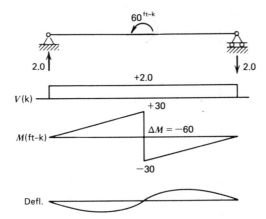

The deflected structure, which could be sketched intuitively from the given load-ing, confirms that there is a change from plus moment to minus moment as one passes from left to right over the point of concentrated moment.

Deflected Structure (for the complete loading): There is positive moment throughout, with sharper curvature in the areas of largest bending moment.

Construct the complete shear and moment diagrams for the beam and loading shown.

4.9.3 Example problem

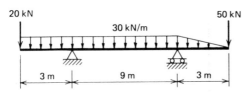

Loading Diagram:

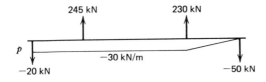

Shear Diagram:

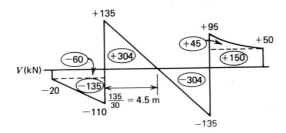

Moment Diagram: The shear diagram areas (shown on shear diagram) are used as moment changes to construct the moment diagram. The total moment change on the cantilever spans results from superimposing the individual areas as shown.

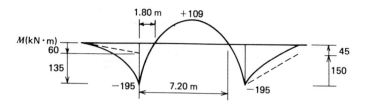

The points of zero moment are determined as follows:

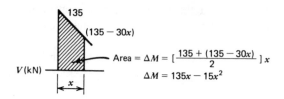

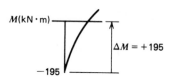

Thus,

$$135 - 15x^2 = +195$$

or

$$x^2 - 9x + 13 = 0$$

from which

$$x = 1.80, \ 7.20 \text{ m}$$

Deflected Structure:

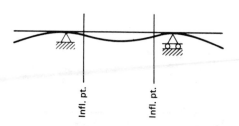

4.9.4 Example problem

Construct shear and moment diagrams for the beam and loading shown. The reactions are given.

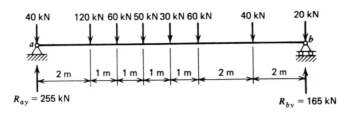

Load Diagram:

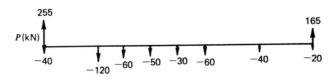

Shear Diagram:

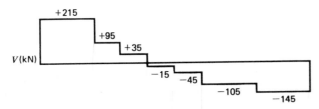

Moment Diagram:

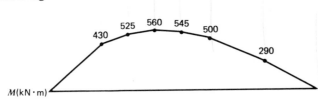

Construct the shear and moment diagrams for the frame structure shown. The column will be viewed from the left, and the normal sign conventions will be employed from that orientation.

4.9.5 Example problem

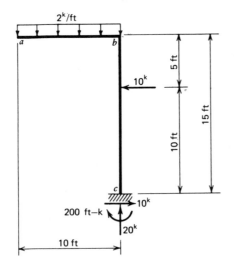

Member Free-Body Diagrams:

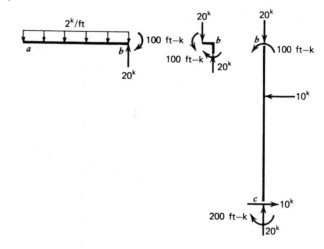

Loading Diagrams:

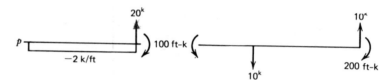

Shear Diagrams:

Moment Diagrams:

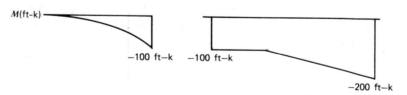

Deflected Structure:

Construct the shear and moment diagrams for the beam structure and loading shown.

4.9.6 Example problem

Tot. 325.

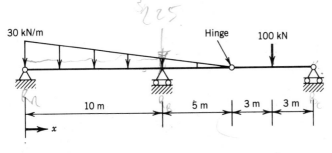

Loading Diagram:

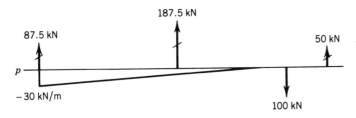

Shear Diagram:

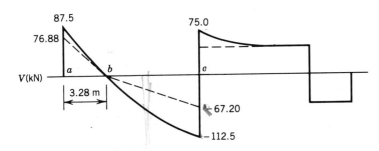

Point of Zero Shear (free-body diagram of section left of x, set $V_x = 0$):

$$\left[\frac{-30 + (-30 + 2x)}{2}\right] x = -87.5; \quad x^2 - 30x + 87.5 = 0$$

$$x = 3.28, 26.73 \text{ m}$$

Shear Diagram Areas: Dashed lines break areas into convenient subsections. Principles of Fig. 4.8 used in sections *ab* and *bc*.

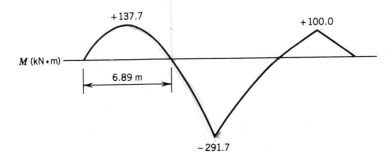

Point of Zero Moment (FBD of section left of x; set $M_x = 0$):

$$87.5x - (30 - 2x)x \cdot \frac{x}{2} - (2x)\frac{x}{2} \cdot \frac{2x}{3} = 0$$

$$x^3 - 45x^2 + 262.5x = 0; \quad x = 0, 6.89, 38.11 \text{ m}$$

Deflected Structure:

4.9.7 Example problem   Construct shear and moment diagrams for the given frame and the designated loading.

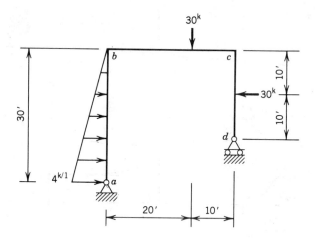

Free-Body Diagrams:

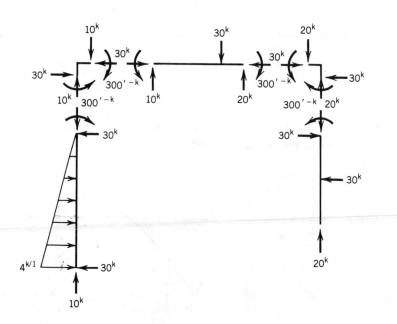

Loading Diagrams:

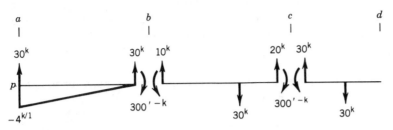

Shear Diagrams:

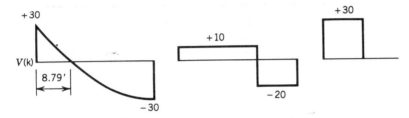

Moment Diagrams:

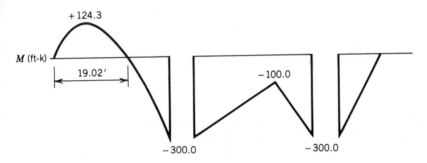

Deflected Structure:

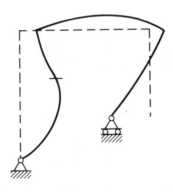

4.10

A convenient tabular method was developed by N. M. Newmark for the construc- **A Tabular** tion of shear and moment diagrams when the member is subjected to concentrated **Method for** transverse loads. For distributed loads, the method can still be used, but the dis- **Concentrated** tributed loads must be replaced by a series of equivalent concentrated loads. The **Transverse** techniques for discretizing loads are described in detail in Newmark's original **Loads** work as cited in Section 4.15 and are also discussed in Section 4.11 of this book.

The method incorporates all the essential features of shear and moment diagram construction, with its major asset being the concise arrangement of the calculations. The following example problem illustrates its application.

4.10.1 Example problem

Determine the ordinates for the shear and moment diagrams for the problem of Section 4.9.4 using the Newmark tabular method.

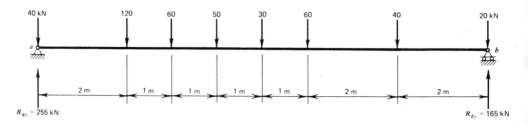

Units

Load P	+215			−120	−60		−50		−30		−60			−40			+145 kN
Shear V	0	+215			+95		+35		−15		−45		−105			−145		0 kN
Mom. Chg. $\Delta M = V \cdot h$		+430			+95		+35		−15		−45		−210			−290		kN · m
Moment M	0			+430	+525		+560		+545		+500			+290			0 kN · m

Notes:

Shear. Begin at point *a* with boundary condition that $V = 0$ just left of support. Proceed left to right, with the changes in shear given by loading diagram. Shear diagram should close with correct boundary condition at point *b*.

Moment change. The moment change in each panel is equal to the area under shear diagram in that panel.

Moment. Begin at point *a* with boundary condition that $M = 0$. Proceed left to right by accumulating the moment changes. Moment diagram should close with correct boundary condition at point *b*.

4.11

Discretization of Forces

Newmark's tabular method is based on the application of concentrated loads; however, this does not preclude the use of this method where distributed loads exist. The technique used is to replace a distributed load with an equivalent set of concentrated loads. This process is known as the *discretization of forces*, and it makes possible the use of the method for any loading condition.

Consider the beam segment shown in Fig. 4.9a, in which the load intensity varies linearly between panel points. This loading is applied directly to the beam; however, let us imagine that the load is supported on a series of imaginary stringers that are then supported by the original beam at the panel points as shown in Fig. 4.9b. The reactions for these stringers form a set of concentrated loads at the panel point as illustrated in Fig. 4.9c.

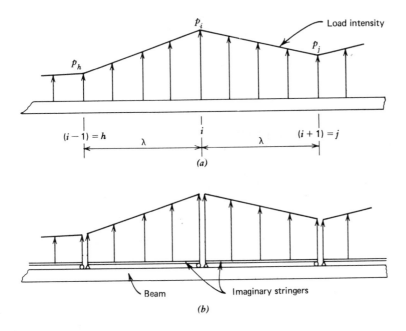

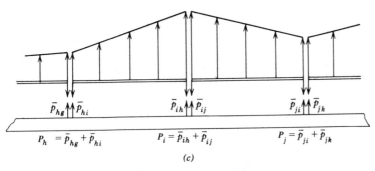

Fig. 4.9 *Discretization of forces.*

The total concentrated load at point i results from the summation of the stringer reactions at the i-end of stringer ih and ij. From Fig. 4.9c, we have

$$P_i = \bar{p}_{ih} + \bar{p}_{ij} \tag{4.10}$$

where $\bar{p}_{ih}$ and $\bar{p}_{ij}$ are the stringer reactions and P_i is the total discrete load at point i.

The reactions for a typical stringer are determined by applying statics to the simply supported stringer and are shown in Fig. 4.10. Making appropriate substitution into Eq. 4.10, we obtain

$$P_i = \frac{\lambda}{6}(2p_i + p_h) + \frac{\lambda}{6}(2p_i + p_j) \tag{4.11}$$

$$= \frac{\lambda}{6}(p_h + 4p_i + p_j)$$

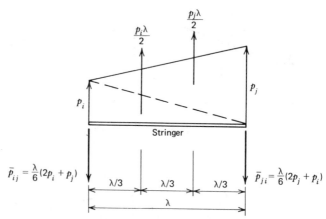

$$\bar{p}_{ij} = \frac{\lambda}{6}(2p_i + p_j)$$

$$\bar{p}_{ji} = \frac{\lambda}{6}(2p_j + p_i)$$

Fig. 4.10 *Stringer reactions.*

This equation is valid only when loading exists for a distance of λ on either side of point i. If point i is at the left end of the member, $P_i = \bar{p}_{ij}$ whereas at the right end, $P_i = \bar{p}_{ih}$. For a beam structure that is divided into n equal panels, there are $n + 1$ panel points. For this case, load discretization can be represented by the following matrix equation:

$$
\begin{Bmatrix} P_1 \\ P_2 \\ P_3 \\ \cdot \\ \cdot \\ \cdot \\ P_{n-1} \\ P_n \\ P_{n+1} \end{Bmatrix} = \frac{\lambda}{6}
\begin{bmatrix}
2 & 1 & 0 & \cdot & \cdot & \cdot & \cdot & \cdot & 0 \\
1 & 4 & 1 & 0 & \cdot & \cdot & \cdot & \cdot & 0 \\
0 & 1 & 4 & 1 & 0 & \cdot & \cdot & \cdot & 0 \\
& & & & & & & & \\
& & & & & & & & \\
& & & & & & & & \\
0 & \cdot & \cdot & \cdot & 0 & 1 & 4 & 1 & 0 \\
0 & \cdot & \cdot & \cdot & \cdot & 0 & 1 & 4 & 1 \\
0 & \cdot & \cdot & \cdot & \cdot & \cdot & 0 & 1 & 2
\end{bmatrix}
\begin{Bmatrix} p_1 \\ p_2 \\ p_3 \\ \cdot \\ \cdot \\ \cdot \\ p_{n-1} \\ p_n \\ p_{n+1} \end{Bmatrix}
\tag{4.12}
$$

For the special case of a uniformly distributed load of intensity p, Eq. 4.12 reduces to

$$
\begin{Bmatrix} P_1 \\ P_2 \\ \cdot \\ \cdot \\ \cdot \\ P_n \\ P_{n+1} \end{Bmatrix} = \lambda p
\begin{Bmatrix} 0.5 \\ 1 \\ \cdot \\ \cdot \\ \cdot \\ 1 \\ 0.5 \end{Bmatrix}
\tag{4.13}
$$

The discretization formulae presented herein are based on a linear variation in load intensity between panel points. For other variations, these equations can be used as an approximation—the smaller λ is, the better the approximation becomes. It is also possible to construct discretization formulae for other load variations, but they will not be developed here.

The shear and moment diagrams that result from using the discretized loads will possess shapes that reflect the concentrated loads and not the original distributed load. Thus, some care must be used in interpreting the results.

Construct the shear and moment diagrams for the beam and loading of Section **4.11.1 Example**
5.6.1. Use a set of discretized loads in which λ = 2 feet and compare the results **problem**
with those obtained earlier.

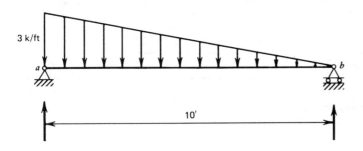

Distributed Load Diagram:

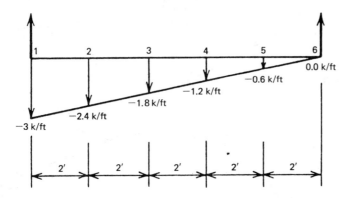

Load Discretization:

$$
\begin{Bmatrix} P_1 \\ P_2 \\ P_3 \\ P_4 \\ P_5 \\ P_6 \end{Bmatrix} = \frac{2}{6} \begin{bmatrix} 2 & 1 & 0 & 0 & 0 & 0 \\ 1 & 4 & 1 & 0 & 0 & 0 \\ 0 & 1 & 4 & 1 & 0 & 0 \\ 0 & 0 & 1 & 4 & 1 & 0 \\ 0 & 0 & 0 & 1 & 4 & 1 \\ 0 & 0 & 0 & 0 & 1 & 2 \end{bmatrix} \begin{bmatrix} -3.0 \\ -2.4 \\ -1.8 \\ -1.2 \\ -0.6 \\ 0 \end{bmatrix} = \begin{Bmatrix} -2.8 \\ -4.8 \\ -3.6 \\ -2.4 \\ -1.2 \\ -0.2 \end{Bmatrix} \qquad (4.12)
$$

Discrete Load Diagram:

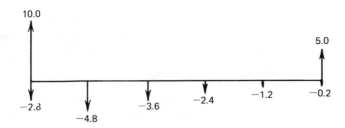

Shear and Moment Diagrams by Newmark's Method:

Load, P		7.2		-4.8		-3.6		-2.4		-1.2		4.8	kips
Shear, V	0	$+7.2$		$+2.4$		-1.2		-3.6		-4.8	0		kips
Mom. Chg., ΔM		$+14.4$		$+4.8$		-2.4		-7.2		-9.6			ft-kips
Moment, M	0		14.4		19.2		16.8		9.6		0		ft-kips

Shear Diagram:

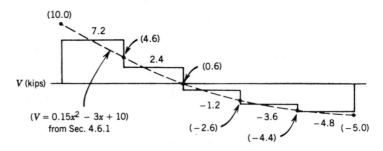

$(V = 0.15x^2 - 3x + 10)$
from Sec. 4.6.1

V (kips)

Moment Diagram:

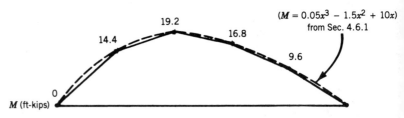

$(M = 0.05x^3 - 1.5x^2 + 10x)$
from Sec. 4.6.1

M (ft-kips)

In this case, the discretized method gives the exact values for moments at the panel points. This occurs because the discretization formulae were exact for the triangular loading. If the discretization process were approximate, there would be some discrepancies in the moment values.

4.12

Statical Determinacy and Stability of Beam and Frame Structures

This chapter has thus far dealt exclusively with statically determinate structures; however, it is necessary to establish criteria for the statical classification of beam and frame structures similar to those for trusses in Section 3.9.

First, the structure can be tested for external determinacy and stability according to the criteria developed in Chapter 2. According to that approach, if r is taken as the least number of independent reaction components that are required for external stability and r_a is the actual number of independent reaction components, then the following criteria hold:

$r_a < r$; structure is statically unstable externally

$r_a = r$; structure is statically determinate externally

$r_a > r$; structure is statically indeterminate externally

The reader is reminded that the conditions $r_a \geq r$ are necessary but not sufficient conditions for the statical classification of the structure. The reaction components must be properly arranged to ensure stability. However, just as was true for trusses, statically determinate reaction components do not ensure that all of the member forces can be determined from statics.

It is not particularly useful to classify beam-type structures internally, as was done with trusses; however, an overall statical classification, which includes both internal and external considerations, is useful. Here, it is necessary to note that for any beam-type member, the internal forces are completely defined throughout the member once the axial force, shear force, and bending moment are determined at any point along the member. Thus, three member actions must be determined for each member of the structure. Thus, if m_a is taken to represent the actual number of members, and r_a continues to represent the actual number of independent reaction components, then there are $(3m_a + r_a)$ unknown quantities that must be determined.

To determine these unknown quantities, Eq. 2.1 can be used to generate three equations of equilibrium at each joint of the structure. In addition, there may be equilibrium equations that can be written by virtue of conditions of construction. Thus, if there are j joints in the structure and n condition equations, then there are $(3j + n)$ equations available for the solution. Thus,

$$(3m_a + r_a) = (3j + n) \tag{4.14}$$

is the relationship that must be satisfied if a beam-type structure is to be statically determinate. This leads to the following criteria:

$(3m_a + r_a) < (3j + n)$; structure is statically unstable
$(3m_a + r_a) = (3j + n)$; structure is statically determinate
$(3m_a + r_a) > (3j + n)$; structure is statically indeterminate

Again, care must be used in applying these criteria. If $(3m_a + r_a) < (3j + n)$, the structure is definitely unstable; however, $(3m_a + r_a) \geq (3j + n)$ are necessary but not sufficient conditions for the classification of the structure. The individual members, the construction conditions, and the reaction components must be properly arranged to ensure stability of the structure.

In applying these criteria, it is recalled from Section 2.10 that for external classification, $r = 3 + n$. Here, n is the number of condition equations that can be written that will involve the external reaction components only at the exclusion of any member forces. However, the n value used for the overall classification in Eq. 4.14 includes all the condition equations that can be written throughout the structure, some of which will be related to internal quantities.

Several structures are classified in Table 4.1 with regard to both the external criteria and the overall criteria. Here, if n for the external classification is different from n for the overall classification, it is so noted. A comparison of the external classification and the overall classification will indicate the degree of internal indeterminacy. For instance, case c is statically indeterminate externally to the fourth degree, and it has an overall indeterminacy to the sixth degree. Thus, the structure is statically indeterminate internally to the second degree.

The construction detail at the center support of case d introduces one condition equation. The pin ensures that there is no support moment; however, a condition of zero moment must be specified at one member end in order to ensure the intended pin end for each member. In fact, if there are p members that frame into

TABLE 4.1 Statical Classification of Beam and Frame Structures

Structure	Structure Characteristics				External Classification		Overall Classification		
	j	n	m_a	r_a	$r = 3 + n$	Classif.	$(3m_a + r_a)$	$(3j + n)$	Classif.
(a) $M = 0$	4	1	3	3	$3 + 1 = 4$	$3 < 4$ Unstable	$(9 + 3) = 12$	$(12 + 1) = 13$	$12 < 13$ Unstable
(b)	8	0	10	3	3	$3 = 3$ Determinate	$(30 + 3) = 33$	$(24 + 0) = 24$	$33 > 24$ Indet., 9th degree, stable
(c) $M = 0$ / $M = 0$	8	3 / 2 (Ext)	8	9	$3 + 2 = 5$	$9 > 5$ Indet., 4th degree, stable	$(24 + 9) = 33$	$(24 + 3) = 27$	$33 > 27$ Indet., 6th degree, stable
(d)	5	1 / 0 (Ext)	5	5	3	$5 > 3$ Indet., 2nd degree, stable	$(15 + 5) = 20$	$(15 + 1) = 16$	$20 > 16$ Indet., 4th degree, stable
(e) $M = 0$	4	1	3	4	$3 + 1 = 4$	$4 = 4$ Determinate stable	$(9 + 4) = 13$	$(12 + 1) = 13$	$13 = 13$ Determinate, stable

a common pin support, $(p - 1)$ conditions must be introduced. Also, note that the free end of a cantilever beam must be treated as a joint. This point is illustrated in the classification of case e.

4.13

Matrix Formulation of V and M Diagrams

In the case of trusses, a single force component (axial force) for each member is adequate to define the force condition at all points along the member. For a planar beam or frame structure, the situation is more complex. Consider the beam structure shown in Fig. 4.11a, where point i is a typical support point. In this case, axial forces are excluded, and only flexural action is considered. A beam segment hi is isolated in Fig. 4.11b, in which the member-end forces are shown to act in positive directions according to the conventions for shear and moment. Here, two force components are necessary to define the member force condition. That is, if the shear and moment are specified at point h, the internal shear and moment can be determined at any point along the segment. Specifically, the member-end forces at the i end can be expressed in terms of those at the h end. That is,

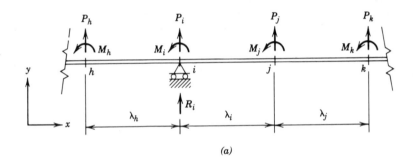

(a)

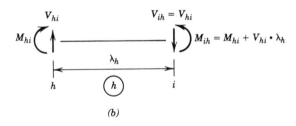

(b)

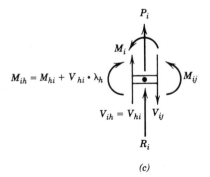

(c)

Fig. 4.11 *Beam structure and loading. (a) Node point and load identification. (b) Beam segment* hi. *(c) Equilibrium of point* i.

$$V_{ih} = V_{hi}$$

$$M_{ih} = M_{hi} + V_{hi}\lambda_h \qquad (4.15)$$

where the double subscript indicates the segment and the order identifies the near and far ends, respectively.

Applying the equations of equilibrium at a typical node point, such as point i in Fig. 4.11c, we obtain

$$V_{ih} - V_{ij} + P_i + R_i = 0 \qquad (4.16)$$

$$-M_{ih} + M_{ij} + M_i = 0$$

Substitution of Eqs. 4.15 into Eqs. 4.16 and reordering of terms gives

$$-V_{hi} + V_{ij} - R_i = P_i \qquad (4.17)$$

$$M_{hi} - M_{ij} + V_{hi}\lambda_h = M_i$$

For a statically determinate structure with no horizontally applied forces, only two reaction components are required for stability. If the structure has n node points, there would be $(n - 1)$ segments, and thus $2(n - 1)$ internal force components are needed to describe the variation in shear and moment throughout the structure. Therefore, there are $2n$ equations of equilibrium available to solve for the $2n$ unknowns—2 reactions and $2(n - 1)$ member forces. In matrix form, with the explicit terms given for points i and j, we have

$$
\begin{bmatrix}
\cdots & \cdots & \cdots & \cdots & \cdots & \cdots \\
\cdots & -1 & 1 & \cdots & -1 & \cdots \\
\cdots & & -1 & 1 & & \cdots \\
\cdots & \lambda_h & \cdots & 1 & -1 & \cdots \\
\cdots & & \lambda_i & & 1 & -1 \\
\cdots & \cdots & \cdots & \cdots & \cdots & \cdots
\end{bmatrix}
\begin{Bmatrix}
\vdots \\ V_{hi} \\ V_{ij} \\ V_{jk} \\ \vdots \\ M_{hi} \\ M_{ij} \\ M_{jk} \\ \vdots \\ R_i \\ \vdots
\end{Bmatrix}
=
\begin{Bmatrix}
\vdots \\ P_i \\ P_j \\ \vdots \\ M_i \\ M_j \\ \vdots \\ \vdots
\end{Bmatrix}
\qquad (4.18)
$$

If point i were not a support point, the R_i term would not be present. However, if a rotational restraint were present at point i, such as at the left end of a structure as shown in Fig. 4.12, then Eq. 4.17 would take the form

$$V_{ij} - R_i = P_i \qquad (4.19)$$

$$-M_{ih} = -MR_i = M_i$$

Of course, for this case, the MR_i term would have to be added to the vector of unknowns in Eq. 4.18, and the coefficient matrix would change to reflect the altered condition at point i. Similarly, if there were condition equations, the coef-

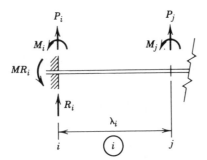

Fig. 4.12 *Fixed-end condition.*

ficient matrix and the vector of unknown forces and reactions would take different forms, but the general concepts would still apply.

Equation 4.18 can be written in the form

$$[C] \left\{ \begin{array}{c} \{V\} \\ \{M\} \\ \{R\} \end{array} \right\} = \left\{ \begin{array}{c} \{P_T\} \\ \{P_M\} \end{array} \right\} \tag{4.20}$$

where $[C]$ is the statics matrix, $\{V\}$, $\{M\}$, and $\{R\}$ are the vectors of shears, moments, and reactions, respectively, and $\{P_T\}$ and $\{P_M\}$ are the applied transverse loads and moments. As was the case in Section 3.10, this equation can be rewritten as

$$[C]\{Q\} = \{P\} \tag{4.21}$$

where $\{Q\}$ is the combined vector of member forces and reactions and $\{P\}$ includes the applied forces. The solution then takes the form

$$\{Q\} = [C]^{-1}\{P\} = [b]_{QP}\{P\} \tag{4.22}$$

In an expanded form, Eq. 4.22 can be written as

$$\left\{ \begin{array}{c} \{V\} \\ \{M\} \\ \{R\} \end{array} \right\} = \left[\begin{array}{cc} [b]_{VPT} & [b]_{VPM} \\ [b]_{MPT} & [b]_{MPM} \\ [b]_{RPT} & [b]_{RPM} \end{array} \right] \left\{ \begin{array}{c} \{P_T\} \\ \{P_M\} \end{array} \right\} \tag{4.23}$$

This form will be useful in subsequent considerations.

The procedure developed here is merely an extension of the matrix formulation developed for trusses in Section 3.10. This general approach is appropriate for any statically determinate structure with the number of member force components dependent on the structure type—one for a planar truss, two for a beam-type structure without axial forces, and six for a full three-dimensional frame structure. It should be noted that this is not a particular good procedure for the analysis of statically determinate structures. However, some of the concepts developed here are important in the treatment of statically indeterminate structures, as will be seen later.

Use the matrix formulation to determine the reactions and the shear and moment values at the designated points along the structure. Plot the resulting shear and moment diagrams. **4.13.1 Example problem**

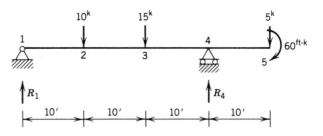

Equations of Equilibrium: The equations of equilibrium are written for each point along the beam. These are collected according to the form of Eq. 4.18 and are expressed as

$$[C]\{Q\} = \{P\} \qquad (4.21)$$

$$
\begin{bmatrix}
1 & & & & & & & & -1 \\
-1 & 1 & & & & & & & \\
& -1 & 1 & & & & & & \\
& & -1 & 1 & & & & & -1 \\
& & & -1 & & & & & \\
\hline
& & & & -1 & & & & \\
10 & & & & 1 & -1 & & & \\
& 10 & & & & 1 & -1 & & \\
& & 10 & & & & 1 & -1 & \\
& & & 10 & & & & 1 & \\
\end{bmatrix}
\begin{Bmatrix}
V_{12} \\ V_{23} \\ V_{34} \\ V_{45} \\ \hline M_{12} \\ M_{23} \\ M_{34} \\ M_{45} \\ R_1 \\ R_4
\end{Bmatrix}
=
\begin{Bmatrix}
0 \\ -10 \\ -15 \\ 0 \\ \hline -5 \\ 0 \\ 0 \\ 0 \\ 0 \\ -60
\end{Bmatrix}
$$

A free-body diagram of each joint would verify the form of the above equations. For example, consider point 4.

$$M_{43} = 10V_{34} + M_{34}$$

$$V_{43} = V_{34}$$

$$\sum P_y = 0: \quad -V_{34} + V_{45} - R_4 = P_4 = 0$$
$$\sum M = 0: \quad 10V_{34} + M_{34} - M_{45} = M_4 = 0$$

Solution for Member Forces and Reactions: The solution of the equation of equilibrium yields the following:

$$
\{Q\} = \begin{Bmatrix} \{V\} \\ \{M\} \\ \{R\} \end{Bmatrix} = [C]^{-1} \begin{Bmatrix} \{P_T\} \\ \{P_M\} \end{Bmatrix} =
\begin{Bmatrix}
V_{12} \\ V_{23} \\ V_{34} \\ V_{45} \\ \hline M_{12} \\ M_{23} \\ M_{34} \\ M_{45} \\ R_1 \\ R_4
\end{Bmatrix}
=
\begin{Bmatrix}
8 \\ -2 \\ -17 \\ 5 \\ \hline 0 \\ 80 \\ 60 \\ -110 \\ 8 \\ 22
\end{Bmatrix}
$$

where V and R quantities have units of kips and M quantities have units of ft-kips.

Shear and Moment Diagrams: Computed values represented by dots (·).

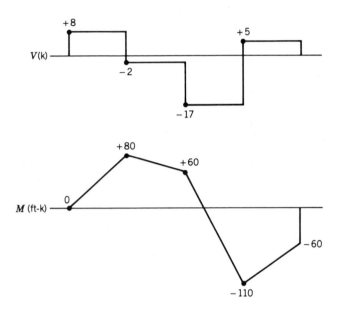

When loads are applied along the member between the node points, a slightly altered format is required to make certain that the effects of these loads are included.

4.13.2 Loads between node points

In this case, the free-body diagram of Fig. 4.11*b* must show the loads acting on the segment *ih,* and the expressions for V_{ih} and M_{ih} must reflect the presence of these loads. Therefore, Eqs. 4.15 will take a modified form, as will Eqs. 4.17. When Eqs. 4.17 are collected in the form of Eq. 4.20 or 4.21, the [*C*] matrix will not be different from that in the case with nodally applied forces only. However, the {*P*} vector will have additional terms—equivalent nodal forces, which are superimposed on the actual nodal forces. The example problem presented in the next section illustrates this application.

Use the matrix approach to determine the reactions and the shear and moment values at the designated points. Plot the final shear and moment diagrams.

4.13.3 Example problem

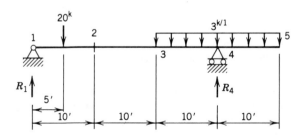

Equations of Equilibrium: The equations of equilibrium are written for each point along the span, and they are expressed in the form of Eq. 4.21:

$$[C]\,\{Q\} = \{P\} \tag{4.21}$$

$$
\begin{bmatrix}
1 & & & & & & & & -1 & & \\
-1 & 1 & & & & & & & & & \\
 & -1 & 1 & & & & & & & & \\
 & & -1 & & & & & & & -1 & \\
 & & & 1 & & & & & & & \\
\hline
 & & & & -1 & & & & & & \\
10 & & & & 1 & -1 & & & & & \\
 & 10 & & & & 1 & -1 & & & & \\
 & & 10 & & & & 1 & -1 & & & \\
 & & & 10 & & & & 1 & & &
\end{bmatrix}
\begin{Bmatrix}
V_{12} \\ V_{23} \\ V_{34} \\ V_{45} \\ \overline{M_{12}} \\ M_{23} \\ M_{34} \\ M_{45} \\ \overline{R_1} \\ R_4
\end{Bmatrix}
=
\begin{Bmatrix}
0 \\ -20 \\ 0 \\ -30 \\ -30 \\ 0 \\ 100 \\ 0 \\ 150 \\ 150
\end{Bmatrix}
$$

The $[C]$ matrix for the above is the same as that for Problem 4.14.1. The $\{P\}$ matrix reflects the new loading, including the effects of the loads between the node points. For example, consider point 2.

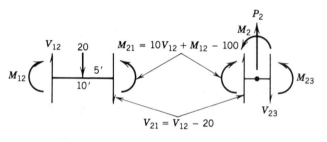

$$\sum P_y = 0: \qquad -(V_{12} - 20) + V_{23} = P_2 = 0$$
$$\sum M = 0: \qquad (10V_{12} + M_{12} - 100) - M_{23} = M_2 = 0$$

Solution for Member Forces and Reactions: The solution of the equation of equilibrium yields

$$
\{Q\} = [C]^{-1}\{P\} =
\begin{Bmatrix}
V_{12} \\ V_{23} \\ V_{34} \\ V_{45} \\ \overline{M_{12}} \\ M_{23} \\ M_{34} \\ M_{45} \\ \overline{R_1} \\ R_4
\end{Bmatrix}
=
\begin{Bmatrix}
16.67 \\ -3.33 \\ -3.33 \\ 30.00 \\ \overline{0} \\ 66.7 \\ 33.3 \\ -150.0 \\ \overline{16.67} \\ 63.33
\end{Bmatrix}
$$

where forces have units of kips and moments have units of ft-kips.

Shear and Moment Diagrams: The computed node point valves are represented by dots (·). The shape of the segments connecting the computed values must reflect the beam loading.

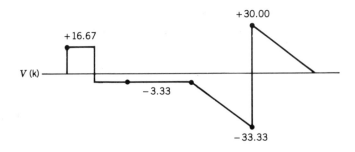

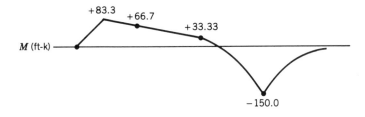

4.14

Although we have limited our treatment thus far to statically determinate structures, it is also necessary to construct shear and moment diagrams for statically indeterminate structures. As will be seen later, most methods of indeterminate analysis yield member-end moments. Once these moments are available, the analyst can proceed to construct the shear and moment diagrams by statics. **Shear and Moment Diagrams for Statically Indeterminate Structures**

For instance, consider span BC of the continuous beam given in Fig. 4.13a. An indeterminate analysis results in the moments M_B and M_C—these are the moments that act on the ends of the beam segment as shown in Fig. 4.13b. Using these moments and the loading on span BC, we proceed to get the end shears by statics and to construct the shear and moment diagrams.

One convenient way to apply statics to beam segment BC is to use superposition in conjunction with a tabular arrangement for the calculations. The procedure is described in Fig. 4.13c, where span BC is shown as a simply supported beam subjected to transverse loading and end moments. All reactions and moments are shown to act in a positive sense. The total reactions on this simply supported beam result from the superposition of two separate effects—the reactions associated with the action of the transverse loads alone (simple beam reactions) and the reactions that result from the action of the end moments alone (moment effect reactions). The total reactions can be related to the end shears by comparing Figs. 4.13b and 4.13c. This comparison shows that a positive (upward) reaction at the left end corresponds to positive shear, whereas a positive reaction at the right end corresponds to negative shear. Thus, once the final reactions are determined, these can be converted to end shears by merely changing the sign of the right reaction. Knowing the shears and moments at the ends, we can now apply any of the methods discussed to complete the shear and moment diagrams.

Where the structure is composed of more than a single member, the tabular procedure is repeated for each span, thus forming the complete shear and moment diagrams for the structure.

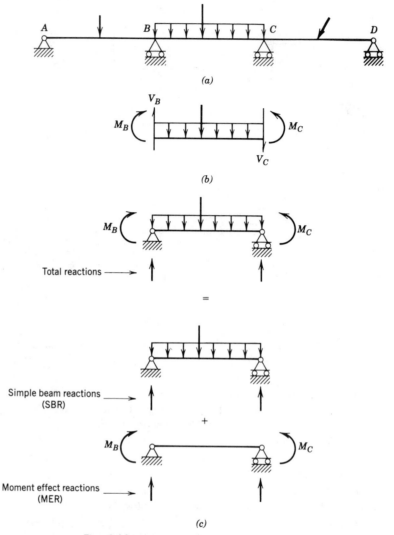

Total reactions →

=

Simple beam reactions
(SBR)

+

Moment effect reactions
(MER)

(c)

Fig. 4.13 *Statically indeterminate structures.*

4.14.1 Example problem   Construct the shear and moment diagrams for the continuous beam given. The end moments for each span have been determined and are given as shown.

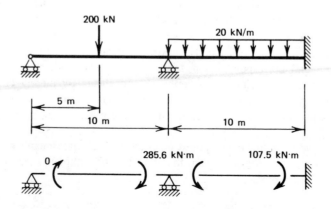

Moments	0		−285.6	−285.6		−107.5	kN · m (concave down is negative)
SBR	+100		+100	+100		+100	kN (upward reaction is positive)
MER	−28.56		+28.56	+17.81		−17.81	kN
Tot. React.	+71.44		+128.56	+117.81		+82.19	kN
Shear	+71.44		−128.56	+117.81		−82.19	kN

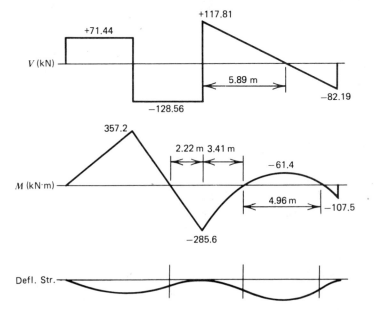

Construct the shear and moment diagrams for the frame shown. The end moments have been determined by an indeterminate analysis and are given as shown.

4.14.2 Example problem

	A	Column	B		B	Beam	C
Moments	+7.7		−15.4		−15.4		−18.0 ft-kips
SBR	0		0		+10.0		+4.0 kips
MER	−1.93		+1.93		−0.19		+0.19 kips
Tot. React.	−1.93		+1.93		+9.81		+4.19 kips
Shear	−1.93		−1.93		+9.81		−4.19 kips

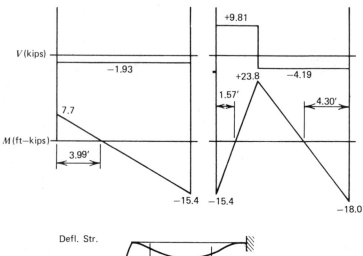

Defl. Str.

4.15

Additional Reading

Beaufait, F.W., *Basic Concepts of Structural Analysis*, Chapter 5, Prentice–Hall, Englewood Cliffs, N.J., 1977.

Hsieh, Yuan-Yu, *Elementary Theory of Structures*, Chapter 3, Prentice–Hall, Englewood Cliffs, N.J., 1970.

Laursen, H.I., *Structural Analysis*, 2nd Ed., Chapter 2, McGraw–Hill, New York, 1978.

Newmark, N.M., "Numerical Procedure for Computing Deflections, Moments, and Buckling Loads," *Trans. ASCE*, **108**, *p*. 1161, 1943.

Norris, C.H., Wilbur, J.B., and Utku, S., *Elementary Structural Analysis*, 3rd Ed., Chapter 3, McGraw–Hill, New York, 1976.

Wang, C. and Salmon, C.G., *Introductory Structural Analysis*, Chapter 4, Prentice–Hall, Englewood Cliffs, N.J., 1984.

4.16

Suggested Problems

1. Use the x coordinate shown to write general expressions for shear and moment for regions A-B and C-D. Evaluate the shear and moment at point B.

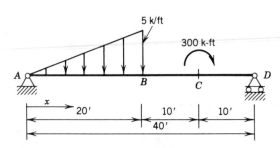

2. Using the given x coordinate, write general expressions for axial load, shear, and moment for regions B-C and C-D. Use the resulting expressions to evaluate these quantities at point C for each of the specified regions. *Note:* Shear acts normal to the member axis.

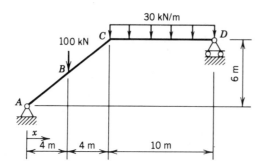

3 through 6. Construct the shear and moment diagrams by the formal integration method for the beams and loading conditions shown.

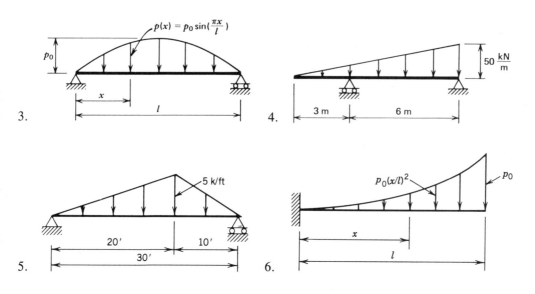

7 through 22. Use the incremental change method to construct the shear and moment diagrams for all members for each of the structures and loadings shown. In each case, *sketch* the deflected structure.

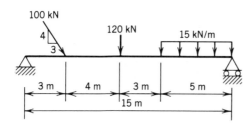

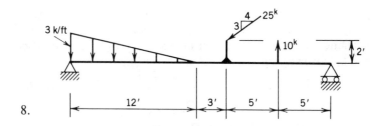

8.

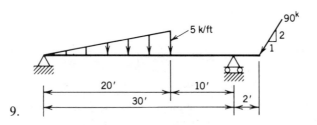

9.

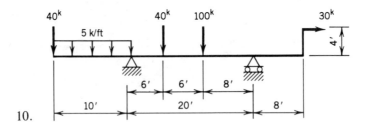

10.

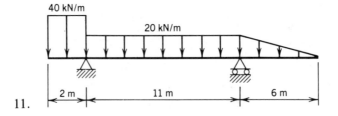

11.

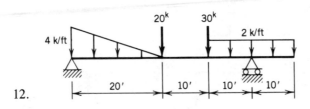

12.

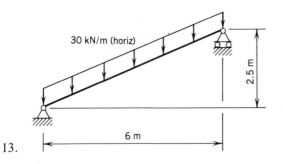

30 kN/m (horiz)

2.5 m

6 m

13.

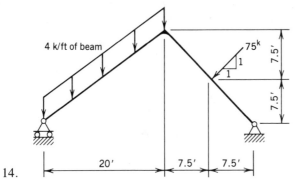

4 k/ft of beam

75^k

7.5'

7.5'

20'

7.5'

7.5'

14.

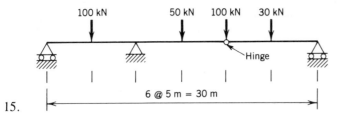

100 kN   50 kN   100 kN   30 kN

Hinge

6 @ 5 m = 30 m

15.

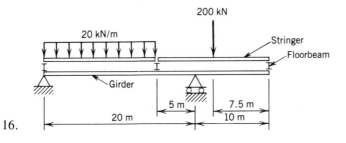

200 kN

20 kN/m

Stringer

Floorbeam

Girder

5 m   7.5 m

20 m   10 m

16.

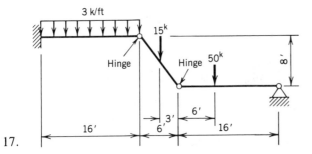

3 k/ft

15^k

Hinge   Hinge 50^k

8'

16'   3'   6'

6'

16'

17.

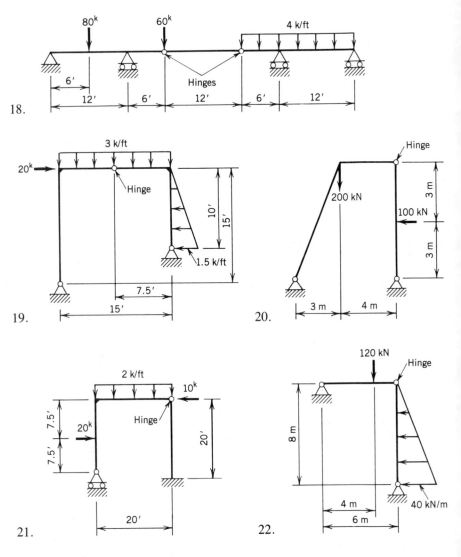

18.

19.

20.

21.

22.

23 through 26. Use the tabular method of Section 4.10 to construct the shear and moment diagrams for the designated structures and loadings. When necessary, discretize distributed loads into an equivalent set of concentrated loads.

23. Structure and loading of Problem 10. Use intervals from left to right of 5, 5, 6, 6, 8, and 8 feet.

24. Structure and loading of Problem 12. Use 10- ft intervals.

25. Structure and loading of Problem 15. Use 5-meter intervals.

26. Structure and loading of Problem 18. Use 6-ft intervals.

27 through 30. Use the matrix method of Section 4.13 to construct the shear and moment diagrams for the designated structures and loadings.

27. Structure and loading of Problem 7. Use intervals from left to right of 3, 4, 3, and 5 meters.

28. Structure and loading of Problem 9. Use intervals of 10, 10, 10, and 2 feet.

29. Structure and loading of Problem 11. Use intervals of 2, 5.5, 5.5, and 6 meters.

30. Structure and loading of Problem 12. Use 10-ft intervals.

31 and 32. Construct the shear and moment diagrams of the statically indeterminate structures shown, and *sketch* the deflected structure for each. The end moments for each span are given.

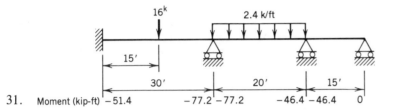

31.   Moment (kip-ft) −51.4   −77.2 −77.2   − 46.4 − 46.4   0

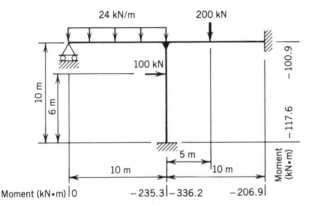

32.   Moment (kN•m) |0   − 235.3|− 336.2   − 206.9|

33. Classify each of the following structures with respect to statical determinacy and stability. Consider both external and overall criteria.

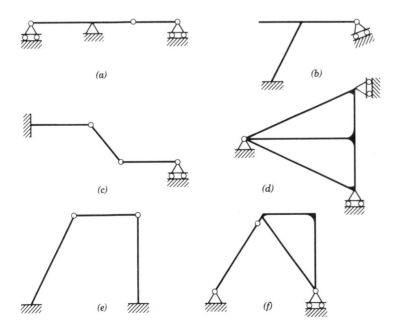

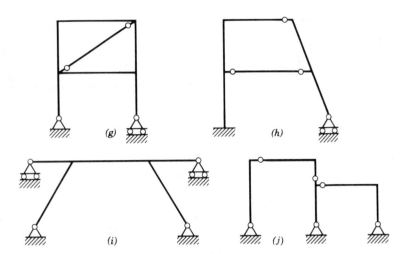

34. Write a computer program for the matrix method developed in Section 4.13. Use the program to solve Problems 27 through 30.

35. Given below are general expressions for shear and moment for beams subjected to concentrated loads and uniformly distributed loads, respectively, based on the force boundary conditions at the left end of the beam.

Concentrated Load:

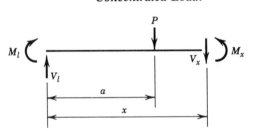

$$V_x = V_l - P \langle x - a \rangle^0$$

$$M_x = M_l + V_l x - P \langle x - a \rangle^1$$

where $\langle x - a \rangle = 0$ for $x < a$

$\qquad\qquad\qquad\ \ = (x - a)$ for $x \geq a$

Uniform Load:

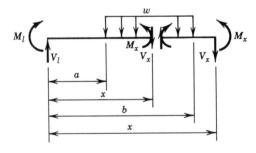

For $x \leq b$:

$$V_x = V_l - w \langle x - a \rangle^1$$

$$M_x = M_l + V_l x - \frac{w \langle x - a \rangle^2}{2}$$

where $\langle x - a \rangle = 0$ for $x < a$

$\qquad\qquad\qquad\ \ = (x - a)$ for $a \leq x \leq b$

For $x > b$:

$$V_x = V_l - w(b - a)$$

$$M_x = M_l + V_l x - \frac{w(b - a)^2}{2} - w(b - a)(x - b)$$

For any x, the total values V_x and M_x are determined from the superposition of individual effects.

a. Verify that the given expressions are correct.

b. Write a computer program to determine the shears and moments at closely spaced intervals along the beam length.

Input data should include the required beam dimensions, the loading (including internal reactions), and the left-end boundary conditions.

The output should include a tabular array that gives shear and moment for each x distance along the beam.

c. Use the program to solve Problems 7, 15, and 18. Intervals of 1 foot or 1 meter should be used, depending on the system of units employed.

CHAPTER 5

*Firth of Forth
Railroad Bridge,
Scotland (photo by
Harry G. West).*

ELASTIC DEFLECTIONS
OF PIN-CONNECTED
FRAMEWORKS

5.1

**Description of
Truss
Deflection
Problem**

When a pin-connected framework or truss is subjected to a system of applied forces, it develops a set of reactions and bar forces that is consistent with the requirements of equilibrium. In response to the member forces, each bar within the truss experiences an axial deformation; however, the assemblage of deformed bars must continue to fit together in accordance with the requirements of compatibility. As a result, the truss joints that are not constrained by the support conditions will experience displacements as the structure adjusts to the deformed member lengths. These joint displacements define the deflected configuration of the truss as a whole.

5.2

**Importance of
Determining
Deflections**

The preceding commentary describes the role that deformations play as a structure seeks an equilibrium configuration in response to a given loading condition. However, deformations are important in their own right.

There are two important reasons for determining deformation quantities. The first has to do with serviceability requirements. Structures generally are required to function within specified limitations regarding deformations if they are to perform in the desired fashion. These serviceability requirements might be related to any stage of erection or to any loading condition. The second reason for determining deformations has to do with the analysis process itself. For statically indeterminate analysis, the equations of equilibrium must be augmented by conditions of compatibility. These conditions result in equations in which deformation quantities are matched, and thus it is necessary to compute specific deformations.

Deformations may be recoverable when the loads are removed or they may be permanent or nonrecoverable in nature. The former type is referred to as an *elastic deformation,* the latter, a *plastic deformation.* Here, we limit our treatment to elastic deformations.

5.3
Axial Force–Deformation Relationships

For any member, there are definite relationships between the member-end forces and the corresponding member-end deformations. In the case of a truss member, we seek the relationship between the axially applied force and the axial deformation.

Consider the member shown in Fig. 5.1a, which is of length l, has a cross-sectional area A, and is subjected to the axial force F. An element dx in length is shown in Fig. 5.1b. Applying the principles of elementary mechanics to this element, we recall that the stress is defined as the force per unit area. Representing this stress by σ, we thus have

$$\sigma = \frac{F}{A} \tag{5.1}$$

The strain is given by the change in length per unit length. If we represent strain by ϵ, we then have

$$\epsilon = \frac{dl}{dx} \tag{5.2}$$

where dl is the change in length or the deformation that accompanies the force F.

For a linearly elastic material, if stress is plotted as ordinate against strain as abscissa, we have the linear relationship shown in Fig. 5.1c. For any given strain, there is a unique stress given by the relationship

$$\sigma = E\epsilon \tag{5.3}$$

where E is Young's modulus or the modulus of elasticity. This modulus is a material property that must be determined from an appropriate test procedure.

Solving Eq. 5.2 for the deformation dl and then substituting Eq. 5.3 and Eq. 5.1, we obtain

$$dl = \frac{F}{EA}\, dx \tag{5.4}$$

The total elongation Δ is given by integrating Eq. 5.4 over the length of the member, and because the axial force and the area are constant throughout the member length, we obtain

$$\Delta = \int_0^l dl = \frac{F}{EA}\int_0^l dx = \frac{Fl}{EA} \tag{5.5}$$

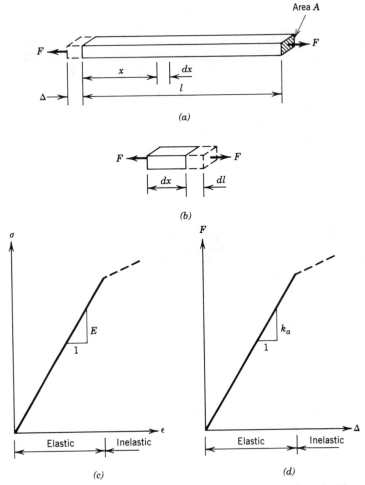

Fig. 5.1 *Axial force–deformation behavior.* (a) *Axially loaded member.* (b) *Element.* (c) *Stress–strain diagram.* (d) *Force–deformation relationship.*

Equation 5.5 is a *deformation–force relationship*, and it can be written in the form

$$\Delta = f_a F \tag{5.6}$$

where f_a, the *axial flexibility*, is given by

$$f_a = \frac{l}{EA} \tag{5.7}$$

In an inverse form, or a *force–deformation relationship*, Eq. 5.5 can be rewritten as

$$F = \frac{EA}{l} \Delta \tag{5.8}$$

or

$$F = k_a \Delta \tag{5.9}$$

where k_a, the *axial stiffness*, is defined as

$$k_a = \frac{EA}{l} \tag{5.10}$$

The relationship between F and Δ is shown in Fig. 5.1d. The shape of this curve is shown to mirror the shape of the stress–strain curve of the constituent material.

In a more general form, consider the member shown in Fig. 5.2. Here the member-end forces and displacements are separately identified—F_1 and δ_1 at end 1, and F_2 and δ_2 at end 2.

Comparing Figs. 5.1a and 5.2, it is clear that $F = F_1 = F_2$ and $\Delta = \delta_1 + \delta_2$. Substitution of these relationships into Eq. 5.8, and a subsequent translation into matrix form, leads to

$$\begin{Bmatrix} F_1 \\ F_2 \end{Bmatrix} = \begin{bmatrix} \dfrac{EA}{l} & \dfrac{EA}{l} \\ \dfrac{EA}{l} & \dfrac{EA}{l} \end{bmatrix} \begin{Bmatrix} \delta_1 \\ \delta_2 \end{Bmatrix} \tag{5.11}$$

In an abbreviated form, this equation can be expressed as

$$\{F\} = [k]_a\{\delta\} \tag{5.12}$$

where $\{F\}$ and $\{\delta\}$ are the member-end force and displacement vectors, respectively, and $[k]_a$ is the *member axial stiffness matrix*. Equation 5.12 is the two-dimensional equivalent of Eq. 5.9, and the individual elements of $[k]_a$ are in terms of EA/l, which reflects the axial stiffness of the member as it is defined by Eq. 5.10.

As noted earlier, Eqs. 5.6 and 5.9 are inverse expressions—the former gives a deformation–force relationship, whereas the latter states a force–deformation relationship. In the two-dimensional form, Eq. 5.12 gives the member-end forces that are associated with prescribed member-end displacements. These are clearly force–deformation relationships. Here, too, it is possible to develop inverse expressions, or displacement–force relationships. These would allow for the determination of certain member-end displacements from prescribed member-end forces, provided that the member is adequately restrained against rigid-body motions. This will be discussed fully in Chapter 16. Our purpose here is merely to emphasize that there are definite relationships between the forces that a member transmits and the deformations that it experiences and that these member relationships are central to determining the deflections of a structure as a whole.

The foregoing discussion is limited to the elastic response of the member. If the member is stressed beyond the elastic range, as shown by the dashed portion of Fig. 5.1c, the resulting force–deformation relationships will reflect this behavior as indicated in Fig. 5.1d. Of course, in the inelastic range, Eqs. 5.6, 5.9, and 5.12 are no longer valid.

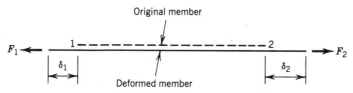

Fig. 5.2 *Member-end forces and displacements.*

5.4

Geometric Approach to Truss Deflection Problems

The first method that will be presented for the determination of truss deflections is based on a geometric approach. This method has the advantages of being easy to understand and of giving the displacements of all truss joints. However, there are disadvantages; it is a graphical method that can be cumbersome to apply, and, unless extreme care is taken, it can be inaccurate. The main reason for including it here is to help the student gain a physical appreciation for the nature of the truss deflection problem.

Consider the simple two-bar truss shown in Fig. 5.3a. Determination of the bar forces enables one to use Eq. 5.5 to compute the corresponding axial deformations ΔL_{AB} and ΔL_{BC} for members AB and BC, respectively—a shortening of AB and a lengthening of BC, as shown in Fig. 5.3b. However, if the two bars are to remain connected at point B, the bars must rotate until they terminate at the common point B'. Under this arrangement, equilibrium and compatibility are both satisfied. Of course, it is assumed that the displacements are small enough that the overall geometry changes are small and thus the bar forces are not altered by the displacements. Also, since the displacements are small, it is assumed that as each bar rotates about its support point, the terminal point moves along the tangent to the arc, or normal to the bar itself. This arrangement is shown in Fig. 5.3c, where the vertical displacement and the horizontal displacement are denoted by v_B and u_B, respectively. This enlarged sketch is referred to as the *Williot diagram*.

This procedure is essentially a process of triangulation in which one uses the location of two points and the deformation of two bars to establish the location of the third point. This technique can readily be applied to a larger truss by working from joint to joint throughout the truss.

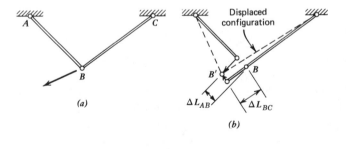

(a)   (b)

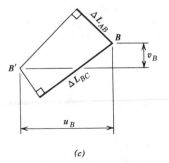

(c)

Fig. 5.3 *Displacement of two-bar truss.* (a) *Truss and loading.* (b) *Bar deformations.* (c) *Williot diagram.*

It should also be noted that the changes in member lengths need not result from applied forces. Instead, they might result from temperature change or errors in fabrication. These cases employ the same general procedure.

For the truss and loading shown below, determine the displacement components at point c. **5.4.1 Example problem**

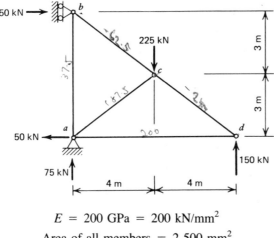

$$E = 200 \text{ GPa} = 200 \text{ kN/mm}^2$$
$$\text{Area of all members} = 2\,500 \text{ mm}^2$$

Bar Forces and Member Deformations: Analysis of the truss and determination of the axial deformations gives the following data:

Bar	Force, F kN	Length, L m	Area, A mm²	Δ = FL/EA mm
ab	+ 37.5	6.0	2 500	0.450
bc	− 62.5	5.0	2 500	− 0.625
cd	− 250.0	5.0	2 500	− 2.500
da	+ 200.0	8.0	2 500	3.200
ac	− 187.5	5.0	2 500	− 1.875

Truss Displacements: The member deformations will be plotted on Fig. 5.4a to an enlarged scale. The original truss is lightly shown, and we begin with member ab, which is restrained against rotation. Since point a cannot displace, Δ_{ab} is accommodated by a vertical movement of point b to b'. Assuming that the directions of bc and ac remain unchanged, and working from their original lengths, we apply the axial deformations Δ_{bc} and Δ_{ac} at the c-end of these members. However, for these two members to remain compatibly attached at point c, they must rotate until they meet at point c'. Now, assuming that the directions of members cd and ad are as originally given, the deformations Δ_{cd} and Δ_{ad} are applied to the original member lengths at the d-end of these members. Again, for compatibility to be satisfied at point d, both members must rotate until they meet at point d'. Connection of the displaced points gives the deformed configuration of the truss.

One should keep in mind that the displacements of the members are very small when compared with the member lengths. Thus, the assumption that the

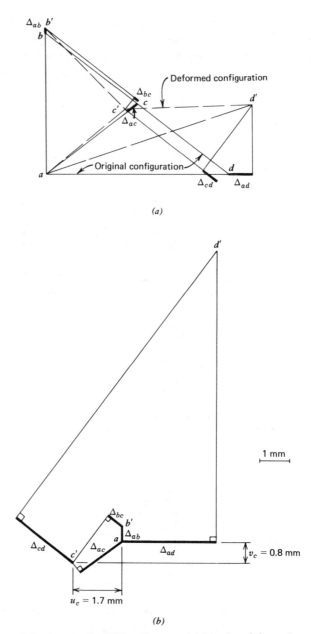

Fig. 5.4 *Truss deflections by the Williot diagram.* (a) *Member deformations.* (b) *Williot diagram.*

member directions do not change is quite valid. This is not evident from Fig. 5.4*a* because the scale for the displacements is much larger than that for the structure. It is also noted that the signs of the deformations are important. A positive deformation is associated with an elongation of the member, whereas a negative deformation indicates a member shortening.

Figure 5.4*a* is used to demonstrate the physical significance of the member elongations; however, it is sufficient simply to sketch the elongations on a Williot diagram as shown in Fig. 5.4*b*. Here, a convenient scale is adopted, and the final displacement of any point is measured directly off the diagram as the distance between the fixed point *a* and the primed position of the point in question. For instance, point *c* displaces vertically downward 0.8 mm and horizontally to the left 1.7 mm.

In this example problem, member *ab* was restrained against rotation, and this provided a reference direction from which the solution could begin.

Consider the truss shown below under the prescribed loading. Determine the displacement of point *c*.

5.4.2 Example problem

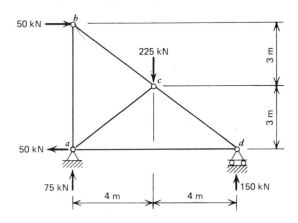

$$E = 200 \text{ kN/mm}^2; \quad \text{member areas} = 2\ 500 \text{ mm}^2$$

This is the same truss and loading that were used in Section 5.4.1, although the boundary conditions are different. In this case, point *b* is free to displace while point *d* is restrained against vertical movement.

We could work this problem by using member *ad* as the reference direction; however, in this case, we will use the solution of Section 5.4.1 with an appropriate adjustment for the change in boundary conditions.

Bar Forces and Member Deformations: The bar forces and member deformations are the same as those given in Section 5.4.1.

Williot Diagram: If we assume initially that member *ab* does not rotate, we get the same Williot diagram as in the previous example. This is shown in Fig. 5.5a. However, it is clear that this set of displacements violates the boundary condition at point *d*, and thus a modification is required.

Mohr Correction Diagram: In order to correct the displacements given by the Williot diagram, a rigid-body rotation of the entire structure about point *a* is employed until the boundary condition at point *d* is satisfied. For this rigid-body motion, the vertical displacement of any point will be proportional to the horizontal distance of the point from point *a*, and the corresponding horizontal displacement sill be proportional to the vertical distance of the point from point *a*.

Thus, the *Mohr correction diagram* will have the shape of the original truss rotated through 90 degrees, with its overall size controlled by the vertical correction displacement required at point *d*. This correction diagram is shown in Fig.

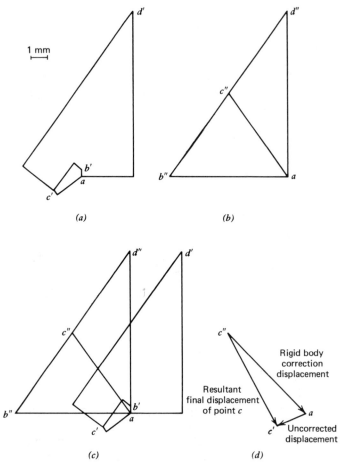

Fig. 5.5 *Truss deflections by Williot–Mohr diagram.* (a) *Williot diagram.* (b) *Mohr correction diagram.* (c) *Williot–Mohr diagram.* (d) *Displacement of point* c.

5.5*b* with the distance from the doubly primed points to point *a* representing the displaced position of any point associated with the rigid-body rotation.

Williot–Mohr Diagram: The Williot and Mohr diagrams are combined in Fig. 5.5*c* into a single diagram by superimposing the Mohr diagram over the Williot diagram. Since point *a* represents a point of zero displacement in each diagram, it becomes a common point on the combined diagram.

On this diagram, the distance from point *a* to a singly primed point gives the displacement of that point when member *ab* is not allowed to rotate, and the distance from a doubly primed point to point *a* gives the displacement associated with the rigid-body rotation. The sum of these displacements, that is, the distance from the doubly primed point to the singly primed point, is the final displacement of the point. This is demonstrated in Fig. 5.5*d* for point *c*.

In this example, the use of the Mohr correction diagram could have been avoided if the Williot diagram had been constructed for zero rotation of member *ad*. However, when no members are restrained against rotation, some member is arbitrarily assumed not to rotate, and then the Mohr correction diagram is used to satisfy the boundary conditions.

The concept of virtual work was introduced in Section 2.14, and the principle of **Virtual Work** virtual displacements was based on virtual work considerations for a rigid body. **for a** We now extend the application of virtual work to a deformable body. As these **Deformable** techniques will be used throughout the text for all types of structures, the devel- **Body** opment here is of a general nature. In Section 5.6, the virtual work method is applied specifically to truss structures.

Consider the deformable body shown in Fig. 5.6a, which is in equilibrium under the P forces. Any element within the body is in equilibrium under a set of σ_P stresses that are caused by the P forces. The body is then subjected to a virtual distortion. This distortion results in a corresponding set of virtual displacements, δD, at each of the external load points and a set of virtual strains, $\delta\epsilon_D$, of the element.

The total virtual distortion of the body is made up of two components. There is a rigid-body motion of the body as a whole and a deformation of the body as illustrated in Fig. 5.6b. Thus, the total virtual work done by the external P forces, δW_e, has two components. There is a work term that couples the P forces with the virtual displacements that result from rigid-body motion, δW_r, and a work

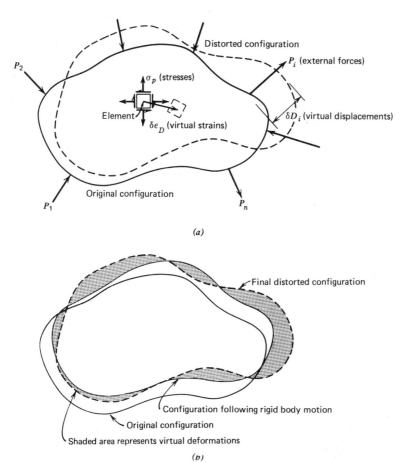

(a)

(b)

Fig. 5.6 *Virtual deformation of deformable body. (a) Deformable body in equilibrium.* (b) *Components of virtual displacements.*

term that couples the P forces with the virtual displacements that result from the distortion of the body, δW_d. Thus,

$$\delta W_e = \delta W_r + \delta W_d$$

or

$$\delta W_r = \delta W_e - \delta W_d \tag{5.13}$$

However, from the principle of virtual displacements, δW_r is zero, and thus

$$\delta W_e = \delta W_d \tag{5.14}$$

It is clear from Fig. 5.6a and from the considerations of Section 2.13 that the external virtual work is given by

$$\delta W_e = \sum_{i=1}^{n} P_i(\delta D)_i \tag{5.15}$$

With respect to the virtual work of distortion, δW_d, consider the element shown in Fig. 5.6a. Each σ_P stress component acts over an elemental area dA to produce an internal force component $F = \sigma_P \cdot dA$. The accumulation of the corresponding $\delta\epsilon_D$ virtual strain component over the elemental length dl yields a virtual displacement $\delta = \delta\epsilon_D \cdot dl$. Thus, the elemental virtual work done by this single stress component is given by

$$d(\delta W_d) = (\sigma_P \cdot dA)(\delta\epsilon_D \cdot dl) \tag{5.16}$$

Noting that the volume of the element is $d\text{Vol} = dA \cdot dl$, and summing the virtual work of all stress components over the entire body, we obtain

$$\delta W_d = \int_{\text{Vol}} \sigma_P(\delta\epsilon_D)\, d\text{Vol} \tag{5.17}$$

Substituting Eqs. 5.15 and 5.17 into Eq. 5.14, we obtain

$$\sum_{i=1}^{n} P_i(\delta D)_i = \int_{\text{Vol}} \sigma_P(\delta\epsilon_D)\, d\text{Vol} \tag{5.18}$$

In summary, Eq. 5.18 states that *if a deformable body is in equilibrium under a set of P forces and it is subjected to a virtual distortion, the external virtual work done by the P forces is equal to the internal virtual work done by the σ_P stresses*.

In applying the method of virtual work, one must think of two separate systems. One system is a *P system* of forces that is in equilibrium, and the other system is a *D system* of virtual deformations that is geometrically compatible. Equation 5.18 thus states the relationship involving the work done by the P system as it undergoes the virtual motions corresponding to the D system.

This formulation is consistent with the development in the preceding section. It is used when the analyst is attempting to determine a selected P force through the introduction of a judiciously selected set of δD virtual displacements. This approach requires the following systems, which are shown in Fig. 5.7.

5.5.1 Virtual work method using virtual displacements

P System: *Actual* force system is equilibrium.

D System: *Virtual* deformation pattern that is geometrically compatible.

Use of Eq. 5.18 gives

$$\underbrace{\sum_{i=1}^{n} P_i(\delta D)_i}_{} = \int_{\text{Vol}} \underbrace{\sigma_P(\delta \epsilon_D) \, d\text{Vol}}_{} \qquad (5.19)$$

P system (actual) — over left term; *D* system (virtual) — under right term.

By selecting a proper set of δD displacements, the desired P force can be isolated on the left-hand side of the equation and thus determined.

When displacement quantities are desired, the virtual work method can still be used; however, a slight modification is required. Here, a fictitious loading system is introduced and the displacements of the actual system are used as displacements for the fictitious loading. The fictitious loading system can be thought of as a set of virtual loads, δP, that are used to determine the actual displacements D. This procedure thus requires the following systems, which are illustrated in Fig. 5.8.

5.5.2 Virtual work method using virtual forces

P System: *Virtual* force system in equilibrium.

D System: *Actual* deformation pattern that is geometrically compatible.

For this form, Eq. 5.14 becomes

$$\underbrace{\sum_{i=1}^{n} (\delta P)_i D_i}_{} = \int_{\text{Vol}} \underbrace{(\delta \sigma_P)\epsilon_D \, d\text{Vol}}_{} \qquad (5.20)$$

P system (virtual) — over left term; *D* system (actual) — under right term.

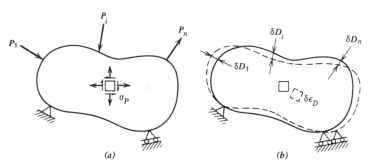

Fig. 5.7 *Virtual displacements.* (a) P *system—actual force system in equilibrium.* (b) D *system—virtual deformation pattern that is geometrically compatible.*

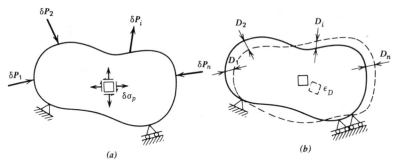

Fig. 5.8 *Virtual forces.* (a) P *system—virtual force system in equilibrium.* (b) D *system—actual deformation system that is geometrically compatible.*

By selecting a proper set of δP forces, the desired D displacement can be isolated on the left side of the equation and thus determined. Generally, the virtual force system involves a unit load at the point and in the direction of the desired displacement. For this reason, the virtual force approach is sometimes referred to as the *dummy unit load method*.

5.6

Truss Deflections by Virtual Work

Truss deflections can be determined by using the virtual work method which uses virtual forces. This method has the disadvantage that we are able to obtain only one displacement quantity at a time for a designated loading condition. In accordance with the procedure outlined in Section 5.5.2, the following systems are employed:

P System: Virtual force system in equilibrium.

D System: Actual deformation pattern that is geometrically compatible.

Using these systems, we equate the external virtual work to the internal virtual work of deformation. Thus, as the *P* system undergoes the displacements of the *D* system, we have

$$\sum_{i=1}^{n} (\delta P)_i D_i = \int_{\text{Vol}} (\delta \sigma_P) \epsilon_D \, d\text{Vol} \tag{5.21}$$

(with annotation "Virtual *P* system" above and "Actual *D* system" below)

Because the right-hand side of this equation represents the internal work of deformation, its form depends on the nature of the deformation. For a truss that is composed of *m* members, this integral represents the summation over all member lengths of the virtual forces of the *P* system that act through the axial deformations of the *D* system. Reference to Fig. 5.9 shows that this can be written in the form

$$\int_{\text{Vol}} (\delta \sigma_P) \epsilon_D \, d\text{Vol} = \sum_{j=1}^{m} \left(\int_l (\delta F_P) \, dL_D \right)_j \tag{5.22}$$

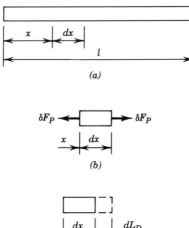

Fig. 5.9 *Virtual P system and actual D system for jth truss member. (a) jth member. (b) Virtual bar force at point x caused by P system of loads. (c) Elongation of element dx caused by D system of deformations.*

where l is the length of the jth member, δF_P is the virtual bar force at some point x on member j corresponding to the P system, and dL_D is the associated member deformation at the same point on member j that occurs over the differential element dx of the D system. Substitution of Eq. 5.22 into Eq. 5.21 gives

$$\sum_{i=1}^{n} (\delta P)_i D_i = \sum_{j=1}^{m} \left(\int_l (\delta F_P) \cdot dL_D \right)_j \qquad (5.23)$$

If dL_D results from a D system of loads on the truss, then Eq. 5.4 gives

$$dL_D = \frac{F_D}{EA} dx \qquad (5.24)$$

where F_D is the bar force at point x on member j corresponding to the D system. Substitution of Eq. 5.24 into Eq. 5.23 gives

$$\sum_{i=1}^{n} (\delta P)_i D_i = \sum_{j=1}^{m} \left(\int_l (\delta F_P) \frac{F_D \, dx}{EA} \right)_j \qquad (5.25)$$

In the present case, since both δF_P and F_D are constant throughout the member lengths, Eq. 5.25 simplifies to

$$\sum_{i=1}^{n} (\delta P)_i D_i = \sum_{j=1}^{m} \left(\delta F_P \frac{F_D l}{EA} \right)_j \qquad (5.26)$$

When dL_D is caused by something other than loads, such as temperature change or fabrication error, Eq. 5.23 is still valid. Again, δF_P is constant throughout the member, and $\int_l dL_D = \Delta L_D$, the total change in length of member j. Thus Eq. 5.23 takes the form

$$\sum_{i=1}^{n} (\delta P)_i D_i = \sum_{j=1}^{m} [(\delta F_P) \cdot \Delta L_D]_j \qquad (5.27)$$

5.7

Application of the Virtual Work Method
When the virtual work method is used to obtain the displacements associated with some D system, the P system is a virtual force system that must be selected with care. Examination of Eq. 5.23 shows that for the determination of any particular displacement quantity D_r, the P system must be composed of a virtual force $(\delta P)_r$ that acts at the point and in the direction of the desired displacement D_r. There can be no other forces in the P system that couple with displacements of the D system. Generally, (δP), is taken as a unit load, and Eq. 5.23 then yields the desired displacement D_r. Thus, a different virtual load system must be used for each displacement quantity that is desired for a given D system.

5.7.1 Example problem

Consider again the truss problem of Section 5.4.1 and determine the horizontal and vertical components of the displacement at point c. All members have a cross-sectional area of $0.002\ 5\ \mathrm{m}^2$ and a modulus of elasticity of 200×10^9 Pa.

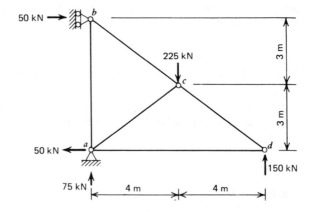

The following virtual work expression is used:

$$\sum_{i=1}^{n} (\delta P)_i D_i = \sum_{j=1}^{m} \left(\delta F_\mathrm{P} \cdot \frac{F_D l}{EA} \right)_j \tag{5.26}$$

Determination of Horizontal Displacement u_c.

   D System: Actual loading system that includes the desired displacement u_c.

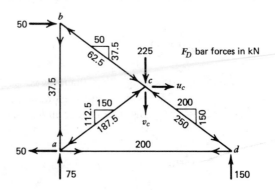

P System: Virtual force system with δP selected to obtain u_c.

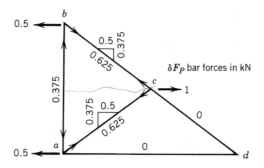

Because the reactions of the P system correspond to nondeflecting points of the D system, they do not contribute to the left-hand side of Eq. 5.26, and therefore

$$\sum_{i=1}^{n} (\delta P)_i D_i = (1)(u_c) \qquad (\text{kN} \cdot \text{m})$$

And, because A and E are the same for all members, the right-hand side of Eq. 5.26 takes the following form:

$$\sum_{j=1}^{m} \left(\delta F_P \frac{F_D l}{EA} \right)_j = \frac{1}{EA} \sum_{j=1}^{m} (\delta F_P \cdot F_D \cdot l)$$

This summation can best be determined by use of the following tabulation:

Bar	l	F_D	δF_P	$\delta F_P \cdot F_D \cdot l$
—	m	kN	kN	$(\text{kN})^2 \cdot$ m
ab	6.0	+37.5	−0.375	−84.38
bc	5.0	−62.5	+0.625	−195.31
cd	5.0	−250.0	0	0
da	8.0	+200.0	0	0
ac	5.0	−187.5	+0.625	−585.94
			$\sum (\delta F_P \cdot F_D \cdot l) =$	−865.63

Thus, the right-hand side of Eq. 5.26 is

$$\frac{1}{EA} \sum_{j=1}^{m} (\delta F_P \cdot F_D \cdot l) = \frac{-865.63}{200 \times 10^6 \times 0.002\,5} = -0.001\,73 \text{ kN} \cdot \text{m}$$

Applying Eq. 5.26, we have

$$(1)(u_c) = -0.001\,73$$

from which

$$u_c = -0.001\,73 \text{ m} = -1.73 \text{ mm}$$

The negative sign simply means that u_c is to the left and not to the right, as assumed.

Determination of Vertical Displacement v_c:

D System: Actual loading system, which includes the desired displacement, v_c. Same system as used for u_c.

P System: Virtual force system with δP selected to obtain v_c.

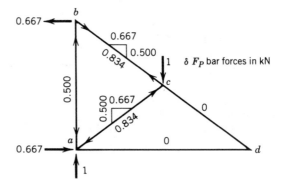

Bar	l	F_D	δF_P	$\delta F_P \cdot F_D \cdot l$
—	m	kN	kN	$(\text{kN})^2 \cdot$ m
ab	6.0	+ 37.5	− 0.500	− 112.50
bc	5.0	− 62.5	+ 0.834	− 260.63
cd	5.0	− 250.0	0	0
da	8.0	+ 200.0	0	0
ac	5.0	− 187.5	− 0.834	+ 781.88

$$\sum (\delta F_P \cdot F_D \cdot l) = +408.75$$

Application of Eq. 5.26 gives

$$(1)(v_c) = \frac{408.75}{200 \times 10^6 \times 0.002\ 5} = 0.000\ 82\ \text{kN} \cdot \text{m}$$

from which

$$v_c = 0.000\ 82\ \text{m} = 0.82\ \text{mm}$$

Here, the positive sign means that the displacement is downward, as assumed.
To avoid unnecessary repetition, the last two columns of the above table could have been appended to the preceding table.

5.7.2 Example problem   Determine the vertical component of the displacement at point d. The areas are given in parentheses (mm²) along each member, and the modulus of elasticity is 200 GPa.

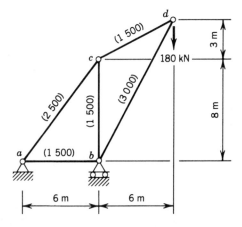

The following virtual work expression is used:

$$\sum_{i=1}^{n} (\delta P)_i D_i = \sum_{j=1}^{m} \left(\delta F_P \cdot \frac{F_D l}{EA} \right)_j \tag{5.26}$$

Determination of Vertical Displacement v_d.

    D System: Actual loading system that includes the desired displacement.

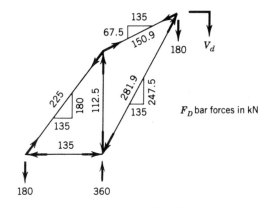

    P System: Virtual force system with δP selected to obtain v_d.

    In this case, a unit vertical load of 1 kN is placed at point d. Therefore, the P system forces can be determined by multiplying the D system forces by $\frac{1}{180}$.

    The summation on the right-hand side of Eq. 5.26 is determined by use of a tabular method. Since member cross-sectional areas differ, they must be included in the tabulation.

Bar	l	A	F_D	δF_P	$\delta F_P F_D l / A$
—	mm	mm^2	kN	kN	kN^2/mm
ab	6 000	1 500	− 135	− 0.75	405.0
ac	10 000	2 500	+ 225	+ 1.25	1 125.0
bc	8 000	1 500	− 112.5	− 0.625	375.0
cd	6 708	1 500	+ 150.9	+ 0.838	565.5
bd	12 530	3 000	− 281.9	− 1.566	1 843.8

$$\sum \left(\delta F_P \cdot \frac{F_D l}{A} \right) = 4\ 314.3$$

Application of Eq. 5.26 yields

$$(1)(v_d) = \frac{1}{E} \sum_{j=1}^{m} \left(\frac{\delta F_p F_D l}{A} \right) = \frac{4\,314.3}{200} \text{ kN} \cdot \text{mm}$$

from which

$$v_c = 21.6 \text{ mm}$$

The positive sign indicates that the deflection is downward.

5.7.3 Example problem

The truss shown below is distorted during fabrication because member bc is 0.5 in. short. What vertical deflection is introduced at point e because of this misfit?

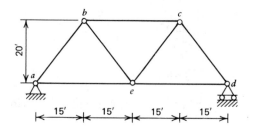

In this application, Eq. 5.27 is used.

$$\sum_{i=1}^{n} (\delta P)_i D_i = \sum_{j=1}^{m} [(\delta F_p) \cdot \Delta L_D]_j \tag{5.27}$$

Determination of Vertical Displacement v_e:

    D System: The actual system that includes the desired displacement v_e.

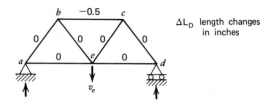

    P System: Virtual force system δP selected to obtain v_e.

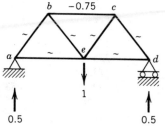

δF_P bar forces in kips.
Only bar bc is determined since it is the only bar that enters into the summation.

Application of Eq. 5.27 gives

$$(1)(v_e) = (-0.75)(-0.5)$$

from which

$$v_e = +0.375 \text{ in.}$$

Note:

The quantities A and E are not required in this problem as ΔL_D is given. Here v_e is not a result of loading but instead a consequence of a geometric adjustment of the structure.

It is suggested that the student determine the horizontal movement induced at point d as a result of the designated misfit.

Determine the vertical displacement that occurs at point b as a result of a temperature change of $+ 50°F$ in members ad and dc. The coefficient of thermal expansion is $\alpha_t = 0.000,006,5$ per $°F$.   **5.7.4 Example problem**

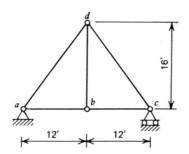

The following virtual work expression is used:

$$\sum_{i=1}^{n}(\delta F)_i D_i = \sum_{j=1}^{m}[(\delta F_P) \cdot \Delta L_D]_j \tag{5.27}$$

D System: The system that includes the desired displacement v_b.

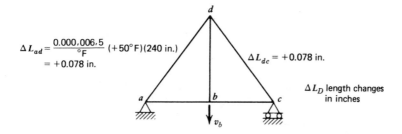

$\Delta L_{ad} = \dfrac{0.000,006,5}{°F}(+50°F)(240 \text{ in.})$
$= +0.078 \text{ in.}$

$\Delta L_{dc} = +0.078 \text{ in.}$

ΔL_D length changes in inches

P System: Virtual force system δP selected to obtain v_b.

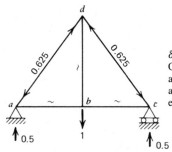

δF_P bar forces in kips. Only bars *ad* and *dc* are given as these are the only bars that enter summation.

Bar	ΔL_D	δF_P	$\delta F_P \cdot \Delta L_D$
—	in.	kip	kip · in.
ad	+0.078	−0.625	−0.048,8
dc	+0.078	−0.625	−0.048,8
		$\sum(\delta F_P)\Delta L_D$	−0.097,6

Applying Eq. 5.27, we have

$$(1)(v_b) = -0.097,6$$
$$v_b = -0.097,6 \text{ in.}$$

Minus sign indicates that point *b* displaces upward, since v_b of *D* system was assumed downward.

Conversion to SI:

$$\alpha_t = (6.5 \times 10^{-6} \text{ per } °F) \times \left(1.8 \frac{°F}{°C}\right) = 11.7 \times 10^{-6} \text{ per } °C$$

$$\Delta T = (50°F) \times \left(\frac{5}{9}\frac{°C}{°F}\right) = 27.8°C$$

D System:

$$l_{ad} = l_{dc} = (20 \text{ ft})(0.3048 \text{ m/ft}) = 6.096 \text{ m} = 6\,096\text{mm}$$
$$\Delta L_{ad} = \Delta L_{dc} = \frac{11.7 \times 10^{-6}}{°C}(27.8°C)(6\,096 \text{ mm}) = 1.98 \text{ mm}$$

P System: 1 kN vertical force at *b* induces forces of −0.625 kN in members *ad* and *dc*.

Bar	ΔL_D	δF_P	$\delta F_P \cdot \Delta L_D$
—	mm	kN	kN · mm
ad	1.98	−0.625	−1.24
dc	1.98	−0.625	−1.24
		$\sum(\delta F_P)\Delta L_D$	−2.48

Equation 5.27 gives

$$(1)(v_b) = -2.48$$
$$v_b = -2.48 \text{ mm}$$

Check: $(-0.0976 \text{ in.} \times 25.4 \text{ mm/in.} = -2.48 \text{ mm})$

5.8 Matrix Representation of the Truss Deflection Problem

Consider the structure shown in Fig. 5.10. The structure coordinate system is defined by the numbered vectors shown on the figure—each vector indicates an admissible load and a corresponding displacement for the system. In matrix form, the load vector $\{P\}$ and the displacement vector $\{\Delta\}$ are given by

$$\{P\} = \begin{Bmatrix} P_1 \\ P_2 \\ \cdot \\ \cdot \\ \cdot \\ P_5 \end{Bmatrix}; \quad \{\Delta\} = \begin{Bmatrix} \Delta_1 \\ \Delta_2 \\ \cdot \\ \cdot \\ \cdot \\ \Delta_5 \end{Bmatrix} \tag{5.28}$$

A *flexibility coefficient* is defined as a quantity that relates the displacement at some point on the structure to a unit load acting at that point or some other point on the structure. That is, a flexibility coefficient f_{ij} gives the value of the ith displacement that occurs as a result of a unit value of the jth load. These flexibility coefficients can be arranged in the form of a flexibility matrix, which relates the truss displacements to the applied loads in the form

$$\begin{Bmatrix} \Delta_1 \\ \Delta_2 \\ \Delta_3 \\ \Delta_4 \\ \Delta_5 \end{Bmatrix} = \begin{bmatrix} f_{11} & f_{12} & f_{13} & f_{14} & f_{15} \\ f_{21} & f_{22} & f_{23} & f_{24} & f_{25} \\ f_{31} & f_{32} & f_{33} & f_{34} & f_{35} \\ f_{41} & f_{42} & f_{43} & f_{44} & f_{45} \\ f_{51} & f_{52} & f_{53} & f_{54} & f_{55} \end{bmatrix} \begin{Bmatrix} P_1 \\ P_2 \\ P_3 \\ P_4 \\ P_5 \end{Bmatrix} \tag{5.29}$$

Equation 5.29 can be abbreviated as

$$\{\Delta\} = [f]\{P\} \tag{5.30}$$

where $[f]$ is the flexibility matrix.

Because flexibility coefficients are displacement quantities that accompany specified loading conditions, they can be obtained by using any of the standard methods for determining truss displacements. The method of virtual work is par-

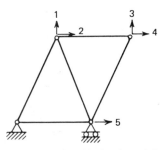

Fig. 5.10 *Truss with admissible loads and displacements.*

ticularly useful because it can be used systematically to determine the flexibility matrix.

Once the flexibility matrix has been established, Eq. 5.30 can be used to obtain the complete set of displacements for any set of loads.

5.8.1 Example problem

Consider the truss shown in which the three admissible load and displacement components are identified.

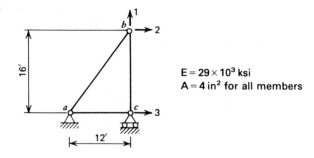

$E = 29 \times 10^3$ ksi
$A = 4$ in^2 for all members

(a) Determine the complete flexibility matrix.
(b) Determine the displacements that result from the load vector

$$\{P\} = \begin{Bmatrix} P_1 \\ P_2 \\ P_3 \end{Bmatrix} = \begin{Bmatrix} -10 \\ +15 \\ 0 \end{Bmatrix} \text{ kips}$$

Flexibility Matrix: Equation 5.26 will be used to determine the individual flexibility coefficients.

$$\sum_{i=1}^{n} (\delta P)_i D_i = \sum_{j=1}^{m} \left[(\delta F_P) \cdot \frac{F_D l}{EA} \right]_j \tag{5.26}$$

D System (above top brace)
P System (below bottom brace)

The following load systems are required:

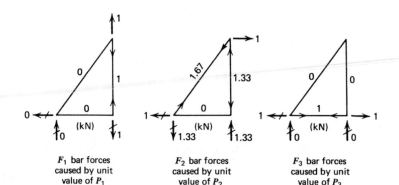

| F_1 bar forces caused by unit value of P_1 | F_2 bar forces caused by unit value of P_2 | F_3 bar forces caused by unit value of P_3 |

The following table describes how these systems are used to obtain the required flexibility coefficients.

Flexibility Coefficient	(δF_P)	F_D	Equation 5.26
f_{11}	F_1	F_1	$1 \cdot f_{11} = \dfrac{1}{EA} \cdot \sum F_1^2 l$
f_{12}	F_1	F_2	$1 \cdot f_{12} = \dfrac{1}{EA} \cdot \sum F_1 F_2 l$
f_{13}	F_1	F_3	$1 \cdot f_{13} = \dfrac{1}{EA} \cdot \sum F_1 F_3 l$
f_{21}	F_2	F_1	$1 \cdot f_{21} = \dfrac{1}{EA} \cdot \sum F_2 F_1 l$
f_{22}	F_2	F_2	$1 \cdot f_{22} = \dfrac{1}{EA} \cdot \sum F_2^2 l$
f_{23}	F_2	F_3	$1 \cdot f_{23} = \dfrac{1}{EA} \cdot \sum F_2 F_3 l$
f_{31}	F_3	F_1	$1 \cdot f_{31} = \dfrac{1}{EA} \cdot \sum F_3 F_1 l$
f_{32}	F_3	F_2	$1 \cdot f_{32} = \dfrac{1}{EA} \cdot \sum F_3 F_2 l$
f_{33}	F_3	F_3	$1 \cdot f_{33} = \dfrac{1}{EA} \cdot \sum F_3^2 l$

This tabulation shows the orderliness of the virtual work method in determining the flexibility coefficients. This is summarized as follows:

$$f_{ij} = \frac{1}{EA} F_i F_j l$$

with P system indicated above and D system indicated below.

The first subscript identifies the unit virtual load of the P system, and the second subscript identifies the given unit load case of the D system. Also, because the order of multiplication does not alter the results, it is clear that $f_{12} = f_{21}, f_{13} = f_{31}$, and $f_{23} = f_{32}$.

The detailed evaluation of the flexibility coefficients is carried out in the following table:

Member	l	F_1	F_2	F_3	$F_1^2 l$	$F_1 F_2 l$	$F_1 F_3 l$	$F_2^2 l$	$F_2 F_3 l$	$F_3^2 l$
—	in.	k	k	k	k² in.	k² in.	k² in.	k² in.	k² in.	k² in.
ab	240	0	+1.67	0	0	0	0	+669.3	0	0
bc	192	+1.00	−1.33	0	+192.0	−255.4	0	+339.6	0	0
ac	144	0	0	+1.00	0	0	0	0	0	+144.0
Σ					+192.0	−255.4	0	+1008.9	0	+144.0

Thus, the flexibility matrix becomes

$$[f] = \frac{1}{EA} \begin{bmatrix} 192.0 & -255.4 & 0 \\ -255.4 & 1008.9 & 0 \\ 0 & 0 & 144.0 \end{bmatrix} \left(\frac{k - in.}{EA}\right)$$

Displacements: The displacements can now be determined for any loading from Eq. 5.30:

$$\{\Delta\} = [f]\{P\} \tag{5.30}$$

For the prescribed load, we have

$$\begin{Bmatrix} \Delta_1 \\ \Delta_2 \\ \Delta_3 \end{Bmatrix} = \frac{1}{(29 \times 10^3 \times 4)} \begin{bmatrix} 192.0 & -255.4 & 0 \\ -255.4 & 1008.9 & 0 \\ 0 & 0 & 144.0 \end{bmatrix} \begin{Bmatrix} -10 \\ +15 \\ 0 \end{Bmatrix} = \begin{Bmatrix} -0.0496 \\ +0.1525 \\ 0 \end{Bmatrix} \text{(inch)}$$

Each element of the flexibility matrix has units of inches. These are the deflections that result from 1 kip loads. The $\{P\}$ vector simply gives the magnitude of the loads, or how much the effects of the unit loads must be amplified. The resulting displacements are in inches. A negative sign simply means that the deflection is opposite from that initially assumed.

5.9

Fundamentals of Energy Methods

Numerous techniques in structural analysis come under the general heading of energy methods. These methods are cast in a variety of forms and each possesses unique advantages and disadvantages. However, it is important to note that each of these methods can be derived from the method of virtual work.

In Section 5.5 it was shown that for a deformable body to be in equilibrium, the virtual work quantities have to satisfy the relationship

$$\delta W_e = \delta W_d \tag{5.31}$$

where δW_e is the external virtual work and δW_d is the virtual work of distortion. Because the work of deforming the system is the strain energy, δW_d represents a virtual change in strain energy δU, and Eq. 5.31 becomes

$$\delta U - \delta W_e = 0 \tag{5.32}$$

or

$$\delta(U - W_e) = 0 \tag{5.33}$$

In energy methods, δW_e is termed the change in potential of the external forces, and the quantity $(U - W_e)$ is called the total *potential energy* of the system and is represented by V. Thus, Eq. 5.33 can be written as

$$\delta V = 0 \tag{5.34}$$

Equation 5.34 is merely a restatement of the method of virtual work. It states that *for a system to be in equilibrium, the variation in the total potential energy must vanish for any virtual deformation*. In other words, *the total potential energy is stationary with regard to variations in the displacements*.

When the change in total potential energy results from a virtual change in the ith displacement quantity, Eqs. 5.32 and 2.20 give

$$\delta V = \delta U - \delta W_e = \frac{\partial U}{\partial \Delta_i} \delta \Delta_i - P_i \delta \Delta_i = 0 \qquad (5.35)$$

where $\delta \Delta_i$ is a virtual displacement corresponding to the force P_i and $\partial U/\partial \Delta_i$ measures the change in strain energy per unit change in Δ_i. Equation 5.35 can be written in the form

$$\left(\frac{\partial U}{\partial \Delta_i} - P_i \right) \delta \Delta_i = 0 \qquad (5.36)$$

and since $\delta \Delta_i \neq 0$, the bracketed term must vanish. This leads to

$$\frac{\partial U}{\partial \Delta_i} = P_i \qquad (5.37)$$

This equation is the mathematical statement of *Castigliano's first theorem*, which states that *the partial derivative of the strain energy with respect to any displacement quantity is equal to the force corresponding to that displacement*. Equation 5.37 is valid for $i = 1, 2, \ldots, n$ and thus represents n equations of equilibrium corresponding to the n displacement degrees of freedom.

As stated earlier, only linearly elastic structures are treated in this book; however, Castigliano's first theorem is not limited to this class of structures.

When the change in total potential energy results from a virtual change in the ith force quantity, Eqs. 5.32 and 2.20 give

$$\delta V = \delta U - \delta W_e = \frac{\partial U}{\partial P_i} \delta P_i - \Delta_i \delta P_i = 0 \qquad (5.38)$$

where δP_i is a virtual force corresponding to the displacement Δ_i and $\partial U/\partial P_i$ measures the change in strain energy per unit change in P_i. Writing Eq. 5.38 in the form

$$\left(\frac{\partial U}{\partial P_i} - \Delta_i \right) \delta P_i = 0 \qquad (5.39)$$

and noting that $\delta P_i \neq 0$, we have

$$\frac{\partial U}{\partial P_i} = \Delta_i \qquad (5.40)$$

This equation is the mathematical expression of *Castigliano's second theorem*, and it states that *the partial derivative of the strain energy with respect to any force quantity is equal to the displacement corresponding to that force*. Equation 5.40 is valid for $i = 1, 2, \ldots, n$, and this leads to n displacement equations. These equations can be used to determine individual displacement quantities or can be used to write compatibility equations in statically indeterminate analysis.

For reasons that are not fully explained in this development, Castigliano's second theorem is limited to linearly elastic structures.

5.10

Truss Deflections by Energy Methods

Castigliano's second theorem, which was given in Section 5.9.2, is particularly well suited for obtaining individual displacement quantities or for generating the complete set of displacement–force equations for the structure as a whole. Both applications are illustrated in the example problems at the end of this section.

In the application of Castigliano's second theorem, it is essential that the total strain energy for the structure, U, be written in terms of the applied structure forces. For each structure force component P_i the corresponding displacement quantity Δ_i can be obtained from the equation

$$\frac{\partial U}{\partial P_i} = \Delta_i \tag{5.41}$$

which was derived in the previous section.

Since the strain energy is equal to the work of deformation, Eq. 2.23 can be used to express the strain energy associated with the elongation of the element shown in Fig. 5.1b. Thus, we have

$$dV = \tfrac{1}{2}F \, dl \tag{5.42}$$

where the factor $\tfrac{1}{2}$ results from the linear relationship between the force, F, and the elongation, dl, as was previously explained through Fig. 2.13. Substituting the expression for dl from Eq. 5.4 into Eq. 5.42, we obtain

$$dV = \frac{F^2}{2EA} \, dx \tag{5.43}$$

where F is the axial force on the element, A is the cross-sectional area, and E is the modulus of elasticity. In this case, since the element force is constant throughout the member length l, the total strain energy is

$$U = \int_{x=0}^{x=l} \frac{F^2 \, dx}{2EA} = \frac{F^2 l}{2EA} \tag{5.44}$$

Thus, the total strain energy stored in a truss that is composed of m members is given by

$$U = \sum_{i=1}^{m} \frac{F_i^2 l_i}{2E_i A_i} \tag{5.45}$$

It is stressed that F_i must be expressed in terms of the externally applied structure forces in order that the required differentiation of Eq. 5.41 can be performed.

5.10.1 Example problem

Determine the horizontal component of displacement at point c. All members have a cross-sectional area of 2 500 mm^2 and $E = 200$ kN/mm^2. This problem was previously worked in Sections 5.4.1 and 5.7.1.

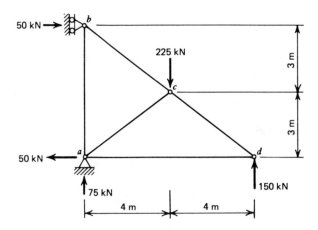

Structure Coordinate System and Member Notation:

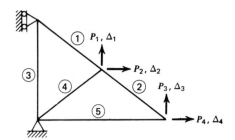

Bar Forces: Determine all bar forces in terms of P_1, P_2, P_3, and P_4 using statics.

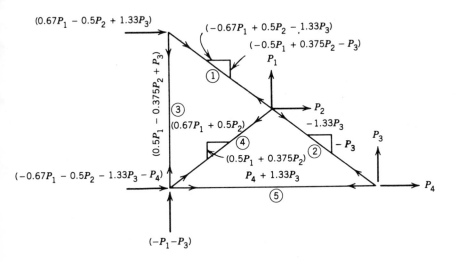

Strain Energy:

$$U = \sum_{i=1}^{5} \frac{F_i^2 l_i}{2E_i A_i} = \frac{1}{2EA} \sum_{i=1}^{5} F_i^2 l_i \qquad (5.45)$$

$$U = \frac{1}{2EA} \{[(-0.67P_1 + 0.5P_2 - 1.33P_3)^2 + (-0.5P_1 + 0.375P_2 - P_3)^2]l_1$$
$$+ [(-1.33P_3)^2 + (-P_3)^2]l_2 + [(0.5P_1 - 0.375P_2 + P_3)^2]l_3$$
$$+ [(0.67P_1 + 0.5P_2)^2 + (0.5P_1 + 0.375P_2)^2]l_4$$
$$+ [(P_4 + 1.33P_3)^2]l_5\}$$

Desired Displacement:

$$\Delta_i = \frac{\partial U}{\partial P_i} \tag{5.41}$$

In this case, $i = 2$ (horizontal displacement at point c)

$$\Delta_2 = \frac{\partial U}{\partial P_2} = \frac{1}{EA} \{[(-0.67P_1 + 0.5P_2 - 1.33P_3)(0.5)$$
$$+ (-0.5P_1 + 0.375P_2 - P_3)(0.375)]l_1$$
$$+ [(0.5P_1 - 0.375P_2 + P_3)(-0.375)]l_3$$
$$+ [(0.67P_1 + 0.5P_2)(0.5) + (0.5P_1 + 0.375P_2)(0.375)]l_4\}$$

but

$$P_1 = -225, \quad P_2 = 0, \quad P_3 = 150, \quad P_4 = 0$$
$$l_1 = 5, \quad l_2 = 5, \quad l_3 = 6, \quad l_4 = 5, \quad l_5 = 8$$

from which,

$$\Delta_2 = \frac{1}{EA} \{-192.05 - 84.38 - 588.00\} = \frac{-864.43 \text{ kN} \cdot \text{m}}{EA}$$

Thus

$$\Delta_2 = u_c = \frac{-864.43}{200 \times 10^6 \times 0.002\ 5} = -0.001\ 73 \text{ m} = -1.73 \text{ mm}$$

5.10.2 Example problem   Use Castigliano's second theorem to generate the displacement–force equations for the structure of Example 5.8.1. From these equations, extract the structure flexibility matrix. The structure coordinate system and the member notation are shown.

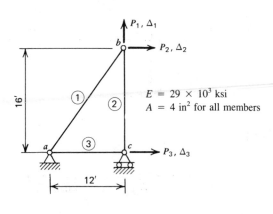

Bar Forces:

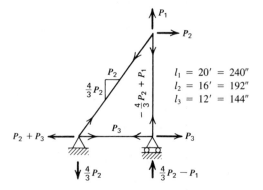

$$l_1 = 20' = 240''$$
$$l_2 = 16' = 192''$$
$$l_3 = 12' = 144''$$

Strain Energy:

$$U = \sum_{i=1}^{3} \frac{F_i^2 l_i}{2E_i A_i} = \frac{1}{2EA} \sum_{i=1}^{3} F_i^2 l_i \qquad (5.45)$$

$$U = \frac{1}{2EA} \{ [(\tfrac{4}{3}P_2)^2 + (P_2)^2] l_1 + (-\tfrac{4}{3}P_2 + P_1)^2 l_2 + (P_3)(^2 l_3) \}$$

Displacement–Force Equations and Flexibility Matrix:

$$\frac{\partial U}{\partial P_i} = \Delta_i \quad (i = 1, 2, 3) \qquad (5.41)$$

$$\frac{\partial U}{\partial P_1} = \Delta_1 \Rightarrow \frac{1}{EA} \{ -\tfrac{4}{3}P_2 + P_1)(+1)l_2 \} = \frac{1}{EA} \{192 P_1 - 256 P_2\} = \Delta_1$$

$$\frac{\partial U}{\partial P_2} = \Delta_2 \Rightarrow \frac{1}{EA} \{ [(\tfrac{4}{3}P_2)\tfrac{4}{3} + (P_2)(1)] l_1 + (-\tfrac{4}{3}P_2 + P_1)(-\tfrac{4}{3}) l_2 \}$$

$$= \frac{1}{EA} \{ -256 P_1 + 1008 P_2 \} = \Delta_2$$

$$\frac{\partial U}{\partial P_3} = \Delta_3 \Rightarrow \frac{1}{EA} \{ (P_3)(1) l_3 \} = \frac{1}{EA} \{144 P_3\} = \Delta_3$$

Or, in matrix form,

$$\frac{1}{EA}
\begin{bmatrix}
192 & -256 & 0 \\
-256 & +1008 & 0 \\
0 & 0 & +144
\end{bmatrix}
\begin{Bmatrix} P_1 \\ P_2 \\ P_3 \end{Bmatrix}
=
\begin{Bmatrix} \Delta_1 \\ \Delta_2 \\ \Delta_3 \end{Bmatrix}$$

From this, it is clear that

$$[f] = \frac{1}{EA}
\begin{bmatrix}
192 & -256 & 0 \\
-256 & 1008 & 0 \\
0 & 0 & 144
\end{bmatrix}
\qquad \left(\frac{\text{k} - \text{in.}}{EA} \right)$$

which agrees with that determined in Example (5.8.1).

5.11

Further Comments on Matrix Methods for Computing Truss Displacements

In Section 5.8, a technique for computing displacements for pin-connected frameworks is cast in a matrix format. The individual flexibility coefficients are calculated by conventional methods and arranged to form a flexibility matrix, which relates the displacements to the loads for as many displaceable points on the structure as are desired. This produces a general formulation, in that the complete array of displacements can be easily determined for as many loading conditions as one wishes to consider. However, there are many cases where the displacements are not explicitly sought apart from other information about the response of the structure. There are, accordingly, many methods available where the displacements, along with the member forces, are natural by-products of a general analysis.

In Chapter 10, a compatibility method is presented for the analysis of statically indeterminate truss structures. This procedure requires the computation of selected displacement quantities in setting up the initial compatibility equations. The solution leads initially to redundant forces, which, when combined with the applied loads, lead to the complete array of displacements for the structure. This technique is best illustrated in Section 10.8. The equations of compatibility can also be generated by Castigliano's second theorem, as illustrated in Section 10.10.

An equilibrium method of analysis for statically indeterminate pin-connected frames is presented in Chapter 12. This method is based on the solution of a set of equilibrium equations that are generated by the matrix synthesis technique of Section 12.5 or by Castigliano's first theorem of Section 12.10. The solution of the governing equilibrium equations leads directly to the truss displacements. An interesting feature of this method is that it is equally applicable to statically determinate and indeterminate structures, and once the displacements are determined, the complete set of member forces can also be determined.

The generalized matrix methods of Part III, which are merely generalized versions of the methods presented in Chapters 10 and 12, provide the complete array of structure displacements along with the member forces.

5.12

Principle of Superposition

The principle of superposition as it is related to force quantities was discussed briefly in Section 2.9. We now expand this principle to include the consideration of displacement quantities.

Consider again the beam structure of Fig. 2.10. If the principle of superposition is applied to displacement quantities, then the individual displacement components at point c that are caused by the two separate loading cases, Δ_{c1} and Δ_{c2}, can be superimposed to obtain the total displacement, Δ_c. That is,

$$\Delta_c = \Delta_{c1} + \Delta_{c2} \tag{5.46}$$

This principle has already been applied in Section 5.8. Consider the first equation from the expanded form of Eq. 5.29. Here we have

$$\Delta_1 = f_{11}P_1 + f_{12}P_2 + f_{13}P_3 + f_{14}P_4 + f_{15}P_5 \tag{5.47}$$

This equation expresses superposition on two levels. On the first level, the deflection at point 1 caused by a unit load at point j, f_{1j}, is amplified by the magnitude of the force P_j to determine that portion of the deflection at point 1 that is caused by P_j. On the second level, the total deflection at point 1 results from the superposition of the contributions from the individual loads, P_j ($j = 1, 2, \ldots, 5$).

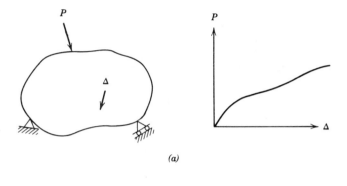

(a)

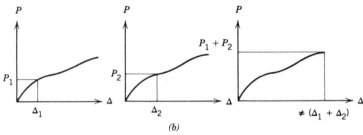

(b)

Fig. 5.11 *Structure with nonlinear response.* (a) *Structure and load–displacement diagram.* (b) *Response of structure for individual loading cases.*

There are two situations in which the principle of superposition does not hold. The first case is associated with instances where the geometry of the structure is appreciably altered under load. In this case, each load that is applied to the structure alters the geometry to the point that subsequent loads are applied to a structure whose stiffness characteristics are changed. Such is the case for the geometrically unstable structures discussed in Section 2.9. The second case in which superposition is not valid is in situations where the structure is composed of a material in which the stress is not linearly related to the strain. Here, the material stiffness changes for each increment of load. For either case, a nonlinear relationship between load and displacement results. This is illustrated in Fig. 5.11a, where a structure is shown along with the governing relationship between the load P and the displacement Δ. Figure 5.11b shows that for two individual values of P, namely P_1 and P_2, the resulting magnitudes of Δ are Δ_1 and Δ_2, respectively— each determined from the governing P–Δ curve. However, it is clear that the combined load of $P = P_1 + P_2$ does not produce a displacement of $\Delta = \Delta_1 + \Delta_2$. Thus, the principle of superposition does not apply.

5.13 Maxwell's and Betti's Laws

Consider the body shown in Fig. 5.12, which is subjected to the forces P_1 and P_2 and experiences the displacements Δ_1 and Δ_2, respectively. Assuming a linearly elastic structure, we can use the principle of superposition to express the total displacements as

$$\Delta_1 = f_{11}P_1 + f_{12}P_2$$
$$\Delta_2 = f_{21}P_1 + f_{22}P_2 \tag{5.48}$$

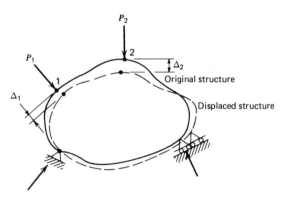

Fig. 5.12 *Forces and corresponding displacements.*

where f_{ij} is the flexibility coefficient, which gives the displacement at point i that results from a unit load at point j.

Two different sequences of loading are now considered:

1. Apply P_1 with $P_2 = 0$, and then apply P_2 with P_1 already on the structure.
2. Apply P_2 with $P_1 = 0$, and then apply P_1 with P_2 already on the structure.

The load–displacement diagrams for the two load points for both loading sequences are shown in Fig. 5.13. Recalling from Section 2.13 that work is given by the area under the load–displacement diagram, we can express the total work done for the first loading sequence, W_1, as

$$W_1 = \tfrac{1}{2} f_{11} P_1^2 + f_{12} P_2 P_1 + \tfrac{1}{2} f_{22} P_2^2 \qquad (5.49)$$

Similarly, the total work for the second loading sequence, W_2, is

$$W_2 = \tfrac{1}{2} f_{11} P_1^2 + \tfrac{1}{2} f_{22} P_2^2 + f_{21} P_1 P_2 \qquad (5.50)$$

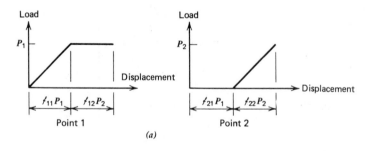

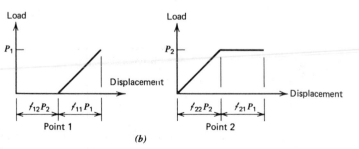

Fig. 5.13 *Load–displacement diagrams. (a) Loading sequence 1. (b) Loading sequence 2.*

Since the final displacements are the same, regardless of the loading sequence, the total work done by the forces P_1 and P_2 must be the same for either path. Thus, $W_1 = W_2$, from which

$$f_{12}P_2P_1 = f_{21}P_1P_2 \tag{5.51}$$

Letting $f_{12}P_2 = \Delta_{12}$, the displacement at point 1 due to the force at point 2, and $f_{21}P_1 = \Delta_{21}$, the displacement at point 2 due to the force at point 1, we obtain

$$P_1\Delta_{12} = P_2\Delta_{21} \tag{5.52}$$

Equation 5.52 states that the work done by the force P_1 as it experiences the displacement caused by the force P_2 is equal to the work done by the force P_2 as it experiences the displacement caused by the force P_1.

If instead of two single loads, two complete loading systems are considered, then the same considerations that led to Eq. 5.52 lead to

$$\sum_{i=1}^{n} (P_i)_1(\Delta_i)_2 = \sum_{j=1}^{m} (P_j)_2(\Delta_j)_1 \tag{5.53}$$

In this case, $(P_i)_1$ is one of the n loads of the first loading system and $(\Delta_i)_2$ is the displacement that this load experiences as a result of the second loading system. Similarly, $(P_j)_2$ is one of the m loads of the second loading condition and $(\Delta_j)_1$ is the displacement that this load experiences as a result of the first loading system.

Equation 5.53 expresses *Betti's law*. That is, *for a linearly elastic structure, the work done by the* n *forces of loading system 1 through the displacements caused by loading system 2 is equal to the work done by the* m *forces of loading system 2 through the displacements caused by loading system 1.*

Returning to the case of two single loads and canceling the load terms from Eq. 5.51, we have $f_{12} = f_{21}$. In a more general form

$$f_{ij} = f_{ji} \tag{5.54}$$

This equation expresses *Maxwell's law*. That is, *for a linearly elastic structure, the displacement at point i due to a unit load at point j is equal to the displacement at point j due to a unit load at point i.*

The development of Betti's and Maxwell's laws as given here are completely general, and they are valid for any linearly elastic structure. As such, they will be used in subsequent portions of the book when other structural systems are considered. However, in our present application to pin-connected structures, the truth of Maxwell's law has already been demonstrated in Section 5.8.1. There, it was shown numerically that $f_{ij} = f_{ji}$, and as a result of this the resulting flexibility matrix was symmetrical.

5.14 Additional Reading

Gerstle, Kurt H., *Basic Structural Analysis*, Section 7.6, Prentice–Hall, Englewood Cliffs, N.J., 1974.

Hoff, N. J., *The Analysis of Structures*, Chapter 2, Wiley, New York, 1956.

McCormac, J. C., *Structural Analysis*, 3rd Ed., Chapter 18, Intext Educational Publishers, New York, 1975.

Norris, C. W., Wilbur, J. B., and Utku, S., *Elementary Structural Analysis*, 3rd Ed., Chapter 8, McGraw–Hill, New York, 1976.

Timoshenko, S. P., and Young, D. H., *Theory of Structures*, Chapter 6, McGraw–Hill, New York, 1965.

5.15

**Suggested
Problems**

1. Use the Williot diagram to determine the horizontal and vertical components of displacement at each joint within the truss shown. Cross-sectional areas are given in parentheses (in.²) on each member. $E = 29 \times 10^3$ ksi.

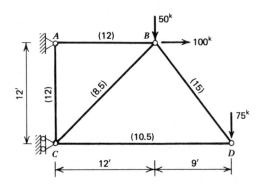

2. Use the Williot–Mohr diagram to determine the horizontal and vertical components of displacement at each joint within the truss shown. Cross-sectional areas are given in parentheses (mm²). $E = 200$ GPa.

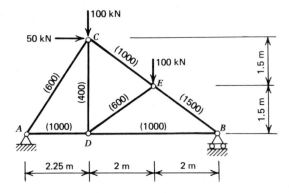

3 through 12. Use the virtual work method to determine the horizontal and vertical displacements at point p for each truss and loading shown.

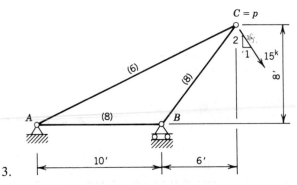

3.

Area in parentheses (in.²) on each member; $E = 29 \times 10^3$ ksi.

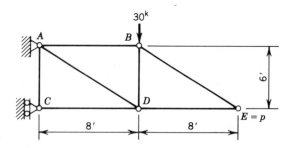

4.

Area $= 2.5$ in.^{2} for each member; $E = 29 \times 10^3$ ksi.

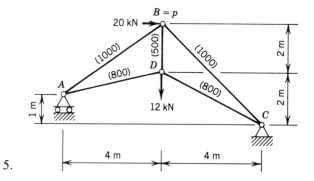

5.

Area in parentheses (mm^2) on each member; $E = 200$ GPa.

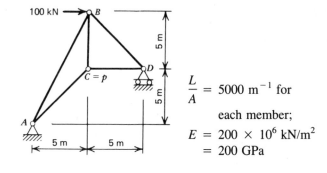

6.

$$\frac{L}{A} = 5000 \text{ m}^{-1} \text{ for}$$

each member;

$$E = 200 \times 10^6 \text{ kN/m}^2$$
$$= 200 \text{ GPa}$$

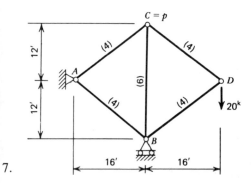

7.

Area in parentheses (in.^{2})
on each member;
$E = 29 \times 10^3$ ksi.

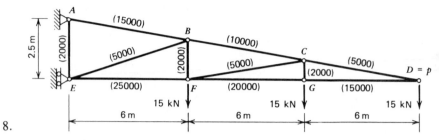

8.

Area in parentheses (mm^2) on each member; $E = 200$ GPa.

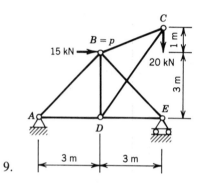

9.

Area $= 1\ 500$ mm^2 for each member; $E = 200$ GPa.

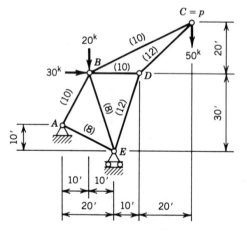

10.

Area in parentheses (in.^{2}) on each member; $E = 29 \times 10^3$ ksi.

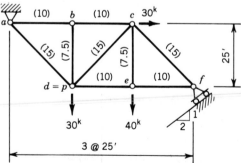

11.

Area in parentheses (in.^{2}) on each member; $E = 29 \times 10^3$ ksi.

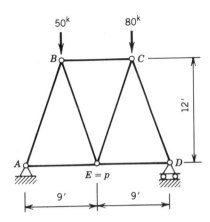

12.

Area in 5.0 in.^{2} for each member; $E = 29 \times 10^3$ ksi.

13. Determine the vertical displacement of point E on the truss of Problem 4 if members AB and BE are each subjected to a temperature change of $+50°F$. Consider the loads of Problem 4 to be removed, and take $\alpha_t = 6.5 \times 10^{-6}$ per °F.

14. Determine the vertical displacement at joint D on the truss of Problem 8 if members AB, BC, and CD are each subjected to a temperature change of $+30°C$. Consider the loads of Problem 8 to be removed, and take $\alpha_t = 1.2 \times 10^{-5}$ per °C.

15. Determine the horizontal displacement of point A on the truss of Problem 5 if members AB and BC are each subjected to a temperature change of $-20°C$ and members AD and DC to a change of $-50°F$. The loads of Problem 5 are to be removed, and $\alpha_t = 1.2 \times 10^{-5}$ per °C.

16. Determine the displacement parallel to the support surface at point f for the structure of Problem 11 if members ad, de, and ef are each subjected to a temperature decrease of $50°F$. Consider the loads of Problem 11 to be removed, and take $\alpha_t = 6.5 \times 10^{-6}$ per °F.

17. Determine the temperature change that must be induced in member AC of Problem 3 to produce an upward deflection of 0.50 inch at point C. $\alpha_t = 6.5 \times 10^{-6}$ per °F.

18. Compute the vertical and horizontal displacements of point E of Problem 4 if the member length changes shown below are induced in the respective members. The loads of Problem 4 are to be removed.

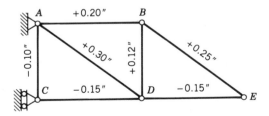

19. Determine the horizontal displacement of joint D on the truss of Problem 6 if member BC is 20 mm short but is forced to fit. Assume that the loads of Problem 6 are removed.

20. Determine the vertical displacement of joint E on the truss of Problem 12 if member BC is 0.5 inch long but is forced into place. Consider the loads of Problem 12 to be removed.

21. Calculate the change in member length that must be induced in member BC of the truss of Problem 7 to produce a horizontal displacement of 0.5 inch to the right at point D.

22. Determine the change in member length for member BD of the truss of Problem 5 that, when combined with the prescribed loading, will produce a net vertical displacement of zero at point D.

23 through 25. For each of the structures indicated, use the virtual work method to generate the complete flexibility matrix for the structure coordinates specified.

  In each case, use the resulting flexibility matrix and the loading of the problem as originally given to determine the complete set of joint displacements.

23. Structure of Problem 1.

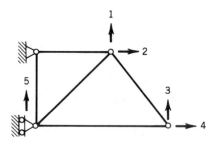

24. Structure of Problem 3.

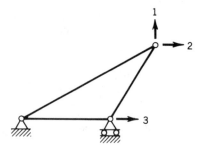

25. Structure of Problem 7.

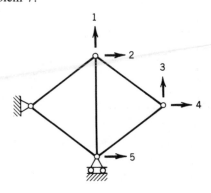

26. Develop a computer program for the generation of the flexibility matrix as defined in Section 5.8 and for the determination of the complete set of joint displacements. The truss analysis program of Problem 3.30 should be incorporated as a subprogram. Use the program to solve Problems 23, 24, and 25.

27. Use Castigliano's second theorem to determine the vertical component of displacement at joint C on the truss of Problem 3. Use the structure coordinates shown in Problem 24.

28. For the truss of Problem 3, use Castigliano's second theorem to generate the displacement–force equations for the structure coordinate system specified in Problem 24. From these equations, extract the structure flexibility matrix, and determine the complete set of joint displacements for the loading of Problem 3.

29. Use Castigliano's second theorem to determine the vertical component of displacement at joint C on the truss of Problem 6. Use the structure coordinates shown below.

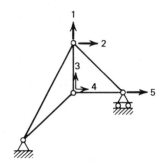

30. For the truss of Problem 6, use Castigliano's second theorem to generate the displacement–force equations for the structure coordinate system specified in Problem 29. From these equations, extract the structure flexibility matrix, and determine the complete set of joint displacements for the loading of Problem 6.

CHAPTER 6

U.S. Steel Building,
Pittsburgh, Pa.
(courtesy American
Bridge Div. of U.S.
Steel).

ELASTIC DEFLECTIONS OF
BEAM AND FRAME STRUCTURES

6.1

**Description
of Flexural
Deformation
Problem**

As a flexural structure responds to loading, it assumes an equilibrium configuration under the combined action of the loads and reactions. Corresponding to this
external state of equilibrium, there is a distribution of internal shears and bending
moments throughout the structure. These internal actions would normally be
shown in the form of shear and moment diagrams for each member. At any given
point within the structure, there is a curvature that is consistent with the moment
at that point. These curvatures accumulate as angle changes along the member
lengths, causing each member to deflect into a flexed or bent configuration. The
individual members of the deformed structure must fit together in a compatible
fashion, and all the displacement boundary conditions must be satisfied.

As was discussed in Section 5.2, it is necessary to compute deflections for evaluating serviceability criteria and for generating compatibility equations for statically indeterminate structures. As was the case in Chapter 5, we will again limit our discussion to elastic deformations.

6.2
Flexural Force– Deformation Relationships

For a flexural member, the force–deformation relationships must relate member-end moments to the corresponding member-end rotations. In a more general formulation, member-end shears and transverse deflections would be included, but they are omitted in our present discussion.

Consider a member under flexural action, and from it isolate the beam element shown in Fig. 6.1a, which is subjected to the moment M. As the member bends, the top fibers of the beam are compressed while the bottom fibers are elongated. In between, there is a longitudinal fiber whose length remains unchanged; this fiber is the so-called neutral fiber of the member. If a tangent is constructed to the neutral fiber at each end of the beam element, it is evident that

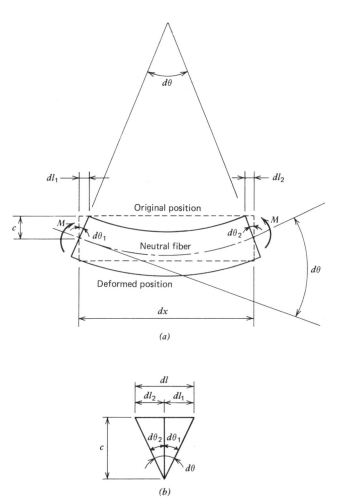

Fig. 6.1 *Flexural deformations of beam element.* (a) *Beam element subjected to moment.* (b) *Strain pattern.*

it experiences an angular deformation of $d\theta$. From the geometric considerations of Figs. 6.1a and 6.1b, it is clear that

$$d\theta = d\theta_1 + d\theta_2 \tag{6.1}$$

and for small angles

$$d\theta = \frac{dl_1}{c} + \frac{dl_2}{c} = \frac{dl}{c} \tag{6.2}$$

where dl is the total shortening of the upper fiber of the member and c is the distance from the neutral fiber to the upper fiber of the beam. Since each longitudinal fiber is essentially an axially loaded element, Eqs. 5.2 and 5.3 can be substituted into Eq. 6.2 to obtain

$$d\theta = \frac{\sigma \, dx}{Ec} \tag{6.3}$$

In this case, σ is a stress that results from flexural action and is given by the familiar expression

$$\sigma = \frac{Mc}{I} \tag{6.4}$$

where M is the moment acting on the element, c is as previously defined, and I is the moment of inertia. Substitution of Eq. 6.4 into Eq. 6.3 gives the final expression for flexural deformation as

$$d\theta = \frac{M}{EI} \, dx \tag{6.5}$$

For a member of length l, the total angle change, ϕ, that would accumulate over the member length is given through the integration of Eq. 6.5. This leads to

$$\phi = \int_0^l d\theta = \int_0^l \frac{M}{EI} \, dx \tag{6.6}$$

This equation is equivalent to Eq. 5.5 for axially loaded members; however, in this case M may vary along the member length. For instance, for the situation shown in Fig. 6.2a, the moment acting on an element at x is given by $M = M_1 (1 - x/l)$, as shown on the moment diagram of Fig. 6.2b. Therefore, since the member is *prismatic* (EI = constant), Eq. 6.6 takes the form

$$\phi = \frac{1}{EI} \int_0^l M_1 (1 - \frac{x}{l}) \, dx = \frac{M_1 l}{2EI} \tag{6.7}$$

Equation 6.7 is a deformation–force relationship that, in this case, relates rotation to moment. It can be written as

$$\phi = f_b M_1 \tag{6.8}$$

where f_b, the *flexural flexibility*, is given by

$$f_b = \frac{l}{\alpha EI} \tag{6.9}$$

The quality α depends on the manner in which M varies along the member length, which, in turn, depends on the end support conditions.

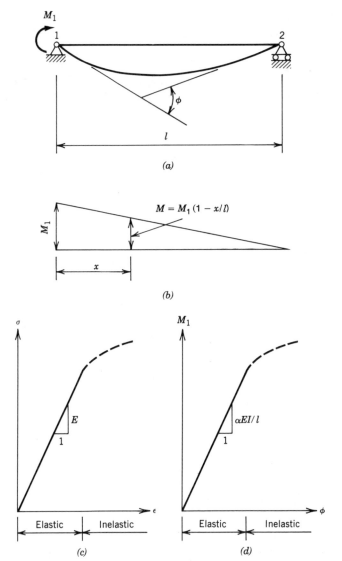

Fig. 6.2 *Flexural force–deformation behavior.* (a) *Flexurally loaded member.*
(b) *Moment diagram.* (c) *Stress–strain diagram.* (d) *Moment–rotation relationship.*

In the form of a force–deformation relationship, Eq. 6.7 can be rewritten as

$$M_1 = \frac{2EI}{l} \phi \tag{6.10}$$

or

$$M_1 = k_b \phi \tag{6.11}$$

where k_b is a measure of the flexural stiffness and is given by

$$k_b = \frac{\alpha EI}{l} \tag{6.12}$$

As noted earlier, α depends on the variation in M along the member length.

Since a linear stress–strain diagram was assumed in deriving Eq. 6.5, the resulting moment–rotation diagram mirrors this linearity, as shown in Figs. 6.2c and 6.2d.

In a more general form, consider the member shown in Fig. 6.3. Here, the end moments and rotations are shown—M_1 and θ_1 at end 1 and M_2 and θ_2 and end 2. The specific relationships between these quantities are subsequently derived in Chapter 13 and are found to be given by

$$\left\{ \begin{matrix} M_1 \\ M_2 \end{matrix} \right\} = \begin{bmatrix} \dfrac{4EI}{l} & \dfrac{2EI}{l} \\ \dfrac{2EI}{l} & \dfrac{4EI}{l} \end{bmatrix} \left\{ \begin{matrix} \theta_1 \\ \theta_2 \end{matrix} \right\} \tag{6.13}$$

In a simple form, this equation can be expressed as

$$\{M = [k]_b \{\theta\} \tag{6.14}$$

where $\{M\}$ and $\{\theta\}$ are the member-end moment and rotation vectors, respectively, and $[k]_b$ is the *member flexural stiffness matrix*.

Equation 6.11 gives a force–deformation relationship for the member as a whole, whereas Eq. 6.14 gives the full interrelationship between individual member-end forces and displacements. The individual elements of $[k]_b$ are in terms of the quantity EI/l, which is consistent with the flexural stiffness of the member as it is given by Eq. 6.12. As noted earlier, Eq. 6.13 could be expanded to include member-end shears and transverse deflections. This is done in Chapters 13 and 16.

Just as Eqs. 6.8 and 6.11 are inverse relationships in a one-dimensional form, Eq. 6.14 has an inverse counterpart. That is, it is possible to develop two-dimensional displacement–force relationships that would provide for the determination of member-end rotations from specified member-end moments. This will be developed in detail in Chapter 16. As was the case earlier for axially loaded members, our purpose here is to stress that there are unique relationships between member forces and deformations, and it is these relationships that form the basis for computing the deflections of structures.

The preceding treatment is based on an elastic response of the member. If the member were stressed beyond the elastic range, as indicated by the dashed line of Fig. 6.2c, the resulting moment–rotation curve would reflect this inelastic behavior as shown in Fig. 6.2d. Equations 6.8, 6.11, and 6.14 would be negated if the member were loaded beyond the elastic range, and they would also be invalid for nonprismatic members in which EI varied along the member length.

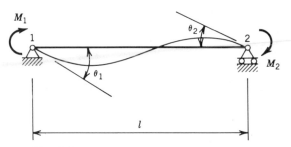

Fig. 6.3 *Member-end moments and rotations.*

6.3

A Geometric
Approach to
Flexural
Deformations:
Moment–Area
Methods

Consider the beam structure of Fig. 6.4a, which is shown in a deflected configuration under the action of the applied loads. A portion of the deflected structure between points A and B is isolated in Fig. 6.4b. Within region AB, an element of length dx is shown with tangents to the deflected member constructed at each end of the element. The angle between these end tangents, which represents the angle change that occurs over the length dx, is denoted $d\theta$. According to Eq. 6.5, this angle change is given by

$$d\theta = \frac{M}{EI}\,dx \qquad (6.15)$$

where M and I are the bending moment and moment of inertia at point x, respectively, and E is the modulus of elasticity of the material. If the M/EI values are plotted as shown in Fig. 6.4c, it is clear that $d\theta$ is given by the shaded area.

The total angle change that occurs between tangents constructed at points A and B is labeled θ_B^A in Fig. 6.4b. This angle is the slope at B relative to the slope at A; it results from the summation of the incremental angle changes between A and B and is given by

$$\theta_B^A = \int_A^B d\theta \qquad (6.16)$$

Substitution of Eq. 6.15 into Eq. 6.16 gives

$$\theta_B^A = \int_A^B \frac{M}{EI}\,dx \qquad (6.17)$$

Equation 6.17 is the basis for the *first moment–area theorem*, which can be stated as follows:

The angle change between points A and B on the deflected structure, or the slope at point B relative to the slope at point A, is given by the area under the M/EI diagram between these two points.

Examination of Fig. 6.4b shows that if the tangents to the element of length dx are extended, they embrace an intercept of $d\Delta$ on a vertical line through point B. For small angles, this intercept is given by

$$d\Delta = \bar{x}\,d\theta \qquad (6.18)$$

Substituting Eq. 6.15 into Eq. 6.18, we obtain

$$d\Delta = \bar{x}\,\frac{M}{EI}\,dx \qquad (6.19)$$

which shows that the intercept $d\Delta$ is given by the static moment of the shaded area taken about an axis through point B. The accumulation of these intercepts for all increments between points A and B gives

$$\Delta_B^A = \int_A^B d\Delta \qquad (6.20)$$

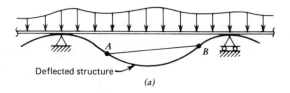

Deflected structure

(a)

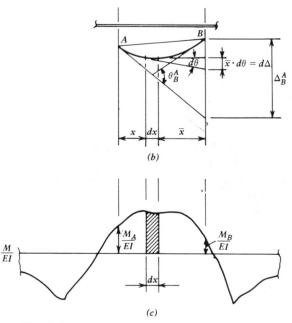

(b)

(c)

Fig. 6.4 *Development of moment–area theorems.*

where Δ_B^A is the vertical displacement of point B on the deflected structure with respect to a line drawn tangent to the structure at point A. Substitution of Eq. 6.19 into Eq. 6.20 yields

$$\Delta_B^A = \int_A^B \frac{M}{EI} \bar{x}\, dx \tag{6.21}$$

Equation 6.21 expresses the essence of the *second moment–area theorem*, which can be stated as follows:

The deflection of point B on the deflected structure with respect to a line drawn tangent to point A on the structure is given by the static moment of the area under the M/EI diagram between points A and B taken about an axis through point B.

It is emphasized that the deflection quantities that are determined by using the second moment–area theorem are normal to the original orientation of the member.

It is to be noted that the deformation quantities determined by using the moment–area theorems are relative quantities, and this fact is emphasized by the notation that is used. The first moment–area theorem is used to get θ_i^j, the slope at i (subscript) relative to the slope at j (superscript), whereas the second moment–area theorem is used to determine Δ_i^j, the displacement at i (subscript) with respect to a line that is tangent at point j (superscript). The relative nature of these quantities must be carefully considered when the moment–area theorems are applied. It is further noted that the moment–area theorems give deformations that result from flexural action only. Thus, these theorems cannot be applied over sections of a member where discontinuities are present. Discontinuities, such as a hinge, allow deformations to occur that are not related to the M/EI diagram and are thus not detected by the moment–area theorems.

6.4

Beam and Frame Deflections by Moment–Area Methods

The application of the moment–area theorems to beam and frame problems is fairly routine; however, because these theorems produce relative deformation quantities, some care has to be exercised. In general, there are two categories of problems. In the first, a displacement boundary condition prescribes a known slope at a support point. This support slope serves as a reference point from which the relative deformation quantities of each moment–area theorem can be applied to determine the actual slope or displacement at any point on the member. In the second situation, there is no point on the structure where the slope is initially known, and thus there is no reference point from which to apply the quantities that are established by the moment–area theorems. Both problem types are among the example problems that follow.

It is possible to develop a detailed sign convention for the moment–area method, but this will not be done here. Instead, an intuitive approach is suggested. For example, consider the structure shown in Fig. 6.5a. A sketch is presented of the deflected structure that satisfies the boundary conditions and possesses member curvatures that are consistent with the moment diagram. Key slopes and deflections are identified, and these are shown in the positively assumed direction as noted in Fig. 6.5a. In applying the moment–area principles, it is convenient to consider two separate cases and superimpose the results as illustrated in Fig. 6.5b. This superposition procedure leads to simpler areas with which to work, and the signs of the individual displacement components (shown parenthetically in Fig. 6.5b) are readily ascertained by comparing their directions with those that were assigned as positive in Fig. 6.5a. For instance, for the positive direction assigned to Δ_c in Fig. 6.5a, the load P_1 produces a negative contribution while the load P_2 induces a positive contribution. After the proper superposition, a positive result for any displacement quantity confirms that the assumed direction is correct.

The areas and centroidal locations for a variety of area segments are given in Fig. 4.7. These are helpful in problem solving and are used in the example problems that follow. These formulae are presented for figures with a horizontal base. They are also valid for skewed base situations when the dimension l is taken as the horizontal projection of the base and the altitude is measured perpendicular to the horizontal. This approach was previously demonstrated in Fig. 4.8.

For computational simplicity in determining the deflection quantities, it is generally most convenient to carry the quantity EI throughout the problem in symbolic form and then substitute the numerical value at the end.

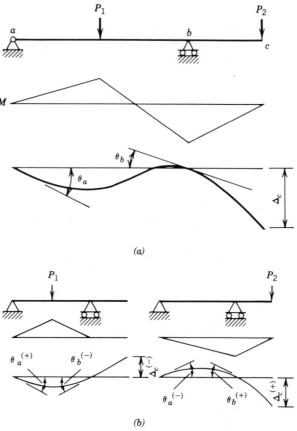

(a)

(b)

Fig. 6.5 *Beam deflections by superposition.* (a) *Loading and positive displacements.* (b) *Displacement contributions from individual loads.*

6.4.1 Example problem

Determine the expressions for the vertical displacement and slope at point B on the cantilever beam shown.

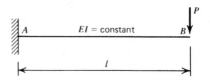

Note:

This problem is of the first type—the slope as point A is specified as zero by the boundary conditions.

M/EI Diagram:

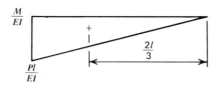

Deflected Structure:

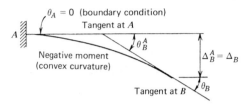

$\theta_A = 0$ (boundary condition)
Tangent at A

A

Negative moment
(convex curvature)

θ_B^A

Tangent at B

$\Delta_B^A = \Delta_B$

θ_B

Deflection Calculations: From first moment–area theorem,

$$\theta_B^A = \frac{1}{2}\left(\frac{Pl}{EI}\right)l = \frac{Pl^2}{2EI}$$

Then,

$$\theta_B = \theta_A + \theta_B^A = 0 + \frac{Pl^2}{2EI} = \frac{Pl^2}{2EI}$$

From second moment–area theorem,

$$\Delta_B^A = \left[\frac{1}{2}\left(\frac{Pl}{EI}\right)l\right] \times \frac{21}{3} = \frac{Pl^3}{3EI}$$

and

$$\Delta_B = \Delta_B^A = \frac{Pl^3}{3EI}$$

6.4.2 Example problem

Determine the slope and deflection at point C on the simply supported beam given.

100 kN

A C B

4 m

10 m

EI = constant: E = 200 GPa = 200×10^6 kN/m^2

$I = 500 \times 10^{-6}$ m^4

$EI = 100\,000$ kN · m^2

Note:
This problem is of the second type; no slope is specified by the boundary conditions.

M/EI Diagram:

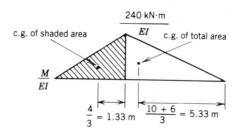

Deflected Structure:

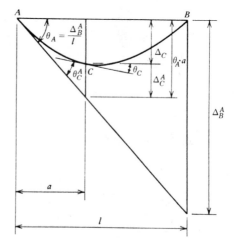

Deflection Calculations: Initially, the slope must be established at some point; point A is selected.

From second moment–area theorem,

$$\Delta_B^A = \frac{1}{2}\left(\frac{240}{EI}\right)10 \times 5.33 = \frac{6\ 396\ \text{kN}\cdot\text{m}^3}{EI}$$

and, for small displacements,

$$\theta_A = \frac{\Delta_B^A}{l} = \frac{6\ 396/EI}{10} = \frac{639.6\ \text{kN}\cdot\text{m}^2}{EI}$$

From first moment–area theorem,

$$\theta_C^A = \frac{1}{2}\left(\frac{240}{EI}\right)4 = \frac{480\ \text{kN}\cdot\text{m}^2}{EI}$$

and thus

$$\theta_C = \theta_A - \theta_C^A = \left(\frac{639.6}{EI} - \frac{480}{EI}\right) = \frac{159.6\ \text{kN}\cdot\text{m}^2}{EI}$$

From second moment–area theorem,

$$\Delta_C^A = \frac{1}{2}\left(\frac{240}{EI}\right)4 \times 1.33 = \frac{638.4\ \text{kN}\cdot\text{m}^3}{EI}$$

from which

$$\Delta_C = \theta_A \cdot a - \Delta_C^A = \left(\frac{639.6}{EI} \times 4\right) - \left(\frac{638.4}{EI}\right) = \frac{1\ 920\ \text{kN} \cdot \text{m}^3}{EI}$$

For the given EI,

$$\theta_C = \frac{159.6}{100\ 000} = 0.001\ 596\ \text{radian}$$

$$\Delta_C = \frac{1\ 920}{100\ 000} = 0.019\ 2\ \text{m} = 19.2\ \text{mm}$$

The positive results for θ_C and Δ_C confirm that these deflection quantities are as shown on the deflection sketch.

Determine the vertical deflection and the slope at point d for the structure shown. **6.4.3 Example problem**

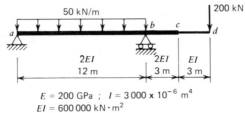

$$E = 200\ \text{GPa}\ ;\ I = 3\ 000 \times 10^{-6}\ \text{m}^4$$
$$EI = 600\ 000\ \text{kN} \cdot \text{m}^2$$

M/EI Diagram: It is convenient to show the M/EI diagram as the superposition of the separate effects of the uniform load and the concentrated load.

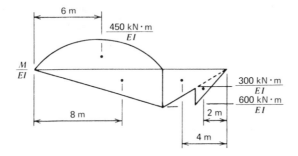

Deflected Structure:

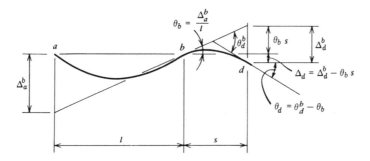

Deflection Calculations: The direction of each displacement quantity shown in the above sketch is taken as positive.

Initially, the slope is established at point B. From second moment–area theorem,

$$\Delta_a^b = \underbrace{\frac{2}{3}\left(\frac{450}{EI}\right)12 \times 6 - \frac{1}{2}\left(\frac{600}{EI}\right)12 \times 8}_{} = \underbrace{\frac{-7\ 200\ \text{kN} \cdot \text{m}^3}{EI}}_{}$$

This term is
taken as positive
because it contributes
positively to Δ_a^b
as shown in
the sketch

Negative simply
indicates that Δ_a^b
is opposite to
the direction shown
in the sketch

For small displacements,

$$\theta_b = \frac{\Delta_a^b}{l} = \frac{-7\ 200/EI}{12} = \frac{-600\ \text{kN} \cdot \text{m}^2}{EI}$$

From first moment–area theorem,

$$\theta_d^b = \frac{1}{2}\left(\frac{600}{EI}\right)6 + \frac{1}{2}\left(\frac{300}{EI}\right)3 = \frac{2\ 250\ \text{kN} \cdot \text{m}^2}{EI}$$

and therefore

$$\theta_d = \theta_d^b - \theta_b = \frac{2\ 250}{EI} - \left(-\frac{600}{EI}\right) = \frac{2\ 850\ \text{kN} \cdot \text{m}^2}{EI}$$

From second moment–area theorem,

$$\Delta_d^b = \frac{1}{2}\left(\frac{600}{EI}\right)6 \times 4 + \frac{1}{2}\left(\frac{300}{EI}\right)3 \times 2$$

$$= \frac{7\ 200}{EI} + \frac{900}{EI} = \frac{8\ 100\ \text{kN} \cdot \text{m}^3}{EI}$$

and then

$$\Delta_d = \Delta_d^b - \theta_b \cdot s = \frac{8\ 100}{EI} - \left(-\frac{600}{EI}\right)6 = \frac{11\ 700\ \text{kN} \cdot \text{m}^3}{EI}$$

For the specified EI,

$$\theta_d = \frac{2\ 850}{600\ 000} = 0.004\ 75\ \text{radian}$$

$$\Delta_d = \frac{11\ 700}{600\ 000} = 0.019\ 5\ \text{m}$$

The positive results for θ_d and Δ_d confirm that these deflection quantities are as shown on the deflection sketch.

6.4.4 Example problem   Determine the slope and deflection under the concentrated load for the structure shown.

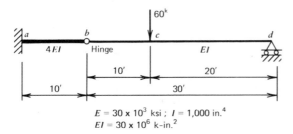

$$E = 30 \times 10^3 \text{ ksi} ; \ I = 1,000 \text{ in.}^4$$
$$EI = 30 \times 10^6 \text{ k-in.}^2$$

M/EI Diagram:

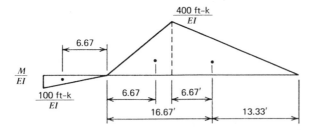

Deflected Structure:

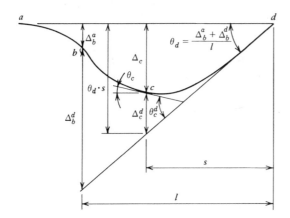

Note:

Solution strategy must treat beam segments on each side of hinge separately since the moment–area theorems cannot be applied over a beam section that includes a hinge.

Deflection Calculations: From second moment–area theorem,

$$\Delta_b^a = \frac{1}{2}\left(\frac{100}{EI}\right) 10 \times 6.67 = \frac{3,335 \text{ ft}^3\text{-k}}{EI}$$

$$\Delta_b^d = \frac{1}{2}\left(\frac{400}{EI}\right) 10 \times 6.67 + \frac{1}{2}\left(\frac{400}{EI}\right) 20 \times 16.67$$

$$= \frac{13,340}{EI} + \frac{66,680}{EI} = \frac{80,020 \text{ ft}^3\text{-k}}{EI}$$

and

$$\theta_d = \frac{\Delta_b^a + \Delta_b^d}{l} = \frac{3,335 + 80,020}{30EI} = \frac{2,778.5 \text{ ft}^2\text{-k}}{EI}$$

From first moment–area theorem,

$$\theta_c^d = \frac{1}{2}\left(\frac{400}{EI}\right) 20 = \frac{4,000 \text{ ft}^2\text{-k}}{EI}$$

Thus,

$$\theta_c = \theta_c^d - \theta_d = \frac{4,000}{EI} - \frac{2,778.5}{EI} = \frac{1,221.5 \text{ ft}^2\text{-k}}{EI}$$

Using second moment–area theorem,

$$\Delta_c^d = \frac{1}{2}\left(\frac{400}{EI}\right) 20 \times 6.67 = \frac{26,680 \text{ ft}^3\text{-k}}{EI}$$

from which

$$\Delta_c = \theta_d \cdot s - \Delta_c^d = \left(\frac{2,778.5}{EI}\right) 20 - \frac{26,680}{EI} = \frac{28,890 \text{ ft}^3\text{-k}}{EI}$$

For the given EI,

$$\theta_c = \frac{1,221.5 \times 144}{30 \times 10^6} = 0.005,86 \text{ radian}$$

$$\Delta_c = \frac{28,890 \times 1,728}{30 \times 10^6} = 1.66 \text{ inches}$$

The positive results for θ_c and Δ_c indicates that the deflection quantities are in the direction shown on the deflection sketch.

6.4.5 Example problem   Determine the vertical deflection under the load for the structure shown.

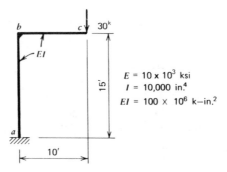

M/EI Diagrams:

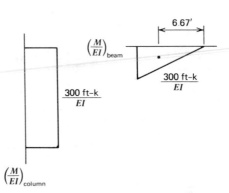

Deflected Structure:

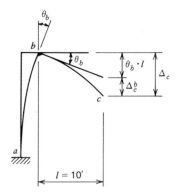

Deflection Calculations: From first moment–area theorem,

$$\theta_b = \theta_b^a = \left(\frac{300}{EI}\right) 15 = \frac{4,500 \text{ ft}^2\text{-k}}{EI}$$

This is the common rotation for the top of column *ab* and the left end of beam *bc*.

From second moment–area theorem,

$$\Delta_c^b = \frac{1}{2}\left(\frac{300}{EI}\right) 10 \times 6.67 = \frac{10,000 \text{ ft}^3\text{-k}}{EI}$$

and thus

$$\Delta_c = \theta_b \cdot l + \Delta_c^b = \left(\frac{4,500}{EI}\right) 10 + \frac{10,000}{EI} = 55,000 \text{ ft}^3\text{-k}$$

For the given *EI*,

$$\Delta_c = \frac{55,000 \times 1,728}{100 \times 10^6} = 0.950 \text{ inch}$$

Determine the horizontal deflection and the slope at point *b* for the frame structure shown.

6.4.6 Example problem

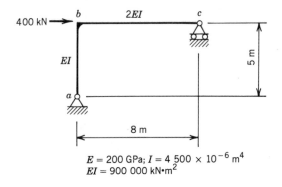

$E = 200 \text{ GPa}; I = 4\,500 \times 10^{-6} \text{ m}^4$
$EI = 900\,000 \text{ kN·m}^2$

M/EI Diagrams:

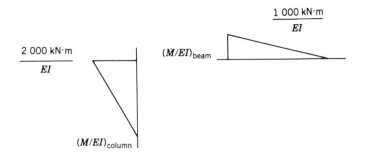

Deflected Structure:

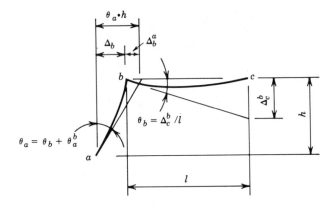

Note:

θ_b is common to the left end of member bc and the top of member ba.

Deflection Calculations: From second moment–area theorem,

$$\Delta_c^b = \frac{1}{2}\left(\frac{1\ 000}{EI}\right) 8 \times 5.34 = \frac{21\ 344\ \text{kN} \cdot \text{m}^3}{EI}$$

from which

$$\theta_b = \frac{\Delta_c^b}{l} = \frac{21\ 344}{8} = \frac{2\ 668\ \text{kN} \cdot \text{m}^2}{EI}$$

Then, from first moment–area theorem

$$\theta_a^b = \frac{1}{2}\left(\frac{2\ 000}{EI}\right) 5 = \frac{5\ 000\ \text{kN} \cdot \text{m}^2}{EI}$$

and, therefore,

$$\theta_a = \theta_b + \theta_a^b = \frac{2\ 668}{EI} + \frac{5\ 000}{EI} = \frac{7\ 668\ \text{kN} \cdot \text{m}^2}{EI}$$

Again, second moment–area theorem gives

$$\Delta_b^a = \frac{1}{2}\left(\frac{2\ 000}{EI}\right) 5 \times 1.67 = \frac{8\ 350\ \text{kN} \cdot \text{m}^3}{EI}$$

and

$$\Delta_b = \theta_a \cdot h - \Delta_b^a = \left(\frac{7\ 668}{EI}\right) 5 - \left(\frac{8\ 350}{EI}\right) = \frac{29\ 990\ \text{kN} \cdot \text{m}^3}{EI}$$

For the given EI,

$$\theta_b = \frac{2\ 668}{900\ 000} = 0.002\ 964\ \text{radian}$$

$$\Delta_b = \frac{29\ 990}{900\ 000} = 0.033\ 322\ \text{m} = 33.3\ \text{mm}$$

The positive signs for θ_b and Δ_b confirm that the deflections are in the directions shown on the sketch.

6.5 The Elastic Load Method

The moment–area method can be expressed as an analogy known as the *elastic load method*. This method is limited, but it provides a transition to the powerful conjugate beam method that will be introduced in Section 6.6. Both the elastic load method and the conjugate beam method allow for the development of systematic sign conventions for determining slopes and deflections.

Consider the simply supported beam shown in Fig. 6.6 along with the resulting M/EI diagram. The systematic application of the moment–area theorems is outlined in Fig. 6.6, and the expressions for θ_A, θ_C, and Δ_C are given by

$$\theta_A = \frac{\Delta_B^A}{l} = \frac{\int_A^B (M/EI)\bar{x}\ dx}{l} \tag{6.22}$$

$$\theta_C = \theta_A - \theta_C^A = \theta_A - \int_A^C \frac{M}{EI}\ dx \tag{6.23}$$

$$\Delta_C = \theta_A \cdot a - \Delta_C^A = \theta_A \cdot a - \int_A^C \frac{M}{EI}\tilde{x}\ dx \tag{6.24}$$

The quantities given by these equations can be determined by considering an analogous problem. Consider an imaginary simply supported beam of the same length as the actual beam, which is acted on by a loading formed by the M/EI diagram (elastic load), as illustrated in Fig. 6.7a. Positive M/EI ordinates are taken as positive (upward) loading, whereas negative M/EI ordinates become negative (downward) loading. By taking moments about point B, we can determine the reaction at A to be

$$R_A = \frac{\int_A^B (M/EI)\bar{x}\ dx}{l} \tag{6.25}$$

This reaction is shown in Fig. 6.7a. A free-body diagram of section AC permits one to determine the shear and moment at point C. With the aid of Fig. 6.7b, we have

$$V_C = R_A - \int_A^C \frac{M}{EI}\ dx \tag{6.26}$$

and

$$M_C = R_A \cdot a - \int_A^C \frac{M}{EI} \tilde{x} \, dx \tag{6.27}$$

Comparing Eqs. 6.22 and 6.25, we see that R_A on the imaginary beam is equivalent to θ_A on the real beam. Furthermore, by comparing Eqs. 6.23 and 6.24 with Eqs. 6.26 and 6.27, we see that V_C and M_C on the imaginary beam are equivalent, respectively, to θ_C and Δ_C on the real beam. This leads to the following general statement concerning the *elastic load method:*

> *The slope and deflection at any point on a simply supported beam segment are given, respectively, by the shear and moment that result from applying the M/EI diagram as loading on an imaginary simply supported beam of the same length as the given real beam.*

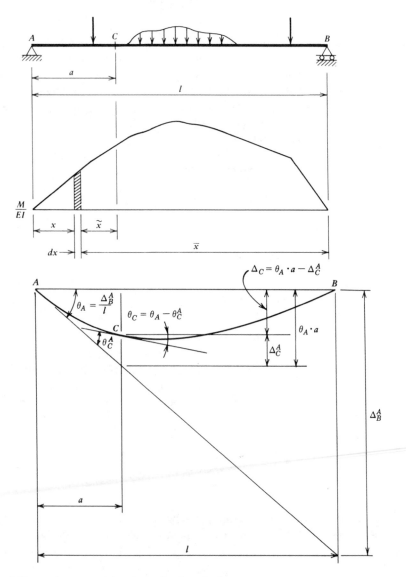

Fig. 6.6 *Development of the elastic-load method.*

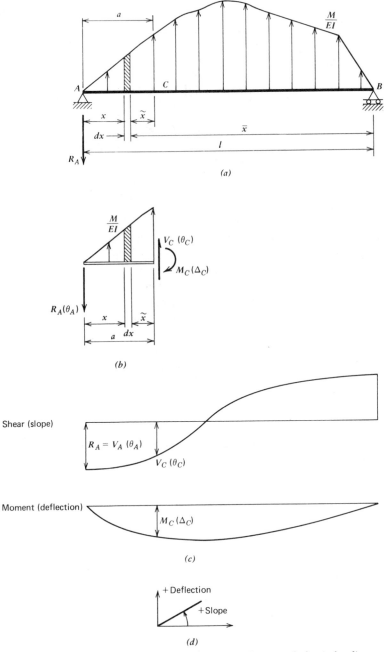

Fig. 6.7 *Details of elastic load analogy.* (a) *Imaginary beam and elastic loading.* (b) *Section* AC. (c) *Shears (slopes) and moments (deflections).* (d) *Sign convention.*

Based on positive upward load, Eqs. 6.25, 6.26, and 6.27 give rise to R_A, V_C, and M_C with directions indicated in Fig. 6.7b. All three of these are negative according to the convention adopted in Chapter 4 for load, shear, and moment, and they form a part of the complete shear and moment diagrams shown in Figs. 6.7c. Because of the elastic load analogy, these shear and moment diagrams for the imaginary beam are the slope and deflection diagrams, respectively, for the

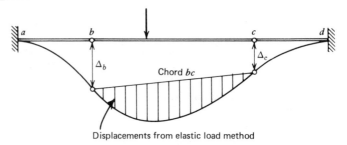

Displacements from elastic load method

Fig. 6.8 *Restrictions of elastic load method.*

real beam. The signed values given by these diagrams are correct if the positive directions for slope and deflection are taken as those shown in Fig. 6.7d.

The beauty of the elastic load method is that the complete signed array of slopes and deflections results from constructing the shear and moment diagrams for the imaginary beam. The major shortcoming of the method is that it can be applied only to a section of the beam in which there is no moment resistance at ends, and the method will give only the rotations and displacements relative to the chord that connects the ends. For instance, consider the beam shown in Fig. 6.8. The elastic load method can be applied to segment *bc* because it is a simply supported section. However, because zero end moments imply zero end displacements, the elastic load method will produce only the shaded displacements illustrated in Fig. 6.8. The end displacements, Δ_b and Δ_c, must be determined separately by the moment–area theorems. These displacements would position the chord for segment *bc*, on which the displacements of the elastic load method must be superimposed in order to get the total displacements.

It should also be noted that, as was the case with the moment–area method, there can be no discontinuities within the section over which the elastic load method is applied.

6.5.1 Example problem

Use the elastic load method to determine the slope and deflection at point *C* for the simply supported beam of Example 6.4.2.

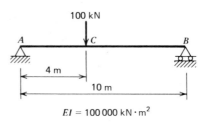

$$EI = 100\,000 \text{ kN} \cdot \text{m}^2$$

Imaginary Beam and Elastic Loading:

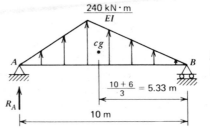

R_A assumed to act in positive direction.

$$\sum M_B = 0: \quad R_A \times 10 + \frac{1}{2}\left(\frac{240}{EI}\right)10 \times 5.33 = 0$$

$$R_A = \frac{-639.6 \text{ kN} \cdot \text{m}^2}{EI}$$

(negative sign indicates that R_A acts downward and is thus negative)

Deflection Quantities: Consider section AC.

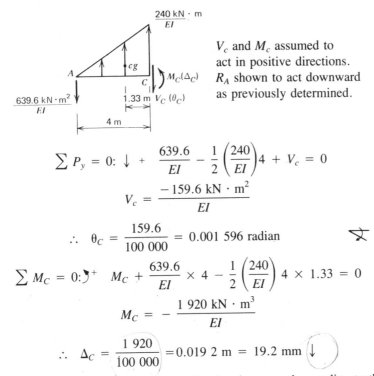

V_c and M_c assumed to act in positive directions. R_A shown to act downward as previously determined.

$$\sum P_y = 0: \downarrow + \quad \frac{639.6}{EI} - \frac{1}{2}\left(\frac{240}{EI}\right)4 + V_c = 0$$

$$V_c = \frac{-159.6 \text{ kN} \cdot \text{m}^2}{EI}$$

$$\therefore \quad \theta_c = \frac{159.6}{100\ 000} = 0.001\ 596 \text{ radian} \qquad \searrow$$

$$\sum M_C = 0: \circlearrowright^+ \quad M_C + \frac{639.6}{EI} \times 4 - \frac{1}{2}\left(\frac{240}{EI}\right)4 \times 1.33 = 0$$

$$M_C = -\frac{1\ 920 \text{ kN} \cdot \text{m}^3}{EI}$$

$$\therefore \quad \Delta_C = \frac{1\ 920}{100\ 000} = 0.019\ 2 \text{ m} = 19.2 \text{ mm } (\downarrow)$$

Both deflection quantities are negative and, when interpreted according to the convention of Fig. 6.7d, have the indicated directions.

6.6 The Conjugate Beam Method

The problem of beam statics is governed by Eqs. 4.1 and 4.3, which state

$$\frac{d^2M}{dx^2} = \frac{dV}{dx} = p \tag{6.28}$$

This is a second-order linear differential equation, and the solution is the classic shear and moment diagram problem: starting with the load, the first integration gives the shear and the second integration gives the moment.

Similarly, the beam deflection problem is governed by Eq. 6.15. Taking y as the beam deflection and noting that $\theta = dy/dx$, we can write

$$\frac{d^2y}{dx^2} = \frac{d\theta}{dx} = \frac{M}{EI} \tag{6.29}$$

This is also a second-order linear differential equation. Here, we start with the curvature, M/EI; the first integration yields the slope, and the second integration gives the deflection.

TABLE 6.1 Basis of Conjugate Beam Method

Real Beam Parameters	Imaginary Beam Parameters	Solution
Load p		Solve $\dfrac{d^2M}{dx^2} = \dfrac{dV}{dx} = p$
Shear V		Start with p and
Moment M		integrate to obtain V and M.
Curvature M/EI	(Load p)	Solve $\dfrac{d^2y}{dx^2} = \dfrac{d\theta}{dx} = \dfrac{M}{EI}$
Slope θ ←───────→	(Shear V)	Start with M/EI and integrate to obtain θ
Deflection y ←─────	(Moment M)	and y.

If the boundary conditions on V and M of the imaginary beam are forced to match the boundary conditions on θ and y of the real beam, then the restrictions of the elastic load method are avoided, and the analogy is complete.

Comparison of Eqs. 6.28 and 6.29 forms the basis of the elastic load method. If M/EI is taken as loading on an imaginary beam, the resulting shears and moments will, in fact, be the slopes and deflections on the real beam. This is shown in Table 6.1 within the box of broken lines.

The deficiency of the elastic load method, which makes it applicable only for a simply supported beam segment, can be corrected if proper attention is given to the boundary conditions. That is, a complete analogy is possible if the boundary conditions on shear and moment on the imaginary beam are made to match the boundary conditions on slope and deflection, respectively, for the real beam. This is highlighted in Table 6.1, and it forms the basis of the *conjugate beam method*.

To use the conjugate beam method, an imaginary beam (conjugate beam) is conceived that has the same length as the real beam and has a set of boundary conditions and internal continuity conditions on shear and moment that match the corresponding real beam boundary conditions and internal continuity conditions on slope and deflection. This transformation procedure is outlined in Table 6.2. For example, a fixed end has geometric boundary conditions of zero slope and deflection. Thus, the conjugate beam must have force boundary conditions of zero shear and moment. These force boundary conditions can occur only at a free end. The final conjugate beam is then subjected to a loading that corresponds to the M/EI diagram of the real beam. The resulting shear and moment diagrams of the conjugate beam give the slope and deflection diagrams, respectively, of the real beam.

It should be emphasized that the same sign convention applies here that was developed for the elastic load method. It is also noted that in the conjugate beam method, the total slopes and deflections are obtained, including those that result from any chord movement that occurs.

TABLE 6.2 Transformation of Real Beam to Conjugate Beam

	Geometric Conditions of Real Beam	Force Conditions of Conjugate Beam
Parameters	Slope Deflection	Shear Moment
	Must Match	
Boundary Conditions	clamped pin roller free	free pin roller clamped
Continuity Conditions	interior support interior roller interior hinge roller transfer	interior hinge roller transfer interior support interior roller

Construct the complete slope and deflection diagrams for the structure shown, and evaluate the slope and deflection at the free end.

6.6.1 Example problem

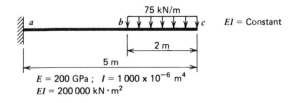

75 kN/m

a b c EI = Constant

2 m

5 m

$E = 200$ GPa ; $I = 1\,000 \times 10^{-6}$ m^4

$EI = 200\,000$ kN $\cdot$ m^2

Moment Diagram:

$M(\text{kN} \cdot \text{m})$

-150

-600

Conjugate Beam and Loading:

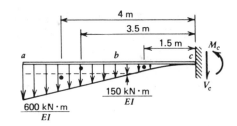

$$\sum P_y = 0: \quad V_c + \left(\frac{150}{EI}\right)3 + \frac{1}{2}\left(\frac{450}{EI}\right)3 + \frac{1}{3}\left(\frac{150}{EI}\right)2 = 0$$

$$V_c = \frac{-1\ 225 \ \mathrm{kN} \cdot \mathrm{m}^2}{EI} = \theta_c \ \text{for real beam}$$

$$\sum M_c = 0: \quad M_c + \left(\frac{150}{EI}\right)3 \times 3.5 + \frac{1}{2}\left(\frac{450}{EI}\right)3 \times 4 + \frac{1}{3}\left(\frac{150}{EI}\right)2 \times 1.5 = 0$$

$$M_c = \frac{-4\ 425 \ \mathrm{kN} \cdot \mathrm{m}^3}{EI} = \Delta_c \ \text{for real beam}$$

Shear Diagram for Conjugate Beam (Slope Diagram for Real Beam):

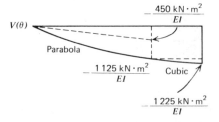

Moment Diagram for Conjugate Beam (Delfection Diagram for Real Beam):

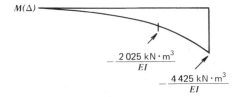

Maximum Slope and Deflection:

$$\theta_c = \frac{-1\ 225}{200\ 000} = -0.006\ 13 \ \text{radian} \left(\text{negative sign indicates} \quad \right)$$

$$\Delta_c = \frac{-4\ 425}{200\ 000} = -0.022\ 1 \ m = -22.1 \ \text{mm (negative sign indicates} \downarrow)$$

6.6.2 Example problem

Determine the slope and deflection under the concentrated load for the structure shown, which is the same structure that was considered in Example 6.4.4.

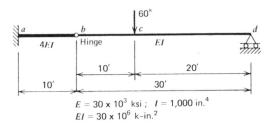

$$E = 30 \times 10^3 \text{ ksi} \; ; \quad I = 1{,}000 \text{ in.}^4$$
$$EI = 30 \times 10^6 \text{ k-in.}^2$$

Moment Diagram:

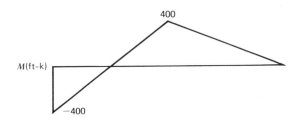

Conjugate Beam and Loading:

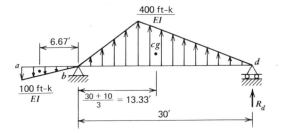

$$\Sigma M_b = 0: \quad \frac{1}{2}\left(\frac{100}{EI}\right) 10 \times 6.67 + \frac{1}{2}\left(\frac{400}{EI}\right) 30 \times 13.33 + R_d \times 30 = 0$$

$$R_d = -\left(\frac{3{,}335 + 79{,}998}{30EI}\right) = \frac{-2{,}777.8 \text{ ft}^2\text{-k}}{EI} \quad \begin{array}{l} \text{(negative sign indicates} \\ \text{that } R_d \text{ acts downward)} \end{array}$$

Deflection Calculations: Consider section cd:

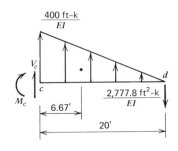

$$\Sigma P_y = 0: \uparrow + \quad V_c + \frac{1}{2}\left(\frac{400}{EI}\right)20 - \frac{2{,}777.8}{EI} = 0$$

$$V_c = \frac{-1{,}222.2 \text{ ft}^2\text{-k}}{EI}$$

$$\therefore \quad \theta_c = \frac{-1{,}222.2 \times 144}{30 \times 10^6} = -0.005{,}867 \text{ rad}$$
(negative sign indicates 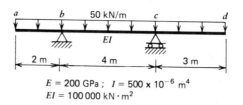)

$$\Sigma M_c = 0: \; \circlearrowleft + \quad M_c - \frac{1}{2}\left(\frac{400}{EI}\right)20 \times 6.67 + \left(\frac{2{,}777.8}{EI}\right)20 = 0$$

$$M_c = \frac{-28{,}876 \text{ ft}^3\text{-k}}{EI}$$

$$\therefore \quad \Delta_c = \frac{-28{,}876 \times 1{,}728}{30 \times 10^6} = -1.66 \text{ inches}$$
(negative sign indicates $\downarrow$)

6.6.3 Example problem

Determine the vertical deflection at point d for the structure shown under the indicated loading.

$E = 200 \text{ GPa}; \quad I = 500 \times 10^{-6} \text{ m}^4$
$EI = 100\,000 \text{ kN} \cdot \text{m}^2$

Moment Diagram:

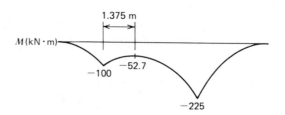

An alternative form of the moment diagram is

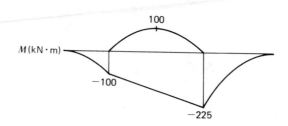

Conjugate Beam and Loadings:

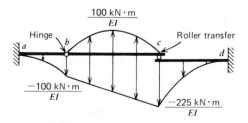

Deflection Calculations: Isolate section *bc*:

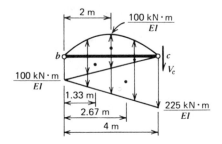

$$\sum M_b = 0: \quad +\curvearrowright$$

$$V_c \times 4 + \frac{1}{2}\left(\frac{100}{EI}\right) 4 \times 1.33 + \frac{1}{2}\left(\frac{225}{EI}\right) 4 \times 2.67 - \frac{2}{3}\left(\frac{100}{EI}\right) 4 \times 2 = 0$$

$$V_c = \frac{-233.55 \text{ kN} \cdot \text{m}^2}{EI}$$

Now isolate section *cd*:

$$\sum M_d = 0: \quad \curvearrowright \quad +$$

$$M_d + \left(\frac{233.55}{EI}\right) 3 + \frac{1}{3}\left(\frac{225}{EI}\right) 3 \times 2.25 = 0$$

$$M_d = \frac{-1\ 206.9 \text{ kN} \cdot \text{m}^3}{EI}$$

$$\therefore \quad \Delta_d = \frac{-1\ 206.9}{100\ 000} = -0.012\ 1 \text{ m} = -12.1 \text{ mm}$$

$$\text{(negative sign indicates } \downarrow)$$

6.7

Numerical Method for Beam Deflections

The conjugate beam method essentially reduces the beam deflection problem to a shear and moment diagram problem. Thus, the tabular method developed by Newmark, which was presented in Section 4.10, can be readily adapted to the determination of beam deflections.

It is recalled that the Newmark method required that concentrated transverse loads be applied to the structure. Where distributed loads were applied, an equivalent set of concentrated loads were determined in accordance with the discretization process of Section 4.11. In the conjugate beam method, the M/EI diagram is taken as the load distribution on the imaginary beam, and this loading must then be discretized into a set of concentrated loads. To do this, we let $M/EI = \alpha$ and assume the α diagram to be composed of a series of straight-line segments between panel points as shown in Fig. 6.9a. This distributed conjugate beam loading must be replaced by a set of equivalent concentrated conjugate beam loads as illustrated by Fig. 6.9b. A typical concentrated load A_i can be determined by adapting Eqs. 4.10 and 4.11, and is given by

$$A_i = \bar{\alpha}_{ih} + \bar{\alpha}_{ij} \tag{6.30}$$

or

$$A_i = \frac{\lambda}{6} (\alpha_h + 4\alpha_i + \alpha_j) \tag{6.31}$$

As noted in Section 4.11, end points must be handled separately. For a beam divided into n equal panels, there are $n + 1$ panel points. For this situation, Eq. 4.12 is applicable, and we have

$$\begin{Bmatrix} A_1 \\ A_2 \\ A_3 \\ \cdot \\ \cdot \\ \cdot \\ A_{n-1} \\ A_n \\ A_{n+1} \end{Bmatrix} = \frac{\lambda}{6} \begin{bmatrix} 2 & 1 & 0 & \cdot & \cdot & \cdot & \cdot & \cdot & 0 \\ 1 & 4 & 1 & 0 & \cdot & \cdot & \cdot & \cdot & 0 \\ 0 & 1 & 4 & 1 & 0 & \cdot & \cdot & \cdot & 0 \\ & & & & & & & & \\ & & & & & & & & \\ 0 & \cdot & \cdot & \cdot & 0 & 1 & 4 & 1 & 0 \\ 0 & \cdot & \cdot & \cdot & \cdot & 0 & 1 & 4 & 1 \\ 0 & \cdot & \cdot & \cdot & \cdot & \cdot & 0 & 1 & 2 \end{bmatrix} \begin{Bmatrix} \alpha_1 \\ \alpha_2 \\ \alpha_3 \\ \cdot \\ \cdot \\ \cdot \\ \alpha_{n-1} \\ \alpha_n \\ \alpha_{n+1} \end{Bmatrix} \tag{6.32}$$

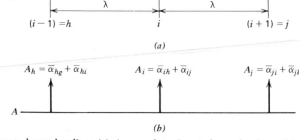

(a)

(b)

Fig. 6.9 Conjugate beam loading. (a) Assumed conjugate beam loading. (b) Discretized conjugate beam loading.

Equations 6.31 and 6.32 are based on a linear variation in the α diagram. As noted in Section 4.11, it is possible to develop discretization formulae for other load variations.

The concentrated loads given by Eqs. 6.31 and 6.32 are used in conjunction with the Newmark method to obtain the shears and moments on the conjugate beam, which are then interpreted as slopes and deflections, respectively, on the real beam.

6.7.1 Example problem

Determine the deflections at 1-meter intervals for the beam of Section 4.10.1. Assume $EI = 300\,000$ kN $\cdot$ m^2 throughout.

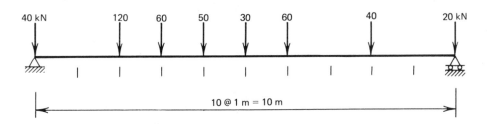

The moments were determined in Example 4.10.1 and are given below.

Conjugate Beam and Loading:

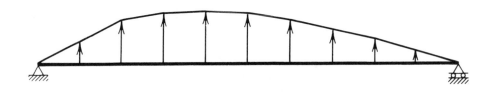

Note: See conjugate beam calculations in the table to follow:

	0	1	2	3	4	5	6	7	8	9	10	Units
Load Distr. α	0	+215	+430	+525	+560	+545	+500	+395	+290	+145	0	$0\ kN \cdot m/EI$
Conc. Lds. A	+215	+1290	+2460	+3090	+3310	+3240	+2940	+2370	+1700	+870	+145	$+145\ kN \cdot m^2/6EI$
Est. Shear V'		−10150	−8860	−6400	−3310	0 (Assumed)	+3240	+6180	+8550	+10250	+11120	$+11120\ kN \cdot m^2/6EI$
Est. Mom. Chg. $\Delta M = V'\lambda$		−10150	−8860	−6400	−3310	0	+3240	+6180	+8550	+10250	+11120	$+11120\ kN \cdot m^3/6EI$
Est. Mom. M'	0 (Start)	−10150	−19010	−25410	−28720	−28720	−25480	−19300	−10750	−500	+10620	$+10620\ kN \cdot m^3/6EI$
Mom. Corr. M''	0	−1062	−2124	−3186	−4248	−5310	−6372	−7434	−8496	−9558	−10620	$-10620\ kN \cdot m^3/6EI$
Final Mom. $M = M' + M''$	0	−11212	−21134	−28596	−32968	−34030	−31852	−26734	−19246	−10058	0	$0\ kN \cdot m^3/6EI$
$M(\Delta)$	0	−0.0062	−0.0117	−0.0159	−0.0183	−0.0189	−0.0177	−0.0149	−0.0107	−0.0056	0	m

Deflections for real beam—negative is downward.

Notes:

Loads: Load distribution, $\alpha = M/EI$; concentrated loads A are given by Eq. 6.32.

Shears (panel slopes on real beam): One could determine reactions for conjugate beam under A loads. Here, estimated shears are determined by assuming $V' = 0$ near the center of the structure and working each way toward supports.

Moments (deflections on real beam): The estimated moments M' are determined by starting at the known point of zero moment at the left support. The right-end moment boundary condition is violated, and thus a linear moment correction is applied. This amounts to a constant shear correction, which is necessary because of an error in the estimated shear diagram. The final moments give the deflections for the original beam.

It is not necessary to use the conjugate beam approach to apply the Newmark method. One can proceed by using geometric concepts instead of the analogy of the conjugate beam approach. The procedure is illustrated in Fig. 6.10. Returning to Eq. 6.15, we have

6.7.2 Physical interpretation of tabular method

$$\alpha = \frac{d\theta}{dx} = \frac{M}{EI} \tag{6.33}$$

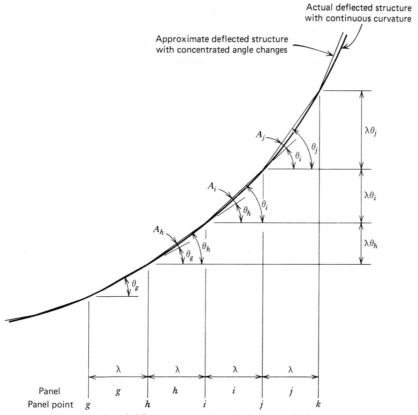

Fig. 6.10 *Interpretation of tabular method.*

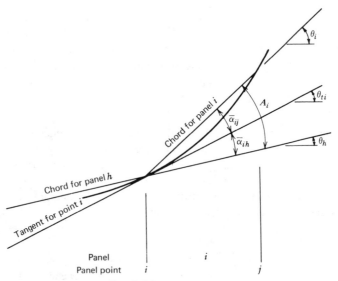

Fig. 6.11 *Angle changes.*

Here it is clear that α represents the rate of change in slope, which is referred to as the curvature. The A values, which come from integrating the α diagram, are concentrated angle changes between panel chords. Beginning with the slope in the end panel (reaction on the conjugate beam), we accumulate the A values to determine the panel chord slopes (panel shears on conjugate beam). The increment of deflection across each panel (moment change on conjugate beam) is given by the product of the panel slope and the panel length. Starting at a boundary point, where the deflection is known, we accumulate the increments of deflection to get the final deflections at the panel points (panel point moments on conjugate beam).

If the slope at a load point rather than a chord slope is desired, it can be determined. Figure 6.10 shows that A_i is the angle change at panel point i. Equation 6.30 shows that this angle change has two components, $\bar{\alpha}_{ih}$ and $\bar{\alpha}_{ij}$. The term $\bar{\alpha}_{ih}$ gives the angle change from the chord in panel h to the tangent at i, and $\bar{\alpha}_{ij}$ gives the angle change from the tangent at i to the chord in panel i. This is shown in Fig. 6.11. Thus, the slope of the tangent at i, θ_{ti}, is given by

$$\theta_{ti} = \theta_h + \bar{\alpha}_{ih} \tag{6.34}$$

6.7.3 Example problem

Determine the deflections at 10-foot intervals and the angle change at the internal hinge for the beam shown.

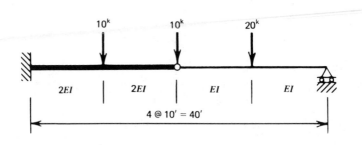

Shears and Moments:

						Units
Load P		−10	−10	−20		kips
Est. Shear V'	20	10	assumed ↘ 0	−20		kips
Est. Mom. Chg. $\Delta M = V'h$	200	100	0	−200		ft-kips
Est. Mom. M'	−300	−100	0 (begin)	0	−200 (error)	ft-kips
Mom. Corr. M''	−200	−100	0	+100	+200	ft-kips
Final Mom. $M = M' + M''$	−500	−200	0	+100	0	ft-kips
Shear Corr. V''	10	10	10	10		kips
Final Shear $V = V' + V''$	30	20	10	−10		kips

Notes:

Loads: The reactions are not initially known.

Shears: The estimated shear values are based on an assumed shear in the panel shown. Shear correction is derived from moment correction.

Moments: The estimated moment values are based on zero moment at the hinge. However, this leads to a violation of the right-end boundary condition. A linear moment correction gives the final moments.

Conjugate Beam and Loading:

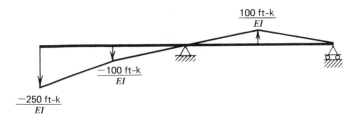

$$\frac{100 \text{ ft-k}}{EI}$$

$$\frac{-100 \text{ ft-k}}{EI}$$

$$\frac{-250 \text{ ft-k}}{EI}$$

Note: See conjugate beam calculations in the table to follow:

Conj. Beam	Real Beam						Units
Load Distr. α	Curv. Distr. α	-250	-100	$-100 \mid +100$	$+100$	0	$\dfrac{\text{ft-kips}}{EI}$
Conc. Load A	Angle Chg. A	-600	-650	0	$+400$	$+100$	$\dfrac{10\ \text{ft}^2\text{-kips}}{6EI}$
Est. Shear V'	Est. Slope θ'	0	-600	-1250	-1250	-850	$\dfrac{10\ \text{ft}^2\text{-kips}}{6EI}$
Est. Mom. Chg. $\Delta M = V'\lambda$	Est. Defl. Chg. $\delta\Delta = \theta'\lambda$	-600	-600	-1250	-1250	-850	$\dfrac{100\ \text{ft}^3\text{-kips}}{6EI}$
Est. Mom. M'	Est. Defl. Δ'	0	-600	-1850	-3100	-3950	$\dfrac{100\ \text{ft}^3\text{-kips}}{6EI}$
Mom. Corr. M''	Defl. Corr. Δ''	0	0	0	$+1975$	$+3950$	$\dfrac{100\ \text{ft}^3\text{-kips}}{6EI}$
Final Mom. $M = M' + M''$	Final Defl. $\Delta = \Delta' + \Delta''$	0	-600	-1850	-1125	0	$\dfrac{100\ \text{ft}^3\text{-kips}}{6EI}$
Shear Corr. V''	Slope Corr. θ''			0	$+1975$	$+1975$	$\dfrac{10\ \text{ft}^2\text{-kips}}{6EI}$
Final Shear $V = V' + V''$	Final Slope $\theta = \theta' + \theta''$	-600		-1250	$+725$	$+1125$	$\dfrac{10\ \text{ft}^2\text{-kips}}{6EI}$
	Tangent Slope θ_t			$-1350 + 625$			$\dfrac{10\ \text{ft}^2\text{-kips}}{6EI}$
	Angle Chg. θ_t			$+1975$			$\dfrac{10\ \text{ft}^2\text{-kips}}{6EI}$

Notes:

A *Quantities:* The unknown conjugate beam reactions, which include the angle change at the hinge, are initially ignored.

Deflection Correction: The conjugate beam moment at the right end, which is the deflection on the real beam, must be zero. The linear correction between hinge and right support amounts to a rigid-body rotation of this beam segment about the hinge.

Angle Change at Hinge: The final panel slopes on either side of the hinge are used in conjunction with the angle changes on each side of the hinge to obtain the slope of the tangent on each side of the hinge. The difference in these tangent slopes gives the angle change at the hinge. This is the conjugate beam reaction that was initially ignored when the A quantities were determined.

6.8

Matrix Formulation of Beam Deflection Problem

The matrix formulation of the shear and moment diagram problem presented in Section 4.13 could be applied to the beam deflection problem by a direct application of the conjugate beam method. The $\{Q\}$ matrix of Eq. 4.21, which would include the shears, moments, and reactions of the conjugate beam, would then be interpreted as displacement quantities for the actual beam. However, the approach here will be to develop governing equations based on a dual understanding of the conjugate beam and the deflection problem.

Figure 6.12a shows a segment of a conjugate beam with the applied P loads,

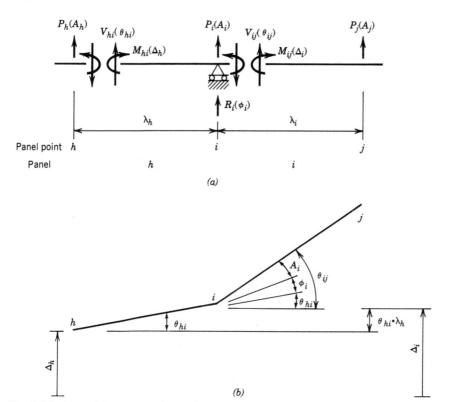

Fig. 6.12 *Beam deflections and associated conjugate beam. (a) Conjugate beam and loading. (b) Beam deflections.*

which are the A concentrated angle changes on the real beam. The R reactions on the conjugate beam correspond to the ϕ angle changes on the actual structure such as would occur at a hinge. The shears and moments, V and M quantities on the conjugate beam, are, respectively, the slopes and deflections on the real beam, which are represented by θ and Δ quantities. Figure 6.12b shows the deflected positions of chords ij and jk along with the pertinent displacement quantities. From the application of statics to Fig. 6.12a or displacement relationships to Fig. 6.12b, it is clear that

$$\theta_{hi} + \phi_i + A_i = \theta_{ij} \tag{6.35}$$
$$\Delta_h + \theta_{hi} \cdot \lambda_h = \Delta_i$$

or, reordering terms,

$$-\theta_{hi} + \theta_{ij} - \phi_i = A_i \tag{6.36}$$
$$\Delta_h - \Delta_i + \theta_{hi} \cdot \lambda_h = 0$$

A comparison of Eqs. 6.36 and 4.17 underscores the analogy of the conjugate beam method. Two variations are evident: no applied moments are present on the conjugate beam and, therefore, the right-hand side of the second of Eqs. 6.36 is zero; and the conjugate beam moments are given a single subscript when they are expressed as deflections for the real beam.

In matrix form, Eqs. 6.36 become

$$\begin{bmatrix} \cdot & \cdot & \cdot & & \cdot & & \cdot & & & & \cdot & \\ \cdots & -1 & 1 & \cdots & & \cdot & & \cdots & -1 & \cdots \\ \cdots & & -1 & 1 & \cdots & & & & \cdots \\ \cdot & & & & & & & & \cdot \\ \cdot & & & & & & & & \cdot \\ \cdot & & & & & & & & \cdot \\ \cdots & \lambda_h & \cdot & \cdots & 1 & -1 & \cdots & & \cdots \\ \cdots & & \lambda_i & \cdots & 1 & -1 & \cdots & & \cdots \\ \cdot & & & & & & & & \cdot \\ \cdot & & & & & & & & \cdot \\ \cdot & & & & & & & & \cdot \\ \cdot & \cdots & & & & & & \cdots & \cdot \end{bmatrix} \begin{Bmatrix} \cdot \\ \cdot \\ \theta_{hi} \\ \theta_{ij} \\ \theta_{jk} \\ \cdot \\ \cdot \\ \Delta_h \\ \Delta_i \\ \Delta_j \\ \cdot \\ \cdot \\ \phi_i \\ \cdot \end{Bmatrix} = \begin{Bmatrix} \cdot \\ \cdot \\ A_i \\ A_j \\ \cdot \\ \cdot \\ \cdot \\ 0 \\ 0 \\ \cdot \\ \cdot \\ \cdot \\ \cdot \\ \cdot \end{Bmatrix} \tag{6.37}$$

Equation 6.37 can be written in the form

$$[G] \begin{Bmatrix} \{\theta\} \\ \{\Delta\} \\ \{\phi\} \end{Bmatrix} = \begin{Bmatrix} \{A\} \\ \{0\} \end{Bmatrix} \tag{6.38}$$

where $[G]$ is the geometric compatibility matrix or the statics matrix for the conjugate beam, $\{\theta\}$ is the panel slope vector, $\{\Delta\}$ is the vector of panel point deflections, and $\{\phi\}$ is the vector of hinge rotations. The $\{A\}$ vector includes the concentrated angle changes that are associated with the member moments.

Equation 6.38 can be written in a condensed form as

$$[G]\{D\} = \{C\} \tag{6.39}$$

where $\{D\}$ is the combined vector of displacement quantities and $\{C\}$ includes the concentrated angle changes. The solution then takes the form

$$\{D\} = [G]^{-1}\{C\} = [b]_{DC}\{C\} \tag{6.40}$$

where $[b]_{DC}$ is the transformation matrix that relates $\{D\}$ to $\{C\}$.

Noting that the lower portion of the $\{C\}$ vector is composed of zeros, Eq. 6.40 can be expressed in an expanded form as

$$\begin{Bmatrix} \{\theta\} \\ \{\Delta\} \\ \{\phi\} \end{Bmatrix} = \begin{bmatrix} [b]_{\theta A} \\ [b]_{\Delta A} \\ [b]_{\phi A} \end{bmatrix} \{A\} \tag{6.41}$$

It should be noted that if the panel point moments are extracted from Eq. 4.23, we can write

$$\{M\} = [b]_{MP_T}\{P_T\} + [b]_{MP_M}\{P_M\} = [b]_{MP}\{P\} \tag{6.42}$$

where $[b]_{MP}$ is the transformation matrix that relates moments to loads. The $\{M\}$ vector gives the moment diagram ordinates. Thus, the ordinates of the load distribution diagram for the conjugate beam, or the ordinates of the curvature diagram for the real beam, are

$$\{\alpha\} = \begin{bmatrix} \diagdown \dfrac{1}{EI} \diagdown \end{bmatrix} \{M\} = \begin{bmatrix} \diagdown \dfrac{1}{EI} \diagdown \end{bmatrix} [b]_{MP}\{P\} \tag{6.43}$$

where $\begin{bmatrix} \diagdown \dfrac{1}{EI} \diagdown \end{bmatrix}$ is a diagonal matrix that gives the $\dfrac{1}{EI}$ values at the panel points.

At points of discontinuity in EI, the average value should be used. Equation 6.43 can be written as

$$[\alpha] = [b]_{\alpha P}\{P\} \tag{6.44}$$

where

$$[b]_{\alpha P} = \begin{bmatrix} \diagdown \dfrac{1}{EI} \diagdown \end{bmatrix} [b]_{MP} \tag{6.45}$$

An equivalent set of concentrated loads $\{A\}$ for the conjugate beam, which are concentrated angle changes for the real beam, is given by

$$\{A\} = [b]_{A\alpha}\{\alpha\} \tag{6.46}$$

where the transformation matrix $[b]_{A\alpha}$ serves to discretize the $\{\alpha\}$ vector to $\{A\}$. If the panel points are equally spaced, Eq. 6.46 is given by Eq. 6.32. If this is not the case, then a slightly different form results.

With these concentrated angle changes, the accompanying deformation quantities can be determined in accordance with Eq. 6.41.

For beam deflections, substitution of Eqs. 6.46 and 6.44 into Eq. 6.41 gives

$$\{\Delta\} = [b]_{\Delta A}[b]_{A\alpha}[b]_{\alpha P}\{P\} \tag{6.47}$$

This can be abbreviated as

$$\{\Delta\} = [b]_{\Delta P}\{P\} \tag{6.48}$$

where

$$[b]_{\Delta P} = [b]_{\Delta A}[b]_{A\alpha}[b]_{\alpha P} \tag{6.49}$$

Equation 6.48 shows that the deflections are directly related to the loads by the transformation matrix $[b]_{\Delta P}$.

6.8.1 Example problem

Determine the panel slopes, the panel point displacements, and the angle change at the hinge for structure and loading given in Section 6.7.3.

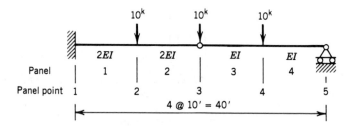

Conjugate Beam and Concentrated Angle Changes: A conventional static analysis, or application of the matrix method of Section 4.13, yields the following moments:

Application of Eqs. 6.44 and 6.46 yields the concentrated angle changes, which act on the conjugate beam as shown below:

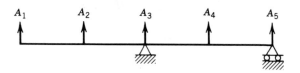

where

$$\{A\} = \begin{Bmatrix} A_1 \\ A_2 \\ A_3 \\ A_4 \\ A_5 \end{Bmatrix} = \frac{10}{6EI} \begin{Bmatrix} -600 \\ -650 \\ 0 \\ 400 \\ 100 \end{Bmatrix} \left(\frac{\text{ft}^2\text{-kips}}{EI} \right)$$

Displacement Equations: The compatibility equations are written for each point along the beam. These are collected according to the form of Eq. 6.37 and are then expressed as Eq. 6.38.

$$[G] \begin{Bmatrix} \{\theta\} \\ \{\Delta\} \\ \{\phi\} \end{Bmatrix} = \begin{Bmatrix} \{A\} \\ \{0\} \end{Bmatrix} \tag{6.38}$$

$$
\begin{bmatrix}
1 & & & & & & & & \\
-1 & 1 & & & & & & & \\
& -1 & 1 & & & & -1 & & \\
& & -1 & 1 & & & & & \\
& & & -1 & & & & -1 & \\
\hline
& & & & -1 & & & & \\
10 & & & & 1 & -1 & & & \\
& 10 & & & & 1 & -1 & & \\
& & 10 & & & & 1 & -1 & \\
& & & 10 & & & & 1 & \\
\end{bmatrix}
\begin{Bmatrix}
\theta_{12} \\
\theta_{23} \\
\theta_{34} \\
\theta_{45} \\
\Delta_1 \\
\Delta_2 \\
\Delta_3 \\
\Delta_4 \\
\phi_3 \\
\phi_5
\end{Bmatrix}
= \frac{10}{6EI}
\begin{Bmatrix}
-600 \\
-650 \\
0 \\
400 \\
100 \\
0 \\
0 \\
0 \\
0 \\
0
\end{Bmatrix}
$$

Solution for Displacement Quantities: The solution of the displacement equations gives the following:

$$
\begin{Bmatrix} \{\theta\} \\ \{\Delta\} \\ \{\phi\} \end{Bmatrix}
= [G]^{-1} \begin{Bmatrix} \{A\} \\ \{0\} \end{Bmatrix}
= \frac{10}{6EI}
\begin{Bmatrix}
-600 \\
-1{,}250 \\
725 \\
1{,}125 \\
0 \\
-6{,}000 \\
-18{,}500 \\
-11{,}250 \\
1{,}975 \\
-1{,}225
\end{Bmatrix}
\begin{array}{l}
\left.\begin{array}{l} \\ \\ \\ \end{array}\right\} \text{Rotation units} \left(\dfrac{\text{ft}^2\text{-kips}}{EI}\right) \\[1em]
\left.\begin{array}{l} \\ \\ \\ \end{array}\right\} \text{Deflection units} \left(\dfrac{\text{ft}^3\text{-kips}}{EI}\right) \\[1em]
\left.\begin{array}{l} \\ \end{array}\right\} \text{Rotation units}
\end{array}
$$

Deflection Diagram: Computed values represented by dots (·).

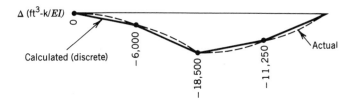

6.9

Deformation quantities for beam and frame structures can also be determined **Beam and** by using the virtual work method. This approach was introduced in Chapter 5 **Frame** and was employed in computing truss deflections. It has the shortcoming that **Deflections by** each application yields a single displacement quantity. It is recalled from Section **Virtual Work** 5.5.2 that two separate systems must be considered. These systems are the following:

P System: *Virtual* force system in equilibrium.

D System: *Actual* deformation configuration which is geometrically compatible.

The P system is subjected to the deformations of the D system, and the external virtual work is equated to the internal virtual work of deformation. This leads to

Virtual P system

$$\sum_{i=1}^{n} (\delta P)_i D_i = \int_{\text{Vol}} (\delta \sigma_P) \epsilon_D \, d \text{ Vol} \tag{6.50}$$

Actual D system

For beam and frame structures, the deformations result from flexural behavior. Thus, the right-hand side of Eq. 6.50, which represents the internal virtual work of deformation, can be expressed as

$$\int_{\text{Vol}} (\delta \sigma_P) \epsilon_D \, d \text{ Vol} = \sum_{j=1}^{m} \left(\int_l (\delta M_P) \, d\theta_D \right)_j \tag{6.51}$$

where l is the length of the jth member, (δM_P) is the virtual moment at some point x along the jth member, and $d\theta_D$ is the angular deformation of the member, at the same point on member j, that occurs over the differential element of length dx for the D system. The summation is over the m members that make up the structure. Substitution of Eq. 6.51 into Eq. 6.50 gives

$$\sum_{i=1}^{n} (\delta P)_i D_i = \sum_{j=1}^{m} \left(\int_l (\delta M_P) \, d\theta_D \right)_j \tag{6.52}$$

If $d\theta_D$ results from a D system of loads on the structure, then Eq. 6.5 gives

$$d\theta_D = \frac{M_D \, dx}{EI} \tag{6.53}$$

where M_D is the moment at point x on member j that results from the loading of the D system. Substitution of Eq. 6.53 into Eq. 6.52 yields

$$\sum_{i=1}^{n} (\delta P)_i D_i = \sum_{j=1}^{m} \left(\int_l (\delta M_P) \frac{M_D \, dx}{EI} \right)_j \tag{6.54}$$

When $d\theta_D$ is caused by something other than direct loading, such as a differential temperature change through the member depth or a fabrication "kink," Eq. 6.51 is still valid with the proper expression being substituted for $d\theta_D$.

6.9.1 Example problem

Determine the vertical deflection and slope at point d for the structure of Section 6.4.3.

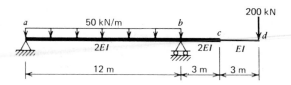

The governing virtual work expression is

$$\sum_{i=1}^{n} (\delta P)_i D_i = \sum_{j=1}^{m} \left(\int_l (\delta M_P) \frac{M_D}{EI} \, dx \right)_j \tag{6.54}$$

Vertical Deflection at Point d:

D System (Actual Load System):

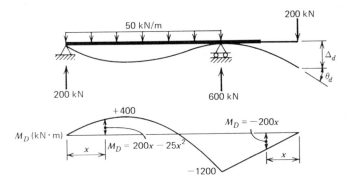

P System (Virtual Load System):

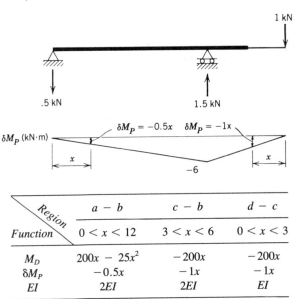

Region / Function	$a - b$	$c - b$	$d - c$
	$0 < x < 12$	$3 < x < 6$	$0 < x < 3$
M_D	$200x - 25x^2$	$-200x$	$-200x$
δM_P	$-0.5x$	$-1x$	$-1x$
EI	$2EI$	$2EI$	EI

Applying Eq. 6.54, we have

$$(1)(\Delta_d) = \frac{1}{2EI} \int_0^{12} (-0.5x)(200x - 25x^2)\,dx + \frac{1}{2EI} \int_3^6 (-1x)(-200x)\,dx$$

$$+ \frac{1}{EI} \int_0^3 (-1x)(-200x)\,dx$$

$$= \frac{1}{2EI}[-33.3x^3 + 3.13x^4]_0^{12} + \frac{1}{2EI}[66.7x^3]_3^6 + \frac{1}{EI}[66.7x^3]_0^3$$

$$= \frac{11\,785 \text{ kN}^2 \cdot \text{m}^3}{EI}$$

$$\Delta_d = \frac{11\,785}{(1)(600\,000)} = 0.019\,6 \text{ m} \quad \text{(the positive sign indicates downward deflection as assumed)}$$

Slope at Point d:

D System (Actual Load System): Same system used to get Δ_d.

P System (Virtual Load System):

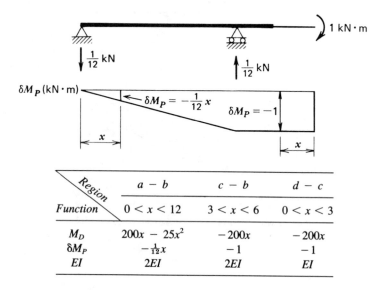

Region Function	$a - b$ $0 < x < 12$	$c - b$ $3 < x < 6$	$d - c$ $0 < x < 3$
M_D	$200x - 25x^2$	$-200x$	$-200x$
δM_P	$-\frac{1}{12}x$	-1	-1
EI	$2EI$	$2EI$	EI

Applying Eq. 6.54, we obtain

$$(1)(\theta_d) = \frac{1}{2EI} \int_0^{12} (-\tfrac{1}{12}x)(200x - 25x^2)\, dx + \frac{1}{2EI} \int_3^6 (-1)(-200x)\, dx$$

$$+ \frac{1}{EI} \int_0^3 (-1)(-200x)\, dx$$

$$= \frac{1}{2EI}[-5.55x^3 - 0.521x^4]_0^{12} + \frac{1}{2EI}[100x^2]_3^6 + \frac{1}{EI}[100x^2]_0^3$$

$$= \frac{2\,856 \text{ kN}^2 \cdot \text{m}^3}{EI}$$

$$\theta_d = \frac{2\,856}{(1)(600\,000)} = 0.004\,76 \text{ radian} \quad \text{(the positive sign indicates clockwise rotation as assumed)}$$

6.9.2 Example problem   Determine the horizontal displacement at point b for the frame structure shown.

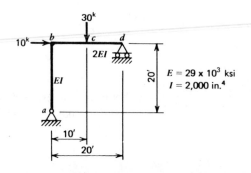

The governing virtual work expression is

$$\sum_{i=1}^{n} (\delta P)_i D_i = \sum_{j=1}^{m} \left(\int_l (\delta M_P)\frac{M_D}{EI} \, dx \right)_j \tag{6.54}$$

Horizontal Deflection at Point b:

D System (Actual Load System):

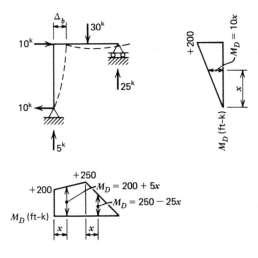

P System (Virtual Load System):

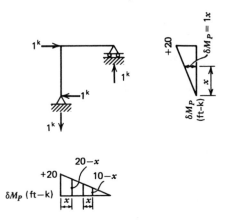

Region Function	a − b 0 < x < 20	b − c 0 < x < 10	c − d 0 < x < 10
M_D	10x	200 + 5x	250 − 25x
δM_P	1x	20 − x	10 − x
EI	EI	2EI	2EI

Using Eq. 6.54, we obtain

$$(1)(\Delta_b) = \frac{1}{EI} \int_0^{20} (1x)(10x) \, dx + \frac{1}{2EI} \int_0^{10} (200 + 5x)(20 - x) \, dx$$

$$+ \frac{1}{2EI} \int_0^{10} (250 - 25x)(10 - x) \, dx = \frac{47{,}500 \text{ k}^2 \cdot \text{ft}^3}{EI}$$

$$\Delta_b = \frac{47{,}500 \times 1{,}728}{1(58 \times 10^6)} = 1.415 \text{ in. (positive } \therefore \text{ to right)}$$

6.10

Flexibility Matrix for Beam- and Frame-Type Structures

Consider the structure shown in Fig. 6.13, in which the structure coordinate system is defined by the numbered vectors shown. Each vector represents an admissible load and its companion displacement for the structural system. In matrix form, the force vector $\{P\}$ and the displacement vector $\{\Delta\}$ are represented by

$$\{P\} = \begin{Bmatrix} P_1 \\ P_2 \\ P_3 \end{Bmatrix}; \quad \{\Delta\} = \begin{Bmatrix} \Delta_1 \\ \Delta_2 \\ \Delta_3 \end{Bmatrix} \tag{6.55}$$

As was described in Section 5.8, a flexibility coefficient measures the displacement at some point on the structure that results from a unit load acting at that or some other point on the structure. Specifically, the flexibility coefficient f_{ij} gives the value of the ith displacement that results from a unit value of the jth load. The complete array of flexibility coefficients can be arranged in the form of a flexibility matrix, which relates the structure displacements to the structure forces in the form

$$\begin{Bmatrix} \Delta_1 \\ \Delta_2 \\ \Delta_3 \end{Bmatrix} = \begin{bmatrix} f_{11} & f_{12} & f_{13} \\ f_{21} & f_{22} & f_{23} \\ f_{31} & f_{32} & f_{33} \end{bmatrix} \begin{Bmatrix} P_1 \\ P_2 \\ P_3 \end{Bmatrix} \tag{6.56}$$

In a simplified form, these equations become

$$\{\Delta\} = [f]\{P\} \tag{6.57}$$

where $[f]$ is the structure flexibility matrix. This equation is of the same form as that previously given as Eq. 5.30. Each flexibility coefficient is a displacement quantity that results from a specific unit loading case and, as such, can be determined by any of the methods presented thus far.

Once the complete flexibility matrix is determined, Eq. 6.57 can be used to establish the complete array of displacements that accompanies any prescribed set of loads.

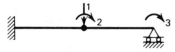

Fig. 6.13 *Simple beam structure with admissible loads and displacements.*

Consider the simple cantilever beam shown below, in which two admissible load **6.10.1 Example** and displacement components are identified.   **problem**

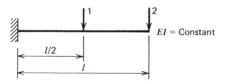

EI = Constant

l/2

l

(a) Determine the complete flexibility matrix.
(b) Determine the displacements that result from the load vector

$$\{P\} = \begin{Bmatrix} P_1 \\ P_2 \end{Bmatrix} = \begin{Bmatrix} 0 \\ 30 \end{Bmatrix} \text{ kN}$$

if $l = 6$ m and $EI = 400\ 000$ kN · m².

Flexibility Matrix: Any of the methods presented in this chapter can be used to obtain the individual flexibility coefficients. The results are as follows:

$P_1 = 1$

f_{11} f_{21}

$P_2 = 1$

f_{12}

f_{22}

$$[f] = \begin{bmatrix} f_{11} & f_{12} \\ f_{21} & f_{22} \end{bmatrix} = \frac{l^3}{EI} \begin{bmatrix} \frac{1}{24} & \frac{5}{48} \\ \frac{5}{48} & \frac{1}{3} \end{bmatrix}$$

Displacements:

$$\{\Delta\} = [f]\{P\} \tag{6.57}$$

For the prescribed loading,

$$\begin{Bmatrix} \Delta_1 \\ \Delta_2 \end{Bmatrix} = \frac{l^3}{EI} \begin{bmatrix} \frac{1}{24} & \frac{5}{48} \\ \frac{5}{48} & \frac{1}{3} \end{bmatrix} \begin{Bmatrix} P_1 \\ P_2 \end{Bmatrix} = \frac{6^3}{400\ 000} \begin{bmatrix} \frac{1}{24} & \frac{5}{48} \\ \frac{5}{48} & \frac{1}{3} \end{bmatrix} \begin{Bmatrix} 0 \\ 30 \end{Bmatrix}$$

$$= \begin{Bmatrix} .001\ 7 \\ .005\ 4 \end{Bmatrix} \text{ m} = \begin{Bmatrix} 1.7 \\ 5.4 \end{Bmatrix} \text{ mm}$$

6.11

The virtual work method is the foundation upon which all energy methods are **Beam and** based; however, other energy-related techniques can be employed to obtain struc- **Frame** tural deflections. Castigliano's second theorem, which was developed in Section **Deflections by** 5.9.2, is particularly well suited for the deflection problem. **Energy**
    In applying Castigliano's second theorem, we must express the total strain **Methods**

energy for the structure U in terms of the applied loads. Then, corresponding to each structure force P_i, the corresponding displacement quantity Δ_i can be determined from the relationship

$$\frac{\partial U}{\partial P_i} = \Delta_i \tag{6.58}$$

which was presented earlier as Eq. 5.40. This equation can be used to determine a single displacement quantity, or it can be used to establish the complete set of displacement–force equations for the structure.

Since the strain energy is equal to the work of deformation, Eq. 2.23 can be used to express the strain energy associated with flexural deformation of the element shown in Fig. 6.1a. This leads to

$$dU = \tfrac{1}{2}M \, d\theta \tag{6.59}$$

where the factor $\tfrac{1}{2}$ results from the linear relationship between the moment, M, and the rotation, $d\theta$, as previously explained in Fig. 2.13. Substituting the expression for $d\theta$ from Eq. 6.5 into Eq. 6.59, we obtain

$$dU = \frac{1}{2EI} M^2 \, dx \tag{6.60}$$

where M is the bending moment on the element, I is the moment of inertia of the element cross section, and E is the modulus of elasticity. Thus, the total strain energy for a moment of length l is

$$U = \frac{1}{2E} \int_{x=0}^{x=l} \frac{M^2}{I} \, dx \tag{6.61}$$

and the total strain energy stored in a structure of m members is given by

$$U = \sum_{i=1}^{m} \frac{1}{2E} \int_{x=0}^{x=l_i} \frac{M_i^2}{I_i} \, dx \tag{6.62}$$

where the i subscript is associated with the ith member.

It is noted that M_i must be expressed in terms of the externally applied loads in order that the differentiation indicated in Eq. 6.58 can be carried out.

6.11.1 Example problem   Use Castigliano's second theorem to develop the displacement–force equations for the structure of Example 6.10.1. From these equations, extract the structure flexibility matrix. The structure coordinate system and the member notation are shown below.

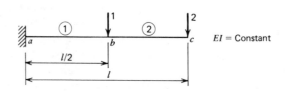

Moment Expressions:

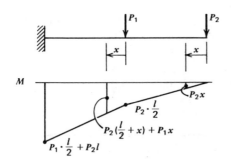

Region Function	Member ①; $a - b$ $0 < x < l/2$	Member ②; $b - c$ $0 < x < l/2$
M	$P_2\left(\dfrac{l}{2} + x\right) + P_1 x$	$P_2 x$

Strain Energy:

$$U = \sum_{i=1}^{m} \frac{1}{2E} \int_{x=0}^{x=l_i} \frac{M_i^2}{I_i}\, dx \qquad (6.62)$$

$i = 1$: $U_1 = \dfrac{1}{2EI}\displaystyle\int_{x=0}^{x=l/2}\left[P_1 x + P_2\left(\dfrac{l}{2} + x\right)\right]^2 dx = \dfrac{1}{EI}\left[\dfrac{P_1^2 l^3}{48} + \dfrac{5P_1 P_2 l^3}{48} + \dfrac{7P_2^2 l^3}{48}\right]$

$i = 2$: $U_2 = \dfrac{1}{2EI}\displaystyle\int_{x=0}^{x=l/2}[P_2 x]^2\, dx = \dfrac{1}{EI}\left[\dfrac{P_2^2 l^3}{48}\right]$

$$U = U_1 + U_2 = \frac{1}{48EI}[P_1^2 l^3 + 5P_1 P_2 l^3 + 8P_2^2 l^3]$$

Displacement–Force Equations and Flexibility Matrix:

$$\frac{\partial U}{\partial P_i} = \Delta_i \qquad (i = 1, 2) \qquad (6.58)$$

$$\frac{\partial U}{\partial P_1} = \Delta_1 \Rightarrow \frac{1}{EI}\left[\frac{P_1 l^3}{24} + \frac{5P_2 l^3}{48}\right] = \Delta_1$$

$$\frac{\partial U}{\partial P_2} = \Delta_2 \Rightarrow \frac{1}{EI}\left[\frac{5P_1 l^3}{48} + \frac{P_2 l^3}{3}\right] = \Delta_2$$

Or, in matrix form,

$$\frac{l^3}{EI}\begin{bmatrix} \frac{1}{24} & \frac{5}{48} \\ \frac{5}{48} & \frac{1}{3} \end{bmatrix}\begin{Bmatrix} P_1 \\ P_2 \end{Bmatrix} = \begin{Bmatrix} \Delta_1 \\ \Delta_2 \end{Bmatrix}$$

From this equation, it is clear that

$$[f] = \frac{l^3}{EI}\begin{bmatrix} \frac{1}{24} & \frac{5}{48} \\ \frac{5}{48} & \frac{1}{3} \end{bmatrix}$$

which agrees with that determined in Example 6.10.1.

6.12

Superposition and Reciprocity

An examination of Sections 5.12 and 5.13 reveals that the development of both the principle of superposition and the reciprocal relationships was not limited to the truss-type structures that were specifically treated in Chapter 5. Indeed, these concepts are equally applicable to beam- and frame-type structures provided that the requisite conditions are satisfied. That is, for both superposition and reciprocity to be valid, the structure must exhibit linear load–deflection relationships.

6.13

Additional Reading

Hseigh, Yuan Yu, *Elementary Theory of Strucutres,* Chapter 8, Prentice–Hall, Englewood Cliffs, N.J., 1970.

McCormac, J. C., *Structural Analysis,* 3rd Ed., Chapters 17 and 18, Intext Educational Publishers, New York, 1975.

Newmark, N. M., "Numerical Procedure for Computing Deflections, Moments, and Buckling Loads," *Trans. ASCE,* **108,** p. 1161, 1943.

Norris, C. H., Wilbur, J. B., and Utku, S., *Elementary Structural Analysis,* 3rd Ed., Chapter 8, McGraw–Hill, New York, 1976.

6.14

Suggested Problems

1 through 22. Use the moment–area method to determine the indicated displacement quantities for each of the indicated structures and loadings.

1. Vertical deflection and rotation at point A.

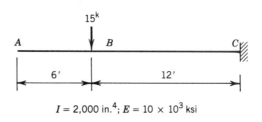

$$I = 2{,}000 \text{ in.}^4; E = 10 \times 10^3 \text{ ksi}$$

2. Vertical deflection and rotation at point B.

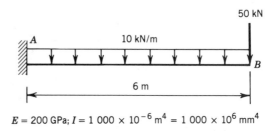

$$E = 200 \text{ GPa}; I = 1\,000 \times 10^{-6} \text{ m}^4 = 1\,000 \times 10^6 \text{ mm}^4$$

3. Vertical deflection and rotation at points C and E.

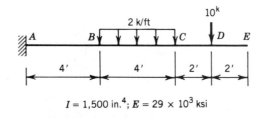

$$I = 1{,}500 \text{ in.}^4; E = 29 \times 10^3 \text{ ksi}$$

4. Vertical deflection at points C and E; rotation at point B.

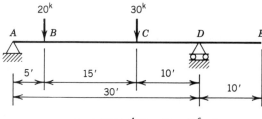

$I = 1,500$ in.^{4}; $E = 29 \times 10^6$ psi

5. Location and magnitude of the maximum vertical deflection.

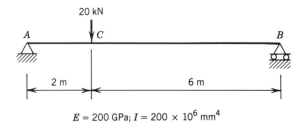

$E = 200$ GPa; $I = 200 \times 10^6$ mm^4

6. Location and magnitude of the maximum vertical deflection. What is the maximum value of P so that this deflection does not exceed 1 inch?

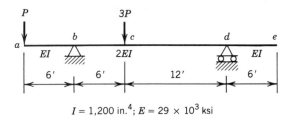

$I = 1,200$ in.^{4}; $E = 29 \times 10^3$ ksi

7. Vertical deflection and rotation at points b and d, and the location and magnitude of maximum vertical deflection.

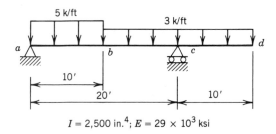

$I = 2,500$ in.^{4}; $E = 29 \times 10^3$ ksi

8. Vertical deflection and rotation at point B.

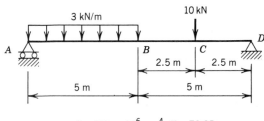

$I = 200 \times 10^6$ mm^4; $E = 70$ GPa

9. Vertical deflection and angle change at point B.

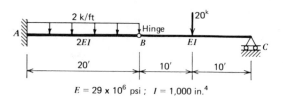

$E = 29 \times 10^6$ psi ; $I = 1,000$ in.^{4}

10. Vertical deflection at point D.

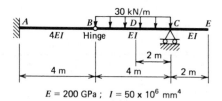

$E = 200$ GPa ; $I = 50 \times 10^6$ mm^4

11. Vertical deflection and rotation at point C.

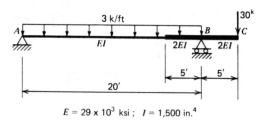

$E = 29 \times 10^3$ ksi ; $I = 1,500$ in.^{4}

12. Vertical deflection and rotation at center of span bc.

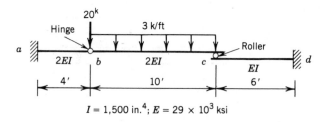

$I = 1,500$ in.^{4}; $E = 29 \times 10^3$ ksi

13. Vertical deflection at point C and the slope at point E.

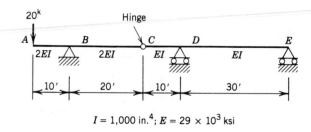

$I = 1,000$ in.^{4}; $E = 29 \times 10^3$ ksi

14. Midspan vertical deflection for spans AB and CD.

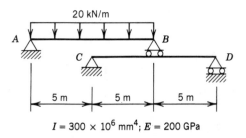

$I = 300 \times 10^6$ mm^4; $E = 200$ GPa

15. Midspan vertical deflection for each stringer and the girder. *Note:* Assume stringers are simply supported between floorbeam support points.

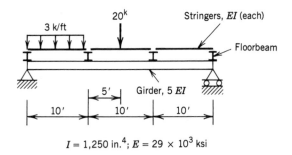

$I = 1,250$ in.^{4}; $E = 29 \times 10^3$ ksi

16. Vertical deflection and rotation at point f.

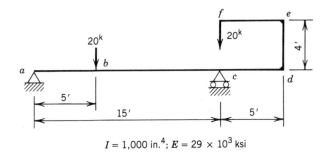

$I = 1,000$ in.^{4}; $E = 29 \times 10^3$ ksi

17. Vertical and horizontal deflections at point C.

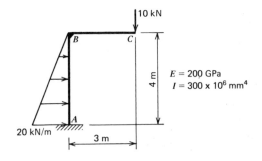

18. Rotation at point A and horizontal deflection at point D.

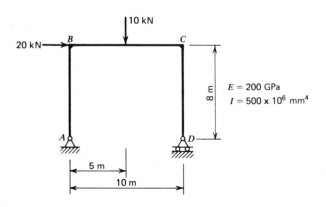

19. Horizontal and vertical deflections of, and the rotation on each side of, point b.

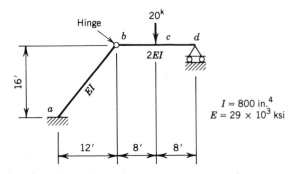

20. Horizontal deflection and rotation at point b and vertical deflection at point c.

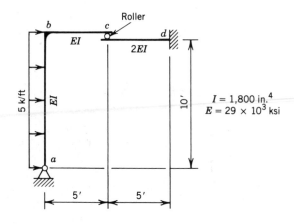

21. Horizontal deflection at point *b* and vertical deflection at point *c*.

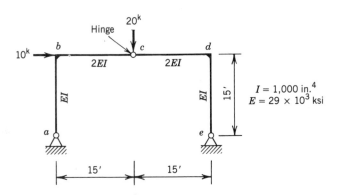

22. Horizontal deflection at points *B* and *C* and the rotation at point *D*.

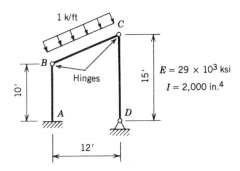

23 through 30. Use the conjugate beam method to determine the indicated displacement quantities for each of the structures and loadings that are designated.

23. Vertical deflection and rotation at point *A* for the structure and loading of Problem 1.

24. Vertical deflection and rotation at point *B* for the structure and loading of Problem 2.

25. Vertical deflections at points *C* and *E*, and the rotation at point *B*, for the structure and loading of Problem 4.

26. Location and magnitude of the maximum vertical deflection for the structure and loading of Problem 5.

27. Vertical deflection and angle change at point *B* for the structure and loading of Problem 9.

28. Vertical deflection at point *D* for the structure and loading of Problem 10.

29. Vertical deflection and rotation at point *C* for the structure and loading of Problem 11.

30. Vertical deflection and rotation at center of span *bc* for the structure and loading of Problem 12.

31 through 36. Use the conjugate beam method to construct the complete slope and deflection diagrams for the prescribed structure and loading.

31. Structure and loading of Problem 1.

32. Structure and loading of Problem 3.

33. Structure and loading of Problem 4.

34. Structure and loading of Problem 5.

35. Structure and loading of Problem 10.

36. Structure and loading of Problem 13.

37. Use the tabular method of Section 6.7 to determine the vertical deflections at 5-m intervals for the structure shown. $EI = 5\,000\,000$ kN $\cdot$ m^2.

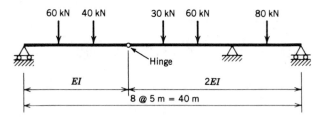

38. Discretize the loading on the following structure into an equivalent set of concentrated loads at 5-ft intervals, and determine the vertical deflections at 5-ft intervals using the tabular method of Section 6.7. $EI = 70 \times 10^6$ k $\cdot$ in.^{2}

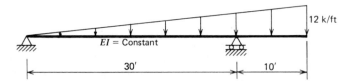

39. Use the tabular method of Section 6.7 to determine the vertical deflections at 1-meter intervals for the structure and loading of Problem 2.

40. Use the tabular method of Section 6.7 to determine the vertical deflections and the angle changes at the hinges for the structure and loading of Problem 12. Use 2-ft intervals.

41. Repeat Problem 37 using the matrix formulation method of Section 6.8.

42. Repeat Problem 38 using the matrix formulation method of Section 6.8.

43. Repeat Problem 39 using the matrix formulation method of Section 6.8.

44. Repeat Problem 40 using the matrix formulation method of Section 6.8.

45 through 49. Use the virtual work method to determine the indicated displacement quantities for each of the structures and loadings that are designated.

45. Vertical deflection and rotation at point B for the structure and loading of Problem 2.

46. Vertical deflection and angle change at point B for the structure and loading of Problem 9.

47. Vertical deflection and rotation at the center of span bc for the structure and loading of Problem 12.

48. Vertical deflection and rotation at point f for the structure and loading of Problem 16.

49. Horizontal and vertical deflection at point C for the structure and loading of Problem 17.

50. For the structure of Problem 1, generate the complete flexibility matrix for the structure coordinate system shown. Use the resulting flexibility matrix and the loading of Problem 1 to determine the vertical displacements at points A and B.

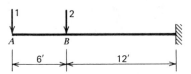

51. For the structure of Problem 9, generate the complete flexibility matrix for the load and displacement coordinates shown. Use the resulting flexibility matrix and the loading of Problem 9 to determine the vertical deflections cooresponding to 1 and 2.

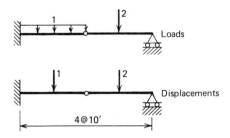

52. Use Castigliano's second theorem to determine the vertical deflection at point C and the slope of point E for the structure and loading of Problem 13.

53. Use Castigliano's second theorem to determine the horizontal and vertical deflections at point B for the structure and loading of Problem 18.

54. Use Castigliano's second theorem to develop the displacement–force equations for the structure of Problem 4 employing the structure coordinate system shown below. From these equations, extract the structure flexibility matrix, and determine the vertical deflections at point B and C for the loading of Problem 4.

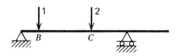

55. Use Castigliano's second theorem to develop the displacement–force equations for the structure of Problem 18 using the coordinate system shown below. From these equations, extract the structure flexibility matrix, and deter-

mine the horizontal deflection at B and the vertical deflection at point D for the loading of Problem 18.

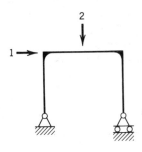

56. Write a computer program for the matrix method developed in Section 6.8. Use the program to solve Problems 41 through 44.

57. Problem 35 of Section 4.16 presented general expressions for shear and moment for beams subjected to concentrated loads and uniformly distributed loads based upon the force boundary conditions at the left end of the beam.

   Given below are additional expressions for slope and deflection based on the force and displacement boundary conditions at the left end of the beam.

Concentrated Load:

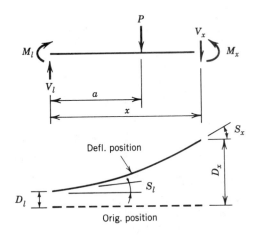

$$S_x = S_l + \frac{M_l x}{EI} + \frac{V_l x^2}{2EI} - \frac{P\langle x - a \rangle^2}{2EI}$$

$$D_x = D_l x + \frac{M_l x^2}{2EI} + \frac{V_l x^3}{6EI} - \frac{P\langle x - a \rangle^3}{6EI}$$

where

$$\langle x - a \rangle = 0 \qquad \text{for } x < a$$
$$= (x - a) \quad \text{for } x \geq a$$

Uniform Load:

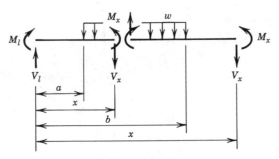

For $x \le b$:

$$S_x = S_l + \frac{M_l x}{EI} + \frac{V_l x^2}{2EI} - \frac{w\langle x - a\rangle^3}{6EI}$$

$$D_x = D_l + S_l x + \frac{M_l x^2}{2EI} + \frac{V_l x^3}{6EI} - \frac{w\langle x - a\rangle^4}{24EI}$$

where

$$\langle x - a\rangle = 0 \qquad \text{for } x < a$$
$$= (x - a) \quad \text{for } x \ge a$$

For $x > b$:

$$S_x = S_l + \frac{M_l x}{EI} + \frac{V_l x^2}{2EI} - \frac{w(b - a)^3}{6EI} - \frac{w(b - a)^2(x - b)}{2EI}$$

$$- \frac{w(b - a)(x - b)^2}{2EI}$$

$$D_x = D_l + S_l x + \frac{M_l x^2}{2EI} + \frac{V_l x^3}{6EI} - \frac{w(b - a)^3(x - 0.75b - 0.25a)}{6EI}$$

$$- \frac{w(b - a)^2(x - b)^2}{4EI} - \frac{w(b - a)(x - b)^3}{6EI}$$

For any x, the total values of S_x and D_x are determined from the superposition of the individual effects.

a. Verify that the given expressions are correct.

b. Expand the computer program written in Problem 35 of Section 4.16 to include the determination of the slopes and deflections at closely spaced intervals along the beam length.

   Input data should include the required beam dimensions and properties, loading (including internal reactions), and the left-end boundary conditions.

   The output should include a tabular array which gives shear, moment, slope, and deflection for each x distance along the beam.

c. Use the program to solve Problems 1, 3, 4, 7, 9, and 12. Intervals of 1 foot or 1 meter should be used, depending on the system of units employed.

CHAPTER 7

*I-81 Interchange,
Harrisburg, Pa.
(courtesy of Gannett
Fleming Engineers
and Planners).*

ANALYSIS FOR VARIABLE LOADING CONDITIONS

7.1

Motivation for Studying Variable Loadings

The design of any structural component requires knowledge of the maximum stress intensities to which the component will be subjected in fulfilling its functional requirements. These stress intensities are determined from the reactions, bar forces, shears, moments, and so forth that result from detailed structural analyses. However, for most structures, there is an array of possible loading conditions, and there are unique loading conditions associated with the maximum stress intensities for each member.

There are, therefore, two major concerns that motivate the study of variable loading conditions. First, of all possible loading conditions, which ones will maximize the stress intensities in a given structural component? And second, what stress intensities correspond to these loading conditions?

In this chapter, only statically determinate structures are considered. The problem of variable loading conditions for statically indeterminate structures is treated later in the book.

7.2

A useful technique for the study of variable loading problems focuses on consideration of the variation in a given response function with the position of load. For instance, consider the simply supported beam of Fig. 7.1a. Selecting the reaction at A as the response function of interest, we will determine its value for various positions of a unit downward load. Figure 7.1b shows the value of R_A for each of

Variation in Response Function with Position of Load

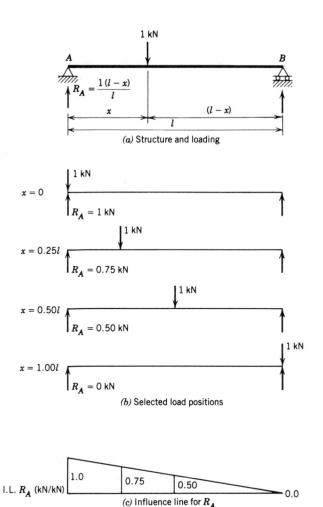

(a) Structure and loading

(b) Selected load positions

(c) Influence line for R_A

Fig. 7.1 *Development of influence line.*

four different positions of the unit load. If the value of R_A is plotted as ordinate along an abscissa that represents the position of the unit load on the structure, we obtain the diagram shown in Fig. 7.1c. A diagram of this type is called an *influence line*—it records the influence that a unit load has on a particular response function as the load transverses the structure. In this case, the response function is R_A, and thus Fig. 7.1c is the influence line for R_A (I.L. R_A).

Each ordinate of Fig. 7.1c gives the value of R_A that results when a unit downward load is applied at a point on the structure corresponding to the abscissa. In this case, it is clear that a positive (upward) value of R_A results for any position of the unit load on the structure and that R_A diminishes linearly as the unit load traverses the structure. In fact, in this case, the equation for the influence line is $1(l - x)/l$, which gives the expression for R_A in terms of the position of the unit load.

In this example, an influence line was developed for a reaction. Actually, an influence line can be constructed for any response function—reaction, bar force, shear, moment, stress, and so forth. Thus, an *influence line is defined as a curve, the ordinate to which at any abscissa equals the value of some particular response function attributable to a unit downward load acting at the point on the structure corresponding to that abscissa.*

The units for an influence line reflect the units of the response function and the unit load. For the example of Fig. 7.1c, the units are force per force (kN/kN). Thus, the ordinate of 0.75 kN/kN indicates that R_A has a value of 0.75 kN per 1 kN load acting downward on the structure at that point.

7.3

Influence Lines by Equilibrium Considerations

The most fundamental approach to the construction of influence lines is to allow the unit load to occupy sequentially several positions on the structure and to compute the value of the desired response function for each load position. The ordinates are plotted to form the influence line for the given response quantity.

The following suggestions are offered for use of this technique in constructing influence lines:

1. Apply the usual analysis procedure to solve for the response function for which the influence line is to be constructed. That is, begin by indicating the positive symbolic representation of the desired response function on the appropriate free-body diagrams.

2. For each position of the unit load, the most convenient free-body diagram is used in conjunction with the equations of static equilibrium to solve for the response function. The sign of the response function will be consistent with the established sign convention and the ordinate of the influence line will reflect the appropriate sign at the abscissa corresponding to the load point.

3. This procedure is carried out for as many positions of the unit load as are needed to establish the shape of the influence line.

4. In solving for an internal response function, the solution generally includes terms involving the reactions. Thus, it is convenient to construct the influence lines for reactions first. These influence lines are then used in the construction of other influence lines.

Construct the influence lines for the shear and moment at point C for the structure **7.3.1 Example** shown.   **problem**

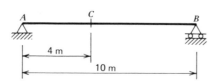

Influence Lines for Reactions: As an initial step, the influence lines for the reactions are constructed. Even though these are not specifically requested, they are invaluable for the construction of the desired influence lines. Taking a unit downward load of 1 kN and assuming positive R_A and R_B to act upward, we obtain the following influence lines:

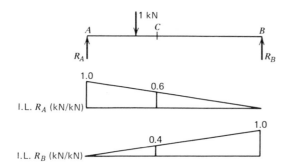

As a check on these influence lines, it is noted that the sum of the ordinates at any point must be 1 kN/kN; that is, the unit downward load must be equilibrated vertically by the reaction components.

Influence Line for Shear at C: To determine the shear at point C, the structure is cut at point C, V_C is shown in the positive sense, and statics is applied to determine the value of V_C. When the unit load acts to the left of point C, it is most convenient to consider the free-body diagram to the right of point C. For this section, $V_C = -R_B$.

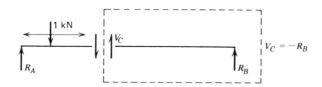

When the unit load acts to the right of point C, it is most convenient to consider the free-body diagram to the left of point C. For this portion, $V_C = R_A$.

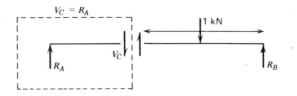

Thus, the influence line is constructed from these two relationships for V_C and the influence lines for reactions.

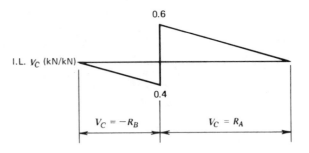

Influence Line for Moment at C: To determine the moment at point C, the structure is cut at point C, M_C is shown in the positive sense, and statics is applied to determine the value of M_C. Again, when the unit load acts to the left of point C, it is most convenient to consider the free-body diagram to the right of point C. For this case, $M_C = 6R_B$.

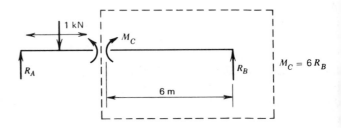

When the unit load is right of point C, the free-body diagram to the left of point C is used. Here, $M_C = 4R_A$.

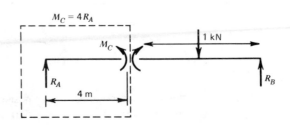

The final influence line is constructed from these two relationships for M_C and the influence lines for reactions.

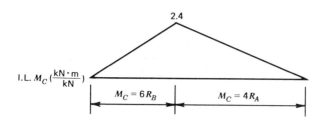

I.L. $M_C (\frac{kN \cdot m}{kN})$

$M_C = 6 R_B$

$M_C = 4 R_A$

2.4

Construct the influence lines for the shear at the hinge at B and for the moment at D for the structure shown below.

7.3.2 Example problem

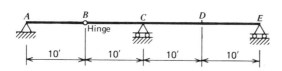

Influence Lines for Reactions:

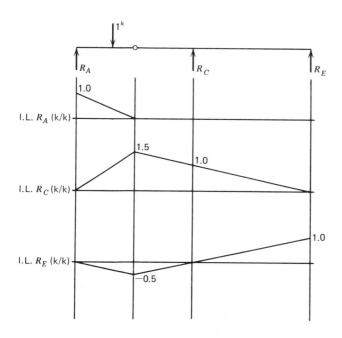

Influence Line for Shear at B:

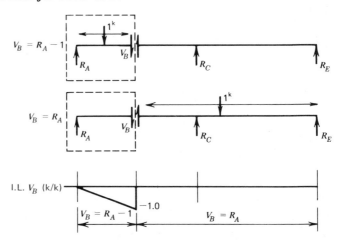

Influence Line for Moment at D:

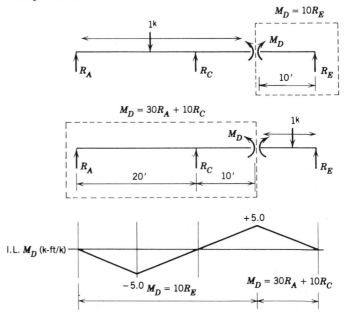

7.3.3 Example problem

Construct the influence lines for the shear in panel 2–3 and the moment at point 5 for the girder shown below.

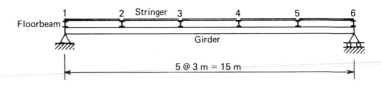

Note:

The girder receives loads at the floorbeam connection points only. For any given loading condition, the shear is constant between load points, and the moment varies linearly between the load points. Thus, it is appropriate to construct the influence lines for shear in a panel and for moment at a panel point.

Influence Lines for Reactions:

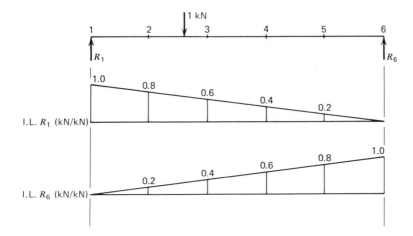

Influence Line for V_{2-3}:   First, consider the unit load to be left of panel 2–3.

Then, consider the unit load to be right of panel 2–3.

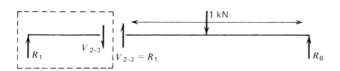

This leads to the following influence line.

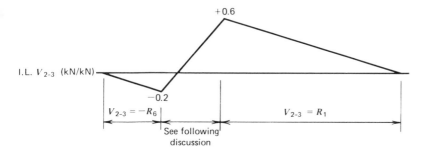

When the unit load is on the stringer between points 2 and 3, it is useful to consider the influence lines for the stringer reactions for stringer 2–3, which in turn are the influence lines for the girder loads at points 2 and 3.

The influence line for V_{2-3} shows that the portion of the unit load that enters the girder as P_2 contributes to V_{2-3} with an influence of -0.2 kN/kN, whereas the portion of the unit load that enters the girder as P_3 contributes to V_{2-3} with an influence of $+0.6$ kN/kN. Since the girder loads P_2 and P_3 vary linearly between the limits of zero and unity as the unit load moves from 2 to 3, their effects on V_{2-3} must also vary linearly between the limits of -0.2 and $+0.6$ as shown.

Influence Line for M_5:

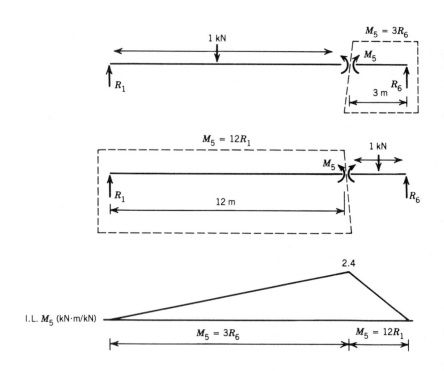

7.3.4 Example problem   Construct the influence lines for the bar forces in members U_1L_1, U_2L_3, and U_3U_4. Consider the unit load to pass along the top chord of the truss.

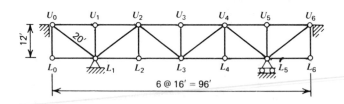

Note:

A truss can receive loads only at the joints. When the unit load is between top chord joints it will enter the truss through adjacent joints in a manner similar to that explained in Section 7.3.3 for panel point loads for girders.

Influence Lines for Reactions:

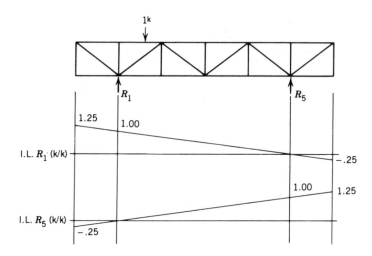

Influence Line for $F_{U_1L_1}$: $F_{U_1L_1}$ is directly related to the load P_1, which enters the truss at joint U_1. An influence line for P_1 leads directly to the influence line for $F_{U_1L_1}$.

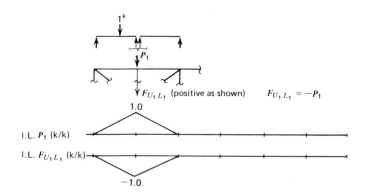

Influence Line for $F_{U_2L_3}$:   The vertical component of $F_{U_2L_3}$ must take the shear in panel 2–3. Thus, we proceed in the same fashion as we did for the girder shear in Section 7.3.3.

For unit load left of joint U_2

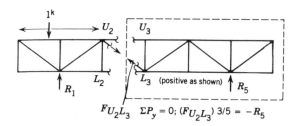

For unit load right of joint U_3

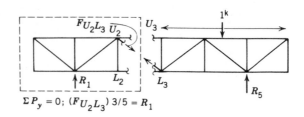

$$\Sigma P_y = 0; \ (F_{U_2L_3}) \ 3/5 = R_1$$

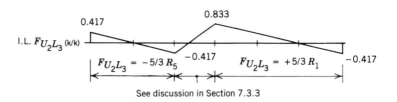

See discussion in Section 7.3.3

Influence Line for $F_{U_3U_4}$:   For unit load left of joint U_3

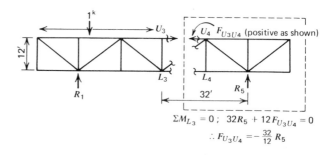

$$\Sigma M_{L_3} = 0; \ 32R_5 + 12F_{U_3U_4} = 0$$
$$\therefore F_{U_3U_4} = -\frac{32}{12}R_5$$

For unit load right of joint U_3

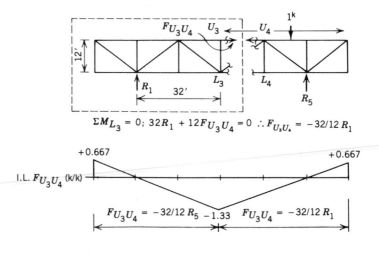

$$\Sigma M_{L_3} = 0; \ 32R_1 + 12F_{U_3U_4} = 0 \ \therefore F_{U_3U_4} = -32/12 \ R_1$$

7.3.5 Example problem   Determine the influence lines for the vertical force transferred at point B, the reactions at D and F, and the shear and moment at point E.

The unit load traverses structure from A to C.

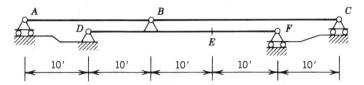

Influence Line for Force at B: Point B is a simple support for spans AB and BC. The vertical reaction at that point, R_B, provides the load at point B for span DF. The influence line for R_B is easily determined.

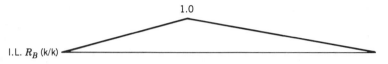

I.L. R_B (k/k)

Influence Lines for Span DF: As the unit load moves from A to C, the load transmitted to span DF at point B is given by the influence line for R_B.

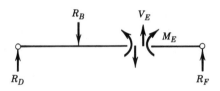

From statics,

$$R_D = \tfrac{2}{3} R_B, \qquad R_F = \tfrac{1}{3} R_B, \qquad V_E = -R_F, \qquad M_E = 10 R_F$$

0.67

I.L. R_D (k/k)

0.33

I.L. R_F (k/k)

I.L. V_E (k/k)

−0.33

3.33

I.L. M_E (ft-k/k)

Determine the influence lines for all reaction components and for the moment at the top of column BD. The unit load moves from A to C.

7.3.6 Example problem

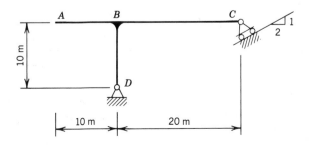

Influence Line Construction: The free-body diagram shows the response quantities of interest, and the application of statics yields the expressions shown.

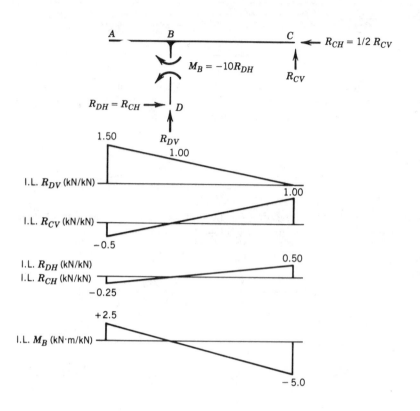

7.4

Influence Lines by Virtual Work: Müller-Breslau Principle

Betti's law, which was presented in Section 5.13, can be used to develop a very useful concept for the construction of influence lines. Consider the structure shown in Fig. 7.2a. Assume that this structure is subjected to a unit downward load at point x. In the process, the structure deforms and develops reactions R_A and R_B as shown in Fig. 7.2b. This situation is referred to as "System 1." Next, assume that the given structure is deformed by removing the restraint associated with the reaction R_A and that a unit displacement is introduced at point A in the positive direction of R_A. This situation is shown in Fig. 7.2c and is referred to as "System 2."

Recalling from Eq. 5.53 that the virtual work done by the forces of System 1 through the displacements of System 2 is equal to the virtual work done by the forces of System 2 through the displacements of System 1, we can write

$$\sum_{i=1}^{n} (P_i)_1(\Delta_i)_2 = \sum_{j=1}^{m} (P_j)_2(\Delta_j)_1 \tag{7.1}$$

where $(P_i)_1$ and $\Delta_i)_2$ represent the force at point i for System 1 and the displacement at point i for System 2, respectively. Similarly, $(P_j)_2$ and $(\Delta_j)_1$ give the force at point j for System 2 and the displacement at point j for System 1, respectively. The above work quantities are virtual in that they exist in essence or effect, but not in fact—that is, they are imagined for the purposes of analysis.

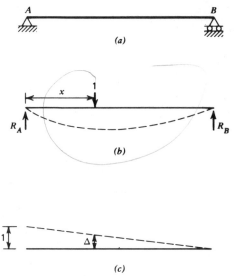

Fig. 7.2 *Müller-Breslau principle.* (a) *Given structure.* (b) *System 1: structure in equilibrium under unit load.* (c) *System 2: structure subjected to unit displacement corresponding to* R_A.

Thus, for the situation in Fig. 7.2, we have

$$(R_A)(1) + (1)(-\Delta) = 0 \tag{7.2}$$

with $(\Delta)_2$ labeled over the second term and $(P)_1$ labeled under the first term.

The negative sign on the Δ quantity simply means that it is in a direction opposite the unit load; thus, negative work results. In the present problem, there are no forces for System 2 since the entire member is merely rotated as a rigid body about point B. However, even if a force were required at point A to impose the deformation pattern of System 2, it would do no work, since there is no displacement at point A in System 1.

Rewriting Eq. 7.2, we have

$$R_A = \Delta \tag{7.3}$$

Equation 7.3 states that the value of R_A for a unit load at point x is given by the displacement at point x when the structure is subjected to a unit displacement corresponding to the positive direction of R_A. Since x can represent any point along the span, the deflection pattern of System 2 traces the influence line for R_A.

Influence lines for other response functions can be determined by the virtual work method in the same general fashion. For instance, consider the influence lines for the shear and moment at point C of the structure shown in Fig. 7.3a. System 1 is given in Fig. 7.3b, in which the unit load is shown at point x and the associated shear and moment are symbolically represented in their positive directions on the cut section at point C.

To determine the influence line for V_C, System 2 requires a unit displacement that will interact with V_C of System 1 to do work, as shown in Fig. 7.3c. This deformation pattern must retain the same slope on each side of the cut section so

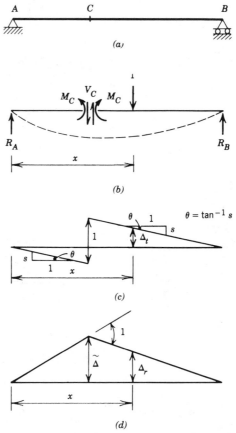

Fig. 7.3 *Müller-Breslau principle for shear and moment influence lines.* (a) *Given structure.* (b) *System 1.* (c) *System 2 used to obtain influence line for* V_C. (d) *System 2 used to obtain influence line for* M_C.

that the rotational displacements of System 2 do not produce any net work as they interact with M_C of System 1.

Thus, applying Eq. 7.1, we have

$$(V_C)(1) \; + \; (M_C)(-\theta) \; + \; (M_C)(\theta) \; + \; (1)(-\Delta_t) = 0 \tag{7.4}$$

Or,

$$V_C = \Delta_t \tag{7.5}$$

The influence line for the shear at point C is thus given by the displacement pattern that results when the structure is broken at point C and a unit displacement is introduced corresponding to positive V_C.

To construct the influence line for M_C, System 2 must reflect a unit rotational displacement that will interact with M_C of System 1 to do work. This situation is shown in Fig. 7.3d. This deformation pattern must retain the same transverse displacement on each side of the cut section so that the displacements of System 2 do not produce any net work as they interact with V_C of System 1.

The application of Eq. 7.1 then leads to

$$(M_C)(1) + (V_C)(-\tilde{\Delta}) + (V_C)(\tilde{\Delta}) + (1)(-\Delta_r) = 0 \qquad (7.6)$$

with $(\Delta)_2$ indicated above and $(P)_1$ indicated below.

from which

$$M_C = \Delta_r \qquad (7.7)$$

Thus, the influence line for the moment at point C is given by the displacement pattern that occurs when the structure is broken at point C and a unit rotational displacement is introduced corresponding to positive M_C.

For all three cases cited, that is, for the construction of influence lines for R_A, V_C, and M_C, the same general pattern was followed. This technique is known as the *Müller-Breslau principle*, and it can be formally stated as follows: *The influence line for any response function is given by the deflection curve that results when the restraint corresponding to that response function is removed and a unit displacement is introduced in its place.*

In applying the Müller-Breslau principle, care must be taken to make certain that the deformation pattern of System 2 interacts with the forces of System 1 such that only the unit load and the response function for which the influence line is being constructed do work. A careful examination of the procedures used to obtain Eqs. 7.3, 7.5, and 7.7 will verify the importance of this fact.

In the present context, the Müller-Breslau principle is most helpful in a qualitative sense for verifying the shape of influence lines. Later in the book, the principle will be used both qualitatively and quantitatively for statically indeterminate structures.

Use the Müller-Breslau principle to construct the influence lines for the reaction at point C, the shear at point B, and the moment at point D for the structure studied in Section 7.3.2. **7.4.1 Example problem**

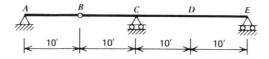

The response functions for which the influence lines are desired are shown below; all functions are directed in the positive sense.

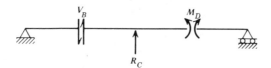

Influence Line for R_C: Remove the restraint corresponding to R_C and introduce a unit vertical displacement at point C.

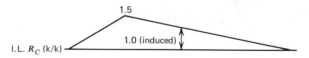

Influence Line for V_B: Remove the restraint associated with V_B and introduce a unit transverse displacement at point B.

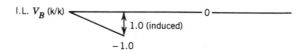

Influence Line for M_D: Remove the restraint corresponding to M_D and introduce a unit angle change at point D.

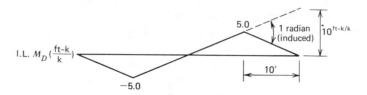

7.4.2 Example problem

Use the Müller-Breslau principle to construct the influence line for the girder shear in panel 2–3 for the structure shown below.

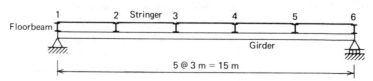

Influence Line for V_{2-3}: The response function for which the influence line is to be constructed is shown in a positive sense on the figure given below.

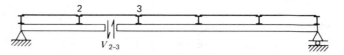

Remove the restraint corresponding to V_{2-3} and introduce a unit transverse displacement, while maintaining the same slope on each side of the cut. Although the displacement is introduced in the girder, the influence line is traced by the stringers, since the loads are applied along this line.

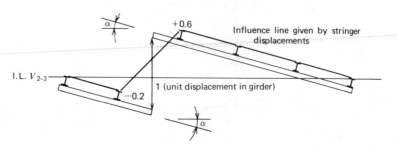

Note:

The unit displacement could be introduced anywhere in panel 2–3, and the same influence line would result.

In Chapters 2, 3, and 4, we developed matrix relationships of the form

$$\{F_R\} = [b]_{F_R P}\{P\} \tag{7.8}$$

where $\{F_R\}$ is a vector representing specific response functions, $\{P\}$ is a load vector, and $[b]_{F_R P}$ is an equilibrium matrix that relates the response functions to the loads. The response function vector may contain reactions, moments, shears, bar forces, or any response quantity of interest.

In an expanded form, Eq. 7.8 can be written as

$$
\begin{Bmatrix}
F_{R1} \\
F_{R2} \\
\vdots \\
F_{Ri} \\
\vdots \\
F_{Rn}
\end{Bmatrix}_{n \times 1}
=
\begin{bmatrix}
b_{11} & b_{12} & \cdots & b_{1j} & \cdots & b_{1m} \\
b_{21} & b_{22} & \cdots & b_{2j} & \cdots & b_{2m} \\
\vdots & & & & & \\
b_{i1} & b_{i2} & \cdots & b_{ij} & \cdots & b_{im} \\
\vdots & & & & & \\
b_{n1} & b_{n2} & \cdots & b_{nj} & \cdots & b_{nm}
\end{bmatrix}_{n \times m}
\begin{Bmatrix}
P_1 \\
P_2 \\
\vdots \\
P_j \\
\vdots \\
P_m
\end{Bmatrix}_{m \times 1}
\tag{7.9}
$$

Examination of the ith row of the $[b]$ matrix reveals that b_{ij} gives the value of F_{Ri} that results from a unit load of P_j. Thus, by definition b_{ij} is the ordinate to the influence line for F_{Ri} corresponding to a unit load at point j. Since j can assume values of 1 to m, the ith row of the $[b]$ matrix gives ordinates of the influence line for F_{Ri} as the unit load occupies successive positions of 1 through m. Accordingly, each row of the $[b]$ matrix represents an influence line for the response function associated with that row. For this reason, the $[b]$ matrix can be appropriately referred to as an *influence matrix*.

For the problems formulated according to the matrix procedures illustrated in earlier chapters, equations of the form of Eq. 7.8 are the natural result. As such, the influence matrix, which contains a complete family of influence lines, is a by-product of the matrix formulation.

A few observations are in order regarding the matrix method for obtaining influence lines, First, the $\{F_R\}$ matrix must be selected to include the response functions for which the influence lines are desired, and the load vector $\{P\}$ must be formulated to include loads corresponding to the desired application points and directions for the placement of unit loads for the generation of the influence lines. In this way, each element of the $[b]$ matrix will represent an ordinate to an influence line for one of the elements of the $\{F_R\}$ vector corresponding to a unit value for one of the loads in the $\{P\}$ vector. It should also be noted that the matrix method is based on a discrete loading arrangement. This means that the relationship $\{V\} = [b]_{VP}\{P\}$ leads to shear intensities between load points and not panel point shears. Thus, the matrix $[b]_{VP}$ produces influence lines for panel shears similar to those determined in Section 7.3.3, as opposed to influence lines for point shears.

7.5.1 Example problem

Determine the influence line ordinates for the reactions for the structure shown.

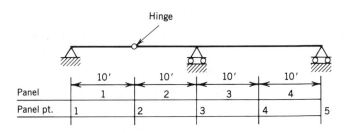

To determine the influence lines, the matrix formulation of Section 2.11 is to be used, and the following loading arrangement is adopted. Note that positive loads are taken as downward, which is in keeping with standard influence line approaches.

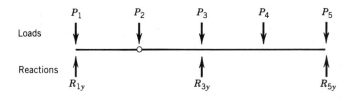

Equations of Equilibrium: The equations of equilibrium are written in the form of Eq. 2.12.

$$
\begin{array}{l}
\sum P_y = 0 \\
\sum M_1 = 0 \\
\sum M_2^{\text{left}} = 0
\end{array}
\begin{bmatrix} 1 & 1 & 1 \\ 0 & 20 & 40 \\ 10 & 0 & 0 \end{bmatrix}
\begin{Bmatrix} R_{1y} \\ R_{3y} \\ R_{5y} \end{Bmatrix}
=
\begin{bmatrix} 1 & 1 & 1 & 1 & 1 \\ 0 & 10 & 20 & 30 & 40 \\ 10 & 0 & 0 & 0 & 0 \end{bmatrix}
\begin{bmatrix} P_1 \\ P_2 \\ P_3 \\ P_4 \\ P_5 \end{bmatrix}
$$

Influence Line Ordinates: The solution of the equations for the reactions yields

$$
\begin{Bmatrix} R_{1y} \\ R_{3y} \\ R_{5y} \end{Bmatrix}
=
\begin{bmatrix} 1.0 & 0 & 0 & 0 & 0 \\ 0 & 1.5 & 1.0 & 0.5 & 0 \\ 0 & -0.5 & 0 & 0.5 & 1.0 \end{bmatrix}
\begin{bmatrix} P_1 \\ P_2 \\ P_3 \\ P_4 \\ P_5 \end{bmatrix}
$$

This is an expanded form of

$$\{R\} = [b]_{RP}\{P\} \tag{2.13}$$

Each row of $[b]_{RP}$ is an influence line for the corresponding component of R_{iy} as the unit load occupies points 1 through 5.

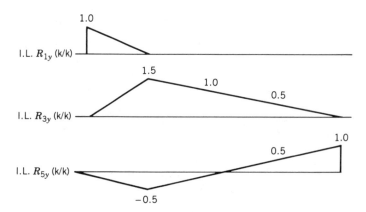

Determine the ordinates for the influence lines for reactions, panel shears, and panel point moments for the structure shown. **7.5.2 Example problem**

To determine the influence lines, the matrix formulation of Section 4.13 is employed. The loading arrangement shown below is used.

Here, positive loads are taken as upward according to the convention of Section 4.13. However, before the influence lines are plotted, an adjustment will be made so that the ordinates will correspond to downward loads.

Equations of Equilibrium: This structure was analyzed in Section 4.13.1, and the equations of equilibrium in the form of Eq. 4.21 were found to be

$$
\begin{bmatrix}
1 & & & & & & & & & & -1 \\
-1 & 1 & & & & & & & & & \\
& -1 & 1 & & & & & & & & \\
& & -1 & 1 & & & & & & & -1 \\
& & & -1 & & & & & & & \\
\hline
& & & & & -1 & & & & & \\
10 & & & & & 1 & -1 & & & & \\
& 10 & & & & & 1 & -1 & & & \\
& & 10 & & & & & 1 & -1 & & \\
& & & 10 & & & & & 1 & &
\end{bmatrix}
\begin{Bmatrix}
V_{12} \\
V_{23} \\
V_{34} \\
V_{45} \\
M_{12} \\
M_{23} \\
M_{34} \\
M_{45} \\
R_1 \\
R_4
\end{Bmatrix}
=
\begin{Bmatrix}
P_1 \\
P_2 \\
P_3 \\
P_4 \\
P_5 \\
0 \\
0 \\
0 \\
0 \\
0
\end{Bmatrix}
$$

Influence Line Ordinates: The solution of the equations of equilibrium yields

$$
\begin{Bmatrix} V_{12} \\ V_{23} \\ V_{34} \\ V_{45} \\ M_{12} \\ M_{23} \\ M_{34} \\ M_{45} \\ R_1 \\ R_4 \end{Bmatrix}
=
\begin{bmatrix}
0 & -0.67 & -0.33 & 0 & 0.33 & 0.033 & 0.033 & 0.033 & 0.033 & 0.033 \\
0 & 0.33 & -0.33 & 0 & 0.33 & 0.033 & 0.033 & 0.033 & 0.033 & 0.033 \\
0 & 0.33 & 0.67 & 0 & 0.33 & 0.033 & 0.033 & 0.033 & 0.033 & 0.033 \\
0 & 0 & 0 & 0 & -1.00 & 0 & 0 & 0 & 0 & 0 \\
0 & 0 & 0 & 0 & 0 & -1.00 & 0 & 0 & 0 & 0 \\
0 & -6.7 & -3.3 & 0 & 3.3 & -0.67 & -0.67 & 0.33 & 0.33 & 0.33 \\
0 & -3.3 & -6.7 & 0 & 6.7 & -0.33 & -0.33 & -0.33 & 0.67 & 0.67 \\
0 & 0 & 0 & 0 & 10.0 & 0 & 0 & 0 & 0 & 0 \\
-1 & -0.67 & -0.33 & 0 & 0.33 & 0.033 & 0.033 & 0.033 & 0.033 & 0.033 \\
0 & -0.33 & -0.67 & -1.0 & -1.33 & -0.033 & -0.033 & -0.033 & -0.033 & -0.033
\end{bmatrix}
\begin{Bmatrix} P_1 \\ P_2 \\ P_3 \\ P_4 \\ P_5 \\ 0 \\ 0 \\ 0 \\ 0 \\ 0 \end{Bmatrix}
$$

The above is an expanded form of

$$
\begin{Bmatrix} \{V\} \\ \{M\} \\ \{R\} \end{Bmatrix}
=
\begin{bmatrix}
[b]_{VPT} & [b]_{VPM} \\
[b]_{MPT} & [b]_{MPM} \\
[b]_{RPT} & [b]_{RPM}
\end{bmatrix}
\begin{Bmatrix} \{P_T\} \\ \{P_M\} \end{Bmatrix}
\tag{4.23}
$$

The rows of $[b]_{VPT}$, $[b]_{MPT}$, and $[b]_{RPT}$ are the influence line ordinates for the shears, moments, and reactions, respectively, as a unit upward load occupies points 1 through 5. Before plotting these ordinates, all signs are changed, since influence lines are customarily constructed for downward loads.

A few sample influence lines are plotted below. A dot $(\cdot)$ is used to identify the calculated ordinate.

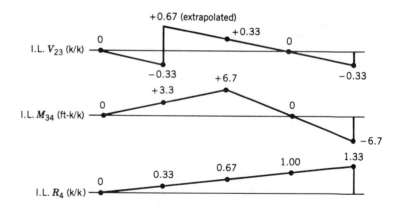

7.6

Use of Influence Lines

Influence lines are useful to the structural analyst in two specific ways. These can be summarized as follows:

1. An influence line for a given response function can be used *qualitatively* to determine how a structure should be loaded in order to maximize the response function.

2. An influence line for a given response function can be used *quantitatively* to determine the value of the corresponding response function for any given loading arrangement.

The first point is useful in establishing the pattern of live loading that will give the maximum value of the response function, whereas the second point is useful for both dead and live loading to establish the value of the response function for a designated loading condition.

It is helpful to consider separately the case of a single concentrated load and that of a uniformly distributed load.

Recalling that each ordinate of an influence line at a given abscissa gives the value **7.6.1** of a response function caused by the placement of a unit load on the structure **Concentrated** corresponding to that abscissa, and noting that superposition is valid, we can state **loads** the following principles for the use of influence lines with concentrated loads:

1. To determine the maximum signed value of a given response function due to a single concentrated load, the load should be placed on the structure at the point where the corresponding signed ordinate to the influence line for that response function is at a maximum.

2. The value of a response function attributable to the action of a single concentrated load equals the product of the magnitude of the load and the ordinate to the influence line for that response function measured at the abscissa corresponding to the point of application of the load.

3. The value of a response function due to a series of concentrated loads equals the summation of the products of the magnitudes of the loads and the respective ordinates to the influence line for the response function at the abscissas corresponding to the points of application of the loads.

The principles stated for concentrated loads in Section 7.6.1 are now used to **7.6.2 Uniformly** develop two additional principles for uniformly distributed loads. **distributed load**

Consider the structural segment shown in Fig. 7.4, which is subjected to a uniformly distributed downward load of intensity p. Also given in this figure is a segment of an influence line for the response function F_R. At section x, an element of the structure dx in length is taken, and an elemental load of $dP = p\,dx$ can be taken as a concentrated load at point x. Using the second principle of Section

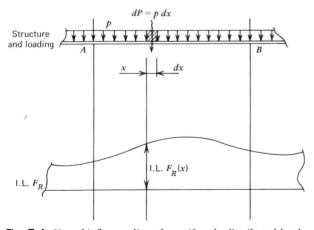

Fig. 7.4 *Use of influence lines for uniformly distributed loads.*

7.6.1, we can express the increment of the response function that results from the load $dP = p \, dx$ as

$$dF_R = p \, dx \, [\text{I.L. } F_R(x)] \qquad (7.10)$$

where $[\text{I.L. } F_R(x)]$ is the ordinate to the influence line at point x.

In accordance with the third principle of Section 7.6.1, the elemental contributions of Eq. 7.10 can be summed between A and B to give the total response F_R as

$$F_R = p \int_{x_A}^{x_B} [\text{I.L. } F_R(x)] \, dx \qquad (7.11)$$

In the above, p has been moved outside the integral since it is a constant. In the form of Eq. 7.11, it is clear that the integral represents the area under the influence line between the limits of points A and B on the structure.

Thus, the following principles can be stated for the use of influence lines with uniformly distributed loads:

1. To determine the maximum signed value of a given response function attributable to a uniformly distributed load, the load should be placed over all sections of the structure where the influence line for that response function has the corresponding signed ordinates.

2. The value of a response function attributable to the action of a uniformly distributed load over a portion of the structure equals the product of the load intensity and the net area under the influence line for that response function between the abscissas corresponding to the limits of the portion of the structure loaded.

7.6.3 Example problem

Consider the simple beam structure shown below. For the loading specified, determine the maximum reaction at point A, the maximum positive shear at point C, and the maximum positive moment at point C. The required influence lines were developed in Example 7.3.1.

Dead load: Uniform load of 10 kN/m.

Live load: Uniform load of 25 kN/m; two concentrated loads of 100 kN that are separated by 3m.

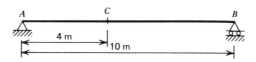

Reaction at Point A: The influence line for R_A is taken from Example 7.3.1.

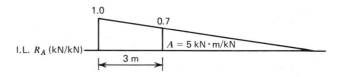

Dead load: 10 kN/m × 5 kN · m/kN   =   50 kN

Live load: 25 kN/m × 5 kN · m/kN   =   125 kN

100 kN × (1.0 + 0.7) kN/kN = 170 kN

Total R_A = 345 kN

Positive Shear at Point C: The influence line for V_C is taken from Example 7.3.1.

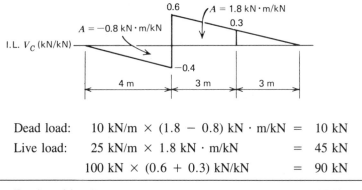

Dead load: 10 kN/m × (1.8 − 0.8) kN · m/kN =   10 kN

Live load: 25 kN/m × 1.8 kN · m/kN   =   45 kN

100 kN × (0.6 + 0.3) kN/kN   =   90 kN

Total positive V_C = 145 kN

Positive Moment at Point C: The influence line for M_C is taken from Example 7.3.1.

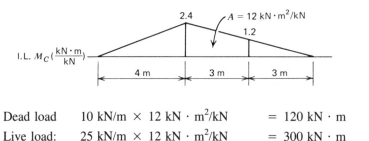

Dead load   10 kN/m × 12 kN · m²/kN   =   120 kN · m

Live load: 25 kN/m × 12 kN · m²/kN   =   300 kN · m

100 kN × (2.4 + 1.2) kN · m/kN =   360 kN · m

Total positive M_C =   780 kN · m

Consider the structure shown below. For the designated loading, determine the maximum upward reaction at point E and the maximum negative moment at point D. The required influence lines were developed in Example 7.3.2.   **7.6.4 Example problem**

Dead load: Uniform load of 1 k/ft.

Live load: Uniform load of 3 k/ft; two concentrated loads of 5^k and 10^k that are separated by 5 ft.

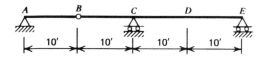

Reaction at Point E: The influence line for R_E is taken from Example 7.3.2.

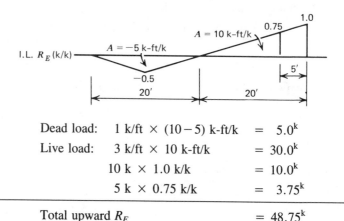

Dead load:	1 k/ft × (10 − 5) k-ft/k	=	5.0^k
Live load:	3 k/ft × 10 k-ft/k	=	30.0^k
	10 k × 1.0 k/k	=	10.0^k
	5 k × 0.75 k/k	=	3.75^k

Total upward R_E	= 48.75^k

Negative Moment at Point D: The influence line for M_D is taken from Example 7.3.2.

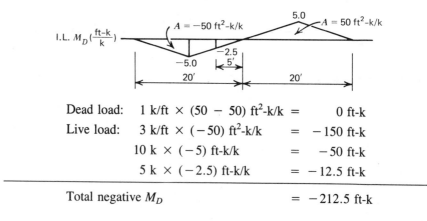

Dead load:	1 k/ft × (50 − 50) ft²-k/k	=	0 ft-k
Live load:	3 k/ft × (− 50) ft²-k/k	=	− 150 ft-k
	10 k × (−5) ft-k/k	=	− 50 ft-k
	5 k × (−2.5) ft-k/k	=	− 12.5 ft-k

Total negative M_D	= − 212.5 ft-k

7.7

Maximum Response Functions in Beams

The sole purpose for considering variable loading conditions is to determine the loading arrangements that will maximize the response functions and to determine the corresponding values of these maximized response functions. Several topics related to the general area of maximum effects are discussed in this section.

7.7.1 Maximum absolute moment

In the example problems of Section 7.6 the maximum moments were computed at designated points on the structure. However, moments computed in this fashion may not include the maximum possible moment on the structure.

On a simply supported beam, the maximum moment occurs at midspan for a uniformly distributed load or at the point of loading for a single concentrated load. One need merely construct the influence line for the moment at the appropriate point and use it in conjunction with the specified loading to determine the maximum moment.

Another case of interest occurs when a simply supported beam is subjected to a series of concentrated loads. For any position of these loads, the moment dia-

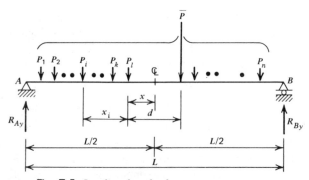

Fig. 7.5 *Loading for absolute maximum moment.*

gram will consist of a series of straight-line segments between the load points; thus, the maximum moment must occur at a load point. However, it is not apparent which load will occupy the position of maximum moment or where the loads should be positioned to create the maximum possible moment.

Consider the simply supported beam shown in Fig. 7.5, which supports a series of n concentrated loads. These loads can be represented by the resultant, $\overline{P}$, which is located at a distance d from the load P_i. The maximum moment will generally occur at one of the loads that is adjacent to $\overline{P}$, and it is assumed that this condition occurs at the load P_l when it it located at a position x from the midspan of the beam. This condition is illustrated in Fig. 7.5.

Summing moments about point B, we determine the reaction at A to be

$$R_{Ay} = \frac{\overline{P}(L/2 + x - d)}{L} \tag{7.12}$$

Thus, the moment at the load P_l is

$$M_{P_l} = R_{Ay}\left(\frac{L}{2} - x\right) - \sum_{i=1}^{k} P_i x_i \tag{7.13}$$

Substituting Eq. 7.12 into Eq. 7.13, we obtain

$$M_{P_l} = \overline{P}\left(\frac{L}{4} - \frac{d}{2} - \frac{x^2}{L} + \frac{xd}{L}\right) - \sum_{i=1}^{k} P_i x_i \tag{7.14}$$

For the maximum value of M_{P_l}

$$\frac{dM_{P_l}}{dx} = \overline{P}\left(-\frac{2x}{L} + \frac{d}{L}\right) = 0 \tag{7.15}$$

from which $x = d/2$.

We thus conclude that the *maximum moment at one of a series of concentrated loads that act on a simply supported beam occurs when the midspan of the structure is located midway between that particular load and the resultant of all loads acting on the span.* This criterion gives the position of the loads for maximum moment to occur at a given load. To obtain the absolute maximum moment, the above location criterion would have to be applied to all loads, and the largest of all maximum moments would then be selected. In general, the largest of the two loads adjacent to the resultant is the load at which the maximum moment occurs.

In many statically determinate beams, the section where the absolute maximum moment occurs cannot be established by inspection, nor can convenient criteria be established. For such cases, it is necessary to compute the maximum moment for sections where the absolute maximum value is likely to occur and then select the controlling maximum value.

7.7.2 Critical loading conditions from influence lines

As was illustrated in Section 7.6, the loading conditions that cause the maximum value for any response function can be determined from the influence line for that function. For complicated loading conditions, however, it is not always apparent how the loads should be arranged for maximum effects. It is possible to develop detailed criteria for the determination of the critical load position for maximizing any response function; however, such criteria will not be presented here. The problems of Sections 7.6 and 7.7 illustrate the essential features for the utilization of influence lines, and the reader is referred to the selected references at the end of the chapter for a complete treatment of load position criteria.

With the computer capabilities that are presently available, it is frequently easier to read into the computer those loading patterns that are feasible critical loading conditions and then analyze the structure for each loading case. A comparison of the results then yields the critical loading case and gives the desired maximum effects.

7.7.3 Highway loadings

Design specifications and building codes usually dictate the loading conditions that must be considered in the design of any particular structure. As an example of a typical situation, we will consider the bridge loadings that are specified in the thirteenth edition of *Standard Specifications for Highway Bridges* of the American Association of State Highway and Transportation Officials (AASHTO). Figure 1.24 gives the designated loading of the so-called HS 20–44 highway loading. According to this loading, a critical condition can result from either of two arrangements, whichever produces the maximum values of the desired response functions. The two arrangements are

1. A lane loading, which consists of a combination of uniform and concentrated loads, as shown in Fig. 1.24*a*.
2. A truck loading, which consists of a single truck–trailer combination, as shown in Fig. 1.24*b*.

Both loading arrangements represent idealizations of the actual situation. The lane loading simulates a line of vehicles with a heavy load in its midst, whereas the truck loading represents a single truck.

Each of the above loading arrangements is for a single lane of traffic, which occupies a 10-foot width of roadway, as is depicted in Fig. 1.24*c*. The number of lanes of traffic that must be supported by each supporting bridge member is determined from a transverse analysis of the structure. For instance, consider a two-lane bridge, of which a cross section is shown in Fig. 7.6. Each traffic lane is 14 feet wide and the two lanes are separated by a 2-foot median strip. If the transverse section of the bridge is assumed to be simply supported by girders A and B, then the influence line for the force between the bridge deck and girder B, R_B, is given in Fig. 7.6*b*. Thus, to maximize R_B, each lane loading should be placed to the extreme edge of the traffic lane, as shown in Fig. 7.6*a*.

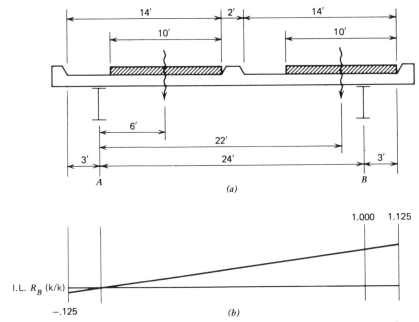

Fig. 7.6 *Position of lane loadings for maximum girder load.* (a) *Bridge cross section.* (b) *Influence line for* R$_B$.

Taking moments about girder A for this position of lane loading, we have

$$\text{(Lane) } 6 + \text{(Lane) } 22 = R_B \cdot 24$$

$$R_B = \frac{28 \text{ (Lane)}}{24} = 1.17 \text{ Lanes}$$

Normally, one would proceed with the analysis of girder B with the normal considerations of maximum effects, using the lane loadings of Fig. 1.24 in conjunction with the required influence lines. Then, the final results would be multiplied by a factor of 1.17 to account for the transverse positioning of the lane loading.

When the bridge deck is transversely supported by more than two longitudinal members, the lane loading taken by each girder would result from a statically indeterminate analysis of a continuous beam on deflecting supports. For this case, the AASHTO specifications provide factors that can be used to determine the loading to be applied to each girder.

7.7.4 Envelope of maximum effects

In the design process, each member must be proportioned to sustain the largest possible values of *all* response quantities at every point along the structure. Thus, the placement of load for the maximization of a response quantity and the corresponding maximum value of that quantity must be determined for a number of points along the structure. The collection of these results for any one response function forms an envelope of maximum effects.

Consider the beam structure shown in Fig. 7.7a along with the prescribed intensities for dead load and live load, p_D and p_L, respectively. The dead load extends over the entire structure, whereas the live load can be applied over any desired section of the structure. Figure 7.7b shows representative influence lines for shear and moment at point x along the span. These influence lines are used in

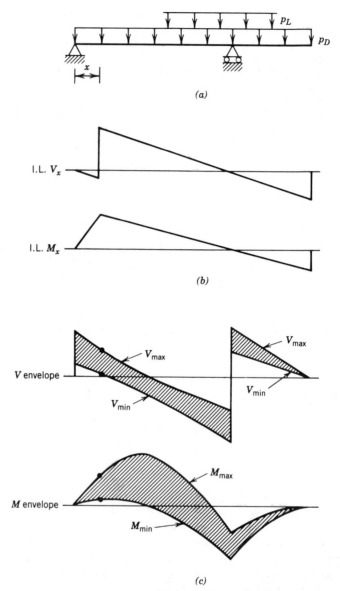

Fig. 7.7 *Envelopes of maximum effects.* (a) *Structure and prescribed loads.* (b) *Representative influence lines.* (c) *Envelopes for shear and moment.*

conjunction with the prescribed loads to determine the extreme values for both shear and moment. These values are plotted as dots in Fig. 7.7c corresponding to the *x* position. If this procedure is followed for several values of *x*, the full envelope of extreme values is produced as shown in Fig. 7.7c.

Problems of this type are best carried out through the systemization provided by tabular calculations. This is illustrated by the example problem of Section 7.7.5.

Of course, an envelope of maximum effects can be generated for any response function coupled with any loading situation. The example problem of Section 7.7.6 involves the determination of the envelope of maximum live load moments associated with the HS 20–44 truck loading.

Construct the envelopes for maximum shears and moments for the structure **7.7.5 Example** shown. Consider 2-meter intervals along the structure. **problem**

$$\text{Dead load:} \quad \text{Uniform load} = p_D = 10 \text{ kN/m}$$
$$\text{Live load:} \quad \text{Uniform load} = p_L = 15 \text{ kN/m}$$

Influence Lines: The influence lines for shear and moment are required at 2-meter intervals along the structure. These are portrayed below in general form.

Shear:

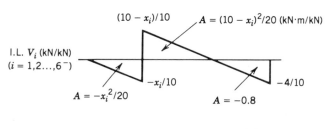

Moment:

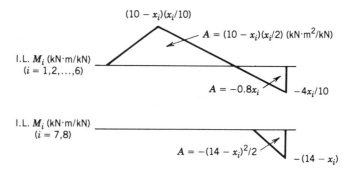

Response Values: The extreme values of the shears and moments are computed in the following tables:

Shear:

Pt i	x_i	Infl. Line Areas (kN · m/kN)			Shear = p × Area (kN)			Envel. Extremes	
					Dead Ld.	Live Ld.		$V_{max} =$	$V_{min} =$
		Pos.	Neg.	Net	V_D	V_L^+	V_L^-	$V_D + V_L^+$	$V_D + V_L^-$
1	0.0	5.0	−0.8	4.2	42.0	75.0	−12.0	117.0	30.0
2	2.0	3.2	−1.0	2.2	22.0	48.0	−15.0	70.0	7.0
3	4.0	1.8	−1.6	0.2	2.0	27.0	−24.0	29.0	−22.0
4	6.0	0.8	−2.6	−1.8	−18.0	12.0	−39.0	−6.0	−57.0
5	8.0	0.2	−4.0	−3.8	−38.0	3.0	−60.0	−35.0	−98.0
6⁻	10.0	0.0	−5.8	−5.8	−58.0	0.0	−87.0	−58.0	−145.0
6⁺	10.0	4.0	0.0	4.0	40.0	60.0	0.0	100.0	40.0
7	12.0	2.0	0.0	2.0	20.0	30.0	0.0	50.0	20.0
8	14.0	0.0	0.0	0.0	0.0	0.0	0.0	0.0	0.0

Moment:

Pt i	x_i	Infl. Line Areas (kN · m²/kN)			Moment = p × Area (kN · m)			Envel. Extremes	
					Dead Ld.	Live Ld.		$M_{max} =$	$M_{min} =$
		Pos.	Neg.	Net	M_D	M_L^+	M_L^-	$M_D + M_L^+$	$M_D + M_L^-$
1	0.0	0.0	0.0	0.0	0.0	0.0	0.0	0.0	0.0
2	2.0	8.0	−1.6	6.4	64.0	120.0	−24.0	184.0	40.0
3	4.0	12.0	−3.2	8.8	88.0	180.0	−48.0	268.0	40.0
4	6.0	12.0	−4.8	7.2	72.0	180.0	−72.0	252.0	0.0
5	8.0	8.0	−6.4	1.6	16.0	120.0	−96.0	136.0	−80.0
6	10.0	0.0	−8.0	−8.0	−80.0	0.0	−120.0	−80.0	−200.0
7	12.0	0.0	−2.0	−2.0	−20.0	0.0	−30.0	−20.0	−50.0
8	14.0	0.0	0.0	0.0	0.0	0.0	0.0	0.0	0.0

Envelopes for Shear and Moment:

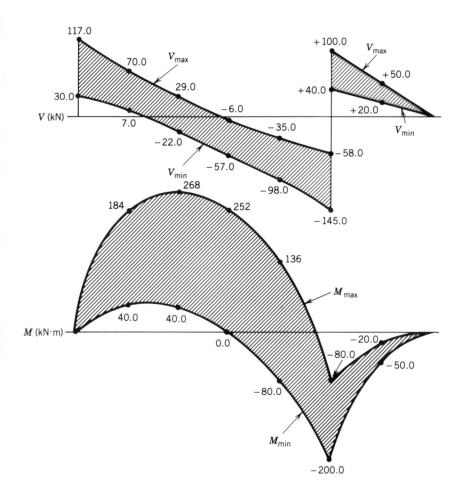

Note:

The shaded portions identify the range of possible shears and moments for the prescribed loads.

Compute the envelope of maximum live load moments for one of the supporting **7.7.6 Example** girders for the single-span bridge shown below. Assume that the structure is a **problem** two-lane bridge with the arrangement shown in Fig. 7.6 and that it supports an HS 20–44 truck loading.

The envelope ordinates should be determined at 5-foot intervals along the structure.

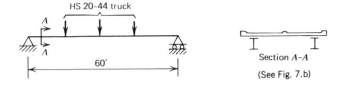

Influence Line Construction: Moment influence lines are required for points at 5-foot intervals along the structure.

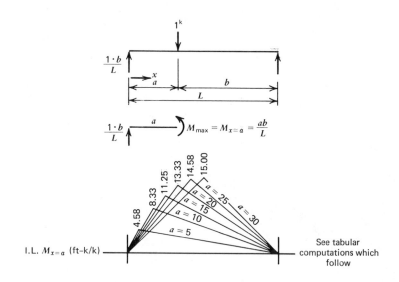

I.L. $M_{x=a}$ (ft-k/k)

See tabular computations which follow

Moments for Single Lane of H 20–44 Loading:

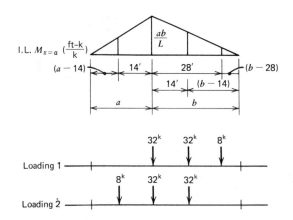

For Loading 1:

$$M_{x=a} = 32\left(\frac{ab}{L}\right) + 32\left(\frac{ab}{L}\right)\left(\frac{b-14}{b}\right) + 8\left(\frac{ab}{L}\right)\left(\frac{b-28}{b}\right)$$

$$= \frac{ab}{L}\left[32 + 32\left(\frac{b-14}{b}\right) + 8\left(\frac{b-28}{b}\right)\right] = \frac{ab}{L}\text{ [Coeff. 1]}$$

For Loading 2:

$$M_{x=a} = 32\left(\frac{ab}{L}\right) + 32\,\frac{ab}{L}\left(\frac{b-14}{b}\right) + 8\,\frac{ab}{L}\left(\frac{a-14}{a}\right)$$

$$= \frac{ab}{L}\left[32 + 32\left(\frac{b-14}{b}\right) + 8\left(\frac{a-14}{a}\right)\right] = \frac{ab}{L}\text{ [Coeff. 2]}$$

a	b	(ab/L)	[Coeff. 1]	[Coeff. 2]	$M_{x=a} = \dfrac{ab}{L} \cdot (\text{Coeff})_{\max}$
(ft)	(ft)	$(ft\text{-}k/k)$	(k)	(k)	$(ft\text{-}k)$
5	55	4.58	59.78	*	273.8
10	50	8.33	58.56	*	487.8
15	45	11.25	57.06	54.58	641.9
20	40	13.33	55.20	55.20	735.8
25	35	14.58	52.80	54.72	797.8
30	30	15.00	49.60	53.33	800.0

*Loading inappropriate since a < 14 ft.

Absolute Maximum Moment: The procedure of Section 7.7.1 is used.

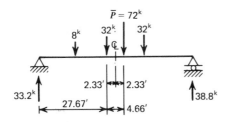

The maximum moment occurs under the wheel positioned at $x = 27.67$ ft.

$$M_{\max} = (33.2 \times 27.67) - (8 \times 14) = 806.6 \text{ ft-k}$$

Envelope of Maximum Live Load Moments:

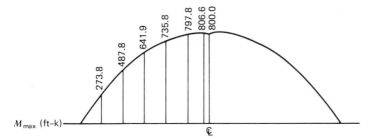

Notes:

1. The above maximum moments are for one lane of loading. These should be multiplied by 1.17 to get the maximum girder moments, as was described in Section 7.7.3, for the arrangement of Fig. 7.6.

2. The moments caused by the HS 20–44 lane loading must also be determined, and they too should be multiplied by 1.17 to get the girder moments.

3. The maximum live load moments (truck or lane) must be combined with the dead load moments to determine the total moments for design.

7.8

Gerstle, K. H., *Basic Structural Analysis*, Chapter 6, Prentice–Hall, Englewood Cliffs, N.J., 1974.

Hsieh, Yuan-Yu, *Elementary Theory of Structures*, Chapters 6 and 7, Prentice–Hall, Englewood Cliffs, N.J., 1970.

Additional Reading

Laursen, H. I., *Structural Analysis,* 2nd Ed., Chapter 10, McGraw–Hill, New York, 1978.

McCormac, J. C., *Structural Analysis,* 3rd Ed., Chapters 11 and 12, Intext Educational Publishers, New York, 1975.

Norris, C. H., Wilbur, J. B., and Utku, S., *Elementary Structural Analysis,* 3rd Ed., Chapter 5, McGraw–Hill, New York, 1976.

Timoshenko, S. P., and Young, D. H., *Theory of Structures,* 2nd Ed., Chapter 3, Mc-Graw–Hill, New York, 1965.

7.9

Suggested Problems

1 through 20. Use the conventional equilibrium approach to construct influence lines for the indicated response functions for each of the structures shown below. In each case, use the Müller-Breslau principle to check the influence lines qualitatively.

1. Reactions at A and B, shear at C, moment at C.

2. Vertical reaction at B, moment at B, shear at C, moment at C.

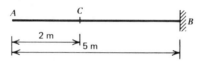

3. Reactions at A and B, shear at C, moment at C, shear at D, moment at D, shear at E, moment at E. *Note:* Point C is just right of support at A; point E is just right of support at B.

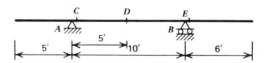

4. Vertical reaction at A, horizontal reaction at A, moment at A, shear at C, moment at C, reaction at B.

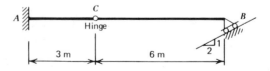

5. All vertical reactions, shear at E, moment at E.

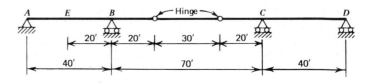

6. Vertical reactions at A and D, horizontal reactions at A and D, shear at F, moment at F. *Note:* Unit load moves from A to C.

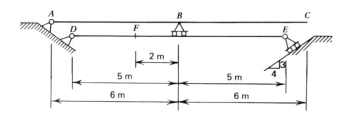

7. Vertical reaction at D, horizontal reaction at D, moment at D, reaction at B, moment at B, shear at C.

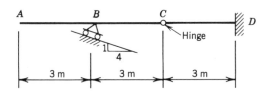

8. Vertical and horizontal reactions at C, shear and moment at B on member CD, shear and moment at E. Unit load mores along A–B–D.

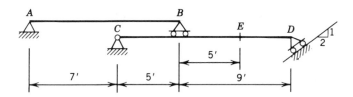

9. Vertical and horizontal reactions at A, reaction at E, shear and moment at G, shear and moment at D. Unit load moves from A to E.

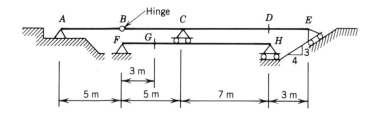

10. Reaction at A, reaction at B, shear in girder at C, moment in girder at C. Unit load moves along stringers.

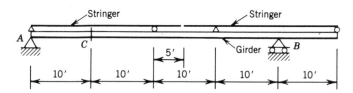

11. Vertical and horizontal reaction at A, vertical reaction at F, shear at C, moment at D. Unit load moves from B to E.

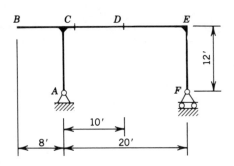

12. All reaction components, shear at C, moment at top of columns BA and DF. Unit load moves from B to E.

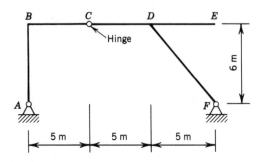

13. All reaction components, shear and moment at E, moment at top of column BD. Unit load moves from A to C.

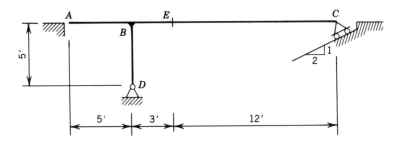

14. All reaction components, shear at e, moment at f, moment at top of column ed. Unit load moves from a to c.

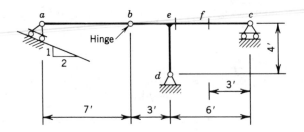

15. Horizontal and vertical reactions at A and B, shear and moment at E. Unit load moves from C to G.

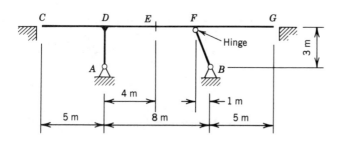

16. Horizontal and vertical reactions at D, members AB and FG. Unit load moves from A to E.

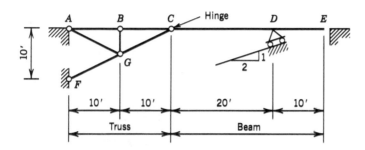

17. Members ab, bc, cg, fg, and bf. Unit load moves from a to e.

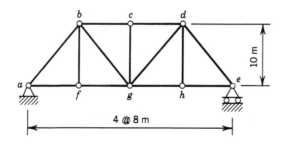

18. Reaction at L_0 and L_3, members U_1L_1, U_2U_3, U_2L_3, L_1L_2, U_4L_4, and U_4L_5.

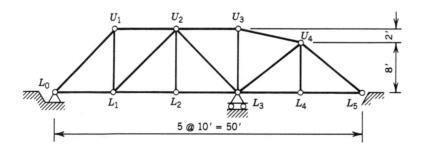

19. Vertical reactions at A, B, and U_4; members U_1U_2, L_1U_2, U_1L_1, L_1L_2, U_2L_2, and L_3U_4. Unit load moves from A to U_4.

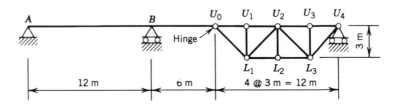

20. All reaction components; members ab, ag, af, bg, fg, and gc. Unit load moves from a to e.

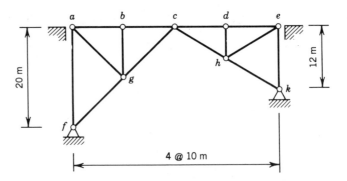

21 through 23. Use the matrix formulation of Section 7.5 to generate the influence lines for the indicated response functions for the structures designated.

21. Structure of Problem 4:

   a. Determine the influence line ordinates at 3-ft intervals for the reactions at A and B using the matrix formulation of Section 2.11.

   b. Determine the influence line ordinates at 3-ft intervals for the reactions at A and B, the panel shears, and the panel point moments.

22. Structure of Problem 4: Determine the influence line ordinates at 3-m intervals for all reaction components, panel shears, and panel point moments.

23. Structure of Problem 5: Determine the influence line ordinates at 10-ft intervals for all vertical reactions, panel shears, and panel point moments.

24. Consider the structure of Problem 3. For the loading specified, determine the maximum vertical reaction at point A, the maximum positive moment at point D, and the maximum positive shear at point D.

   Dead load: Uniform load of 2 k/ft.
   Live load: Uniform load of 4 k/ft and a roving concentrated load of 25 k.

25. Consider the structure of Problem 5. For the loading specified, determine the maximum vertical reaction at B, the maximum negative shear at E, and the maximum negative moment at E.

   Dead load: Uniform load of 2.5 k/ft.
   Live load: Uniform load of 4.0 k/ft and a roving concentrated load of 30 kips.

26. Consider the structure of Problem 13. For the loading specified, determine the

maximum horizontal reaction at D, the maximum positive and negative shear at E, and the maximum moment at the top of column BD.

Dead load: Uniform load of 1.5 k/ft.

Live load: Uniform load of 4.0 k/ft and two roving concentrated loads of 20 kips each.

27. Consider the structure of Problem 15. For the loading specified, determine the maximum horizontal reaction at point A, the maximum positive moment at point E, and the maximum positive shear at point E.

Dead load: Uniform load of 14 kN/m.

Live load: Uniform load of 24 kN/m and two concentrated loads of 72 kN that are separated by 2 m.

28. Consider the structure of Problem 18. For the loading specified, determine the maximum positive and negative axial forces in members U_1L_1, U_2U_3, U_2L_3, and U_4L_5.

Dead load: Uniform load of 4.0 k/ft.

Live load: Uniform load of 2.5 k/ft and two concentrated loads of 20 kips that are separated by 8 feet.

29. Construct the envelopes for maximum shears and moments for the structure of Problem 3. Consider 2.5-ft intervals along the structure.

Dead load: Uniform load $= p_D = 2.0$ k/ft.

Live load: Uniform load $= p_L = 3.5$ k/ft.

30. Construct the envelopes for maximum shears and moments for the structure of Problem 7. Consider 1-meter intervals along the structure.

Dead load: Uniform load $= p_D = 5.0$ kN/m.

Live load: Uniform load $= p_L = 8.0$ kN/m.

31. Construct the envelopes for maximum shears and moments in the beam elements of the structure of Problem 13. Consider 2.5-ft intervals along the structure and include points both left and right of point B.

Dead load: Uniform load $= p_D = 1.5$ k/ft.

Live load: Uniform load $= p_L = 2.5$ k/ft.

32. Develop the envelope of maximum live-load moments for one of the supporting girders for the single-span bridge structure shown. The structure can carry two lanes of traffic, and the loading for a single lane is given. The envelope ordinates should be determined at 2-m intervals.

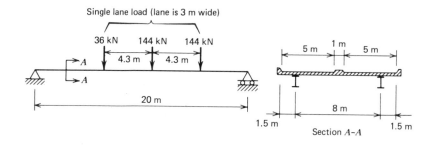

CHAPTER 8

*Hampton Roads
Coliseum, Hampton
Roads, Va. (courtesy
Bethlehem Steel
Corporation).*

NONPLANAR STRUCTURES

8.1

Equilibrium Requirements

As was stated earlier, equilibrium is a condition that prevails when a body is initially at rest and remains at rest when it is acted upon by a set of forces. For a general three-dimensional structure, this condition is ensured by the satisfaction of the equations of static equilibrium. These equations were first given in Chapter 2 for a planar structure, and they are expended here for nonplanar structures.

$$\sum P_x = 0; \quad \sum P_y = 0; \quad \sum P_z = 0$$
$$\sum M_x = 0; \quad \sum M_y = 0; \quad \sum M_z = 0 \tag{8.1}$$

The first set of three equations gives the summation of all force components in the directions of the three mutually perpendicular x, y, and z axes, and the second set represents the summation of moments about these three axes, respectively.

These equations must include the combined effects of the applied forces and the reaction forces that develop at the support points.

Since we are presently studying statically determinate structures, these equations will be sufficient for the complete analysis of the structure.

8.2

As was the case with planar structures, a force in space is completely specified **Specification** when its magnitude, direction, and line of action are given. Figure 8.1 shows a **of a Force** typical force P_i with its components P_{ix}, P_{iy}, and P_{iz}, which act along the three mutually perpendicular coordinate axes. The magnitude of the resultant force is given by

$$P_i = \sqrt{P_{ix}^2 + P_{iy}^2 + P_{iz}^2} \qquad (8.2)$$

The individual components are related to the resultant force by the expressions

$$\begin{aligned} P_{ix} &= P_i \cos \alpha \\ P_{iy} &= P_i \cos \beta \\ P_{iz} &= P_i \cos \gamma \end{aligned} \qquad (8.3)$$

where α, β, and γ are the angles that the resultant force P_i makes with the x, y, and z coordinate axes, respectively. These angles are not mutually independent in that if any two are specified, the third can be determined, as can be evidenced by combining Eqs. 8.2 and 8.3. The quantities $\cos \alpha$, $\cos \beta$, and $\cos \gamma$ are called the direction cosines since these angles fix the direction of the force P_i.

The line of action, indicated by l in Fig. 8.1, must still be specified. This is done by designating a set of coordinates, or a specific point on the structure, at which the force acts.

Thus, there are six characteristics that completely define a force in space—the magnitude, two direction cosines, and the three coordinates that define the line of action.

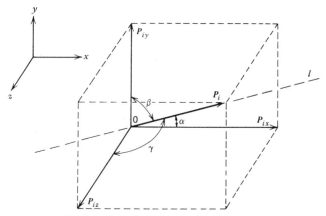

Fig. 8.1 *Specification of a force in space.*

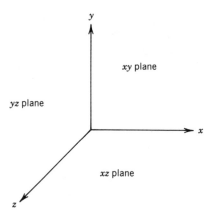

Fig. 8.2 *Mutually perpendicular planes.*

8.3

Support Conditions; Reaction Forces

For planar structures, the restraints provided by the support in the xy plane were described in detail in Sections 2.5 and 2.6. For nonplanar structures, the same considerations must be taken into account in the three mutually perpendicular xy, xz, and yz planes as shown in Fig. 8.2.

For instance, consider the restraint conditions that are provided by the support at point a in Fig. 8.3. As was the case with a planar structure, the three reaction force components of R_{ax}, R_{ay}, and M_{az} define the restraints within the xy plane as shown in Fig. 8.3, where M_{az} is the moment about the z axis. Similarly, the reaction components R_{ax}, R_{az}, M_{ay} and R_{ay}, R_{az}, M_{ax} define the support restraints in the xz and yz planes, respectively. Thus, a total of three force components and three moment components completely define the resultant reaction. These six quantities are equivalent to the six force characteristics described in the previous section.

In most cases, the support mechanism is such that fewer than six reaction components are developed. The considerations here are similar to those discussed in Section 2.5 for planar structures and summarized in Table 2.1. In this case, however, the restraints in three separate support planes must be considered. Table

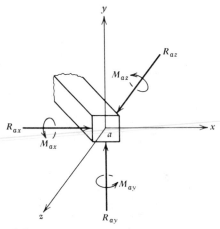

Fig. 8.3 *Components of reaction force at point* a.

TABLE 8.1 Types of Supports

Type Support	Symbolic Representation		Reaction Components	
	Plan[a]	Elevation[a]	Known	Unknown
Fixed			Nothing	R_{ax}, R_{ay}, R_{az} M_{ax}, M_{ay}, M_{az}
Sleeve (pinned about z axis)			$M_{az} = 0$	R_{ax}, R_{ay}, R_{az} M_{ax}, M_{ay}
Universal joint			$M_{ax} = M_{ay}$ $= M_{az} = 0$	R_{ax}, R_{ay}, R_{az}
Roller (free to roll in x direction)			$R_{ax} = 0$ $M_{ax} = M_{ay}$ $= M_{az} = 0$	R_{ay}, R_{az}
Ball			$R_{ax} = R_{az} = 0$ $M_{ax} = M_{ay}$ $= M_{az} = 0$	R_{ay}

[a]*Key Sketch*
Plan: xz plane
Elevation: xy plane

8.1 gives some representative support situations in which the restraints are enumerated. In these cases, two members are assumed to frame into a support at point a as shown in the key sketch. One of these members is in the xz plane, and the other member is inclined to this plane, but in the same vertical plane. In each case, the unknown reaction components are identified. Of course, many other possibilities beyond the representative cases of Table 8.1 are possible when all the possible combinations in the three planes of restraint are considered.

8.4 External Statical Determinacy and Stability

Based on the considerations of the previous section, each support point can be evaluated with regard to the number of unknown reaction components. For the entire structure, the total number of unknown reaction components r_a is given by the summation of the unknown at all support points. The statical classification of a structure is dependent on the value of r_a and the arrangement of the corresponding reaction components.

The unknown reaction components are determined by the systematic application of the equations of static equilibrium, which are given as Eq. 8.1. Since six equations are available for the solution, it is possible to solve for six independent reaction components. Thus, if $r_a = 6$, the structure is classified as statically de-terminate externally. However, if $r_a > 6$, there are more unknowns than there are equations available for their determination, and thus the structure is classified as statically indeterminate externally. Where $r_a < 6$, there are fewer unknowns than there are equations, and thus a solution is not possible. Structures in this latter category are statically unstable since there is no solution that will satisfy the re-quirements of equilibrium.

As was true with planar structures, when the structure is statically indetermi-nate, the equations of static equilibrium remain a necessary part of the solution. They do not form the basis for a solution in themselves, but when augmented with equations of compatibility, a solution is possible.

Summarizing the criteria, we have the following:

$r_a < 6$; structure is statically unstable externally

$r_a = 6$; structure is statically determinate externally

$r_a > 6$; structure is statically indeterminate externally

The criteria summarized above must be applied with some care. If $r_a < 6$, the structure is definitely unstable; however, $r_a \geq 6$ are necessary but not sufficient criteria for structural classification. The reaction components must be arranged to resist rigid-body translation along, and rotation about, each of the coordinate axes. That is, the reactions must be arranged to ensure external stability. If this is not satisfied, the structure is classified as geometrically unstable.

As was true for planar structures, it is possible to have condition equations that will increase the number of reactions that can be determined from statics. If there are n condition equations, then $r_a = 6 + n$ becomes the condition for statical determinacy. Again, the structure may be unstable if the reaction compo-nents are not properly arranged.

8.5

Computation of Reactions Using Equations of Equilibrium

In determining the reactions for nonplanar structures, the same general techniques are used as were used in Chapter 2 for planar structures. The initial step is to construct appropriate free-body diagrams and to show each unknown reaction component to act in a specified direction. The equations of equilibrium are then written in consistency with the arrangements shown on the free-body diagrams, and these equations are simultaneously solved for the unknown reactions. When the solution produces a positive reaction component, then the assumed direction is correct; however, a negative reaction component signifies that the wrong direc-tion was assumed on the free-body diagram.

This section will treat only those structures that are statically determinate ex-ternally. Thus, the six equations of static equilibrium provide all that is needed for the determination of the reactions. For statically indeterminate structures, the equations of equilibrium would still play a necessary role in the solution, but they would not be sufficient.

8.5.1 Example problem

Determine the reactions for the structure given below. Each support is a roller-type mechanism that is capable of developing the reaction components shown.

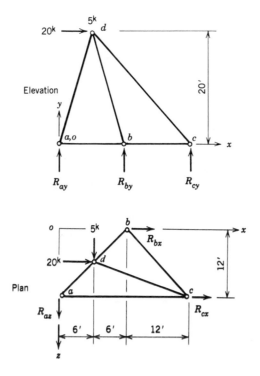

Free-Body Diagrams: The entire structure is taken as a free-body diagram and the given sketches (with the assumed directions of the reaction components) will be used to write the equations of equilibrium.

Determination of Reactions: Since $r_a = 6$, and the arrangement is stable, the structure is statically determinate. One could write the six equations of equilibrium and solve them simultaneously for the six unknowns. Indeed, this is necessary in some cases; however, in the present case, an effort is made to solve for the unknowns in a sequence that will uncouple the equations.

$\sum M_x = 0$

$$(R_{by} \times 12) + (5 \times 20) = 0; \quad R_{by} = -8.33^k$$

$\sum M_z = 0$

$$(R_{cy} \times 24) + (R_{by} \times 12) - (20 \times 20) = 0$$
$$R_{cy} = \frac{400 - (-8.33 \times 12)}{24} = +20.83^k$$

$\sum F_y = 0$

$$R_{ay} + R_{by} + R_{cy} = 0$$
$$R_{ay} = 8.33 - 20.83 = -12.50^k$$

$\sum M_y = 0$

$$(R_{cx} \times 12) + (20 \times 6) - (5 \times 6) = 0$$
$$R_{cx} = \frac{-120 + 30}{12} = -7.5^k$$

$\sum F_x = 0$

$$R_{bx} + R_{cx} + 20 = 0$$
$$R_{bx} = -20 + 7.5 = -12.5^k$$

$\sum F_z = 0$

$$R_{az} + 5 = 0; \quad R_{az} = -5^k$$

Note:

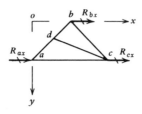

If the roller at point a had been oriented to resist R_{ax}, as shown here, then $r_a = 6$ would still have been satisfied. However, the structure would be unstable because the 5^k load in the z direction could not be resisted. The structure would be conditionally stable if no loads were applied in the z direction, but it would then be classified as geometrically unstable, and the three horizontal reaction components would be statically indeterminate.

8.6

Uni-Connected Space Frameworks

A common three-dimensional structure that frequently can be analyzed on the basis of statics alone is the *uni-connected space framework*. This class of structure results when the characteristics of a planar truss are extended to three dimensions, with the pin connections of a planar truss being replaced by universal joints that are not capable of sustaining a member-end moment. For such structures to be stable, there can be no relative motion between the joints beyond that which occurs as a result of member deformations. Such stability generally results from a combination of tetrahedral units that are interconnected so as to include all points that are to be embraced within the structure space. The details concerning member arrangements for stable frameworks are treated in Sections 8.7 and 8.12.

Uni-connected space frameworks can be statically indeterminate either externally or internally. In the former case, there are more reaction components than can be determined from statics, whereas in the latter case, there are more member forces than can be determined by statics. It is possible for the structure to be statically indeterminate externally but have an overall classification of statically determinate.

This section considers only statically determinate structures, and it is limited to a presentation of the fundamental concepts and procedures needed for the analysis of a uni-connected framework. Complicated frameworks, whether statically determinate or indeterminate, would be treated by the more general matrix techniques that are developed in Part III.

The key assumptions necessary for the analysis of uni-connected space frameworks are mere extensions of those employed in the analysis of planar trusses. Since these assumptions were justified in detail in Chapter 3, they are stated here simply in the context of a three-dimensional structure as follows:

1. Members are connected at their ends with frictionless universal joints.
2. All loads and reactions are applied to the framework at the joints only.
3. The centroidal axis of each member is straight and coincides with the line connecting the joint centers at each end of the member.

When these assumptions are satisfied, each member of the structure will carry pure axial load, with no bending moment or shear present.

The simplest form of a stable uni-connected framework is the tetrahedral arrange- **Variations in** ment of members shown in Fig. 8.4a. In this case, to the basic planar triangular **Framework** unit *abc*, point *d* is attached by the addition of three members. The resulting **Configurations** tetrahedron includes four separate stable triangular units, and the relative move- ments between points result only from the member elongations that accompany the material strains. Since these small movements are ignored in the present con- text, the tetrahedral unit *abcd* is taken as a rigid unit.

This simple tetrahedral framework can be enlarged to include an additional point *e* by adding three members as shown in Fig. 8.4b. This procedure can be repeated to enlarge the system further to any size and configuration, and the struc- ture formed by this approach is called a *simple uni-connected space framework*. For example, Fig. 8.4c shows an enlargement of the structure shown in Fig. 8.4b by the ordered addition of joints *f, g,* and *h.* The resulting structure is composed of five tetrahedra. An alternative arrangement is shown in Fig. 8.4d—here the original tetrahedron of Fig. 8.4a is enlarged by a different strategy, through the ordered addition of points *e, f, g,* and *h.* This example serves to illustrate that not all stable structures are composed of a nested array of tetrahedra. In this case, there are four tetrahedra and a five-sided polygon.

It must be noted that in forming a simple framework, the original triangular

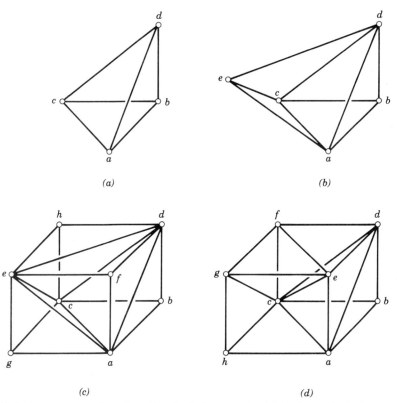

(a)   (b)

(c)   (d)

Fig. 8.4 *Arrangement of members for simple frameworks.* (a) *Basic tetrahedral arrangement.* (b) *Enlargement of framework to include joint at* e. (c) *Further enlargement to include joints at* f, g, *and* h. (d) *Alternative enlargement of original tetrahedron.*

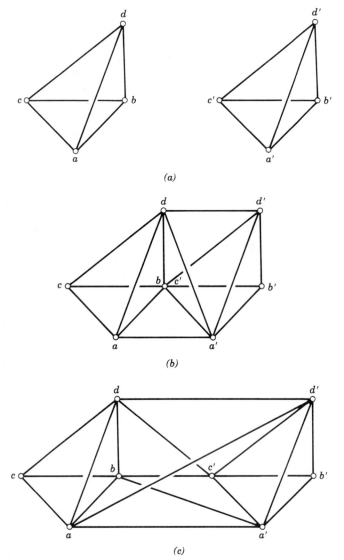

Fig. 8.5 *Member arrangement for compound frameworks.* (a) *Two simple frameworks.*
(b) *Compound framework formation using common joint and three members.*
(c) *Compound framework formation using six members.*

unit must connect points that are not located along a straight line, and each added
joint must not lie in the plane of the triangular unit to which it is being connected.

Other stable framework configurations are possible that do not evolve from
the procedure outlined above. Consider the case where two or more simple frameworks are connected to form a *compound uni-connected space framework*. For
example, take the two simple frameworks shown in Fig. 8.5*a*. As a first step,
these two frameworks are joined by bringing together joints *b* and *c'*. This arrangement is not stable because rotation could occur about any axis through the
common point. If a member were added between points *a* and *a'*, the resulting
structure would be unstable with respect to rotations about axes along members
ab and *a'c'*. The addition of a member between joints *d* and *a'* would inhibit
rotation about an axis along member *ab*, and the addition of a member between

joints d and d' would remove the remaining instability associated with rotation about an axis along member $a'c'$. The compound framework is now stable—the attachment at the common points prevents relative translations, and the addition of three members restricts relative rotations.

An alternative connection would be the addition of six members between the original simple frameworks as shown in Fig. 8.5c. Here, member bc' prevents overall translation between the original two tetrahedra, but it takes five additional members to restrict all possible relative rotations.

It should be noted that although the structure of Fig. 8.5b was formed as a compound framework, it is actually a simple framework that could have been formed by the sequential addition of joints a', d', and b' to the original tetrahedron $abcd$. However, the structure of Fig. 8.5c is truly a compound framework in that it could not be formed from the procedure used to form a simple framework.

In this discussion, only simple cases are considered for the purposes of definition. When more complicated simple frameworks are connected, such as those shown in Figs. 8.4c and d, the connecting sections may contain polyhedra. For additional information on more complicated frameworks, the reader is directed to the references in Section 8.17.

8.8 Joint Identification, Bar Force Notation, and Sign Convention

The matters of joint identification, bar force notation, and sign convention are important in establishing a consistent method of analysis. The techniques adopted here are direct extensions of what was done for planar trusses.

Each joint of the structure is labeled with a letter as shown in Fig. 8.6a. A given member force is symbolically represented by the letter F with an appended double subscript indicating the terminal points of the member. For instance, F_{ij} represents the force in the member that connects joints i and j of the structure. This force, which is an internal force, becomes an external force for any free-body diagram that cuts through member ij, as shown in Fig. 8.6b. Since the member must carry an axial force, the force acts in the same direction as the member itself, and the relationships between the member length and its geometric components are similar to those between the member force and its force compo-

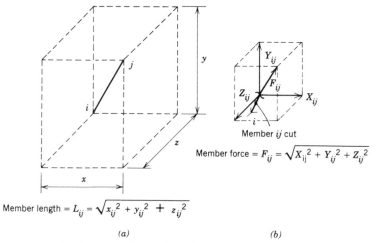

Member ij cut

Member force $= F_{ij} = \sqrt{X_{ij}^2 + Y_{ij}^2 + Z_{ij}^2}$

Member length $= L_{ij} = \sqrt{x_{ij}^2 + y_{ij}^2 + z_{ij}^2}$

(a)   (b)

Fig. 8.6 *Member and force identification.* (a) *Member.* (b) *Force.*

nents. Thus, if member ij is of length L_{ij} and it has projections of x_{ij}, y_{ij}, and z_{ij} along the three coordinate axes, and if the bar force F_{ij} has components X_{ij}, Y_{ij}, and Z_{ij} along the same axes, then

$$\frac{F_{ij}}{L_{ij}} = \frac{X_{ij}}{x_{ij}} = \frac{Y_{ij}}{y_{ij}} = \frac{Z_{ij}}{z_{ij}} \tag{8.4}$$

From this relationship, any bar force component can be determined from any other. That is,

$$F_{ij} = X_{ij}\left(\frac{L}{x}\right)_{ij} = Y_{ij}\left(\frac{L}{y}\right)_{ij} = Z_{ij}\left(\frac{L}{z}\right)_{ij}$$

$$X_{ij} = F_{ij}\left(\frac{x}{L}\right)_{ij} = Y_{ij}\left(\frac{x}{y}\right)_{ij} = Z_{ij}\left(\frac{x}{z}\right)_{ij}$$

$$Y_{ij} = F_{ij}\left(\frac{y}{L}\right)_{ij} = X_{ij}\left(\frac{y}{x}\right)_{ij} = Z_{ij}\left(\frac{y}{z}\right)_{ij} \tag{8.5}$$

$$Z_{ij} = F_{ij}\left(\frac{z}{L}\right)_{ij} = X_{ij}\left(\frac{z}{x}\right)_{ij} = Y_{ij}\left(\frac{z}{y}\right)_{ij}$$

As was done in planar trusses, a tensile member force is taken to be positive and a compressive member force, negative. In the analysis process, the analyst must isolate a free-body diagram in which the desired member force is exposed as an unknown. This unknown force should be assumed to act in the positive direction (tension) on the cut face of the member. The subsequent analysis leads to the desired member force; a positive result indicates that the assumed direction is correct and that the member is in tension, whereas a negative result means that the assumed direction is incorrect and that the member is in compression.

8.9

Overall Statical Determinacy and Stability

The discussion of determinacy and stability in Section 8.4 was based solely on the application of the six equations of static equilibrium to the entire structure. In some cases, it turns out that the reaction components are not statically determinate based on external considerations. However, the total analysis of the structure (reaction components and member forces) is statically determinate when both external and internal conditions are considered.

Each joint of a uni-connected structure forms a concurrent force system in which the three moment equations of Eq. 8.1 are automatically satisfied. Thus, only three independent equations of static equilibrium can be written at each joint. It must be noted that the static relations written externally for the entire structure will not provide further independent equations when this approach is used. Therefore, if both external reaction components and internal member forces are considered, a necessary, but not sufficient, condition for the overall determinacy of the structure is that the total number of unknown member forces m, plus the total number of unknown reaction components r_a, must not exceed three times the number of joints j. That is,

$$m + r_a = 3j$$

or

$$m = 3j - r_a \tag{8.6}$$

In this form, m is the number of members required to form a statically determinate structure that will connect j joints and possess r_a reaction components. If m_a is the actual number of members in the structure, then the following criteria hold:

$m_a < m$; structure is statically unstable

$m_a = m$; structure is statically determinate

$m_a > m$; structure is statically indeterminate

It must be emphasized that $m_a \geq m$ are necessary but not sufficient conditions. Also, if $m_a > m$, it is not clear from this test whether the indeterminacy stems from redundant reaction components or redundant members. Of course, the criteria of Section 8.4 could be used to classify the structure externally, and those results could be coupled with the overall classification of this section to determine the degrees of external and internal redundancies.

The criteria enumerated above clearly apply to simple frameworks, but it is not immediately evident that they apply to compound frameworks. The two simple frameworks of Fig. 8.5a would each require a minimum of six reaction components for external stability. Thus, according to the requirements of Eq. 8.6, with $r_a = 6$ for each structure, we have

$$m_1 = 3j_1 - 6 \tag{8.7}$$

and

$$m_2 = 3j_2 - 6 \tag{8.8}$$

where the subscript simply corresponds to the two separate cases. Since $j_1 = j_2 = 4$, these equations lead to $m_1 = m_2 = 6$, and since $m_a = 6$ for each structure, both are statically determinate.

Addition of Eqs. 8.7 and 8.8 yields

$$m_1 + m_2 = 3j_1 + 3j_2 - 12 \tag{8.9}$$

and it is reasonable to assume that this equation would reflect the requirements for statical determinacy for any compound framework that would be formed by combining these two simple frameworks.

Rearranging the terms of Eq. 8.9, we have

$$(m_1 + m_2 + 3) = 3(j_1 + j_2 - 1) - 6 \tag{8.10}$$

If the two simple frameworks are combined according to the scheme of Fig. 8.5b, then $m = (m_1 + m_2 + 3)$ and $j = (j_1 + j_2 - 1)$ for the new structure. Therefore, Eq. 8.10 correctly expresses the requirements of Eq. 8.6 for the compound structure, which is externally stabilized through six reaction components.

In an alternative rearrangement of terms, Eq. 8.9 could be written in the form

$$(m_1 + m_2 + 6) = 3(j_1 + j_2) - 6 \tag{8.11}$$

The compound framework of Fig. 8.5c has $m = (m_1 + m_2 + 6)$ and $j = (j_1 + j_2)$, and thus Eq. 8.11 again reflects the requirements of Eq. 8.6 for the properly stabilized compound structure.

These examples sustain the conclusion that Eq. 8.6 is a valid requirement for the statical determinacy of compound frameworks and that the resulting criteria apply to both simple and compound frameworks.

The treatment given here is somewhat brief, and it does not address some subtle points regarding structure classification for complicated structures with unusual arrangements of members. Section 8.12 briefly treats the topic of *complex structures,* but for a more complete treatment of advanced topics, the reader is directed to the references given at the end of this chapter.

8.10

Analysis of Uni-Connected Space Frameworks

Two slightly different techniques can be employed in the analysis of statically determinate space frameworks. The first approach is used for structures that are statically determinate externally. Here, all the reaction components can be determined by applying the six equations of static equilibrium to the entire structure. For these cases, the reactions are determined first, and then the individual member forces are determined by taking a sequence of free-body diagrams that permit the systematic determination of all member forces. The second technique involves situations where the reactions are statically indeterminate externally and cannot be determined by the external application of the equations of static equilibrium. In these cases, once it has been determined that the overall structure is statically determinate, one proceeds directly to the free-body diagrams of the various parts of the structure that lead to the requisite number of equations for the combined solution of the reactions and the member forces.

By either approach, a free-body diagram may involve the isolation of a joint of the structure, which is a simple extension of the method of joints that is used in planar truss analysis, or it may treat any convenient portion of the structure that allows the analyst to isolate and solve for the desired unknown forces. There are several special theorems that result from considering a free-body diagram of a joint under prescribed conditions of member arrangement and loading. The first theorem follows from Fig. 8.7 and is given as follows:

1. If a joint is arranged so that all members except one lie in the same plane, then the force component normal to the plane in the one member not in the plane is equal to the resultant force component normal to the plane of the external loads applied at the joint.

The remaining theorems are actually corollaries to the first:

2. For the same arrangement of bars, if the resultant normal force compo-

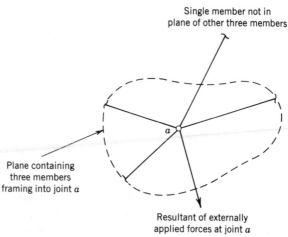

Fig. 8.7 *Special arrangement of members at a joint.*

nent of the external loads at the joint is zero, then the force in the member not in the plane is zero.

3. At a joint where three noncoplanar members meet, and at which no external loads act, all members have zero force unless two of the members are colinear.

4. If all but two of the members framing into a joint have zero force, and these two members are not colinear, and if, further, there are no external forces acting at the joint, then the force in each of these two bars is zero.

It is not necessary that these theorems be employed in the analysis process; however, their use frequently simplifies the analysis by permitting certain bar forces to be determined by inspection.

Determine the complete set of bar forces for the structure and loading specified in **8.10.1 Example** Example 8.5.1. The structure is shown below along with the reactions that were **problem** previously determined.

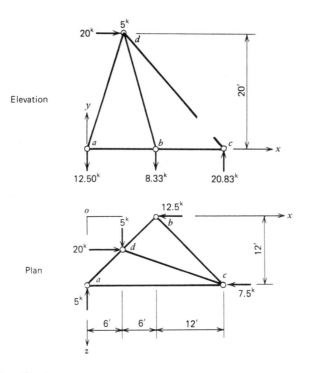

Structure Classification:

$$j = 4; \qquad r_a = 6; \qquad m_a = 6$$
$$m = 3j - r_a = 12 - 6 = 6$$
$m_a = m$ ∴ Structure is statically determinate. Arrangement is stable.

Determination of Member Forces: It is convenient to arrange the geometric components and the force components in a table that permits easy access to the data as they are needed. The geometric data are entered in the table immediately, whereas the force data are added as the analysis proceeds. At the conclusion of the analysis, the complete solution is within the table.

Member	Geometry (ft)				Force (kips)			
	x	y	z	L	X	Y	Z	F
ab	12.0	0	12.0	17.0	-8.75	0	-8.75	-12.40
bc	12.0	0	12.0	17.0	$+6.25$	0	$+6.25$	$+8.85$
ac	24.0	0	0	24.0	$+5.00$	0	0	$+5.00$
ad	6.0	20.0	6.0	21.7	$+3.75$	$+12.50$	$+3.75$	$+13.56$
bd	6.0	20.0	6.0	21.7	$+2.50$	$+8.33$	$+2.50$	$+9.04$
cd	18.0	20.0	6.0	27.6	-18.75	-20.83	-6.25	-28.75

Isolate Joint a as *FBD*:

$$\Sigma\, P_y = 0 \;\uparrow+$$

$$Y_{ad} - 12.50 = 0; \quad Y_{ad} = +12.50^k$$

$$X_{ad} = Y_{ad}\left(\frac{x}{y}\right)_{ad} = (+12.50)\left(\frac{6}{20}\right) = +3.75^k$$

$$Z_{ad} = Y_{ad}\left(\frac{z}{y}\right)_{ad} = (+12.50)\left(\frac{6}{20}\right) = +3.75^k \tag{8.5}$$

$$F_{ad} = Y_{ad}\left(\frac{L}{y}\right)_{ad} = (+12.50)\left(\frac{21.7}{20}\right) = 13.56^k$$

These results are added to the table for member ad.

$$\Sigma\, P_z = 0 \;\uparrow+$$

$$5 + Z_{ad} + Z_{ab} = 0$$

$$Z_{ab} = -5 - 3.75 = -8.75^k$$

$$X_{ab} = Z_{ab}\left(\frac{x}{z}\right)_{ab} = (-8.75)\left(\frac{12}{12}\right) = -8.75^k$$

$$Y_{ab} = Z_{ab}\left(\frac{y}{z}\right)_{ab} = (-8.75)\left(\frac{0}{12}\right) = 0 \tag{8.5}$$

$$F_{ab} = Z_{ab}\left(\frac{L}{z}\right)_{ab} = (-8.75)\left(\frac{17.0}{12}\right) = -12.40^k$$

$$\Sigma\, P_x = 0 \;\xrightarrow{+}$$

$$X_{ad} + X_{ab} + X_{ac} = 0$$

$$X_{ac} = -3.75 + 8.75 = +5.00^k$$

$$Y_{ac} = X_{ac}\left(\frac{y}{x}\right)_{ac} = (+5.00)\left(\frac{0}{24}\right) = 0$$

$$Z_{ac} = X_{ac}\left(\frac{z}{x}\right)_{ac} = (+5.00)\left(\frac{0}{24}\right) = 0 \tag{8.5}$$

$$F_{ac} = X_{ac}\left(\frac{L}{x}\right)_{ac} = (+5.00)\left(\frac{24}{24}\right) = +5.00^k$$

These results are added to the table for members ab and ac.

Isolate Joint b as *FBD:*

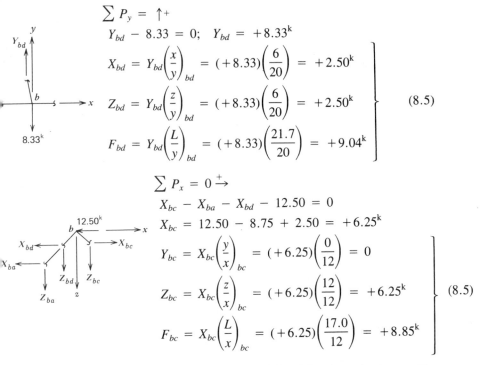

$$\Sigma P_y = \uparrow +$$

$$Y_{bd} - 8.33 = 0; \quad Y_{bd} = +8.33^k$$

$$\left.\begin{array}{l}X_{bd} = Y_{bd}\left(\dfrac{x}{y}\right)_{bd} = (+8.33)\left(\dfrac{6}{20}\right) = +2.50^k \\[2mm] Z_{bd} = Y_{bd}\left(\dfrac{z}{y}\right)_{bd} = (+8.33)\left(\dfrac{6}{20}\right) = +2.50^k \\[2mm] F_{bd} = Y_{bd}\left(\dfrac{L}{y}\right)_{bd} = (+8.33)\left(\dfrac{21.7}{20}\right) = +9.04^k\end{array}\right\} \quad (8.5)$$

$$\Sigma P_x = 0 \xrightarrow{+}$$

$$X_{bc} - X_{ba} - X_{bd} - 12.50 = 0$$

$$X_{bc} = 12.50 - 8.75 + 2.50 = +6.25^k$$

$$\left.\begin{array}{l}Y_{bc} = X_{bc}\left(\dfrac{y}{x}\right)_{bc} = (+6.25)\left(\dfrac{0}{12}\right) = 0 \\[2mm] Z_{bc} = X_{bc}\left(\dfrac{z}{x}\right)_{bc} = (+6.25)\left(\dfrac{12}{12}\right) = +6.25^k \\[2mm] F_{bc} = X_{bc}\left(\dfrac{L}{x}\right)_{bc} = (+6.25)\left(\dfrac{17.0}{12}\right) = +8.85^k\end{array}\right\} \quad (8.5)$$

These results are added to the table for members bd and bc. As a check, consider

$$\Sigma P_z = 0 \downarrow +$$

$$Z_{ba} + Z_{bd} + Z_{bc} = 0$$

$$-8.75 + 2.50 + 6.25 = 0$$

Isolate Joint c as *FBD:*

$$\Sigma P_y = 0 \uparrow +$$

$$Y_{cd} + 20.83 = 0; \quad Y_{cd} = -20.83^k$$

$$\left.\begin{array}{l}X_{cd} = Y_{cd}\left(\dfrac{x}{y}\right)_{cd} = (-20.83)\left(\dfrac{18}{20}\right) = -18.75^k \\[2mm] Z_{cd} = Y_{cd}\left(\dfrac{z}{y}\right)_{cd} = (-20.83)\left(\dfrac{6}{20}\right) = -6.25^k \\[2mm] F_{cd} = Y_{cd}\left(\dfrac{L}{y}\right)_{cd} = (-20.83)\left(\dfrac{27.6}{20}\right) = -28.75^k\end{array}\right\} \quad (8.5)$$

These results are added to the table for member cd. As a check, consider

$$\Sigma P_x = 0 \xrightarrow{+}$$

$$-X_{ca} - X_{cd} - X_{cb} - 7.5 = 0$$

$$-5.00 + 18.75 - 6.25 - 7.5 = 0$$

Also, as a check, consider $\sum P_z = 0 \uparrow +$

$$Z_{cd} + Z_{cb} = 0$$
$$-6.25 + 6.25 = 0$$

Isolate Joint d as *FBD*:

Here, we check $\sum P_x = 0$, $\sum P_y = 0$, $\sum P_z = 0$ and find that these conditions are satisfied.

8.10.2 Example problem Determine the reactions and member forces for the structure and loading shown. The support mechanisms include ball supports at c and d, a roller at a, and a universal joint at b.

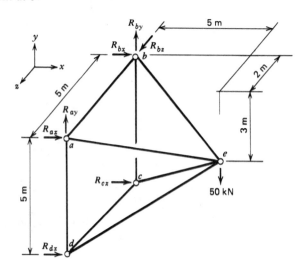

Structure Classification:

External: $r_a = 7 > 6$; thus, structure is statically indeterminate externally.

Overall: $j = 5$; $r_a = 7$; $m_a = 8$; $m = 3j - r_a = 15 - 7 = 8$. $m_a = m$; thus, structure is statically determinate. Arrangement is stable.

Analysis: One could write three equations of equilibrium at each joint and solve the resulting 15 equations for the 15 unknowns (7 reactions and 8 member forces). This is essentially the procedure used in the matrix method presented in Section 8.11. For hand calculations, we proceed as follows:

Reactions: Proceed as if R_{ax} were known, and solve for all other reaction components in terms of R_{ax}.

$\sum M_z = 0$
$$-(R_{ax} \times 5) - (R_{bx} \times 5) - (50 \times 5) = 0$$
$$R_{bx} = -(50 + R_{ax})$$

$\sum M_y = 0$
$$-(R_{bx} \times 5) - (R_{cx} \times 5) = 0$$
$$R_{cx} = -R_{bx} = 50 + R_{ax}$$

$$R_{ax} + R_{bx} + R_{cx} + R_{dx} = 0$$

$\sum P_x = 0 \quad \xrightarrow{+}$

$$R_{ax} - (50 + R_{ax}) + (50 + R_{ax}) + R_{dx} = 0$$

$$R_{dx} = -R_{ax}$$

$\sum M_x = 0$

$$(50 \times 2) - (R_{ay} \times 5) = 0; \quad R_{ay} = 20 \text{ kN}$$

$\sum P_z = 0$

$$R_{bz} = 0$$

$\sum P_y = 0$

$$R_{ay} + R_{by} - 50 = 0; \quad R_{by} = 50 - 20 = 30 \text{ kN}$$

The actual value of R_{ax} is determined by establishing the force in member ab at joints a and b and equating the results.

Member Forces:

Member	Geometry (m)				Force (kN)			
	x	y	z	L	X	Y	Z	F
ab	0	0	5.0	5.0	0	0	-12.0	-12.0
bc	0	5.0	0	5.0	0	$+12.0$	0	$+12.0$
cd	0	0	5.0	5.0	0	0	$+12.0$	$+12.0$
da	0	5.0	0	5.0	0	$+8.0$	0	$+8.0$
ae	5.0	3.0	3.0	6.56	$+20.0$	$+12.0$	$+12.0$	$+26.24$
be	5.0	3.0	2.0	6.16	$+30.0$	$+18.0$	$+12.0$	$+36.96$
ce	5.0	2.0	2.0	5.74	-30.0	-12.0	-12.0	-34.44
de	5.0	2.0	3.0	6.16	-20.0	-8.0	-12.0	-24.64

Joint a: $\quad X_{ae} = -R_{ax}; \quad Z_{ae} = X_{ae}\left(\dfrac{z}{x}\right)_{ae} = (-R_{ax})\left(\dfrac{3}{5}\right) = -0.6R_{ax}$

$$Z_{ab} = -Z_{ae} = 0.6R_{ax}$$

Joint b: $\quad X_{be} = -R_{bx} = (50 + R_{ax})$

$$Z_{be} = X_{be}\left(\dfrac{z}{x}\right)_{be} = (50 + R_{ax})\left(\dfrac{2}{5}\right) = 20 + 0.4R_{ax}$$

$$Z_{ba} = -Z_{be} = -(20 + 0.4R_{ax})$$

but

$$Z_{ab} = Z_{ba}; \quad 0.6R_{ax} = -20 - 0.4R_{ax}; \quad R_{ax} = -20 \text{ kN}$$

The reactions can now be summarized:

$$R_{ax} = -20 \text{ kN}$$
$$R_{bx} = -(50 + R_{ax}) = -30 \text{ kN}$$
$$R_{cx} = (50 + R_{ax}) = +30 \text{ kN}$$
$$R_{dx} = -R_{ax} = +20 \text{ kN}$$
$$R_{ay} = +20 \text{ kN}$$
$$R_{bz} = 0$$
$$R_{by} = +30 \text{ kN}$$

Continuing with the member forces:

$Z_{ab} = 0.6 R_{ax} = -12$ kN (X_{ab}, Y_{ab}, F_{ab} in table from Eq. 8.5)

$X_{ae} = -R_{ax} = -(-20) = +20$ kN (Y_{ae}, Z_{ae}, F_{ae} in table from Eq. 8.5)

$X_{be} = (50 + R_{ax}) = (50 - 20) = +30$ kN (Y_{be}, Z_{be}, F_{be} in table from Eq. 8.5)

Joint d:

$X_{de} = -R_{dx} = -20$ kN (Y_{de}, Z_{de}, F_{de} in table from Eq. 8.5)

$Z_{dc} = -Z_{de} = -(-12) = +12$ kN (X_{dc}, Y_{dc}, F_{dc} in table from Eq. 8.5)

$Y_{da} = -Y_{de} = -(-8) = +8$ kN (X_{da}, Z_{da}, F_{da} in table from Eq. 8.5)

Joint c:

$X_{ce} = -R_{cx} = -30$ kN (Y_{ce}, Z_{ce}, F_{ce} in table from Eq. 8.5)

$Y_{cb} = -Y_{ce} = -(-12) = +12$ kN (X_{cb}, Z_{cb}, F_{cb} in table from Eq. 8.5)

8.10.3 Example problem   Determine the reactions and member forces for the structure and loading shown. Each support is provided by a roller with the restraints shown.

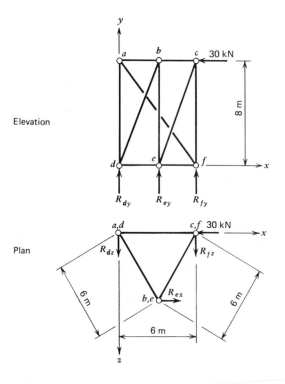

Structure Classification:

External: $r_a = 6$; thus, structure is statically determinate externally. Arrangement is stable.

Overall: $j = 6$; $r_a = 6$; $m_a = 12$; $m = 3j - r_a = 18 - 6 = 12$. $m_a = m$; thus, structure is statically determinate. Arrangement is stable.

Reactions:

$$\sum M_x = 0 \quad d \xrightarrow{\quad} f \quad R_{ey} = 0$$

$$\sum M_z = 0 \quad +\quad (30 \times 8) + (R_{ey} \times 3) + (R_{fy} \times 6) = 0$$

$$R_{fy} = \frac{-(30 \times 8)}{6} = -40 \text{ kN}$$

$$\sum P_y = 0 \quad \uparrow_+ \quad R_{dy} + R_{ey} + R_{fy} = 0; \quad R_{dy} = +40 \text{ kN}$$

$$\sum P_x = 0 \quad \xrightarrow{+} \quad R_{ex} - 30 = 0; \quad R_{ex} = +30 \text{ kN}$$

$$\sum M_y = 0 \quad + \quad R_{ex} \times 6\left(\frac{\sqrt{3}}{2}\right) - R_{fz} \times 6 = 0$$

$$R_{fz} = 30\left(\frac{\sqrt{3}}{2}\right) = +25.98 \text{ kN}$$

$$\sum P_z = 0 \quad R_{dz} + R_{fz} = 0; \quad R_{dz} = -R_{fz} = -25.98 \text{ kN}$$

Member Forces:

| Member | Geometry (m) | | | | Force (kN) | | | |
	x	y	z	L	X	Y	Z	F
ab	3.0	0	5.2	6.0	0	0	0	0
bc	3.0	0	5.2	6.0	0	0	0	0
ca	6.0	0	0	6.0	−30.0	0	0	−30.0
de	3.0	0	5.2	6.0	+15.0	0	+25.98	+30.0
ef	3.0	0	5.2	6.0	−15.0	0	−25.98	−30.0
fd	6.0	0	0	6.0	−15.0	0	0	−15.0
ad	0	8.0	0	8.0	0	−40.0	0	−40.0
be	0	8.0	0	8.0	0	0	0	0
cf	0	8.0	0	8.0	0	0	0	0
bd	3.0	8.0	5.2	10.0	0	0	0	0
ce	3.0	8.0	5.2	10.0	0	0	0	0
af	6.0	8.0	0	10.0	+30.0	+40.0	0	+50.0

To assist in visualizing the analysis, much of which is done by inspection, it is suggested that the student cut a piece of paper and label, fold, and orient it as shown by the following.

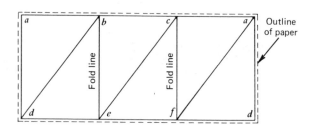

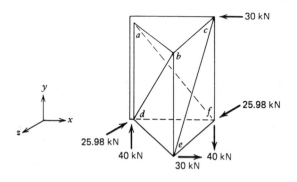

The analysis proceeds as follows:

Joint b: $F_{bc} = 0$ by Theorem 2; $X_{bc} = Y_{bc} = Z_{bc} = 0$

Joint a: $F_{ab} = 0$ by Theorem 2; $X_{ab} = Y_{ab} = Z_{ab} = 0$

Joint b: $F_{bd} = F_{be} = 0$ by Theorem 4; $X_{bd} = Y_{bd} = Z_{bd} = 0$
$$X_{be} = Y_{be} = Z_{be} = 0$$

Joint c: $\sum P_z = 0$; $Z_{ce} = 0$; $X_{ce} = Y_{ce} = F_{ce} = 0$

$\sum P_x = 0$; $X_{ca} = -30$ kN; Y_{ca}, Z_{ca}, F_{ca} from Eq. 8.5

$\sum P_y = 0$; $Y_{cf} = 0$; $X_{cf} = Z_{cf} = F_{cf} = 0$

Joint e: $\sum P_y = 0$;   verify $Y_{ec} = 0$

Joint a: $\sum P_x = 0$; $X_{af} = +30$ kN; Y_{af}, Z_{af}, F_{af} from Eq. 8.5

$\sum P_y = 0$; $Y_{ad} = -40$ kN; X_{ad}, Z_{ad}, F_{ad} from Eq. 8.5

Joint f: $\sum P_y = 0$;   verify $Y_{af} = +40$ kN

$\sum P_z = 0$; $Z_{fe} = -25.98$ kN; X_{fe}, Y_{fe}, F_{fe} from Eq. 8.5

$\sum P_x = 0$; $X_{fd} = -15$ kN; Y_{fd}, Z_{fd}, F_{fd} from Eq. 8.5

Joint d: $\sum P_x = 0$; $X_{de} = +15$ kN; Y_{de}, Z_{de}, F_{de} from Eq. 8.5

$\sum P_y = 0$;   verify $Y_{da} = -40$ kN

$\sum P_z = 0$;   verify $Z_{de} = +25.98$ kN

Joint e: $\sum P_x = 0$ and $\sum P_z = 0$ as checks on F_{de} and F_{ef}

8.11

Matrix Formulation of Space Framework Analysis

A matrix method of analysis for statically determinate uni-connected space frames is based on the satisfaction of the equations of equilibrium at each joint of the structure. Each joint forms a concurrent force system for which three independent equations of equilibrium can be written. If the conditions of determinacy and stability developed in Section 8.9 are satisfied, then the method of joints approach

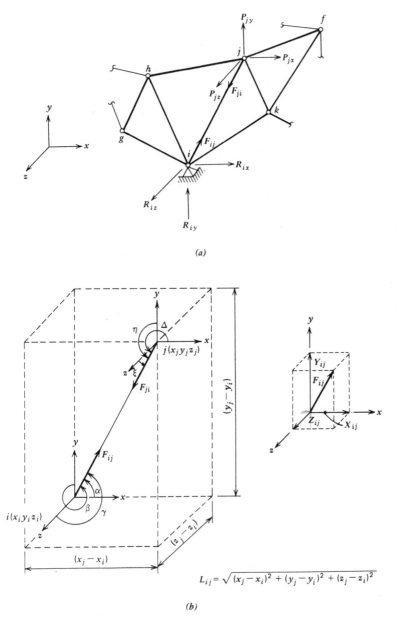

Fig. 8.8 *Generalized analysis of uni-connected space frame. (a) Section of statically determinate space frame. (b) Member* ij.

will provide the requisite number of equations for the determination of the unknown member forces and reactions.

For the general development of the method, consider the portion of a statically determinate frame shown in Fig. 8.8a. Joint i is a typical support point in
which the reaction R_i is shown to be resolved into its three components. Joint j is
a typical joint within the structure with the applied load P_j resolved into its three
components. The member force F_{ij} is shown to act in tension with the member
actions on joints i and j shown in the figure. All forces are shown to act in the
positive sense.

Figure 8.8b isolates member ij with the coordinates of the member ends given in terms of the structure coordinate system. At the i end of the member, the force F_{ij} is resolved into its components X_{ij}, Y_{ij}, and Z_{ij} in the x, y, and z directions, respectively. These can be expressed in the form

$$X_{ij} = F_{ij} \cos \alpha = F_{ij} \left(\frac{x_j - x_i}{L_{ij}} \right) = F_{ij} l_{ij}$$

$$Y_{ij} = F_{ij} \cos \beta = F_{ij} \left(\frac{y_j - y_i}{L_{ij}} \right) = F_{ij} m_{ij} \tag{8.12}$$

$$Z = F_{ij} \cos \gamma = F_{ij} \left(\frac{z_j - z_i}{L_{ij}} \right) = F_{ij} n_{ij}$$

In these expressions, the quantities l_{ij}, m_{ij}, and n_{ij} are the direction cosines, which define the angles between the direction of member ij and the x, y, and z axes, respectively. At end j, the force F_{ji} is similarly resolved into its components X_{ji}, Y_{ji}, and Z_{ji}, and these are given by

$$X_{ji} = F_{ji} \cos \Delta = F_{ji} \left(\frac{x_i - x_j}{L_{ij}} \right) = F_{ji} l_{ji}$$

$$Y_{ji} = F_{ji} \cos \eta = F_{ji} \left(\frac{y_i - y_j}{L_{ij}} \right) = F_{ji} m_{ji} \tag{8.13}$$

$$Z_{ji} = F_{ji} \cos \xi = F_{ji} \left(\frac{z_i - z_j}{L_{ij}} \right) = F_{ji} n_{ji}$$

where l_{ji}, m_{ji}, and n_{ji} are the direction cosines for member ji.

An examination of Eqs. 8.12 and 8.13 reveals that $l_{ij} = -l_{ji}$, $m_{ij} = -m_{ji}$, and $n_{ij} = -n_{ji}$. Also, since the member is in pure tension, $F_{ij} = F_{ji}$, and thus we have

$$X_{ji} = F_{ij} \left(-l_{ij} \right)$$
$$Y_{ji} = F_{ij} \left(-m_{ij} \right) \tag{8.14}$$
$$Z_{ji} = F_{ij} \left(-n_{ij} \right)$$

When the equations of equilibrium are applied to joints i and j, we obtain

$$X_{ig} + X_{ih} + X_{ij} + X_{ik} + R_{ix} = 0$$
$$Y_{ig} + Y_{ih} + Y_{ij} + Y_{ik} + R_{iy} = 0$$
$$Z_{ig} + Z_{ih} + Z_{ij} + Z_{ik} + R_{iz} = 0$$
$$X_{jh} + X_{ji} + X_{jk} + X_{jf} + P_{jx} = 0 \tag{8.15}$$
$$Y_{jh} + Y_{ji} + Y_{jk} + Y_{jf} + P_{jy} = 0$$
$$Z_{jh} + Z_{ji} + Z_{jk} + Z_{jf} + P_{jz} = 0$$

Or, using the form suggested by Eqs. 8.12 and 8.14, we can rewrite Eq. 8.15 in the form

$$F_{ig}l_{ig} + F_{ih}l_{ih} + F_{ij}l_{ij} + F_{ik}l_{ik} + R_{ix} = 0$$
$$F_{ig}m_{ig} + F_{ih}m_{ih} + F_{ij}m_{ij} + F_{ik}m_{ik} + R_{iy} = 0$$
$$F_{ig}n_{ig} + F_{ih}n_{ih} + F_{ij}n_{ij} + F_{ik}n_{ik} + R_{iz} = 0 \qquad (8.16)$$
$$F_{jh}l_{jh} - F_{ij}l_{ij} + F_{jk}l_{jk} + F_{jf}l_{jf} + P_{jx} = 0$$
$$F_{jh}m_{jh} - F_{ij}m_{ij} + F_{jk}m_{jk} + F_{jf}m_{jf} + P_{jy} = 0$$
$$F_{jh}n_{jh} - F_{ij}n_{ij} + F_{jk}n_{jk} + F_{jf}n_{jf} + P_{jz} = 0$$

For a structure that has a total of n joints, there are $3n$ equations of equilibrium that can be written in the following matrix form:

$$
\begin{bmatrix}
\cdot & \cdot & \cdot & \cdot & & \cdot & & \cdot & & & \cdot & & & & & & \cdot & & & \cdot \\
\cdots & -l_{ig} & -l_{ih} & -l_{ik} & -l_{ij} & \cdot & \cdot & & & \cdots & -1 & \cdot & \cdot & \cdot & \cdots \\
\cdots & -m_{ig} & -m_{ih} & -m_{ik} & -m_{ij} & \cdot & \cdot & & & & & -1 & \cdot & \cdot \\
\cdots & -n_{ig} & -n_{ih} & -n_{ik} & -n_{ij} & \cdot & \cdot & & & & \cdot & & -1 & \cdot & \cdot \\
\cdots & \cdot & \cdot & \cdot & l_{ij} & -l_{jh} & -l_{jk} & -l_{jf} & \cdot & \cdot & & & & \cdot & \cdots \\
\cdots & \cdot & \cdot & \cdot & m_{ij} & -m_{jh} & -m_{jk} & -m_{jf} & \cdot & \cdot & & & & \cdot & \cdots \\
\cdots & \cdot & \cdot & \cdot & n_{ij} & -n_{jh} & -n_{jk} & -n_{jh} & \cdot & \cdot & & & & \cdot & \cdots \\
\end{bmatrix}
\begin{Bmatrix}
\cdot \\
F_{ig} \\
F_{ih} \\
F_{ik} \\
F_{ij} \\
F_{jh} \\
F_{jk} \\
F_{jf} \\
\cdot \\
R_{ix} \\
R_{iy} \\
R_{iz} \\
\cdot
\end{Bmatrix}
=
\begin{Bmatrix}
\cdot \\
\\
\\
\\
P_{jx} \\
P_{jy} \\
P_{jz} \\
\\
\\
\\
\end{Bmatrix}
\qquad (8.17)
$$

$$3n \times 3n \qquad 3n \times 1 \qquad 3n \times 1$$

This equation clearly shows that a total of $3n$ unknown member forces and reactions can be determined from the solution of $3n$ simultaneous equations. If there are more than $3n$ unknowns, the system is statically indeterminate. If there are fewer than $3n$ unknowns, there is not a unique solution. As was true with planar trusses, this latter case corresponds to an unstable structure in which there are not enough members and reactions to satisfy all of the equilibrium requirements.

In an abbreviated matrix form, Eq. 8.17 can be written as

$$[C] \begin{Bmatrix} \{F\} \\ \{R\} \end{Bmatrix} = \{P\} \qquad (8.18)$$

in which $[C]$ is the coefficient matrix of direction cosines or the overall statics matrix, $\{F\}$ and $\{R\}$ are the vectors of unknown member forces and reaction components, respectively, and $\{P\}$ is the load vector, which includes the components of the applied forces. In a condensed form, Eq. 8.18 may be written as

$$[C]\{Q\} = \{P\} \qquad (8.19)$$

where $\{Q\}$ is the vector of unknown member forces and reactions. The solution is then given by

$$\{Q\} = [C]^{-1}\{P\} = [b]_{QP}\{P\} \tag{8.20}$$

where $[b]_{QP}$ relates the member forces and reactions to the applied forces.

It is recalled from Section 8.9 that a structure can be geometrically unstable if it has the requisite number of members but if these members are not stably arranged. For such structures, which are of critical form, the $[C]$ matrix does not possess an inverse and thus the solution expressed by Eq. 8.20 is not defined.

A modification of the above procedure is based on a subtle variation that is credited to Southwell. According to this technique, Eq. 8.12 is rewritten in the form

$$X_{ij} = \frac{F_{ij}}{L_{ij}}(x_j - x_i) = T_{ij} \cdot \Delta x_{ij}$$

$$Y_{ij} = \frac{F_{ij}}{L_{ij}}(y_j - y_i) = T_{ij} \cdot \Delta y_{ij} \tag{8.21}$$

$$Z_{ij} = \frac{F_{ij}}{L_{ij}}(z_j - z_i) = T_{ij} \cdot \Delta z_{ij}$$

Here, T_{ij} is referred to as the tension coefficient or force coefficient for member ij—that is, the force per unit of length for the member. The quantities Δx_{ij}, Δy_{ij}, and Δz_{ij} are the directed projections of members ij along the three coordinate axes, respectively. If this approach is used, Eq. 8.19 is still valid, but $[C]$ contains the directed member projections and $\{Q\}$ contains the tension coefficients. The solution, which is given by Eq. 8.20, leads to the tension coefficients, and the final member forces result from multiplying these tension coefficients by the corresponding member lengths. Although this method is completely equivalent to the method that employs the direction cosines, it has the advantage that the $[C]$ matrix is easier to generate and the member lengths are used only at the end to obtain the member forces.

It is stressed that the methods given here are for statically determinate structures in which the entire formulation is based on statics. Later chapters will present matrix approaches for statically indeterminate structures that will take into account the deformation characteristics of the individual members. It turns out that these methods will be equally applicable to statically determinate structures, and they will have the advantage of providing the joint displacements as well as the member forces.

8.11.1 Example problem Determine the reactions and the member forces for the structure and loading of Example 8.10.2. The structure is determinate and stable as shown earlier. The figure shows the structure, the joint coordinates, and the assumed directions of the reactions and member forces.

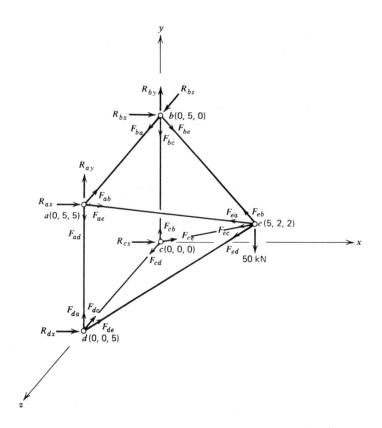

Direction Cosines:

$$l_{ij} = \frac{x_j - x_i}{L_{ij}}; \quad m_{ij} = \frac{y_j - y_i}{L_{ij}}; \quad n_{ij} = \frac{z_j - z_i}{L_{ij}}$$

$$L_{ij} = \sqrt{(x_j - x_i)^2 + (y_j - y_i)^2 + (z_j - z_i)^2}$$

Member ij	$(x_j - x_i)$ m	$(y_j - y_i)$ m	$(z_j - z_i)$ m	L_{ij} m	l_{ij}	m_{ij}	n_{ij}
ae	5.0	−3.0	−3.0	6.56	0.762	−0.457	−0.457
ea	−5.0	3.0	3.0	6.56	−0.762	0.457	0.457
be	5.0	−3.0	2.0	6.16	0.812	−0.487	0.325
eb	−5.0	3.0	−2.0	6.16	−0.812	0.487	−0.325
ce	5.0	2.0	2.0	5.74	0.871	0.348	0.348
ec	−5.0	−2.0	−2.0	5.74	−0.871	−0.348	−0.348
de	5.0	2.0	−3.0	6.16	0.812	0.325	−0.487
ed	−5.0	−2.0	3.0	6.16	−0.812	−0.325	0.487
ab	0	0	−5.0	5.00	0	0	−1.000
ba	0	0	5.0	5.00	0	0	1.000
bc	0	−5.0	0	5.00	0	−1.000	0
cb	0	5.0	0	5.00	0	1.000	0
cd	0	0	5.0	5.00	0	0	1.000
dc	0	0	−5.0	5.00	0	0	−1.000
da	0	5.0	0	5.00	0	1.000	0
ad	0	−5.0	0	5.00	0	−1.000	0

Equations of Equilibrium: The equations of equilibrium are written at each joint and expressed in the form of Eq. 8.19.

$$[C]\{Q\} = \{P\}$$

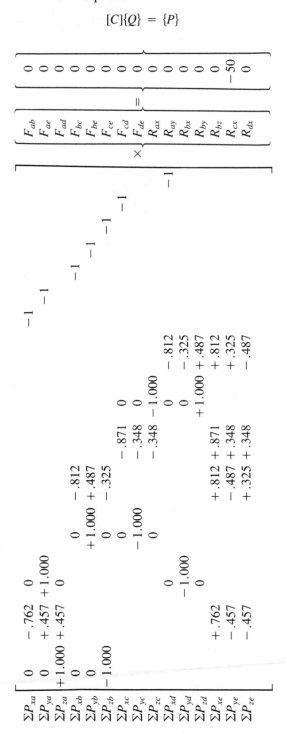

Solution for Member Forces and Reactions:

$$\{Q\} = [C]^{-1}\{P\} = \begin{Bmatrix} F_{ab} \\ F_{ae} \\ F_{ad} \\ F_{bc} \\ F_{be} \\ F_{ce} \\ F_{cd} \\ F_{de} \\ R_{ax} \\ R_{ay} \\ R_{bx} \\ R_{by} \\ R_{bz} \\ R_{cx} \\ R_{dx} \end{Bmatrix} = \begin{Bmatrix} -12.0 \\ +26.24 \\ +8.0 \\ +12.0 \\ +36.96 \\ -34.44 \\ +12.0 \\ -24.64 \\ -20.0 \\ +20.0 \\ -30.0 \\ +30.0 \\ 0 \\ +30.0 \\ +20.0 \end{Bmatrix} kN$$

These agree with the results of Example 8.10.2.

8.12

Complex Frameworks

As was the case with planar trusses, there are uni-connected space frameworks that are neither simple nor compound. For instance, consider the structure of Fig. 8.4c, which is redrawn in Fig. 8.9a along with the six reaction components that are needed for external stability. This structure is a simple framework that was constructed by starting with the basic tetrahedral unit *abcd* and systematically adding joints *e, f, g,* and *h* through the addition of three members to secure each

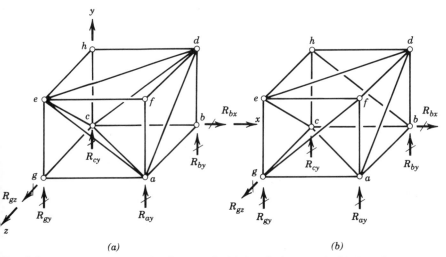

(a)   (b)

Fig. 8.9 *Formation of a complex framework.* (a) *Simple framework.* (b) *Complex framework.*

new joint. A count of the members, joints, and reaction components ($m_a = 18$, $j = 8$, $r_a = 6$) and the application of Eq. 8.6 ($m = 3 \times 8 - 6 = 18$) reveals that $m_a = m$, and therefore the structure is statically determinate.

An examination of the framework shows that the diagonals slope in opposite directions on opposing faces of the cube. If, however, members *ae* and *cd* are removed and are replaced by members *gf* and *hb*, the framework shown in Fig. 8.9*b* results. This structure remains stable and statically determinate, but it is neither simple nor compound. Such a structure is referred to as a *complex frame-work*.

It is noted again that the treatment given here is brief. Much more could be stated regarding special cases and more complicated structures. Such topics are beyond the scope of this treatment, and the reader is encouraged to examine the works cited in Section 8.17.

8.13

Deflections of Uni-Connected Space Frameworks

The virtual work method will be employed to determine framework deflections. The method evolves from a direct application of the procedures that were employed in determining planar truss deflections in Sections 5.6 and 5.7. According to this technique, the following two systems are employed:

P System: Virtual force system in equilibrium.

D System: Actual deformation pattern that is geometrically compatible.

The key relationship for a joint loaded framework is

$$(\delta P)_i D_i = \sum_{j=1}^{m} \left(\delta F_P \frac{F_D l}{EA} \right)_j \tag{8.22}$$

where D_i represents a desired deflection at point i corresponding to the actual loading on the framework, which causes the set of F_D member forces, and $(\delta P)_i$ is a virtual load that is introduced at the point and in the direction of the desired deflection, which introduces the virtual set of δF_P member forces. Of course, in the present context, the F_D and δF_P member forces are determined through two separate framework analyses. In the above summation, j ranges over the m members of the framework.

For cases where the deflection is caused by something other than joint loading, the required formula is

$$(\delta P)_i D_i = \sum_{j=1}^{m} [(\delta F_P) \cdot \Delta L_D]_j \tag{8.23}$$

where ΔL_D represents the imposed length changes of the members, such as might result from fabrication errors or temperature changes.

In applying either of the above expressions, it is computationally convenient to designate $(\delta P)_i$ as a unit load in determining D_i. Of course, a separate virtual loading must be employed for each deflection quantity that is desired.

8.13.1 Example problem

Determine the horizontal displacement in the x direction at point d for the structure and loading of Example 8.10.1. In addition to the data presented earlier, the following material and member properties are specified.

Cross-sectional areas:

Members *ab*, *bc*, and *ac*: $A = 5.0$ in.^{2}
Members *ad*, *bd*, and *cd*: $A = 8.0$ in.^{2}

$$E = 29 \times 10^3 \text{ ksi}$$

The following virtual work expression is used:

$$(\delta P)_i D_i = \sum_{j=1}^{m} \left(\delta F_P \frac{F_D l}{EA} \right)_j \tag{8.22}$$

Determination of Horizontal Displacement u_d.

D System: Actual loading system that includes the desired displacement, u_d. The corresponding F_D member forces were determined in Example 8.10.1 and are shown in the summary table that follows.

P System: Virtual force system with δP selected to obtain u_d. An analysis procedure similar to that outlined in Examples 8.5.1 and 8.10.1 yields the reactions and member forces given below, and the member forces are entered in the summary table.

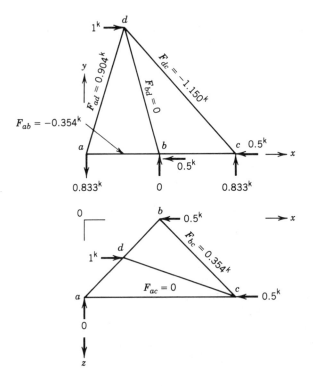

Summary Table:

Member	L in.	A in.^{2}	F_D kips	δF_P kips	$\delta F_P F_D l/A$ kips^2/in.
ab	204.0	5.0	−12.40	−0.354	179.10
bc	204.0	5.0	8.85	0.354	127.82
ac	288.0	5.0	5.00	0	0
ad	260.4	8.0	13.56	0.904	399.01
bd	260.4	8.0	9.04	0	0
cd	331.2	8.0	−28.75	−1.150	1,368.79

$$\sum \left(\delta F_P \frac{F_D l}{A} \right) = 2,074.72$$

Application of Eq. 8.22 yields

$$(1) \quad u_d = \frac{1}{E} \left(\delta F_P \frac{F_D l}{A} \right) = \frac{2,074.72}{29 \times 10^3} = 0.072 \text{ kip-in.}$$

from which

$$u_d = 0.072 \text{ in.}$$

The positive sign indicates that the deflection is in the positive x direction—the direction of the applied unit load.

8.14

Nonplanar Beam and Frame Structures

The emphasis in this chapter has been on uni-connected space frameworks, which represent a three-dimensional version of a planar truss. Of course, not all three-dimensional structures are composed of members whose ends are attached to universal joints. Here, a few brief observations are made concerning nonplanar beam and frame structures. In most instances, the analysis concepts are direct extensions of those employed in planar systems, but the member actions are more complicated, as will be explained in the next section.

We will limit discussion to statically determine structures. Therefore, the determination of the reactions is accomplished through a systematic application of the equations of equilibrium as expressed by Eqs. 10.1. It is thus clear that a statically determinate system has six independent reaction components, which must be arranged to ensure external stability. Additional reaction components are necessary for stability, and they are determinate, when conditions of construction allow for the expression of condition equations that would augment the equations of equilibrium.

The example problems that follow show some representative cases.

Compute the reactions for the structure shown below.

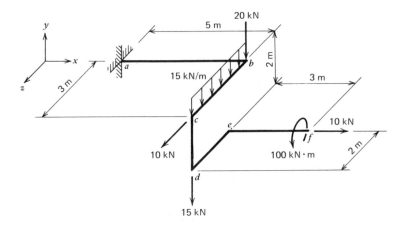

Free-Body Diagram: There are six unknown reactive components at point a.

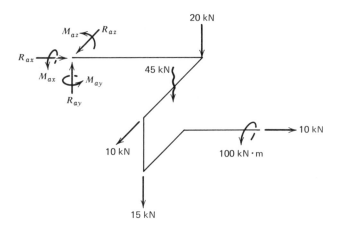

Determination of Reactions: Since $r_a = 6$ and the arrangement is stable, the structure is statically determinate. The reaction components are determined from the equations of equilibrium.

$\sum P_x = 0 \xrightarrow{+} R_{ax} + 10 = 0; \quad R_{ax} = \underline{\underline{-10 \text{ kN}}}$

$\sum P_y = 0 \uparrow_{+} \quad R_{ay} - 20 - 45 - 15 = 0; \quad R_{ay} = \underline{\underline{+80 \text{ kN}}}$

$\sum P_z = 0 \quad R_{az} + 10 = 0; \quad R_{az} = \underline{\underline{-10 \text{ kN}}}$

$\sum M_x = 0 \quad M_{ax} + (45 \times 1.5) + (15 \times 3) + 100 = 0$

$\qquad\qquad M_{ax} = \underline{\underline{-212.5 \text{ kN} \cdot \text{m}}}$

$\sum M_y = 0 \quad M_{ay} - (10 \times 5) + (10 \times \{3 - 2\}) = 0$

$\qquad\qquad M_{ay} = \underline{\underline{+40 \text{ kN} \cdot \text{m}}}$

$\sum M_z = 0 \quad M_{az} - (20 \times 5) - (45 \times 5) - (15 \times 5) + (10 \times 2) = 0$

$\qquad\qquad M_{az} = \underline{\underline{+380 \text{ kN} \cdot \text{m}}}$

8.14.2 Example problem Determine the reaction components for the structure shown. The admissible reaction components are shown symbolically.

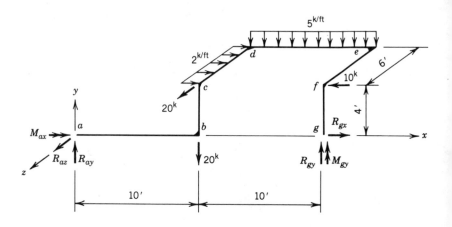

Free-Body Diagram: The uniformly distributed loads are replaced with equivalent concentrated loads.

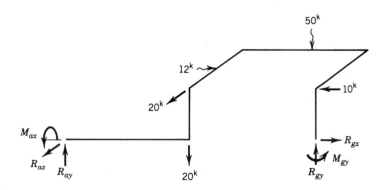

Determination of Reactions: Since $r_a = 6$ and the arrangement is stable, the structure is statically determinate. The reaction components are determined from the equations of equilibrium.

$\sum P_x = 0 \quad \xrightarrow{+} \quad R_{gx} + 12 - 10 = 0; \quad R_{gx} = \underline{\underline{-2^k}}$

$\sum P_z = 0 \quad \nearrow^+ \quad R_{az} + 20 = 0; \quad R_{az} = \underline{\underline{-20^k}}$

$\sum M_x = 0 \quad \overset{a}{\curvearrowright}_{\downarrow}^+ \quad M_{ax} + (20 \times 4) - (50 \times 6) = 0; \quad M_{ax} = \underline{\underline{220^{\text{ft-k}}}}$

$\sum M_z = 0 \quad \overset{a}{\underset{+}{\curvearrowright}} \quad \begin{array}{l} -(20 \times 10) - (12 \times 4) - (50 \times 15) + (10 \times 4) \\ + (R_{gy} \times 20) = 0 \end{array}$

$$R_{gy} = \underline{\underline{47.9^k}}$$

$\sum P_y = 0 \quad \uparrow^+ \quad R_{ay} - 20 - 50 + R_{gy} = 0; \quad R_{ay} = 70 - R_{gy} = \underline{\underline{22.1^k}}$

$\sum M_y = 0 \quad \overset{\curvearrowright}{\underset{g}{\raisebox{-1ex}{}}}_+ \quad M_{gy} - (12 \times 3) + (20 \times 10) + (R_{az} \times 20) = 0$

$$M_{gy} = -164 - 20R_{az} = \underline{\underline{+236^{\text{ft-k}}}}$$

A variety of boundary conditions can be employed to provide external stability for **8.14.3** a given structure. Consider again the structure of Example 8.14.2 but with full **Variations in** fixity at point a and complete freedom at point g as shown in Fig. 8.10a. Here, **boundary** the six reaction components at point a are determined by applying the six equilib- **conditions** rium equations in the fashion employed for Example 8.14.1.

If the restraint associated with M_{az} were to be removed, the structure would be unstable with respect to rotation about the z axis, but this can be restored by introducing either an M_{gz} restraint or an R_{gy} restraint at point g. The latter selection is shown in Fig. 8.10b. Similarly, if M_{ay} is removed, R_{gz} or M_{gy} is needed to

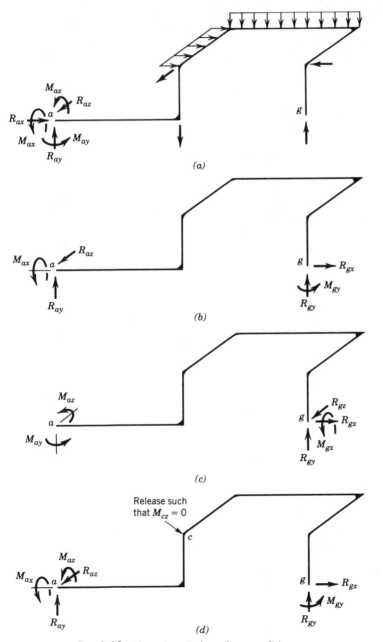

Fig. 8.10 *Alterations in boundary conditions.*

stabilize the system; M_{gy} is shown in Fig. 8.10b. If M_{ax} were to be removed, M_{gx} would have to be introduced; in this case, rotational stability about the x axis cannot be ensured by introducing a force reaction at point g. However, if R_{ax} is removed, translational stability can be ensured by introducing R_{gx} as shown in Fig. 8.10b. The set of reactions shown in Fig. 8.10b corresponds to those used in Example 8.14.2. Other boundary restraint trade-offs are possible; Fig. 8.10c provides yet another example. In each case, the reaction components must be arranged so as to preclude rigid-body instabilities.

Now, consider again the case shown in Fig. 8.10b; however, in this instance, incorporate a release at point c that requires the moment about the z axis to be zero. Since this restricts the capacity of the structure to carry moment about the z axis, the reaction component M_{az} must be added to ensure rotational stability. The solution now requires the six equilibrium equations and the condition equation that arises from taking a free-body diagram of segment ca and setting M_{cz} equal to zero.

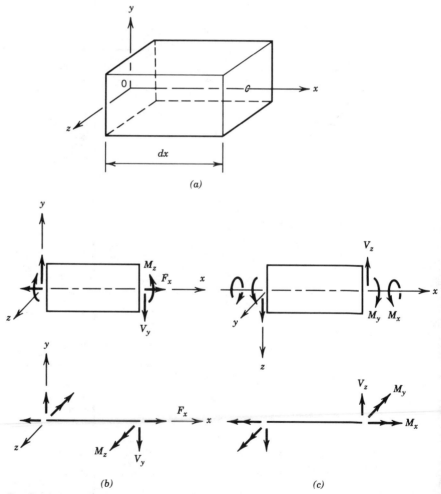

Fig. 8.11 *Notation and sign convention.* (a) *Member element.* (b) *Member forces for planar case—xy plane.* (c) *Additional member forces for nonplanar case—xz plane.*

Member Force Notation and Sign Convention

The member force notation for nonplanar beam and frame structures is a bit more complicated than the one employed for planar structures, which was described in Section 4.3; but it is a mere extension of that simpler case.

Consider the element shown in Fig. 8.11a. For planar structures, axial force was considered, as was flexure in the xy plane. Figure 8.11b shows the member forces associated with these modes of behavior. The member actions and notation are as they were in Section 4.3, with the addition of some clarifying subscripts: F_x is the axial force along the x axis, V_y is the shear directed along the y axis, and M_z is the bending moment about the z axis. For the nonplanar case, flexure must be added in the xz plane and torsion about the x axis must be included. The member forces associated with bending, the shear along the z axis, V_z, and the moment about the y axis, M_y, are shown in Fig. 8.11c, as is the moment about the x axis, M_x, which is the torsional moment.

With respect to sign convention, the general concepts considered in planar structures are carried forward. Positive member forces are as follows: Positive F_x produces tension on the element; positive M_z causes compression on the positive y fibers in the xy plane; positive V_y acts in the negative y direction on the right-exposed face and in the positive y direction of a left-exposed face; positive M_y causes compression on the positive z fibers in the xz plane; positive V_z acts in the negative z direction on a right-exposed face and in the positive z direction on the left-exposed face; and positive M_x acts according to the right-hand rule on the right-exposed face and in the opposite sense on the left-exposed face. All of the member forces shown on the element of Fig. 8.11 are positive, with vector representations for the moments shown according to the right-hand rule.

The sign convention described above lacks elegance and generality, but it will suffice for our present considerations. In the formal development of matrix methods in Chapters 17 and 18, a superior sign convention will be employed that has greater generality for three-dimensional structures.

Deflections for Nonplanar Beam and Frame Structures

Several methods were introduced in Chapter 6 for computing the deflections in planar beam and frame structures. Most of these methods focused on the deflections associated with flexural action because axial deformations are normally ignored in structures of this type. However, for nonplanar frame structures, the effects of axial and torsional deformations are normally included along with the flexural actions in two mutually perpendicular planes.

The most versatile method for the inclusion of the combined contributions of axial, torsional, and flexural deformations is the virtual work method. The specific application that uses virtual forces, which was introduced in Section 5.5.2, will be employed. The following two systems are used:

P System: Virtual force system in equilibrium.

D System: Actual deformation pattern that is geometrically compatible.

Using these systems, the external virtual work is equated to the internal virtual work of deformation. This leads to

$$\overset{\text{Virtual } P \text{ system}}{\sum_{i=1}^{n} (\delta P)_i D_i = \int_{\text{Vol}} (\delta \sigma_P) \epsilon_D \, d\text{Vol}} \tag{8.24}$$

Actual D system

The reader is directed to Section 5.5 for the detailed development of the above equation. However, it is recalled that D_i represents an actual deflection at point i that is geometrically compatible with the internal strain ϵ_D. The external virtual force, δP_i, is introduced at the point and the direction of D_i, and each of these forces contributes to the internal virtual stress, $\delta \sigma_P$. The summation ranges over the n external load points.

The right-hand side of Eq. 8.24 is composed of three portions—the internal virtual works associated with axial effects, flexural effects, and torsional effects.

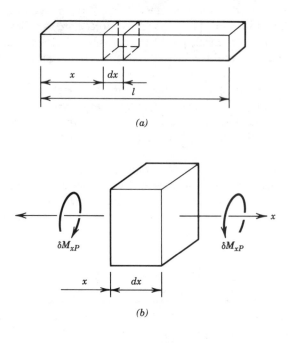

(a)

(b)

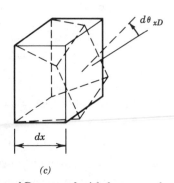

(c)

Fig. 8.12 *Virtual P system and actual D system for jth frame member.* (a) *jth member.* (b) *Virtual torsion at point x caused by P system of loads.* (c) *Twist of element dx caused by D system of deformations.*

The axial contribution has already been developed in Section 5.6. Likewise, the flexural portion was formulated in Section 6.9, and it must be applied in the present context for bending in both the xy and xz planes. The torsional internal virtual work is developed using the methods that were employed earlier in Sections 5.6 and 6.9. Reference to Fig. 8.12 shows that the portion of the right-hand side of Eq. 8.24 that is related to torsion can be expressed

$$\left(\int_{\text{Vol}} (\delta \sigma_P) \epsilon_D \, d\text{Vol} \right)_{\text{Torsion}} = \sum_{j=1}^{m} \left(\int_l (\delta M_{xP}) \, d\theta_{xD} \right)_j \qquad (8.25)$$

where l is the length of the jth member, which is one of a total of m members; δM_{xP} is the virtual torsion about the x axis at some point x on member j corresponding to the P system; and $d\theta_{xD}$ is the associated twisting rotation at the same point on member j that occurs over the differential element dx of the D system. From elementary mechanics, using procedures similar to those discussed in Sections 5.3 and 6.2 for axial and flexural deformations, respectively, we obtain

$$d\theta_{xD} = \frac{M_{xD} \, dx}{GJ} \qquad (8.26)$$

where M_{xD} is the torsion about the x axis at some point x on the jth member, G is the shear modulus, and J is a torsional section property. The shear modulus is given by

$$G = \frac{E}{2(1 + v)} \qquad (8.27)$$

where E is the modulus of elasticity and v is Poisson's ratio. The torsional property, J, depends on the nature of the cross section and is discussed in strength of materials texts. Thus, including all deformation contributions, Eq. 8.24 becomes

$$\sum_{i=1}^{n} (\delta P)_i D_i = \sum_{j=1}^{m} \left(\int_l \delta F_{xP} \frac{F_{xD} \, dx}{EA} + \int_l \delta M_{xP} \frac{M_{xD} \, dx}{GJ} \right.$$
$$\left. + \int_l \delta M_{yP} \frac{M_{yD} \, dx}{EI_y} + \int_l \delta M_{zP} \frac{M_{zD} \, dy}{EI_z} \right) \qquad (8.28)$$

In the above, the four terms on the right-hand side represent the four separate deformational contributions—axial, torsion, bending about the y axis, and bending about the z axis.

The notation for the axial and flexural components is more complicated than was evidenced in Eqs. 5.25 and 6.54 because of the need for axis subscripts.

Of course, as was pointed out earlier in the application of the virtual work method, Eq. 8.28 is used to find one displacement quantity at a time. For the displacement D_r, a single virtual load of $\delta P_r = 1$ is used. Since there are no other forces in the virtual system that couple with displacements of the D system, the desired deflection is isolated on the left-hand side of Eq. 8.28.

8.16.1 Example problem Determine the vertical deflection at point d for the structure shown.

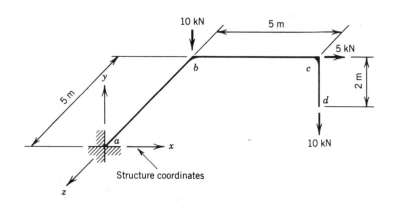

Structure coordinates

Member and Material Properties (in terms of member coordinates):

$$A = 5\ 000\ \text{mm}^2 = 5\ 000 \times 10^{-6}\ \text{m}^2$$
$$I_x = J = 1\ 000 \times 10^6\ \text{mm}^4 = 1\ 000 \times 10^{-6}\ \text{m}^4$$
$$I_y = I_z = 1\ 500 \times 10^6\ \text{mm}^4 = 1\ 500 \times 10^{-6}\ \text{m}^4$$
$$E = 200\ \text{GPa} = 200 \times 10^6\ \text{kN/m}^2$$
$$v = 0.3$$
$$G = E/2(1 + v) = 77 \times 10^6\ \text{kN/m}^2$$

Vertical Deflection at Point d:

D System (Actual Load System): Application of Eqs. 8.1 yields the reactions, and the member forces are shown on appropriate free-body diagrams.

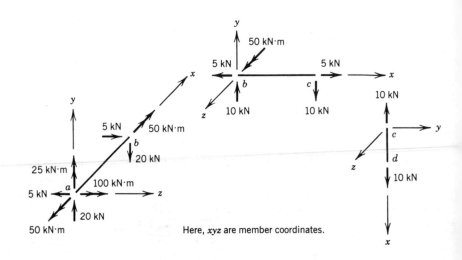

Here, xyz are member coordinates.

The member force diagrams for the D system are as follow:

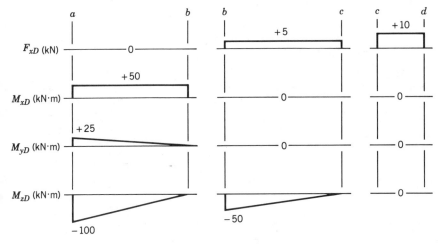

P System (Virtual Load System): The virtual load system is composed of a unit downward load at point d. The resulting member forces are shown on the individual free-body diagrams.

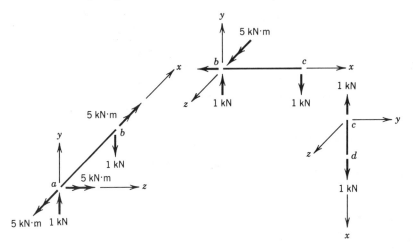

The member force diagrams for the P system are as follows:

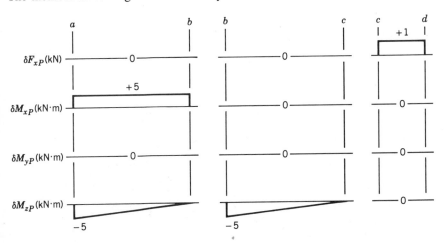

Deflection Calculation: Equation 8.28 will be used, and the following table gives a summary of the required member force expressions.

Region / Function	$a - b$ $0 < x < 5$	$b - c$ $0 < x < 5$	$c - d$ $0 < x < 2$
F_{xD}	0	+5	+10
M_{xD}	+50	0	0
M_{yD}	$+25 - 5x$	0	0
M_{zD}	$-100 + 20x$	$-50 + 10x$	0
δF_{xP}	0	0	+1
δM_{xP}	+5	0	0
δM_{yP}	0	0	0
δM_{zP}	$-5 + 1x$	$-5 + 1x$	0

The governing virtual expression is

$$\sum_{i=1}^{n} (\delta P)_i D_i = \sum_{j=1}^{n} \left(\int_l \delta F_{xP} \frac{F_{xD}\, dx}{EA} + \int_l \delta M_{xP} \frac{M_{xD}\, dx}{GJ} \right.$$
$$\left. + \int_l \delta M_{yP} \frac{M_{yD}\, dx}{EI_y} + \int_l \delta M_{zP} \frac{M_{zD}\, dx}{EI_z} \right)_j \quad (8.28)$$

Substitution from the table of member force quantities gives

$$(1)(v_d) = \frac{1}{EA}\int_0^2 (1)(10)\, dx + \frac{1}{GJ}\int_0^5 (5)(50)\, dx + \frac{1}{EI_z}\int_0^5 (-5 + x)(-100 + 20x)\, dx$$
$$+ \frac{1}{EI_z}\int_0^5 (-5 + x)(-50 + 10x)\, dx$$

$$= \frac{20}{EA} + \frac{1\,250}{GJ} + \frac{1\,250}{EI_z}$$

$$= \frac{20}{200 \times 10^6 \times 5\,000 \times 10^{-6}} + \frac{1\,250}{77 \times 10^6 \times 1\,000 \times 10^{-6}}$$
$$+ \frac{1\,250}{200 \times 10^6 \times 1\,500 \times 10^{-6}}$$

$$= 0.000\,020 + 0.016\,234 + 0.004\,167 \;(\text{kN} \cdot \text{m})$$

$$v_d = 0.020\,421 \text{ m (positive sign confirms that deflection is downward)}$$

8.17

Additional Reading

Kinney, J. S., *Indeterminate Structural Analysis*, Chapter 4, Addison–Wesley, Reading, Mass., 1959.

Michalos, J., and Wilson, E. N., *Structural Mechanics and Analysis*, Chapter 4, Macmillian, New York, 1965.

Norris, C. H., Wilbur, J. B., and Utku, S., *Elementary Structural Analysis,* Chapter 4, McGraw–Hill, New York, 1976.

Sack, Ronald L., *Structural Analysis,* Chapter 8, McGraw–Hill, New York, 1984.

Timoshenko, S. P., and Young, D. H., *Theory of Structures,* Chapter 4, McGraw–Hill, New York, 1965.

8.18

1 through 3. Determine the member forces in all members for each of the struc- **Suggested**
tures for the indicated loads.   **Problems**

1.

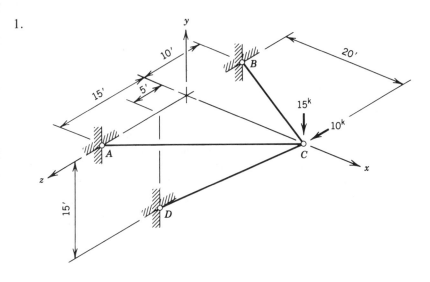

2.

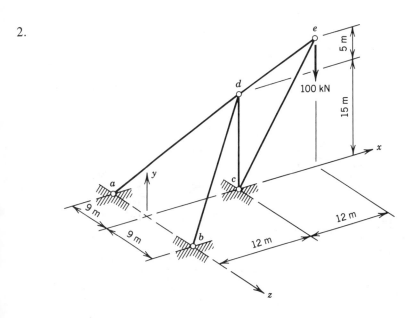

3.

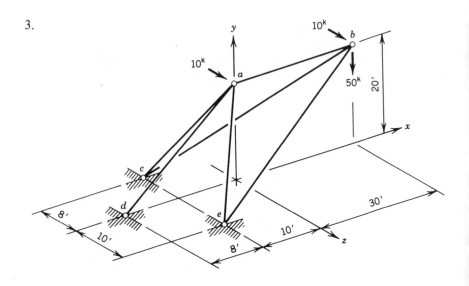

4 through 6. Determine the reaction components and all member forces for the given structures with the indicated loadings. In each case, classify the structure with respect to statical determinacy and stability before proceeding with the analysis. The symbol ↛ represents an unknown reaction component.

4.

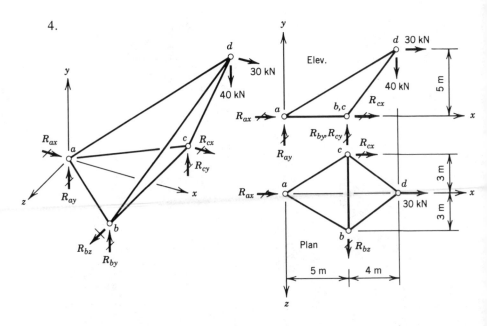

5.

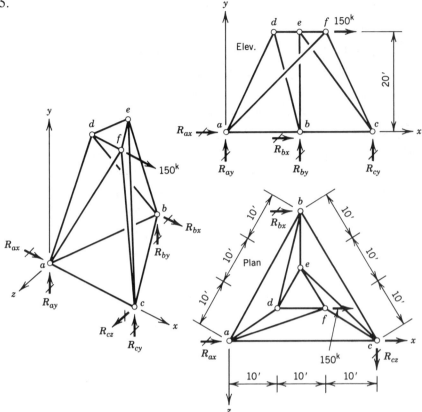

6.

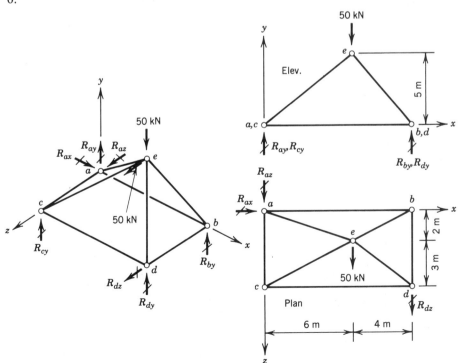

7 through 9. Set up each of the following in the form of $[C]\{Q\} = \{P\}$.

7. Structure and loading of Problem 3.

8. Structure and loading of Problem 4.

9. Structure and loading of Problem 6.

10. Write a computer program for the method of framework analysis developed in Section 8.11. Use the program to solve Problems 3, 4, 5, and 6.

11. Determine the vertical and horizontal deflections at point C for the structure and loading of Problem 1. The cross-sectional area for all members is 4 in.^{2}, and $E = 29 \times 10^3$ ksi.

12. Determine the vertical deflection under the applied load for Problem 2. The cross-sectional areas are given below, and $E = 200 \times 10^6$ kN/m^2.

$$\text{Members } ad, bd, de: \quad A = 1\,500 \text{ mm}^2$$
$$\text{Members } dc, ec: \quad A = 3\,000 \text{ mm}^2$$

13. Determine the deflections in the x and y directions at point e for the structure and loading of Problem 6. The cross-sectional areas are given below, and $E = 200 \times 10^6$ kN/m^2.

$$\text{Members } ae, be, ce, de: \quad A = 2\,000 \text{ mm}^2$$
$$\text{Members } ab, bd, dc, ca: \quad A = 1\,500 \text{ mm}^2$$

14. Determine the reactions for the structure shown.

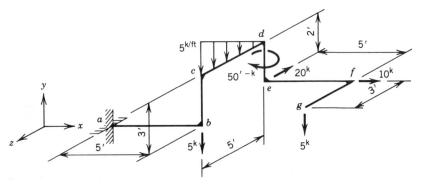

15. Compute the reaction for the grid structure shown. The symbol $\rightarrow$ represents an unknown reaction component.

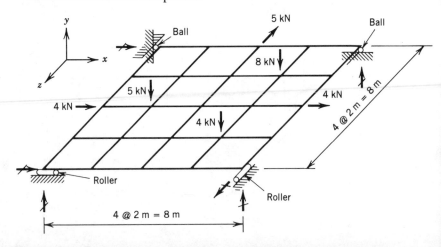

16 through 21. Determine the reaction components for each of the structures shown.

    →   represents an unknown force reaction.

    ↠   represents an unknown moment reaction.

16.

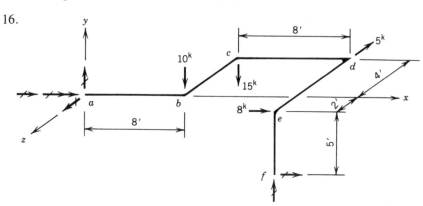

17. The same structure and loading as Problem 16, but with the altered set of boundary conditions shown.

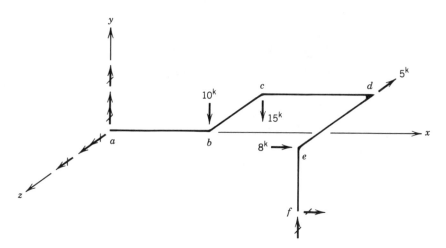

18. The same structure and loading as Problem 14, but with universal-type joint supports at points a and g. Explain the significance of your results.

19.

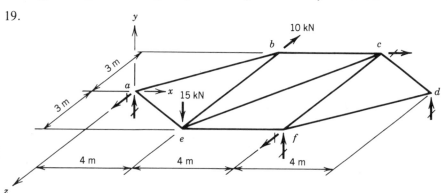

Note: Structure in xz plane

20.

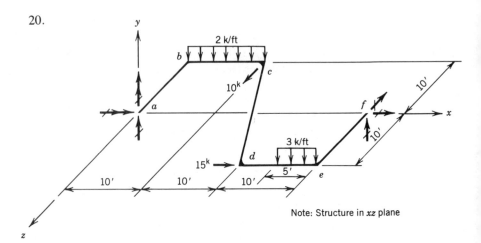

Note: Structure in *xz* plane

21. The same structure and loading as Problem 19, but with the altered set of boundary conditions shown.

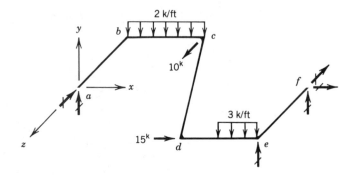

22. Determine the rotation about the *x* axis at point *b* and the vertical deflection at point *g* for the structure and loading of Problem 14. The following section and material properties apply to all members:

$$A = 4 \text{ in.}^2$$
$$I_x = 200 \text{ in.}^4$$
$$I_y = I_z = 500 \text{ in.}^4 \qquad \left(\begin{array}{c}\text{in terms of}\\ \text{member coordinates}\end{array}\right)$$
$$E = 29 \times 10^3 \text{ ksi}$$
$$\nu = 0.3$$

23. Determine the displacements in the *x*, *y*, and *z* directions at point *e* for the structure and loading of Problem 20. The following section and material properties apply to all members:

$$A = 6 \text{ in.}^2$$
$$I_x = 400 \text{ in.}^4$$
$$I_y = I_z = 1,000 \text{ in.}^4 \qquad \left(\begin{array}{c}\text{in terms of member coordinates;}\\ \text{member } y \text{ axis normal to structure } xz \text{ plane}\end{array}\right)$$
$$E = 29 \times 10^3 \text{ ksi}$$
$$\nu = 0.3$$

PART II

STATICALLY INDETERMINATE STRUCTURES

Curved box girder railroad bridge over Latah Creek, Spokane, Washington (courtesy of Bethlehem Steel Corporation).

FUNDAMENTALS OF STATICALLY INDETERMINATE ANALYSIS

9.1
Requirements and Limitations of Equilibrium

As was stated in Section 2.3, equilibrium refers to the condition of a structure when it is at rest and remains at rest during the application of loading. For equilibrium to be realized, there must be a balance of the force tendencies that would act to disrupt the structure in any way. For planar structures, equilibrium is ensured by satisfying the three equations of static equilibrium as they were expressed in Eq. 2.1.

It is to be emphasized that equilibrium is a requirement that must be satisfied. Thus, the analysis must lead to a set of reactions and internal forces that satisfy the equations of static equilibrium. If these equations are sufficient for the complete structural analysis, the structure is said to be *statically determinate*. However, if there are more independent reaction components and/or internal forces

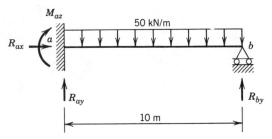

Fig. 9.1 *Statically indeterminate beam structure.*

than can be determined from the application of the equations of equilibrium, then the structure is said to be *statically indeterminate*. This does not imply that a solution does not exist. As long as the structure is stable, a solution will exist; however, the conditions of equilibrium are insufficient for completing the solution.

Consider the beam structure shown in Fig. 9.1. According to the criteria developed in Section 2.7, this structure is statically indeterminate since there is one

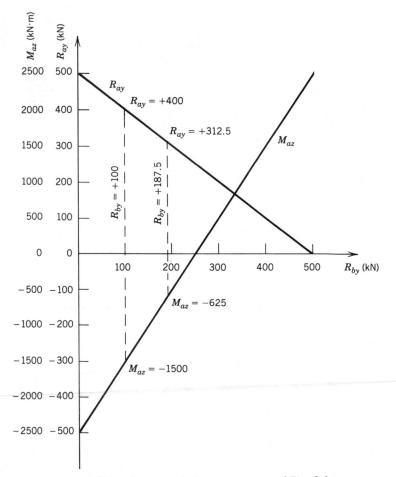

Fig. 9.2 *Equilibrium solutions to structure of Fig. 9.1.*

more reaction component than there are equations of equilibrium available for the solution. Application of these equations leads to the obvious result that $R_{ax} = 0$ and that

$$R_{ay} + R_{by} = 500 \qquad (9.1)$$
$$M_{az} - 10R_{by} = -2500$$

A unique solution for the reactions is not possible in this case because there are three unknown reactions and only two equations. However, equilibrium solutions can be determined by rewriting the equations in the form

$$R_{ay} = 500 - R_{by} \qquad (9.2)$$
$$M_{az} = -2500 + 10R_{by}$$

Here, for any assigned value at R_{by}, the two equations can be solved for R_{ay} and M_{az}. Figure 9.2 shows the results of several such solutions as R_{by} is allowed to take on values ranging from 0 kN to $+500$ kN. For instance, if $R_{by} = +100$ kN, then R_{ay} and M_{az} are $+400$ kN and -1500 kN · m, respectively. This plot shows that there is an infinite number of solutions—each resulting in values for R_{ay} and M_{az} that, along with the assigned value of R_{by}, satisfy the equilibrium requirements.

Thus, it can be concluded that equilibrium is a requirement that the solution must satisfy. However, there is a limitation in that equilibrium considerations alone give no clue regarding which one of the infinite array of possible equilibrium solutions is the correct one.

9.2 Static Indeterminacies; Redundancies

Before studying the detailed methods of analysis for statically indeterminate structures, it is essential that the criteria for the statical classification of structures be clearly understood. This topic has already been given careful consideration, and the reader is urged to review Sections 2.7, 2.10, 3.9, 4.12, 8.4, and 8.9. These sections develop the statical classification criteria for several different types of structures with regard to external, internal, and overall static determinacy and stability.

In essence, all of the criteria that are developed in these sections involve a comparison between the number of independent unknown force quantities and the number of independent equations of equilibrium that are available for the solution of the unknowns. The criteria always take the following forms:

1. If there are more equations than there are unknowns, the structure is statically unstable.
2. If there is the same number of equations as unknowns, the structure is statically determinate.
3. If there are fewer equations than unknowns, the structure is statically indeterminate.

The first of these criteria is absolute. The second and third criteria give conditions that are necessary, but not sufficient, for the statical classification of the structure. The unknown force quantities must be arranged so as to ensure the stability of the structure.

Thus, the criteria that have been developed indicate that a structure is indeterminate when there are more reaction components available and/or members

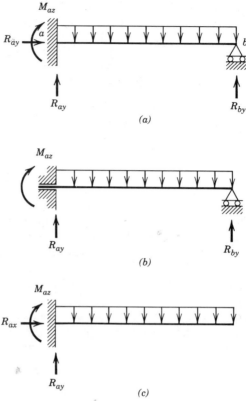

Fig. 9.3 *Selection of redundant reaction.* (a) *Statically indeterminate structure.*
(b) *Unstable and statically indeterminate primary structure.* (c) *Stable and statically
determinate primary structure.*

present than are necessary for the stability of the structure. The degree of external
indeterminacy is equal to the number of reaction components that are available in
excess of the number that is required for external stability. These excess reaction
components are called *redundants* because they are unnecessary for the stability
of the structure. The degree of internal indeterminacy is given by the number of
internal force components that are present in excess of those that are needed for
internal stability. These are also called *redundants* since they are not required for
a stable structure.

In certain methods of statically indeterminate analysis, it is necessary for analysts to identify explicitly the reaction components or internal force components
that they wish to select as the redundants. These are then conceptually removed
from the structure, and the statically determinate structure that remains is called
the *primary structure*. However, it is essential that the redundants be selected so
that the primary structure is stable.

As an example, consider the beam structure of Fig. 9.3a, which is the same
structure that was considered in the previous section. Since $r_a = 4 > 3$, the
structure is statically indeterminate externally to the first degree. Thus, there is
one redundant, or unnecessary, reaction component. If R_{ax} is selected as the redundant, the primary structure is as shown in Fig. 9.3b. This structure is both
unstable and indeterminate. However, if R_{by} is taken as the redundant, then the
primary structure is statically determinate and stable, as shown in Fig. 9.3c.

As was described in Chapter 2, compatibility places constraints on the displacements of a structure to ensure that the individual elements of the structure fit together properly and that the structure conforms to the displacement boundary conditions prescribed at the support points. Certain internal compatibility conditions are automatically satisfied through the formulation of the method of analysis. Other internal conditions are explicitly introduced whenever the structure is separated into free-body diagrams. Here, care must be taken to make certain that the equations that are formulated for the analysis retain the required compatibility at the points where the structure has been severed. External compatibility conditions require compliance between the displacements that the structure undergoes at its support points and the displacements that the support mechanisms permit to occur.

It is important to note that compatibility is a requirement that must be satisfied. Thus, the analysis must lead to a deformed configuration that will adhere to all of the deformation constraints imposed on the structure. However, in general, there is an infinite number of deformed patterns that will satisfy the compatibility conditions. Therefore, satisfaction of compatibility is not enough to ensure the correct solution.

As an example, consider again the beam of Section 9.1. At that point, the analysis could not be completed by statics alone because of the statically indeterminate nature of the structure. However, if the reaction component R_{by} is removed, as shown in Fig. 9.3c, the resulting primary structure is statically determinate. The reactions can now be determined from statics, and the deflections, which include a deflection of point b, Δ_{b1}, of $-62\,500/EI$ (downward)[1], can be determined by any of the methods presented in Chapter 6. Comparing this solution, which is shown in Fig. 9.4a, with the given structure of Fig. 9.1, we see that the designated boundary condition at point b is violated. That is, the vertical restraint required by the support at point b is not maintained.

To remedy this problem, we allow the primary structure to be acted upon by the redundant reaction R_{by}, as shown in Fig. 9.4b. The solution for this case includes a deflection of point b, Δ_{b2}, equal to $+333R_{by}/EI$ (upward)[1], which can also be determined by the methods of Chapter 6.

For a solution that includes the proper loading and also satisfies the designated boundary conditions at point b, the solutions shown in Figs. 9.4a and 9.4b must be superimposed so that the final vertical displacement at point b, Δ_b, is equal to zero. Thus, we have the compatibility equation

$$\Delta_b = \Delta_{b1} + \Delta_{b2} = 0 \tag{9.3}$$

or, upon substitution and rearrangement,

$$\frac{333R_{by}}{EI} = \frac{62\,500}{EI} \tag{9.4}$$

from which

$$R_{by} = +187.5 \text{ kN} \tag{9.5}$$

Returning to Eq. 9.2, and using $R_{by} = +187.5$ kN, we obtain $R_{ay} = 312.5$ kN and $M_{az} = -625$ kN $\cdot$ m, as shown in Fig. 9.2.

---

[1]Deflection in meters if EI has units of kN $\cdot$ m^2.

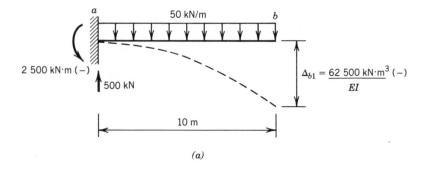

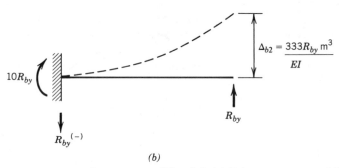

Fig. 9.4 *Primary structure for structure of Fig. 9.1.* (a) *Primary structure subjected to given loading.* (b) *Primary structure subjected to redundant reaction.*

In this example, it is clear that compatibility is a requirement that the final solution must satisfy. However, there is the limitation that compatibility considerations alone cannot extract the correct solution from the infinite array of deformed structures that satisfy the compatibility conditions.

The final solution clearly satisfies the requirements of equilibrium given in Eq. 9.2 and the requirements of compatibility stated as Eq. 9.5. These equations collectively form a set of three equations that must be solved simultaneously for the unknown reaction components.

9.4

Kinematic Indeterminacies; Redundancies

It is convenient at this point to introduce the notation of *kinematic indeterminacy* in a fashion that is analogous to the concept of static indeterminacy. As static indeterminacy deals with the number of force quantities that must be determined in order to render the equilibrium solution complete, kinematic indeterminacy refers to the number of displacement quantities (kinematic degrees of freedom) that are necessary to define the deformational response of the structure.

To illustrate the concept, consider the structure previously studied in Section 9.1. This structure is shown again in Fig. 9.5a along with a qualitative deflected shape. Since the structure is fixed at point a and vertically restrained at point b and the axial deformation is zero in this case, there is only one kinematic degree of freedom—θ_b, the rotation at point b. Therefore, the structure is kinematically indeterminate to the first degree.

Extending the analogy with static determinacy, we identify the rotation, θ_b,

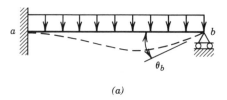

(a)

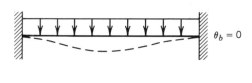

(b)

Fig. 9.5 *Selection of redundant displacement. (a) Kinematically indeterminate structure with redundant displacement, θ_b. (b) Kinematically determinate primary structure.*

as a *kinematic redundancy*, and if it is removed from the structure ($\theta_b = 0$), the resulting *primary structure* is said to be *kinematically determinate*. Figure 9.5b shows the kinematically determinate primary structure for our example case. It should be noted that a structure is rendered kinematically determinate by setting the displacements equal to zero or some predetermined values. The latter case would be appropriate, for example, if there were prescribed settlement displacements.

9.5

In Eq. 9.1, the conditions of equilibrium were applied to the structure of Fig. 9.1. **Alternative** These conditions were found to be inadequate in determining a unique solution, **Form of** and the compatibility condition of Eq. 9.3 was used to complete the solution. **Analysis**

As an alternative approach, let's begin with compatibility considerations. The structure of Fig. 9.1 is shown again in Fig. 9.6a. In this case, an end moment of M_{bz} has been added which must eventually be set equal to zero. It is clear that this structure is statically indeterminate. Next, the statically determinate primary structure of Fig. 9.3c is separately loaded with w, R_{by}, and M_{bz} as shown in the successive figures of Fig. 9.6b. The displacement quantities in these figures were determined by the methods of Chapter 6.

The separate solutions of Fig. 9.6b must be superimposed to obtain the correct boundary conditions for the given structure of Fig. 9.6a. Thus, we have

$$\Delta_{b1} + \Delta_{b2} + \Delta_{b3} = \Delta_b = 0 \tag{9.6}$$
$$\theta_{b1} + \theta_{b2} + \theta_{b3} = \theta_b$$

Substitution of the displacement quantities from Fig. 9.6b into Eq. 9.6 yields

$$\frac{R_{by}l^3}{3EI} + \frac{M_{bz}l^2}{2EI} = \frac{wl^4}{8EI} \tag{9.7}$$
$$\frac{R_{by}l^2}{2EI} + \frac{M_{bz}l}{EI} = \frac{wl^3}{6EI} + \theta_b$$

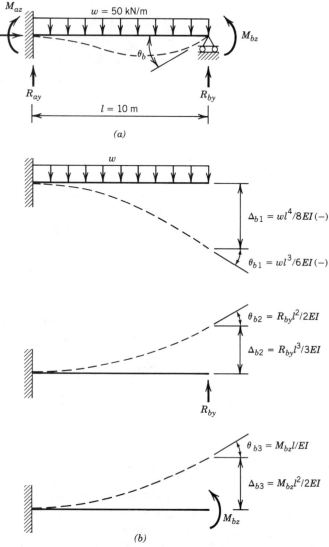

Fig. 9.6 *Statically indeterminate beam structure* (a) *Loading and deformation.* (b) *Statically determinate primary structure with individual loadings.*

Solving Eq. 9.7 for M_{bz} and R_{by} and then applying statics to the structure of Fig. 9.6a, we obtain

$$R_{ay} = \frac{wl}{2} + \frac{3k_b}{2l}\,\theta_b$$

$$M_{az} = -\frac{wl^2}{12} - \frac{k_b}{2}\,\theta_b$$

$$R_{by} = \frac{wl}{2} - \frac{3k_b}{2l}\,\theta_b$$

$$M_{bz} = -\frac{wl^2}{12} + k_b\theta_b$$

(9.8)

where $k_b = 4EI/l$, as defined in Section 6.2.

Equations 9.8 express all of the response quantities in terms of the single kinematic degree of freedom θ_b. The solution represented by these equations satisfies the required compatibility conditions and the overall equilibrium requirements for the structure. Figure 9.7 gives the member-end forces associated with a range of $EI\theta_b$ values from zero to 1 200 kN · m² for the structure shown in Fig. 9.6a. For example, for $EI\theta_b = 400$ kN · m², $R_{ay} = +274$ kN, $M_{az} = -496.7$ kN · m, $R_{by} = +226$ kN, and $M_{bz} = -256.7$ kN · M.

Equations 9.8 also provide the member-end forces associated with the kinematically determinate primary structure shown in Fig. 9.5b. With $\theta_b = 0$, these member-end forces have the values shown in Fig. 9.8a, and these are consistent with the values shown in Fig. 9.7 for $EI\theta_b = 0$. In lieu of the negative signs for M_{az} and M_{bz}, the moments are shown in the opposite directions from the positive directions of Fig. 9.6a. Although this solution is a valid equation solution, it violates the required force boundary condition on the moment at point b.

To remedy this problem, the primary structure is subjected to the redundant displacement θ_b. Based upon Eq. 6.13, and noting that $k_b = 4EI/l$, the associated end moments and reactions take on the values shown in Fig. 9.8b.

For a solution that reflects the proper loading and satisfies the designated force boundary conditions at point b, the solution shown in Figs. 9.8a and 9.8b must

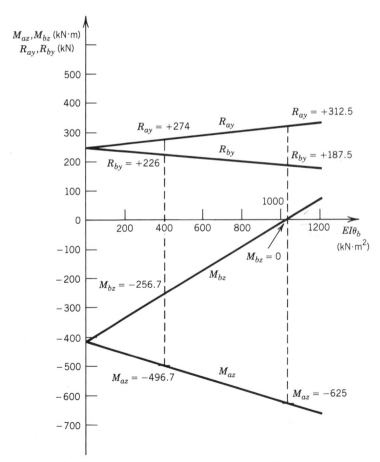

Fig. 9.7 *Compatible solution to structure of Fig. 9.6a.*

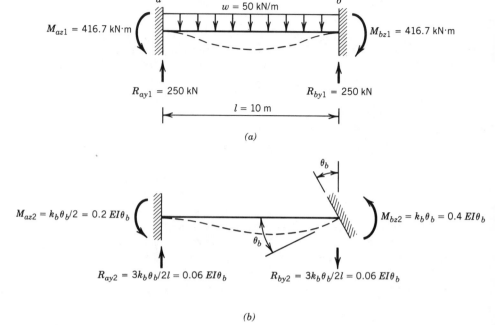

Fig. 9.8 *Kinematic primary structure of Fig. 9.5b. (a) Primary structure subjected to given loading. (b) Primary structure subjected to redundant rotation.*

be superimposed so that the final moment at point b is zero. Thus, we have the equilibrium equation

$$M_{bz1} + M_{bz2} = 0 \tag{9.9}$$

or, upon substitution,

$$-416.7 + 0.4EI\theta_b = 0 \tag{9.10}$$

from which

$$EI\theta_b = 1\,041.7 \text{ kN} \cdot \text{m}^2 \tag{9.11}$$

Returning to Eqs. 9.8 and taking $EI\theta_b = 1\,041.7$ kN $\cdot$ m^2, we obtain $R_{ay} = +312.5$ kN, $M_{az} = -625$ kN $\cdot$ m, $R_{by} = 187.5$ kN, and $M_{bz} = 0$ kN $\cdot$ m, and these values are shown in Fig. 9.7.

In this formulation of the problem, compatibility considerations gave rise to an infinite array of possible solutions, as demonstrated in Fig. 9.7. However, compatibility alone could not sort out the correct solution. Instead, the full satisfaction of equilibrium was necessary to extract the appropriate solution.

The final solution clearly satisfies the partial requirements of equilibrium and the compatibility conditions expressed through Eq. 9.8 and the augmenting equilibrium condition of Eq. 9.9.

9.6

Static versus Kinematic Indeterminacy

The example problem treated in the previous sections of this chapter had one static redundancy, R_{by}, and one kinematic redundancy, θ_b. This agreement in the order of static and kinematic indeterminacy is not usually the case. Table 9.1 shows

TABLE 9.1 Comparison of Static and Kinematic Indeterminacies

Structure	Static Classification (Redundant Forces Shown)	Kinematic Classification (Redundant Displs. Shown)	Degree of Indeterminacy	
			Static	Kinematic
			1	5
			4	2
			1	8
			3	6

Note: In evaluating kinematic redundants, axial deformations are frequently omitted for flexural structures. This would reduce the kinematic indeterminacy to 3, 1, and 3 for structures *a*, *b*, and *d*, respectively.

several examples of the determination of indeterminacy. Within the scope of these simple examples, no clear trends are evident; however, for larger frame structures, the degree of static indeterminacy is normally greater than the degree of kinematic indeterminacy.

One additional observation is crucial. In establishing the statically determinate primary structure, there are a number of alternative selection patterns for the redundant force quantities. However, in developing the kinematically determinate primary structure, no selection process is necessary. Instead, all of the displacement quantities that are necessary to describe the structure's response automatically become redundant quantities.

9.7

Compatibility Methods of Analysis

Two different methods of analysis are available for the solution of statically indeterminate structures. This section will deal with the first—the so-called *compatibility method*. This method is straightforward and easy to understand, and it serves as an effective method for certain classes of problems. As the name suggests, it is based on the solution of a set of equations that express compatibility relationships throughout the structure. This method is also referred to as the *force method* or the *flexibility method,* since the unknowns in the governing equations are forces and the coefficients of these unknown forces are flexibility quantities.

It is important to note that the final solution must satisfy both the conditions of compatibility and the requirements of equilibrium. However, when the method is properly formulated, the compatibility equations represent a superposition of a set of partial solutions, each of which satisfies the requirements of equilibrium.

The illustrative problem discussed earlier in Sections 9.1 and 9.3 serves as an example of the compatibility method. Equation 9.3 states a condition of compatibility concerning the vertical displacement at point b. The individual displacement quantities that are required are determined from the individual analyses shown in Fig. 9.4, each of which satisfies equilibrium. In the final form of the compatibility equation, Eq. 9.4, the coefficient of R_{by} is a flexibility coefficient, and this equation allows for the solution of the redundant reaction R_{by}. This equation, along with the overall equilibrium equations, Eqs. 9.1, provides the three equations needed for the determination of the reactions R_{ay}, M_{az}, and R_{by}.

For large structures, there are many redundant force quantities and, therefore, a number of compatibility equations must be simultaneously solved for the corresponding number of redundant forces. Since there are several acceptable patterns for the selection of the redundant forces, there is no unique formulation of the problem. This variability is a detriment to computer formulation because it is not possible to automate the procedure for generating the compatibility equations.

Some classical compatibility methods of analysis are presented in Chapters 10 and 11. In addition to the classical formulations, these methods can also be cast in a matrix formulation. The formal matrix compatibility approach, the flexibility method, is presented in Chapter 18.

9.8

Equilibrium Methods of Analysis

The second method of analysis of statically indeterminate structures, which is equally applicable for statically determinate structures, is the *equilibrium method.* This method is also straightforward conceptually and has advantages for certain types of structures. As the name indicates, this method is based on the solution of a set of equations that express the equilibrium requirements for the structure. This approach is also called the *displacement method* or the *stiffness method,* because the unknown quantities in the governing equations are displacements and the coefficients of these unknowns are stiffness quantities.

Again, it is stressed that the final solution must satisfy both the requirements of equilibrium and the conditions of compatibility. The solution of the governing equations of equilibrium will ensure the former, but care must be exercised in developing these equations to make certain that none of the compatibility conditions are violated.

The problem treated in Section 9.5 provides an example of the equilibrium method. Equation 9.9 gives the equilibrium requirement regarding the moment at point b. The individual moment quantities that are needed in this equation are determined from the individual analyses shown in Fig. 9.8, each of which fully satisfies the conditions of compatibility. In the final form of the equilibrium equation, Eq. 9.10, the coefficient of θ_b is a stiffness coefficient, and this equation allows for the solution of the redundant displacement, θ_b. This equation, along with Eqs. 9.8, which represent overall equilibrium and compatibility, provides the relationships needed to evaluate all the reaction components.

For larger structures, there are, of course, many redundant displacement quantities—one corresponding to each kinematic degree of freedom. Therefore, a number of equilibrium equations must be solved simultaneously for the displacement quantities. However, unlike the compatibility method, there is no selectivity in choosing the redundants—all the kinematic degrees of freedom are taken as redundant displacements. This is a major advantage for the automation that is needed in computer programming.

Several classical equilibrium methods are developed in Chapters 12, 13, and 14. These methods are given in their classical formulations, and in some cases the matrix representations are presented. The formal matrix equilibrium procedure, which is referred to as the stiffness method, is developed in Chapter 17.

9.9 Advantages and Disadvantages of Indeterminate Structures

Courses of instruction in structural analysis generally treat first the area of statically determinate structures as was done in Part I of this book. This treatment typically includes the stress analysis, which leads to the determination of the forces that act on any element of the structure, and deformation analysis, which leads to the determination of the displacements of any point on the structure. With these tools mastered, the student is ready to consider statically indeterminate structures where the stress analysis and the deformation analysis are integral parts of a solution that satisfies the required equilibrium and compatibility conditions, as was illustrated in the earlier sections of this chapter.

One might legitimately ask, however, why indeterminate structures are used. They are obviously more difficult to analyze than determinate ones, so why aren't statically determinate structures used exclusively? And, if there are specific advantages in statically indeterminate structures, why are determinate structures used in certain cases? As it turns out, specific advantages and disadvantages are associated with each type of structure.

The major advantages of statically indeterminate structures are manifested in three ways. First, a statically indeterminate structure displays greater stiffness in resisting load than does a comparable statically determinate structure. For example, consider the two structures shown in Figs. 9.9a and 9.9b, in which the indi-

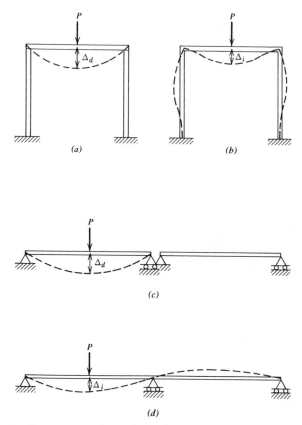

Fig. 9.9 *Comparative responses for statically determinate and indeterminate structures.* (a) *Post and lintel.* (b) *Portal frame.* (c) *Two simply supported beams.* (d) *Two-span continuous beam.*

vidual elements of each structure have the same cross-sectional dimensions and lengths. The post and lintel arrangement shows a beam member (lintel) that is supported atop two vertical struts (posts). When the load P is applied at midspan of the beam, the beam deflects with a vertical displacement of Δ_d at the load point. If the beam is integrally connected to the column, a portal frame is formed. In this case, the same load P will cause a vertical displacement of $\Delta_i < \Delta_d$ at the load point. Computing the stiffness for each case, that is, the load per unit displacement, we see $P/\Delta_i > P/\Delta_d$. That is, the stiffness of the indeterminate portal frame is greater than that of the determinate post and lintel system. The increased stiffness of the portal frame stems from the fact that the ends of the beam are restrained by the columns. Thus, as the beam deflects, the columns assist in resisting the load. This behavior differs from the post and lintel construction, where the beam simply deflects and the struts are merely passive supports for the beam.

A similar situation exists when the two simply supported beams of Fig. 9.9c are compared with the two-span continuous beam of Fig. 9.9d. Again, if the individual elements in each system have the same lengths and cross-sectional dimensions, then $\Delta_i < \Delta_d$, and thus $P/\Delta_i > P/\Delta_d$. Again, the statically determinate case reduces to a simply supported beam, with the loaded span carrying the entire load. However, the statically indeterminate beam provides an end restraint for the loaded span, and the unloaded span participates in resisting the load.

The second advantage of a statically indeterminate structure is that it will have lower stress intensities than would a comparable statically determinate structure. Consider Fig. 9.10, which shows the moment diagrams for the structures and loading conditions, respectively, given in Fig. 9.9. For both statically determinate cases, the full static moment M_s acts on the midspan section of the loaded span. However, for the statically indeterminate cases, the midspan moment on the loaded span is less than the static moment because of the effects of the negative restraining moments at the ends of the loaded span. The reduced moments for the indeterminate structures mean that the stress levels are lower, and thus smaller beam sections can be selected than would be required for the determinate structures. This, of course, leads to economies in the design of the structure.

The third advantageous feature of a statically indeterminate structure is related to safety. Recall from the nature of the criteria for statical classification that statically indeterminate structures have redundant force quantities. That is, there are either internal forces or external reactions that are not needed for stability, and if these force components are removed, the structure will not become unstable. This redundancy represents safety in that upon failure of a joint, member, or support to carry an assigned force, the structure will not necessarily collapse. For example, consider the frame of Fig. 9.9b. The frame and the associated moment diagram are redrawn in Fig. 9.11a. Again, M_s is the full static moment. Now, suppose that because of repeated cyclical-type loading, a crack develops in the beam under the load such that the moment capacity is severely decreased. The resulting deflected structure and the redistributed moment diagram are shown in Fig. 9.11b.

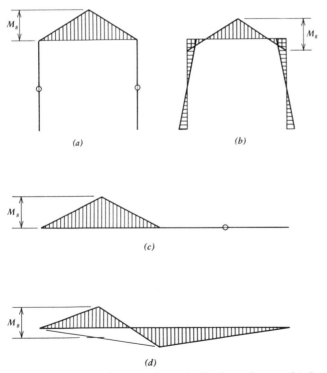

(a) (b) (c) (d)

Fig. 9.10 *Comparative moment diagrams for statically determinate and indeterminate structures.* (a) *Post and lintel.* (b) *Portal frame.* (c) *Two simply supported beams.* (d) *Two-span continuous beam.*

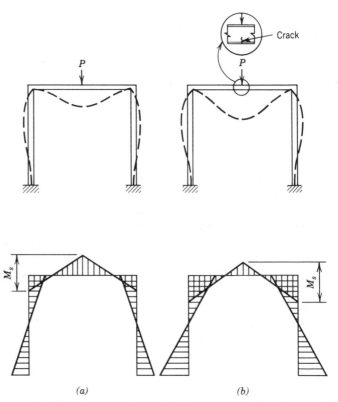

Fig. 9.11 *Redistribution of moment in a redundant structure.* (a) *Statically indeterminate frame.* (b) *Partially collapsed frame.*

The deflected structure reflects the fact that the structure is under distress; however, collapse does not occur. Therefore, if the difficulty is observed and the structure is taken out of service, injuries can be prevented and repairs can be made. If a similar crack developed in the beam of Fig. 9.9a, failure would be immediate.

For complex structures, the redistribution process that accompanies partial failure can be complicated to trace; however, again, it is the presence of redundant quantities that precludes failure.

Actually, all of the advantages cited above are a result of the inherent differences between the determinate and indeterminate structures. That is, the integral attachment of the individual structural members acts to reduce the deflections and stress intensities and to preclude failure. This property is referred to as *continuity* because the separate members are continuously attached where they connect.

The primary disadvantage of statically indeterminate structures results from the same feature that underlies its advantages—continuity. As an example, consider the statically indeterminate three-span continuous beam shown in Fig. 9.12a. Should this structure undergo a support settlement of Δ_b at point b relative to the other supports, it would assume the deformed pattern in Fig. 9.12b, which in turn would induce the bending moments shown in Fig. 9.12c. Thus, without any loading on the structure, sizable moments would be introduced at the interior support points. With these moments superimposed upon those that result from the design loads, it is possible that the structure will be overstressed at one of the interior

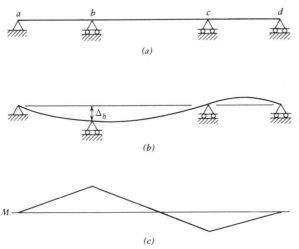

Fig. 9.12 *Relative settlement in three-span continuous beam. (a) Three-span continuous beam. (b) Deformations caused by relative settlement. (c) Moments induced by settlement.*

supports. For this reason, a statically indeterminate beam bridge cannot be used at a site where the foundation conditions are such that relative support settlements cannot be precluded.

In situations where such settlements can occur, a statically determinate structure should be employed, such as the cantilever structure shown in Fig. 9.13a. When the support at point b settles in this case, the structure merely realigns itself as shown in Fig. 9.13b. Since the members remain straight, no moments are induced.

The potential disadvantage of a statically indeterminate structure has been illustrated here for a specific structure and a distinct deformation. However, the same phenomenon occurs for any statically indeterminate structure. That is, any induced deformations, such as relative settlements, fabrication errors, or member length changes caused by temperature variations, will induce internal forces throughout the structure. Deformations that can be predicted, such as those that result from temperature change, can be included as a loading condition in the analysis; however, those that cannot be predicted may cause serious structural distress.

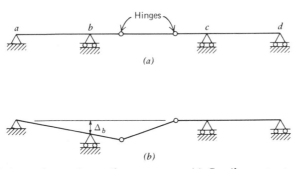

Fig. 9.13 *Relative settlement in cantilever structure. (a) Cantilever structure. (b) Relative settlement pattern.*

9.10

Additional Reading

Cross, H., and Morgan, N. D., *Continuous Frames of Reinforced Concrete,* Wiley, New York, 1932.

Ghali, A., and Neville, A. M., *Structural Analysis,* Chapter 1, Intext Educational Publishers, Scranton, Pa., 1972.

Kinney, J. S., *Indeterminate Structural Analysis,* Chapter 3, Arts. 1–6, Addison–Wesley, Reading, Mass., 1959.

McCormac, J. C., *Structural Analysis,* 3rd Ed., Chapter 15, Intext Educational Publishers, New York, 1975.

Parcel J. I., and Moorman, R. B. B., *Analysis of Statically Indeterminate Structures,* Chapter 1, Wiley, New York, 1955.

The Newburgh-Beacon Bridge, New York (courtesy Modjeski and Masters, Consulting Engineers).

ANALYSIS OF PIN-CONNECTED FRAMEWORKS BY COMPATIBILITY METHODS

10.1
Compatibility Methods for Trusses

Any method of structural analysis must lead to a solution that satisfies all of the requisite conditions of equilibrium and compatibility. The methods that are broadly classified as *compatibility methods* are those in which the key relationships that are used in the solution are compatibility equations. The conditions of equilibrium are inherently satisfied throughout the solution as the individual compatibility equations are generated. Since the final equations are composed of flexibility quantities and lead to the solution of redundant forces, the compatibility method is sometimes referred to as the *fexibility method* or the *force method*.

There are many methods that can be classified as compatibility methods, and, as will be seen later, these methods are not limited to the analysis of pin-connected frameworks. In this chapter, we consider the so-called method of *consistent deformations*. This technique is a system flexibility approach that requires that certain displacements, which are determined from considering the structural system as a whole, are made consistent with the requirements of compatibility. Element flexibility approaches, which consider the compatibility of the individual structural elements, are considered in Chapter 18.

10.2

External versus Internal Redundancies

As was described in Chapter 9, redundant forces are those that can be removed from the structure without impairing the stable integrity of the structure. These redundant forces may be either external or internal. In the former case, the redundants are reaction forces, whereas in the latter, the redundant forces are member forces. In each case, these forces are unnecessary for the stability of the structure.

This chapter considers only planar trusses, and thus the nature and number of redundancies can be determined by examining the provisions of Section 3.9.

10.3

Method of Consistent Deformations for Determining Redundant Reactions

The simple continuous beam solution that was conceptualized in Sections 9.1 through 9.3 was actually an application of the method of consistent deformations. In this section, the method is formally applied to the problem of externally redundant trusses.

Consider the simple truss shown in Fig. 10.1a. The objective of the analysis is to determine the four independent reaction components, R_1 through R_4, and the F set of bar forces. The considerations of Section 3.9 indicate that the structure is externally indeterminate to the first degree. That is, there is one redundant reaction component. In this case, R_1 is taken as the redundant reaction, with the remaining reactions being sufficient in number and arrangement to ensure the external stability of the structure.

Upon removal of R_1, the statically determinate *primary structure* of Fig. 10.1b remains. Since this structure is statically determinate, the reactions R_{20} through R_{40} can be determined along with the F_0 set of bar forces, where the 0 subscript indicates that these quantities are associated with the actual loading on the primary structure. Corresponding to this arrangement, there is a displacement Δ_{10} at the point and in the direction of the released redundant. This displacement is, of course, in violation of the prescribed boundary condition for point a of the original structure, which requires that $\Delta_1 = 0$, as shown in Fig. 10.1a. Thus, the solution of the primary structure must be altered to meet the boundary conditions.

This alteration is accomplished by introducing a unit value of the redundant reaction on the primary structure, as is shown in Fig. 10.1c. Here, the reactions R_{21} through R_{41} and the F_1 set of bar forces result from a static analysis of the primary structure, with the 1 subscript indicating that these values are associated with a unit value of R_1. The displacement corresponding to the released redundant in this case is identified as f_{11}, which is the flexibility coefficient that expresses the deflection at the point and in the direction of R_1 that is caused by a unit value of R_1.

Since f_{11} is the displacement that results from a unit value of R_1, superposition can be used to obtain the displacement for any other value of R_1 by simply multiplying f_{11} by the magnitude of R_1. Thus, the deflection at the point and in the direction of the released redundant that is caused by the redundant reaction R_1 is identified as Δ_{1R} and is given by

$$\Delta_{1R} = f_{11}R_1 \tag{10.1}$$

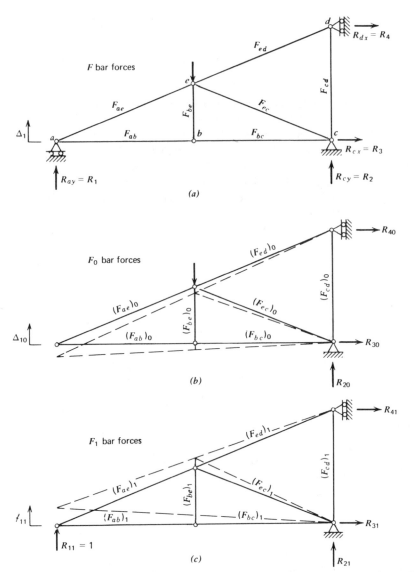

Fig. 10.1 *Consistent deformation analysis of externally indeterminate truss.* (a) *Statically indeterminate truss.* (b) *Statically determinate primary structure.* (c) *Unit value of* R_1.

For the desired solution of Fig. 10.1*a*, the solutions of Figs. 10.1*b* and 10.1*c* must be superimposed. Specifically, the displacements Δ_{10} and Δ_{1R} are combined to give the final displacement, Δ_1. Thus, we have

$$\Delta_{10} + \Delta_{1R} = \Delta_1 \tag{10.2}$$

Or, using Eq. 10.1, we have

$$\Delta_{10} + f_{11}R_1 = \Delta_1 \tag{10.3}$$

Solving for R_1, we have

$$R_1 = \frac{(\Delta_1 - \Delta_{10})}{f_{11}} \tag{10.4}$$

In this case, all displacements are considered to be positive when upward. Thus, Δ_{10} is actually negative, as shown in Fig. 10.1b. Also, Δ_1 is zero, as shown in Fig. 10.1a, because of the nonyielding constraint associated with the redundant reaction.

It should be noted that Eq. 10.3 is a compatibility equation that has units of displacement. Since f_{11} also has units of displacement, the quantity R_1 is unitless. That is, Eq. 10.4 merely gives the magnitude of the redundant reaction, or the factor by which the unit load solution of Fig. 10.1c must be multiplied so that when it is combined with the primary solution of Fig. 10.1b, the correct solution of Fig. 10.1a is obtained.

Once R_1 has been determined, statics can be applied to determine the nonredundant reactions. There is, however, a more general approach. The superposition pattern expressed in Eq. 10.3 for displacements holds for all other aspects of the solution. Thus, to determine one of the nonredundant reactions, such as R_q, we have

$$R_q = R_{q0} + R_{q1}R_1 \tag{10.5}$$

Or, a typical bar force, such as F_{rs}, is determined from

$$F_{rs} = (F_{rs})_0 + (F_{rs})_1 R_1 \tag{10.6}$$

Consider next the structure of Fig. 10.1a with one additional vertical reaction component at point b. This structure is shown in Fig. 10.2a. In this case, the criteria of Section 3.9 lead to the conclusion that the structure is externally redundant to the second degree. Removing R_1 and R_2, the statically determinate structure of Fig. 10.2b is obtained, which is the same primary structure that was employed in Fig. 10.1b. In this case, however, there are two reaction points at which the compatibility of the boundary conditions has been violated. Thus, we must systematically modify the primary structure until the boundary conditions of the original structure are satisfied. This is done by introducing, in turn, unit values of the redundant reactions on the primary structure and determining the effects of each of these unit load cases. These are shown in Fig. 10.2c.

It should be noted that in Fig. 10.2 the 0 subscript is used to identify the quantities associated with the actual loading on the primary structure, whereas the subscripts 1 and 2 denote the quantities associated with the application of unit values of R_1 and R_2, respectively, on the primary structure. The deflection f_{ij} is the flexibility coefficient that expresses the deflection at the point and in the direction of the redundant reaction R_i that is caused by a unit value of R_j.

In this case, there are two compatibility equations—one associted with the displacement corresponding to R_1 and the other associated with the displacement corresponding to R_2. Thus, we have

$$\Delta_{10} + \Delta_{1R} = \Delta_1 \tag{10.7}$$
$$\Delta_{20} + \Delta_{2R} = \Delta_2$$

where Δ_{1R} and Δ_{2R} are the displacements that result from the combined effects of the redundant reactions R_1 and R_2. From superposition, these are given by

$$\Delta_{1R} = f_{11}R_1 + f_{12}R_2$$
$$\Delta_{2R} = f_{21}R_1 + f_{22}R_2 \tag{10.8}$$

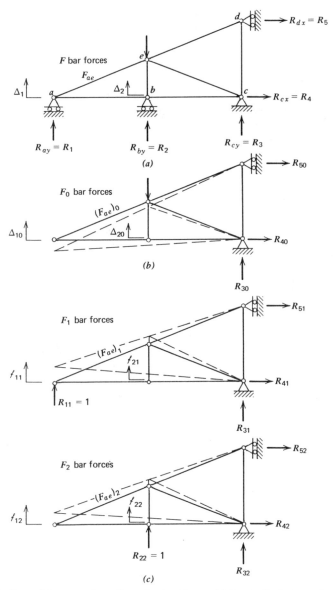

Fig. 10.2 *Second-degree externally redundant truss. (a) Statically indeterminate truss. (b) Statically determinate primary structure. (c) Unit values of redundant reactions.*

Substitution of Eqs. 10.8 into Eqs. 10.7 leads to

$$f_{11}R_1 + f_{12}R_2 = \Delta_1 - \Delta_{10}$$
$$f_{21}R_1 + f_{22}R_2 = \Delta_2 - \Delta_{20}$$

(10.9)

or in matrix form

$$\begin{bmatrix} f_{11} & f_{12} \\ f_{21} & f_{22} \end{bmatrix} \begin{Bmatrix} R_1 \\ R_2 \end{Bmatrix} = \begin{Bmatrix} \Delta_1 - \Delta_{10} \\ \Delta_2 - \Delta_{20} \end{Bmatrix}$$

(10.10)

In this form, the square matrix of f terms is the structure flexibility matrix involving the displacements on the primary structure that are associated with the directions of R_1 and R_2. Again, since upward displacements are considered positive, Δ_{10} and Δ_{20} are negative, as indicted in Fig. 10.2b.

The solution of Eq. 10.10 gives the magnitudes of the reactions R_1 and R_2. The superposition pattern expressed in Eqs. 10.7 and 10.8 can be used to determine the final values of the nonredundant reactions and the bar forces. For instance, for the reaction R_q, we have

$$R_q = R_{q0} + R_{q1}R_1 + R_{q2}R_2 \tag{10.11}$$

and for bar force F_{rs}, we have

$$F_{rs} = (F_{rs})_0 + (F_{rs})_1R_1 + (F_{rs})_2R_2 \tag{10.12}$$

It is noted that superposition plays a vital role in the development of the method of consistent deformations. Since, as was pointed out in Section 5.12, superposition is valid only for structures that display a linear response to loading, it is concluded that the method of consistent deformations can be applied only to this class of structures.

For more redundant reactions, the same procedure is followed. Each redundant requires a matching compatibility equation, and these collectively lead to a set of simultaneous equations in the form of Eq. 10.10. The solution of these equations gives the magnitudes of the redundant reactions, and the final solution is then formed by superposition similar to that expressed in Eqs. 10.11 and 10.12.

It is stressed that the final solution satisfies all of the requirements of compatibility and equilibrium. The solution for the redundant forces stems directly from a satisfaction of the compatibility equations, and each individual solution of the primary structure results from a statically determinate analysis that satisfies the requirements of equilibrium. Thus, the superposition indicated in Eqs. 10.11 and 10.12 is consistent with the requirements of compatibility and equilibrium.

10.4

Application of the Method of Consistent Deformations

In applying the method of consistent deformations, it is necessary to generate compatibility equations in the form of Eq. 10.3 or 10.9. This requires the determination of the Δ_{i0} displacements and the f_{ij} flexibility coefficients. All of these quantities result from elementary displacement calculations for statically determinate trusses, which are determined in accordance with the methods presented in Chapter 5.

The example problems presented in this section are designed to illustrate the application of the method.

10.4.1 Example problem

Determine the reactions and bar forces for the statically indeterminate truss shown. The quantity EA is the same for each member.

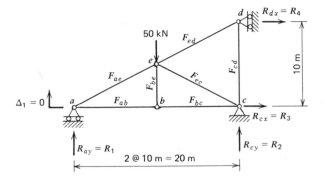

Structure Classification:

$j = 5$	Internal:	$m = 2j - r = 10 - 3 = 7$	
$m_a = 7$		$m_a = m$ ∴ Statically determinate	
$r_a = 4$	External:	$r_a > r;\quad 4 > 3$	
$r = 3$		∴ Statically indeterminate to first degree	

Primary Structure and Loadings: Select R_1 as the redundant reaction.

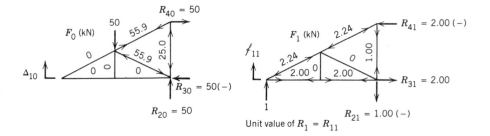

Displacement Calculations: The virtual work method is used to obtain the required displacement quantities. The fundamental equation is

$$\sum_{i=1}^{n} (\delta P)_i D_i = \sum_{j=1}^{m} \left(\delta F_P \cdot \frac{F_D l}{EA} \right)_j \tag{5.26}$$

The following table describes how this equation is applied.

Displ. Quantity	δF_P	F_D	Equation 5.26
Δ_{10}	F_1	F_0	$1 \cdot \Delta_{10} = \dfrac{1}{EA} \sum F_1 F_0 l$
f_{11}	F_1	$\cdot F_1$	$1 \cdot f_{11} = \dfrac{1}{EA} \sum F_1^2 l$

The desired summations are carried out in the following table:

Member	l m	F_0 kN	F_1 kN	$F_1 F_0 l$ $(\text{kN})^2 \cdot \text{m}$	$F_1^2 l$ $(\text{kN})^2 \cdot \text{m}$
ab	10.0	0	+2.00	0	40.0
bc	10.0	0	+2.00	0	40.0
ae	11.2	0	−2.24	0	56.2
ed	11.2	+55.9	−2.24	−1 402.4	56.2
eb	5.0	0	0	0	0
ec	11.2	−55.9	0	0	0
cd	10.0	−25.0	+1.00	−250.0	10.0
Σ				−1 652.4	+202.4

$$1 \cdot \Delta_{10} = \frac{1}{EA} \sum_{j=1}^{m} (F_1 F_0 l)_j = \frac{-1\,652.4}{EA} \quad \left(\frac{\text{kN}^2 \cdot \text{m}}{EA}\right)$$

$$\therefore \Delta_{10} = \frac{-1\,652.4}{EA} \quad \left(\frac{\text{kN} \cdot \text{m}}{EA}\right)$$

$$1 \cdot f_{11} = \frac{1}{EA} \sum_{j=1}^{m} (F_1^2 l)_j = \frac{202.4}{EA} \quad \left(\frac{\text{kN}^2 \cdot \text{m}}{EA}\right)$$

$$\therefore f_{11} = \frac{202.4}{EA} \quad \left(\frac{\text{kN} \cdot \text{m}}{EA}\right)$$

Note:
In the above, the units are given in parentheses. Positive displacement is upward since the unit load of P system acts upward.

Calculation of Redundant Reaction:

$$R_1 = \left(\frac{\Delta_1 - \Delta_{10}}{f_{11}}\right) \tag{10.4}$$

Since

$$\Delta_1 = 0; \quad R_1 = -\frac{(-1\,652.4)}{202.4} = +8.16$$

Note:
In this case, the quantity EA cancels. Thus, the final results are not dependent on the actual values of E and A.

Member Forces and Reactions: The member forces can be determined by extending the table used above and by applying Eq. 10.6.

Member rs		$(F_{rs})_1 \cdot R_1$ kN	$F_{rs} = (F_{rs})_0 + (F_{rs}) \cdot R_1$ kN
ab		+16.32	+16.32
bc		+16.32	+16.32
ae		−18.28	−18.28
ed		−18.28	+37.62
eb		0	0
ec		0	−55.90
cd		+8.16	−16.84

where

$(F_{rs})_1 = F_1$ force in member rs
$(F_{rs})_0 = F_0$ force in member rs

The reactions are then determined from Eq. 10.5.

Reaction q	R_{q0} kN	R_{q1} kN	$R_{q1} \cdot R_1$ kN	$R_q = R_{q0} + R_{q1} \cdot R_1$ kN
1	—	1.00	+8.16	+8.16
2	50	−1.00	−8.16	+41.84
3	−50	2.00	+16.32	−33.68
4	50	−2.00	−16.32	+33.68

where

$R_{q1} = R_q$ for F_1 system
$R_{q0} = R_q$ for F_0 system

Final Results:

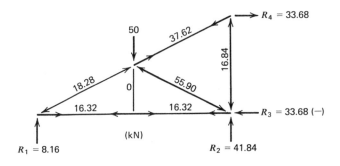

Determine the bar forces and the reactions for the statically indeterminate truss **10.4.2 Example** shown below. Member cross-sectional areas are given in square inches in paren- **problem** theses on each member, and E is the same for all members.

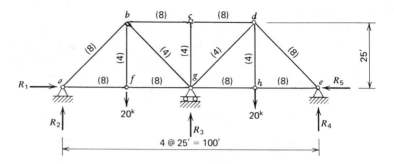

Structural Classification:

$$j = 8$$ Internal: $m = 2j - r = 16 - 3 = 13$

$$m_a = 13$$ $m_a = m$ $\therefore$ Statically determinate

$$r_a = 5$$

$$n = 0$$ External: $r_a > r;$ $5 > 3$

$$r = 3$$ Statically indeterminate to second degree

Primary Structure and Loadings: Select R_1 and R_2 as the redundant reactions.

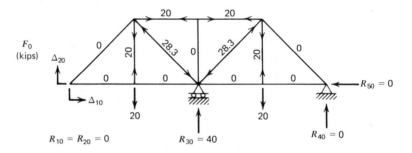

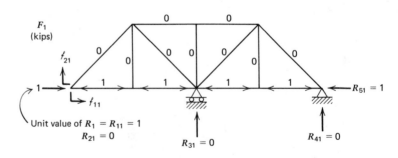

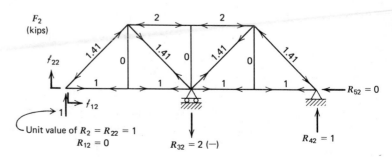

Displacement Calculations: The virtual work method is used to obtain the required displacement quantities. The fundamental equation is

$$\sum_{i=1}^{n} (\delta P_i) D_i = \sum_{j=1}^{m} \left(\delta F_P \cdot \frac{F_D l}{EA} \right)_j \tag{5.26}$$

which is applied in the manner outlined in the following table:

Displacement Quantity	δF_P	F_D	Equation 5.26
Δ_{10}	F_1	F_0	$1 \cdot \Delta_{10} = \dfrac{1}{E} \sum F_1 F_0 \dfrac{l}{A}$
Δ_{20}	F_2	F_0	$1 \cdot \Delta_{20} = \dfrac{1}{E} \sum F_2 F_0 \dfrac{l}{A}$
f_{11}	F_1	F_1	$1 \cdot f_{11} = \dfrac{1}{E} \sum F_1^2 \dfrac{l}{A}$
f_{21}	F_2	F_1	$1 \cdot f_{21} = \dfrac{1}{E} \sum F_2 F_1 \dfrac{l}{A}$
f_{12}	F_1	F_2	$1 \cdot f_{12} = \dfrac{1}{E} \sum F_1 F_2 \dfrac{l}{A}$
f_{22}	F_2	F_2	$1 \cdot f_{22} = \dfrac{1}{E} \sum F_2^2 \dfrac{l}{A}$

The desired summations are carried out in the table on page 382.

Thus, from Eq. 5.26.

$$\Delta_{10} = 0$$

$$\Delta_{20} = \frac{-956.2}{E} \quad \text{(units of } k \cdot \text{ft/in.}^2 \cdot E\text{)}$$

$$f_{11} = \frac{+12.52}{E}$$

$$f_{12} = f_{21} = \frac{-12.52}{E}$$

$$f_{22} = \frac{+90.30}{E}$$

Calculation of Redundants, Member Forces, and Reactions:

$$f_{11}R_1 + f_{12}R_2 = -\Delta_{10}$$

$$f_{21}R_1 + f_{22}R_2 = -\Delta_{20} \tag{10.9}$$

$$12.52R_1 - 12.52R_2 = 0$$

$$-12.52R_1 + 90.30R_2 = +956.2$$

From which

$$R_1 = R_2 = +12.29 \text{ kips}$$

Member rs	A in.²	l ft	F_0 kip	F_1 kip	F_2 kip	$\frac{F_1 F_0 l}{A}$ k²·ft/in.²	$\frac{F_2 F_0 l}{A}$ k²·ft/in.²	$\frac{F_1^2 l}{A}$ k²·ft/in.²	$\frac{F_2 F_1 l}{A}$ k²·ft/in.²	$\frac{F_2^2 l}{A}$ k²·ft/in.²	$(F_{rs})_1 \cdot R_1$ kip	$(F_{rs})_2 \cdot R_2$ kip	$(F_{rs})_0 + (F_{rs})_1 \cdot R_1 + (F_{rs})_2 \cdot R_2$ kips
ab	8	35.4	0	0	−1.41	0	0	0	0	8.80	0	−17.33	−17.33
bc	8	25	+20.0	0	−2	0	−125.0	0	0	12.50	0	−24.58	−4.58
cd	8	25	+20.0	0	−2	0	−125.0	0	0	12.50	0	−24.58	−4.58
de	8	35.4	0	0	−1.41	0	0	0	0	8.80	0	−17.33	−17.33
af	8	25	0	−1	+1	0	0	3.13	−3.13	3.13	−12.29	12.29	0
fg	8	25	0	−1	+1	0	0	3.13	−3.13	3.13	−12.29	12.29	0
gh	8	25	0	−1	+1	0	0	3.13	−3.13	3.13	−12.29	12.29	0
he	8	25	0	−1	+1	0	0	3.13	−3.13	3.13	−12.29	12.29	0
bf	4	25	+20.0	0	0	0	0	0	0	0	0	0	20.0
bg	4	35.4	−28.3	0	+1.41	0	−353.1	0	0	17.59	0	17.33	−10.97
cg	4	25	0	0	0	0	0	0	0	0	0	0	0
dg	4	35.4	−28.3	0	+1.41	0	−353.1	0	0	17.59	0	17.33	−10.97
dh	4	25	+20.0	0	0	0	0	0	0	0	0	0	20.0
Σ						0	−956.2	+12.52	−12.52	+90.30			

The final bar forces are determined from

$$F_{rs} = (F_{rs})_0 + (F_{rs})_1 \cdot R_1 + (F_{rs})_2 \cdot R_2 \qquad (10.12)$$

and are tabulated in the last three columns of the table.
The reactions are determined from

$$R_q = R_{q0} + R_{q1} \cdot R_1 + R_{q2} \cdot R_2 \qquad (10.11)$$

and are given in the following table:

Reaction q	R_{q0} kip	R_{q1} kip	R_{q2} kip	$R_{q0} + R_{q1} \cdot R_1 + R_{q2} \cdot R_2$ kip
1	0	1	0	12.29
2	0	0	1	12.29
3	40	0	−2	15.42
4	0	0	1	12.29
5	0	1	0	12.29

10.5 Support Settlements and Elastic Supports

It was illustrated in Chapter 9 that support settlements in statically indeterminate structures induce forces throughout the structure. The problems of Section 10.4 represented situations in which the supports were nonyielding; however, the development of Section 10.3 included provisions for treating the problem of support settlements.

For instance, consider the structure shown in Fig. 10.1. The compatibility equation given as Eq. 10.3 includes the displacement Δ_1, which is the upward vertical displacement at point a for the actual structure. This displacement appears on the right-hand side of Eq. 10.4, which is the expression for the magnitude of the redundant reaction, R_1. In Fig. 10.1, the vertical support at point a is a nonyielding support for which $\Delta_1 = 0$. If there is a downward settlement of Δ_{1S}, then $\Delta_1 = -\Delta_{1S}$, and Eq. 10.4 would have the form

$$R_1 = \frac{-(\Delta_{1S} + \Delta_{10})}{f_{11}} \qquad (10.13)$$

Thus, it is clear that the magnitude of the redundant reaction is affected by the support settlement. Once R_1 has been established, the final reactions and bar forces are determined in accordance with Eqs. 10.5 and 10.6, respectively.

In a similar manner, if downward settlements of Δ_{1S} and Δ_{2S} are prescribed for the structure of Fig. 10.2, then $\Delta_1 = -\Delta_{1S}$ and $\Delta_2 = -\Delta_{2S}$, and Eq. 10.10 becomes

$$\begin{bmatrix} f_{11} & f_{12} \\ f_{21} & f_{22} \end{bmatrix} \begin{Bmatrix} R_1 \\ R_2 \end{Bmatrix} = -\begin{Bmatrix} \Delta_{1S} + \Delta_{10} \\ \Delta_{2S} + \Delta_{20} \end{Bmatrix} \qquad (10.14)$$

The solution of Eq. 10.14 gives the magnitudes of the redundant reactions R_1 and R_2 and the final reactions and bar forces are determined from Eqs. 10.11 and 10.12, respectively.

Frequently, structures are mounted on supports that are not rigid; however, there is not a prescribed settlement. Instead, the support movement depends on the magnitude of the reaction force. The simplest representation of this kind of support is called an *elastic support,* in which settlement is a linear function of the reaction force. This arrangement is shown in Fig. 10.3a for the truss that was previously considered in Fig. 10.1. Here, it is clear that

$$R_1 = k_s (-\Delta_1) \tag{10.15}$$

where k_s is the elastic spring constant, as shown in Fig. 10.3b. Solving Eq. 10.15 for Δ_1 and substituting the result in Eq. 10.3, we obtain

$$\Delta_{10} + f_{11}R_1 = -\frac{R_1}{k_s} \tag{10.16}$$

from which

$$R_1 = \frac{-\Delta_{10}}{(f_{11} + 1/k_s)} \tag{10.17}$$

For a structure with multiredundant reactions, each reaction provided by an elastic support must be represented by an expression of the type given in Eq. 10.15.

In the example problems of Section 10.4, where there were no support displacements, common EA values were factored out of the summation quantities used to determine the displacements Δ_{i0} and f_{ij}. Since $\Delta_i = 0$ in these cases, the common EA terms canceled uniquely when Eqs. 10.4 and 10.9 were solved for the redundant reactions. However, when support displacements are admitted, and

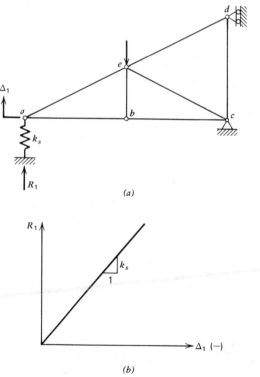

Fig. 10.3 *Elastic support. (a) Indeterminate truss with elastic support at point a. (b) Elastic support stiffness.*

$\Delta_i \neq 0$, then these common EA terms no longer cancel, and they must be included in the determination of the redundant reactions.

Thus, in the absence of support movement, the member forces and reactions of a statically indeterminate truss are dependent on the relative EA values. However, when support movement exists, the member forces and reactions depend on the absolute EA values. This fact is illustrated in the following problem.

Determine the reactions and bar forces for the truss shown in Example 10.4.1 if, in addition to the 50 kN load, there is a downward settlement of point a such that $\Delta_1 = -\Delta_{1S} = -10$ mm. In this case, $A = 3\,000$ mm^2 and $E = 200$ GN/m^2. **10.5.1 Example problem**

Structure Classification: Same as Example 10.4.1.

Primary Structure and Loadings: Again, R_1 is selected as the redundant and the loadings are the same as those of Example 10.4.1.

Displacement Calculations: Δ_{10} and f_{11} are required. These are the same as those of Example 10.4.1.

Calculation of Redundant Reaction:

$$R_1 = \frac{-(\Delta_{1S} + \Delta_{10})}{f_{11}} \tag{10.13}$$

$$EA = (200 \times 10^6 \text{ kN/m}^2) \times (0.003\,000 \text{ m}^2)$$
$$= 0.6 \times 10^6 \text{ kN}$$

$$\Delta_{10} = \frac{-1\,652.4 \text{ kN} \cdot \text{m}}{EA} = \frac{-1\,652.4}{0.6 \times 10^6} = -0.002\,754 \text{ m}$$
$$= -2.75 \text{ mm}$$

$$f_{11} = \frac{202.4 \text{ kN} \cdot \text{m}}{EA} = \frac{202.4}{0.6 \times 10^6} = 0.000\,337 \text{ m} = 0.337 \text{ mm}$$

$$\Delta_{1S} = 10 \text{ mm}$$

$$R_1 = \frac{-(10 - 2.75)}{0.337} = -21.5 \text{ kN}$$

Bar Forces and Reactions: The bar forces are determined by using the bar forces for the primary structure loadings from Example 10.4.1 and by applying Eq. 10.6.

Member rs	$(F_{rs})_1 \cdot R_1$ kN	$F_{rs} = (F_{rs})_0 + (F_{rs})_1 \cdot R_1$ kN
ab	−43.00	−43.00
bc	−43.00	−43.00
ae	+48.16	+48.16
ed	+48.16	+104.06
eb	0	0
ec	0	−55.90
cd	−21.50	−46.50

where

$(F_{rs})_1 = F_1$ force in member rs

$(F_{rs})_0 = F_0$ force in member rs

The reactions are determined by using the reactions for the primary structure loadings from Example 10.4.1 and by applying Eq. 10.5.

Reaction q	R_{q0} kN	R_{q1} kN	$R_{q1} \cdot R_1$ kN	$R_q = R_{q0} + R_{q1} \cdot R_1$ kN
1	—	1.00	−21.5	−21.5
2	50	−1.00	+21.5	+71.5
3	−50	2.00	−43.0	−93.0
4	50	−2.00	+43.0	+93.0

where

$R_{q1} = R_q$ for F_1 system

$R_{q0} = R_q$ for F_0 system

Final Results:

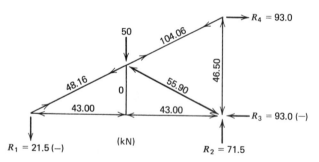

10.6

Method of Consistent Deformations for Determining Redundant Member Forces

Section 10.3 deals with the problem of determining the redundant reactions for a structure that is statically indeterminate externally. The method of consistent deformations is equally useful for the determination of redundant member forces for structures that are statically indeterminate internally.

Consider the truss shown in Fig. 10.4a. The objective is to determine the independent reaction components and the F set of bar forces. Application of the provisions of Section 3.9 reveals that this structure is internally indeterminate to the first degree. Thus, there is one redundant member. In the present case, the member bf is taken as the redundant member. When the load-carrying capacity of this member is released, the remaining members are sufficient in number and arrangement to ensure the internal stability of the structure. This release is accomplished by envisioning a telescoping sleevelike fixture along the member length that permits the member to retain its position and alignment within the structure without its being able to transmit any load.

Upon release of member bf, the statically determinate primary structure of Fig. 10.4b is formed. Since this structure is determinate, the reactions and the F_0 set of bar forces can be determined. Again, the 0 subscript indicates that these

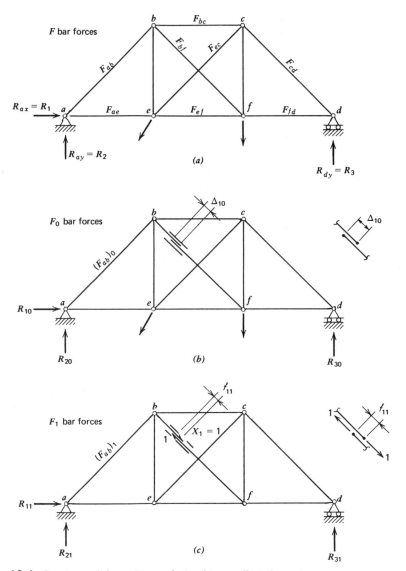

Fig. 10.4 *Consistent deformation analysis of internally indeterminate truss.*
(a) *Internally statically indeterminate truss.* (b) *Statically determinate primary structure.*
(c) *Unit value of* X_1.

quantities are associated with the actual loading on the primary structure. For this loading, there is an overlap, Δ_{10}, of the cut member within the sleeve, which is in violation of the prescribed continuity for this member in the original structure. Thus, this solution of the primary structure must be modified in order to restore the required continuity to member *bf*.

The required modification is brought about by introducing a unit value of the redundant member force on the primary structure, as is shown in Fig. 10.4c. Here, the F_1 set of bar forces results from a static analysis of the primary structure, with the subscript 1 indicating that these values are associated with a unit value of the redundant $X_1 = (F_{bf})_1$. Here, too, there is an overlap within the sleeve of the cut member that is denoted by f_{11}.

Since f_{11} is the overlapping displacement associated with a unit value of X_1, superposition can be used to determine the corresponding displacement that occurs when the cut member is subjected to a force equal to the actual value of the redundant member force X_1. This is represented by Δ_{1R} and is given by

$$\Delta_{1R} = f_{11}X_1 \tag{10.18}$$

For the final solution of Fig. 10.4a, the solutions of Figs. 10.4b and 10.4c must be superimposed. Specifically, the displacements Δ_{10} and Δ_{1R} must be combined to restore the continuity of the severed member. Thus,

$$\Delta_{10} + \Delta_{1R} = 0 \tag{10.19}$$

Substituting Eq. 10.18 into Eq. 10.19, and solving for X_1, we obtain

$$X_1 = \frac{-\Delta_{10}}{f_{11}} \tag{10.20}$$

The final member forces and reactions can now be determined from statics, or they can be determined by superposition. In the latter case, the final member force in member rs is given by

$$F_{rs} = (F_{rs})_0 + (F_{rs})_1 X_1 \tag{10.21}$$

and the final value of R_q is

$$R_q = R_{q0} + R_{q1}X_1 \tag{10.22}$$

If there are two redundant member forces, th: continuity equations take the form

$$\begin{bmatrix} f_{11} & f_{12} \\ f_{21} & f_{22} \end{bmatrix} \begin{Bmatrix} X_1 \\ X_2 \end{Bmatrix} = \begin{Bmatrix} -\Delta_{10} \\ -\Delta_{20} \end{Bmatrix} \tag{10.23}$$

where f_{ij} is a flexibility coefficient, which gives the displacement along the release of the ith redundant member that is associated with a unit load in the jth redundant member of the primary structure and Δ_{i0} is the displacement along the ith redundant member that is caused by the actual loading on the primary structure.

Solution of Eq. 10.23 gives the magnitudes of the redundant forces X_1 and X_2. The final force in member rs and the final value of R_q are then determined by superposition to be

$$F_{rs} = (F_{rs})_0 + (F_{rs})_1 X_1 + (F_{rs})_2 X_2 \tag{10.24}$$

$$R_q = R_{q0} + R_{q1}X_1 + R_{q2}X_2 \tag{10.25}$$

If there are more than two redundant reactions, the same general procedures is followed. Each redundant member requires a continuity equation. These equations collectively provide a set of simultaneous equations in the form of Eq. 10.23. The solution of these equations gives the magnitudes of the redundant member forces, and the final solution is then completed by superposition similar to that expressed in Eqs. 10.24 and 10.25. As explained earlier, the final solution will inherently satisfy all of the requirements of compatibility and equilibrium.

In the example problem that follows, the required displacement quantities are determined from the virtual work method. It is noted that the summations must include the terms corresponding to the cut member, since internal virtual work is done on this member in the determination of some displacement quantities.

Determine the member forces for the structure given. The quantity EA is the same **10.6.1 Example** for each member of the structure.   **problem**

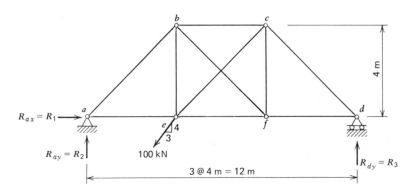

Structure Classification:

$j = 6$ Internal: $m = 2j - r = 12 - 3 = 9$

$m_a = 10$ $m_a > m$; $10 > 9$

$r_a = 3$ ∴ Statically indeterminate to first degree

$n = 0$ External: $r_a = r$; $3 = 3$

$r = 3$ ∴ Statically determinate

Primary Structure and Loadings: Select F_{bf} as a redundant member force.

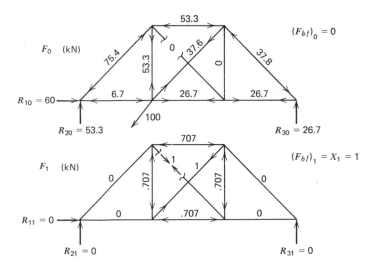

Displacement Calculations: The fundamental virtual work expression is used to obtain the required displacement quantities.

$$\sum_{i=1}^{n} (\delta P)_i D_i = \sum_{j=1}^{m} \left(\delta F_P \cdot \frac{F_D l}{EA} \right)_j \qquad (5.26)$$

This equation is applied, as described in the following table:

Displacement Quantity	δF_P	F_D	Equation 5.26
Δ_{10}	F_1	F_0	$1 \cdot \Delta_{10} = \dfrac{1}{EA} \sum F_1 F_0 l$
f_{11}	F_1	F_1	$1 \cdot f_{11} = \dfrac{1}{EA} \sum F_1^2 l$

The required summations are formed in the following table:

Bar rs	l m	F_0 kN	F_1 kN	$F_1 F_0 l$ kN$^2 \cdot$ m	$F_1^2 l$ kN$^2 \cdot$ m	$(F_{rs})_1 \cdot X_1$ kN	$F_{rs} = (F_{rs})_0 + (F_{rs})_1 \cdot X_1$ kN
ab	5.66	−75.4	0	0	0	0	−75.4
bc	4	−53.3	−0.707	150.7	2.0	5.0	−48.3
cd	5.66	−37.8	0	0	0	0	−37.8
ae	4	−6.7	0	0	0	0	−6.7
ef	4	26.7	−0.707	−75.5	2.0	5.0	31.7
fd	4	26.7	0	0	0	0	26.7
be	4	53.3	−0.707	−150.7	2.0	5.0	58.3
bf	5.66	0	1	0	5.66	−7.1	−7.1
ce	5.66	37.6	1	212.8	5.66	−7.1	30.5
cf	4	0	−0.707	0	2.0	5.0	5.0
Σ				137.3	19.32		

Note:
The redundant member *bf* must be included in the tabulation.

$$1 \cdot \Delta_{10} = \frac{1}{EA} \sum_{j=1}^{m} (F_1 F_0 l)_j = \frac{137.3}{EA}; \quad \Delta_{10} = \frac{137.3}{EA} \quad \left(\frac{\text{kN} \cdot \text{m}}{EA}\right)$$

$$1 \cdot f_{11} = \frac{1}{EA} \sum_{j=1}^{m} (F_1^2 l)_j = \frac{19.32}{EA}; \quad f_{11} = \frac{19.32}{EA} \quad \left(\frac{\text{kN} \cdot \text{m}}{EA}\right)$$

Calculation of Redundant and Member Forces:

$$X_1 = \frac{-\Delta_{10}}{f_{11}} \tag{10.20}$$

from which

$$X_1 = \frac{-137.3}{19.32} = -7.1$$

The final bar forces are determined from

$$F_{rs} = (F_{rs})_0 + (F_{rs})_1 \cdot X_1 \tag{10.21}$$

and are calculated in the last two columns of the table.

Consider again the structure of Fig. 10.4. If member bf has a specified length **Temperature** change Δ_1 that is a result of *temperature change* or *fabrication error*, then the **Change or** continuity requirement that was previously expressed as Eq. 10.19 becomes **Fabrication Error**

$$\Delta_{10} + \Delta_{1R} = \Delta_1 \qquad (10.26)$$

Substitution of Eq. 10.18 into Eq. 10.26 yields

$$X_1 = \frac{\Delta_1 - \Delta_{10}}{f_{11}} \qquad (10.27)$$

If there are two redundant member forces, the continuity equations take the form

$$\begin{bmatrix} f_{11} & f_{12} \\ f_{21} & f_{22} \end{bmatrix} \begin{Bmatrix} X_1 \\ X_2 \end{Bmatrix} = \begin{Bmatrix} \Delta_1 - \Delta_{10} \\ \Delta_2 - \Delta_{20} \end{Bmatrix} \qquad (10.28)$$

where, in addition to the terms defined after Eq. 10.23, Δ_i is the specified change in member length for the ith redundant member. Equations 10.27 and 10.28 give the magnitudes of the redundant member forces. The final member forces and reactions are given by superposition using Eqs. 10.21 and 10.22 for a single redundant or Eqs. 10.24 and 10.25 for the case with two redundants.

A specified length change that occurs as a result of temperature variation or fabrication error plays the same role in internally indeterminate structures as does support movement in externally indeterminate structures. In the absence of such length changes, it is sufficient to use relative EA values in the computations. When length changes are included, however, the absolute EA values must be used.

It should be noted that length changes due to temperature effects or fabrication errors are not necessarily limited to the redundant members. The effects of length changes for the nonredundant members are included in the calculation of the Δ_{i0} quantities for the primary structure.

Consider a statically indeterminate truss that has k redundant reaction components **Matric** and m redundant member forces. In addition, there are n independent forces that **Formulation** correspond to the n independent displacements that the structure can undergo. **of Method of** Thus, the statically determinate primary structure is subjected to $(k + m + n)$ **Consistent** forces. These forces can be arranged in a single $\{P\}$ vector, in which the first k **Deformations** elements represent the unknown redundant reactions, the next m elements correspond to the unknown redundant member forces, and the remaining n elements are the known forces that are applied to the structure. In this arrangement, it is convenient to specify that $k + m = q$ and $k + m + n = s$.

Corresponding to the $\{P\}$ vector, there is a $\{\Delta\}$ vector that includes the displacements on the primary structure. The first k elements of this array are the displacements that correspond to the redundant reactions. These must be zero for all points where there are nonyielding supports, or they are specified at points where there are support settlements. The next m elements are associated with the member releases. These are zero except for members that have a designated temperature change or fabrication error. The final n elements of $\{\Delta\}$ are the unknown displacements at the displaceable joints of the structure.

In accordance with the discussion in Section 5.8, there is an interactive relationship between the forces and displacements of the primary structure. In terms of the flexibility coefficients of the structure, this interactive relationship is given in the form

$$
\begin{bmatrix}
f_{11} & \cdots & f_{1k} & f_{1,k+1} & \cdots & f_{1q} & f_{1,q+1} & \cdots & f_{1s} \\
 & & & & & & & & \\
f_{k1} & \cdots & f_{kk} & f_{k,k+1} & \cdots & f_{kq} & f_{k,q+1} & \cdots & f_{ks} \\
f_{k+1,1} & \cdots & f_{k+1,k} & f_{k+1,k+1} & \cdots & f_{k+1,q} & f_{k+1,q+1} & \cdots & f_{k+1,s} \\
 & & & & & & & & \\
f_{q1} & \cdots & f_{qk} & f_{q,k+1} & \cdots & f_{qq} & f_{q,q+1} & \cdots & f_{q,s} \\
f_{q+1,1} & \cdots & f_{q+1,k} & f_{q+1,k+1} & \cdots & f_{q+1,q} & f_{q+1,q+1} & \cdots & f_{q+1,s} \\
 & & & & & & & & \\
f_{s1} & \cdots & f_{sk} & f_{s,k+1} & \cdots & f_{sq} & f_{s,q+1} & \cdots & f_{ss}
\end{bmatrix}
\begin{Bmatrix}
P_1 \\ \vdots \\ P_k \\ P_{k+1} \\ \vdots \\ P_q \\ P_{q+1} \\ \vdots \\ P_s
\end{Bmatrix}
=
\begin{Bmatrix}
\Delta_1 \\ \vdots \\ \Delta_k \\ \Delta_{k+1} \\ \vdots \\ \Delta_q \\ \Delta_{q+1} \\ \vdots \\ \Delta_s
\end{Bmatrix}
\tag{10.29}
$$

where f_{ij} is the flexibility coefficient that gives the displacement Δ_i that results from a unit value of the load P_j. An examination of Eq. 10.29 shows that the first k equations are compatibility equations for the displacements that are associated with the redundant reactions. The next m equations are continuity equations for the displacements that correspond to the cut redundant members. Finally, the last n equations give the displacements corresponding to the applied loads on the structure. A study of the load vectors shows that the first $(k + m) = q$ load terms are unknown redundant forces, whereas the remaining n load terms are known. However, in the displacement vector, the first q terms are known, while the remaining n terms are unknown. This is consistent with the requirements of not overspecifying boundary conditions, as was explained in Section 2.17.

If Eq. 10.29 is partitioned along the bold broken lines shown in the expression of that equation, we have

$$
\begin{bmatrix}
[f]_{11} & [f]_{12} \\
[f]_{21} & [f]_{22}
\end{bmatrix}
\begin{Bmatrix}
\{P\}_1 \\ \{P\}_2
\end{Bmatrix}
=
\begin{Bmatrix}
\{\Delta\}_1 \\ \{\Delta\}_2
\end{Bmatrix}
\tag{10.30}
$$

In this form, $[f]_{11}$ relates the known displacements $\{\Delta\}_1$ to the unknown redundant forces $\{P\}_1$, $[f]_{12}$ relates the known displacements $\{\Delta\}_1$ to the known applied loads $\{P\}_2$, $[f]_{21}$ relates the unknown displacements $\{\Delta\}_2$ to the unknown redundant forces $\{P\}_1$, and $[f]_{22}$ relates the unknown displacements $\{\Delta\}_2$ to the known applied loads $\{P\}_2$. Expansion of Eq. 10.30 leads to

$$
[f]_{11}\{P\}_1 + [f]_{12}\{P\}_2 = \{\Delta\}_1
\tag{10.31}
$$

$$
[f]_{21}\{P\}_1 + [f]_{22}\{P\}_2 = \{\Delta\}_2
\tag{10.32}
$$

In Eq. 10.31, it is clear that $[f]_{12}\{P\}_2$ gives the displacements associated with the redundant forces that are caused by the applied loads. Representing these as $\{\Delta_0\}_1$, we can rewrite Eq. 10.31 as

$$
[f]_{11}\{P\}_1 = \{\Delta\}_1 - \{\Delta_0\}_1
\tag{10.33}
$$

This has the same form as was previously expressed by Eqs. 10.4 and 10.10 for externally indeterminate structures and by Eqs. 10.20 and 10.23 for internally indeterminate structures.

Equations 10.33 can readily be solved for the unknown redundant forces in the form

$$\{P\}_1 = [/]_{11}^{-1}(\{\Delta\}_1 - \{\Delta_0\}_1) \tag{10.34}$$

The vector $\{P\}_1$ can now be substituted into Eq. 10.32 to obtain the unknown displacements $\{\Delta\}_2$ in the form

$$\{\Delta\}_2 = [/]_{21}[/]_{11}^{-1}(\{\Delta\}_1 - \{\Delta_0\}_1) + [/]_{22}\{P\}_2 \tag{10.35}$$

Once the redundant reactions and member forces have been determined, the remaining member forces and reactions for the statically determinate primary structure can be found. For this purpose, Eq. 3.17 is rewritten in the form

$$[C]_P \begin{Bmatrix} \{F\}_P \\ \{R\}_P \end{Bmatrix} = \begin{Bmatrix} \{P\}_1 \\ \{P\}_2 \end{Bmatrix} \tag{10.36}$$

where $[C]_P$ is the overall statics matrix for the primary structure, $\{F\}_P$ and $\{R\}_P$ are the unknown member forces and reaction components, respectively, for the primary structure, and $\{P\}_1$ and $\{P\}_2$ are the applied loads for the primary structure—$\{P\}_1$ being the redundant forces as determined from Eq. 10.34 and $\{P\}_2$ being the specified applied forces. Solving Eq. 10.36, we obtain

$$\begin{Bmatrix} \{F\}_P \\ \{R\}_P \end{Bmatrix} = [C]_P^{-1} \begin{Bmatrix} \{P\}_1 \\ \{P\}_2 \end{Bmatrix} \tag{10.37}$$

Of course, the $[C]_P$ matrix could be developed as explained in Section 3.10 and its inverse determined according to the requirements of Eq. 10.37. However, it is clear that each column of $[C]_P^{-1}$ is composed of elements of $\{F\}_P$ and $\{R\}_P$ that are associated with a unit value of the corresponding element of $\{P\}_1$ or $\{P\}_2$ that multiplies that column. Since the bar forces and reactions for these unit load cases are needed to generate the flexibility coefficients required in Eq. 10.30, the explicit development of $[C]_P$, and the determination of its inverse, is not necessary.

Therefore, Eq. 10.37 yields the final bar forces and reactions for the primary structure through a superposition of contributions from each load component applied to the primary structure in which $\{P\}_2$ are the externally applied loads and $\{P\}_1$ are the solved redundant forces. This parallels the format of Eq. 10.29, in which the final displacements are expressed as a superposition of the contributions from the same load components.

10.9 Application of Matrix Formulation

It is clear from the development of the matrix formulation that the most crucial step is in the generation of the required flexibility matrix. The individual flexibility coefficients, which are actually displacement quantities, can be determined from any of the methods developed in Chapter 5. This approach, in which the structure flexibility coefficients are determined from considering the structure as a whole, is a *system approach* to the problem. This is contrasted to the *element approach* developed in Chapter 18, in which the structure flexibility coefficients are generated through a synthesis of the individual element flexibilities.

This chapter is limited to the system approach. In the example problem that follows, the virtual work method is used to determine the individual flexibility coefficients from which the structure flexibility matrix is formed.

10.9.1 Example problem Determine the reactions, the member forces, and the displacements at the free joints for the truss shown. Assume EA is the same for each member.

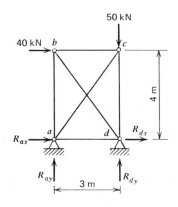

Structure Classification:

$j = 4$ Internal: $m = 2j - r = 8 - 3 = 5$

$m_a = 6$ $m_a > m; \quad 6 > 5$

$r_a = 4$ $\therefore$ Statically indeterminate to first degree

$r = 3$ External: $r_a > r; \quad 4 > 3$

$\therefore$ Statically indeterminate to first degree

Primary Structure: Select R_{dx} as the redundant reaction component and F_{bd} as the redundant member force.

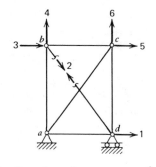

$$
\{P\} =
\begin{Bmatrix}
P_1 \\
P_2 \\
\hline
P_3 \\
P_4 \\
P_5 \\
P_6
\end{Bmatrix}
=
\begin{Bmatrix}
R_{dx} \\
F_{bd} \\
\hline
40 \\
0 \\
0 \\
-50
\end{Bmatrix};
\qquad
\{\Delta\} =
\begin{Bmatrix}
\Delta_1 \\
\Delta_2 \\
\hline
\Delta_3 \\
\Delta_4 \\
\Delta_5 \\
\Delta_6
\end{Bmatrix}
=
\begin{Bmatrix}
0 \\
0 \\
\hline
\Delta_3 \\
\Delta_4 \\
\Delta_5 \\
\Delta_6
\end{Bmatrix}
$$

$$
\{P\} = \begin{Bmatrix} \{P\}_1 \\ \hline \{P\}_2 \end{Bmatrix};
\qquad
\{\Delta\} = \begin{Bmatrix} \{\Delta\}_1 \\ \hline \{\Delta\}_2 \end{Bmatrix}
$$

Loading on Primary Structure:

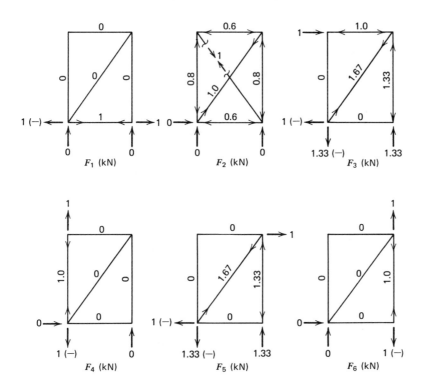

Determination of Flexibility Coefficients: The standard virtual work expression is used to determine the flexibility coefficients. That is,

$$\sum_{i=1}^{n} (\delta P)_i D_i = \sum_{j=1}^{m} \left(\delta F_P \frac{F_D L}{EA} \right)_j \qquad (5.26)$$

Specifically, for f_{pq}, we have

$$1 \cdot f_{pq} = \frac{1}{EA} \sum F_p \cdot F_q L$$

The necessary summations are formed in the following table:

Member	L	F_1	F_2	F_3	F_4	F_5	F_6	$F_1^2 L$	$F_2 F_1 L$
—	m	kN	kN	kN	kN	kN	kN	kN²·m	kN²·m
ab	4.0	0	−0.8	0	1.0	0	0	0	0
ac	5.0	0	1.0	1.67	0	1.67	0	0	0
ad	3.0	1.0	−0.6	0	0	0	0	3.00	−1.80
bc	3.0	0	−0.6	−1.0	0	0	0	0	0
cd	4.0	0	−0.8	−1.33	0	−1.33	1.0	0	0
bd	5.0	0	1.0	0	0	0	0	0	0
Σ								3.00	−1.80

F_2^2L kN²·m	F_3F_1L kN²·m	F_3F_2L kN²·m	F_3^2L kN²·m	F_4F_1L kN²·m	F_4F_2L kN²·m	F_4F_3L kN²·m	F_4^2L kN²·m	F_5F_1L kN²·m	F_5F_2L kN²·m
2.56	0	0	0	0	−3.20	0	4.00	0	0
5.00	0	8.35	13.94	0	0	0	0	0	8.35
1.08	0	0	0	0	0	0	0	0	0
1.08	0	1.80	3.00	0	0	0	0	0	0
2.56	0	4.26	7.08	0	0	0	0	0	4.26
5.00	0	0	0	0	0	0	0	0	0
17.28	0	14.41	24.02	0	−3.20	0	4.00	0	12.61

F_5F_3L kN²·m	F_5F_4L kN²·m	F_5^2L kN²·m	F_6F_1L kN²·m	F_6F_2L kN²·m	F_6F_3L kN²·m	F_6F_4L kN²·m	F_6F_5L kN²·m	F_6^2L kN²·m
0	0	0	0	0	0	0	0	0
13.94	0	13.94	0	0	0	0	0	0
0	0	0	0	0	0	0	0	0
0	0	0	0	0	0	0	0	0
7.08	0	7.08	0	−3.20	−5.32	0	−5.32	4.00
0	0	0	0	0	0	0	0	0
21.02	0	21.02	0	−3.20	−5.32	0	−5.32	4.00

Note:

Cut bar must be included in tabulation.

$$[/] = \frac{1}{EA}\left[\begin{array}{cc|cccc} 3.00 & -1.80 & 0 & 0 & 0 & 0 \\ -1.80 & 17.28 & 14.41 & -3.20 & 12.61 & -3.20 \\ \hline 0 & 14.41 & 24.02 & 0 & 21.02 & -5.32 \\ 0 & -3.20 & 0 & 4.00 & 0 & 0 \\ 0 & 12.61 & 21.02 & 0 & 21.02 & -5.32 \\ 0 & -3.20 & -5.32 & 0 & -5.32 & 4.00 \end{array}\right] = \left[\begin{array}{c|c} [/]_{11} & [/]_{12} \\ \hline [/]_{21} & [/]_{22} \end{array}\right]$$

Redundant Reactions and Member Forces:

$$[/]_{11}\{P\}_1 + [/]_{12}\{P\}_2 = \{\Delta\}_1$$
$$\{P\}_1 = [/]_{11}^{-1}(\{\Delta\}_1 - [/]_{12}\{P\}_2) \tag{10.31}$$

$$\{P\}_1 = \begin{Bmatrix} R_{dx} \\ F_{bd} \end{Bmatrix} = \frac{EA}{48.60}\begin{bmatrix} 17.28 & 1.80 \\ 1.80 & 3.00 \end{bmatrix}$$

$$\times \left(\begin{Bmatrix} 0 \\ 0 \end{Bmatrix} - \frac{1}{EA}\begin{bmatrix} 0 & 0 & 0 & 0 \\ 14.41 & -3.20 & 12.61 & -3.20 \end{bmatrix}\begin{Bmatrix} 40 \\ 0 \\ 0 \\ -50 \end{Bmatrix} \right)$$

$$\begin{Bmatrix} R_{dx} \\ F_{bd} \end{Bmatrix} = \begin{Bmatrix} -27.27 \\ -45.46 \end{Bmatrix} \text{ kN}$$

Displacements:

$$[/]_{21}\{P\}_1 + [/]_{22}\{P\}_2 = \{\Delta\}_2 \tag{10.32}$$

$$\{\Delta\}_2 = \begin{Bmatrix} \Delta_3 \\ \Delta_4 \\ \Delta_5 \\ \Delta_6 \end{Bmatrix} = \frac{1}{EA} \begin{bmatrix} 0 & 14.41 \\ 0 & -3.20 \\ 0 & 12.61 \\ 0 & -3.20 \end{bmatrix} \begin{Bmatrix} -27.27 \\ -45.46 \end{Bmatrix}$$

$$+ \frac{1}{EA} \begin{bmatrix} 24.02 & 0 & 21.02 & -5.32 \\ 0 & 4.00 & 0 & 0 \\ 21.02 & 0 & 21.02 & -5.32 \\ -5.32 & 0 & -5.32 & 4.00 \end{bmatrix} \begin{Bmatrix} 40 \\ 0 \\ 0 \\ -50 \end{Bmatrix}$$

$$\begin{Bmatrix} \Delta_3 \\ \Delta_4 \\ \Delta_5 \\ \Delta_6 \end{Bmatrix} = \frac{1}{EA} \begin{Bmatrix} 571.72 \\ 145.47 \\ 533.55 \\ -267.33 \end{Bmatrix} \left(\frac{\text{kN} \cdot \text{m}}{EA} \right)$$

Member Forces and Reactions for Primary Structure:

$$\begin{Bmatrix} \{F\}_P \\ \{R\}_P \end{Bmatrix} = [C]_P^{-1} \begin{Bmatrix} \{P\}_1 \\ \{P\}_2 \end{Bmatrix} \tag{10.37}$$

$$\begin{Bmatrix} F_{ab} \\ F_{ac} \\ F_{ad} \\ F_{bc} \\ F_{cd} \\ R_{ax} \\ R_{ay} \\ R_{dy} \end{Bmatrix} = \begin{bmatrix} 0 & -0.8 & 0 & 1.0 & 0 & 0 \\ 0 & 1.0 & 1.67 & 0 & 1.67 & 0 \\ 1.0 & -0.6 & 0 & 0 & 0 & 0 \\ 0 & -0.6 & -1.0 & 0 & 0 & 0 \\ 0 & -0.8 & -1.33 & 0 & -1.33 & 1.0 \\ -1.0 & 0 & -1.0 & 0 & -1.0 & 0 \\ 0 & 0 & -1.33 & -1.0 & -1.33 & 0 \\ 0 & 0 & 1.33 & 0 & 1.33 & -1.0 \end{bmatrix} \begin{Bmatrix} -27.27 \\ -45.46 \\ \hline 40.00 \\ 0 \\ 0 \\ -50.00 \end{Bmatrix} = \begin{Bmatrix} 36.67 \\ 21.34 \\ 0 \\ -12.73 \\ -66.83 \\ -12.73 \\ -53.20 \\ 103.20 \end{Bmatrix} \text{kN}$$

Notes:

- Each column of $[C]_P^{-1}$ includes the bar forces and reactions for one of the unit loading cases for the primary structure.
- If there were a temperature change or fabrication error for member *bd*, then Δ_2 would no longer be zero. Instead, it would reflect the induced change in member length.
- If there were a horizontal support movement at point *d*, then Δ_1 would no longer be zero. Instead, it would have to be set equal to the induced support displacement.

Final Results:

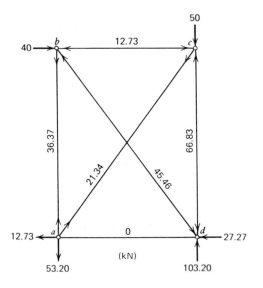

10.9.2 Example problem For the structure of Example 10.9.1, determine the displacements at points b and c and the member forces that occur as a result of a temperature change of $-25°C$ in member bd.

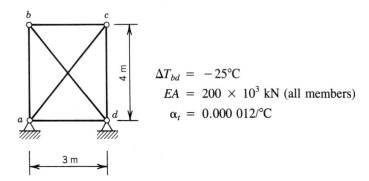

$$\Delta T_{bd} = -25°C$$
$$EA = 200 \times 10^3 \text{ kN (all members)}$$
$$\alpha_t = 0.000\ 012/°C$$

Primary Structure and Coordinate Systems: The same selections are made as were employed in Example 10.9.1. In this case, however,

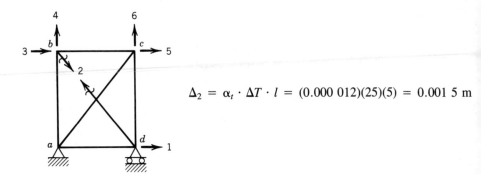

$$\Delta_2 = \alpha_t \cdot \Delta T \cdot l = (0.000\ 012)(25)(5) = 0.001\ 5 \text{ m}$$

Note:

Δ_2 is plus since a member shortening produces a positive Δ for the coordinate system adopted.

$$\{P\} = \begin{Bmatrix} \{P\}_1 \\ \{P\}_2 \end{Bmatrix} = \begin{Bmatrix} P_1 \\ P_2 \\ \hline P_3 \\ P_4 \\ P_5 \\ P_6 \end{Bmatrix} = \begin{Bmatrix} R_{dx} \\ F_{bd} \\ \hline 0 \\ 0 \\ 0 \\ 0 \end{Bmatrix}; \qquad \{\Delta\} = \begin{Bmatrix} \{\Delta\}_1 \\ \{\Delta\}_2 \end{Bmatrix} = \begin{Bmatrix} \Delta_1 \\ \Delta_2 \\ \hline \Delta_3 \\ \Delta_4 \\ \Delta_5 \\ \Delta_6 \end{Bmatrix} = \begin{Bmatrix} 0 \\ 0.001\,5 \\ \hline \Delta_3 \\ \Delta_4 \\ \Delta_5 \\ \Delta_6 \end{Bmatrix}$$

Redundant Reaction and Member Force: All flexibility quantities are the same as those determined in Example 10.9.1

$$[f]_{11}\{P\}_1 + [f]_{12}\{P\}_2 = \{\Delta\}_1$$
$$\{P\}_1 = [f]_{11}^{-1}(\{\Delta\}_1 - [f]_{12}\{P\}_2) \tag{10.31}$$

Since $\{P\}_2 = \{0\}$, we have

$$\{P\}_1 = \begin{Bmatrix} R_{dx} \\ F_{bd} \end{Bmatrix} = \frac{EA}{48.60}\begin{bmatrix} 17.80 & 1.80 \\ 1.80 & 3.00 \end{bmatrix}\begin{Bmatrix} 0 \\ 0.001\,5 \end{Bmatrix} = \begin{Bmatrix} 5.55 \times 10^{-5}EA \\ 9.25 \times 10^{-5}EA \end{Bmatrix}$$

For $EA = 200 \times 10^3$ kN, we obtain

$$\begin{Bmatrix} R_{dx} \\ F_{bd} \end{Bmatrix} = \begin{Bmatrix} 11.1 \\ 18.5 \end{Bmatrix} \text{ kN}$$

Displacements:

$$[f]_{21}\{P\}_1 + [f]_{22}\{P\}_2 = \{\Delta\}_2$$

For $\{P\}_2 = \{0\}$,

$$\{\Delta\}_2 = \begin{Bmatrix} \Delta_3 \\ \Delta_4 \\ \Delta_5 \\ \Delta_6 \end{Bmatrix} = \frac{1}{EA}\begin{bmatrix} 0 & 14.41 \\ 0 & -3.20 \\ 0 & 12.61 \\ 0 & -3.20 \end{bmatrix}\begin{Bmatrix} 5.55 \times 10^{-5}EA \\ 9.25 \times 10^{-5}EA \end{Bmatrix} = \begin{Bmatrix} 0.001\,333 \\ -0.000\,296 \\ 0.001\,167 \\ -0.000\,296 \end{Bmatrix} \text{m}$$

Member Forces and Reactions for Primary Structure:

$$\begin{Bmatrix} \{F\}_P \\ \{R\}_P \end{Bmatrix} = [C]_P^{-1}\begin{Bmatrix} \{P\}_1 \\ \{P\}_2 \end{Bmatrix} \tag{10.37}$$

$[C]_P^{-1}$ is the same as that for Example 10.9.1, and since $\{P\}_2 = \{0\}$,

$$\begin{Bmatrix} F_{ab} \\ F_{ac} \\ F_{ad} \\ F_{bc} \\ F_{cd} \\ \hline R_{ax} \\ R_{ay} \\ R_{dy} \end{Bmatrix} = \begin{bmatrix} 0 & -0.8 \\ 0 & 1.0 \\ 1.0 & -0.6 \\ 0 & -0.6 \\ 0 & -0.8 \\ \hline -1.0 & 0 \\ 0 & 0 \\ 0 & 0 \end{bmatrix}\begin{Bmatrix} 11.1 \\ 18.5 \end{Bmatrix} = \begin{Bmatrix} -14.8 \\ 18.5 \\ 0 \\ -11.1 \\ -14.8 \\ \hline -11.1 \\ 0 \\ 0 \end{Bmatrix}$$

Final Results:

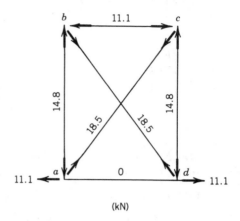

(kN)

10.10

Compatibility Equations by Energy Methods

Castigliano's second theorem was developed in Section 5.9.2, and it was stated there that this theorem is useful in determining individual displacement quantities or in writing compatibility equations for statically indeterminate analysis. The application concerning the determination of individual displacements was illustrated for trusses in Section 5.10. The use of the theorem is generating the governing compatibility equations for the analysis of statically indeterminate trusses is illustrated in this section.

It is recalled from Section 5.9.2 that the mathematical expression for Castigliano's second theorem has the form

$$\frac{\partial U}{\partial P_i} = \Delta_i \tag{10.38}$$

where U is the total strain energy for the structure and P_i and Δ_i represent the ith externally applied load and the corresponding displacement, respectively. The strain energy must first be expressed in terms of the member forces. This is accomplished according to the procedure developed in Section 5.10, where it was shown that for a truss of m members

$$U = \sum_{i=1}^{m} \frac{F_i^2 l_i}{2E_i A_i} \tag{10.39}$$

In this expression, F_i is the member force, which must be expressed in terms of the P loads that are applied to the primary structure, l_i is the member length, A_i is the member cross-sectional area, and E_i is the modulus of elasticity. It is noted that the total strain energy must include the strain energy stored in any cut members.

In the present problem, Eq. 10.38 must be applied for each displacement degree of freedom on the primary structure. This leads to a system of interactive displacement–force relationships that are identical to those given in Eq. 10.29.

These equations are partitioned to separate the redundant forces from the applied forces as shown in Eq. 10.30, and the solution proceeds in accordance with the method explained in Section 10.8.

Consider the structure given in Example 10.9.1. Show that Castigliano's second **10.10.1** theorem can be used to generate the compatibility equations in the form of Eq. **Example** 10.29. **problem**

See Section 10.9.1 for the given structure, the structural classification, and the selection of primary structure.

Member Forces for Primary Structure:

$R_{dx} = P_1 =$ redundant reaction

$F_{bd} = P_2 =$ redundant member force (this is an internal force for the original structure, but it is external force for the primary structure)

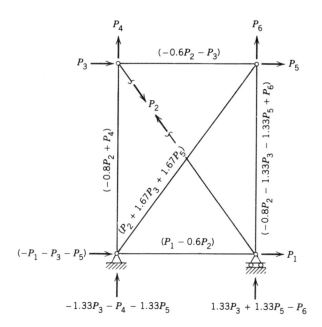

Strain Energy:

$$U = \sum_{i=1}^{m} \frac{F_i^2 l_i}{2E_i A_i} = \frac{1}{2EA} \sum_{i=1}^{m} F_i^2 l_i \qquad (10.39)$$

$$U = \frac{1}{2EA}[(-0.8P_2 + P_4)^2(4) + (-0.6P_2 - P_3)^2(3)$$

$$+ (P_2 + 1.67P_3 + 1.67P_5)^2(5) + (P_2)^2(5)$$

$$+ (0.8P_2 - 1.33P_3 - 1.33P_5 + P_6)^2(4)$$

$$+ (P_1 - 0.6P_2)^2(3)]$$

Note:
Cut bar must be included in strain energy expression.

Compatibility Equations from Castigliano's Second Theorem:

$$\frac{\partial U}{\partial P_1} = \Delta_1 \Rightarrow \frac{1}{EA}[(P_1 - 0.6P_2)(1)(3)] = \Delta_1$$

$$\frac{1}{EA}(3.00P_1 - 1.80P_2) = \Delta_1$$

$$\frac{\partial U}{\partial U_2} = \Delta_2 \Rightarrow \frac{1}{EA}[(-0.8P_2 + P_4)(-0.8)(4) + (-0.6P_2 - P_3)(-0.6)(3)$$

$$+ (P_2 + 1.67P_3 + 1.67P_5)(1)(5) + (P_2)(1)(5)$$
$$+ (-0.8P_2 - 1.33P_3 - 1.33P_5 + P_6)(-0.8)(4)$$
$$+ (P_1 - 0.6P_2)(-0.6)(3)]$$

$$\frac{1}{EA}(-1.80P_1 + 17.28P_2 + 14.41P_3 - 3.20P_4$$

$$+ 12.61P_5 - 3.20P_6) = \Delta_2$$

etc. for

$$\frac{\partial U}{\partial P_i} = \Delta_i \qquad \text{for } i = 3, 4, 5, 6$$

The resulting equations are

$$\frac{1}{EA}\begin{bmatrix} 3.00 & -1.80 & 0 & 0 & 0 & 0 \\ -1.80 & 17.28 & 14.41 & -3.20 & 12.6 & -3.20 \\ \hline & & & & & \\ & & \text{etc.} & & & \\ & & & & & \\ & & & & & \end{bmatrix}\begin{Bmatrix} P_1 \\ P_2 \\ \hline P_3 \\ P_4 \\ P_5 \\ P_6 \end{Bmatrix} = \begin{Bmatrix} \Delta_1 \\ \Delta_2 \\ \hline \Delta_3 \\ \Delta_4 \\ \Delta_5 \\ \Delta_6 \end{Bmatrix}$$

These equations agree with those developed in Example 10.9.1

10.11

Multiple Loading Cases

It is frequently necessary in structural design to analyze the structure for a number of separate loading conditions. Each of these loading conditions would represent a possible load pattern for the structure, and all conditions would have to be considered in order to determine the most severe loading for each member in the structure.

The method of consistent deformations lends itself readily to the study of multiple loading cases. This is most evident when one considers Eq. 10.34. Each loading condition produces its own $(\{\Delta\}_1 - \{\Delta_0\}_1)$ vector, where $\{\Delta\}_1$ includes any specified displacements and $\{\Delta_0\}_1$ reflects the applied loading. However, the loading condition has no effect on $[f]_{11}^{-1}$, since the flexibility coefficients are solely dependent on the properties of the structure. Thus, for each loading condition, Eq. 10.34 yields a set of redundant forces that are represented by the vector $\{P\}_1$. Allowing $(\{\Delta\}_1 - \{\Delta_0\}_1)$ for the ith loading case to be represented by $\{\Delta\}_1^i$, and the corresponding redundant forces to be represented by $\{P\}_1^i$, we can generalize Eq. 10.34 for n loading conditions in the form

$$[\{P\}_1^1 \cdots \{P\}_1^i \cdots \{P\}_1^n] = [f]_{11}^{-1}[\{\Delta\}_1^1 \cdots \{\Delta\}_1^i \cdots \{\Delta\}_1^n] \quad (10.40)$$

10.12 Selection of Redundants

An important consideration in using the method of consistent deformations deals with the selection of the redundants. This topic is discussed in greater detail in Chapter 18. Here, it is sufficient to offer a few observations.

As a general rule, the redundants should be selected so that the resulting primary structure will possess an $[/]_{11}$ matrix that is well conditioned. This will minimize the effect of round-off errors when $[/]_{11}$ is inverted in accordance with Eq. 10.34.

10.13 Additional Reading

Beaufait, F. W., *Basic Concepts of Structural Analysis*, Sections 8.1 through 8.3, Prentice–Hall, Englewood Cliffs, N.J., 1977.

Hsieh, Yuan-Yu, *Elementary Theory of Structures*, Chapter 9, Prentice–Hall, Englewood Cliffs, N.J., 1970.

McCormac, J. C., *Structural Analysis*, 3rd Ed., Chapter 20, Intext Educational Publishers, New York, 1975.

Wang, Chu-Kia, *Matrix Methods of Structural Analysis*, 2nd Ed., Chapter 8, International Textbook, Scranton, Pa., 1970.

10.14 Suggested Problems

1 through 9. Use the method of consistent deformations to determine the reactions and member forces for each of the statically indeterminate trusses given, subject to the prescribed loading.

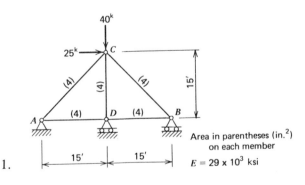

1.

Area in parentheses (in.^{2}) on each member

$E = 29 \times 10^3$ ksi

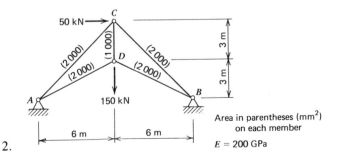

2.

Area in parentheses (mm^2) on each member

$E = 200$ GPa

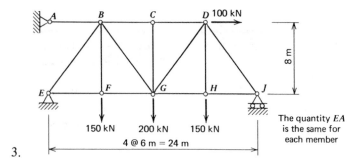

150 kN   200 kN   150 kN

4 @ 6 m = 24 m

The quantity EA
is the same for
each member

3.

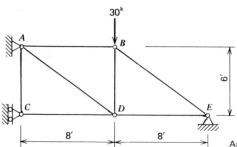

30^k

8'   8'

Area = 2.5 in.^{2} for
each member

$E = 29 \times 10^3$ ksi

4.

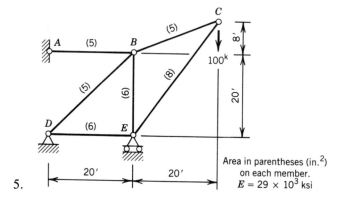

(5)

(5)

(5)

(8)

(6)

(6)

100^k

20'   20'

Area in parentheses (in.^{2})
on each member.
$E = 29 \times 10^3$ ksi

5.

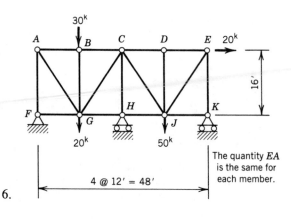

30^k

20^k 50^k

20^k

4 @ 12' = 48'

The quantity EA
is the same for
each member.

6.

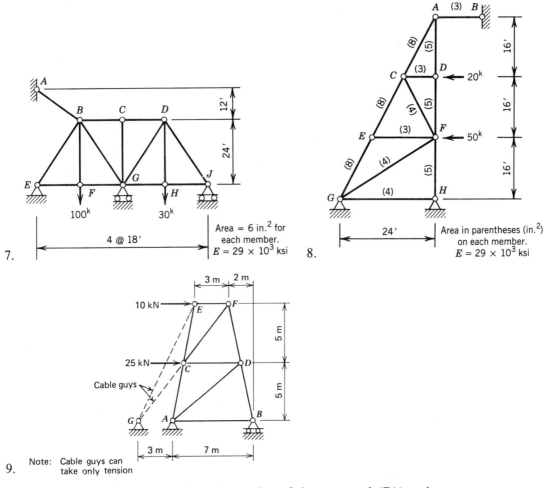

7.

Area = 6 in.² for each member.
E = 29 × 10³ ksi

8.

Area in parentheses (in.²) on each member.
E = 29 × 10³ ksi

9.

Note: Cable guys can take only tension

The quantity EA is the same for each member of the truss, and (EA)ᵤₐᵦₗₑ is one-tenth of the EA values for the truss members.

10. Determine the reactions and member forces for the truss of Problem 2 if the support at point B settles downward by 25 mm and to the right by 15 mm. Disregard the loads given in Problem 2.

11. Determine the reactions and member forces for the truss of Problem 4 if, in addition to the given load, the support at point E settles downward by 0.75 inch.

12. Determine the reactions and member forces for the truss of Problem 1 if the following vertical support settlements occur: point A, 0.50 inch upward; point B, 0.50 inch downward; point D, 0.75 inch downward. Disregard the loads given in Problem 1.

13. Determine the reactions and member forces for the structure of Problem 7 if, in addition to the given loads, the support at point A shifts 0.5 inch toward point B, and point J settles 0.65 inch downward.

14. Determine the reactions and member forces for the structure of Problem 8 if, in addition to the given loads, point G settles 0.75 inch downward and point H is uplifted 0.25 inch.

15. Determine the reactions and the member forces for the truss and loading of Problem 1 if the rigid support of point D is replaced by the elastic support shown below. Compare your solution with that obtained in Problem 1.

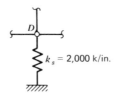

$k_s = 2,000$ k/in.

16. Determine the reactions and the member forces for the truss and loading of Problem 2 if the rigid support of point B is replaced by the elastic support shown. Compare your solution with that obtained in Problem 2.

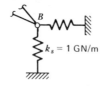

$k_s = 1$ GN/m

17. Determine the reactions and member forces for the truss and loading of Problem 7 if the rigid supports at points G and J are replaced by the elastic supports shown. Compare your solution with the one obtained in Problem 7.

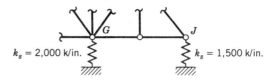

$k_s = 2,000$ k/in. $k_s = 1,500$ k/in.

18. Determine the reactions and member forces for the structure of Problem 1 if members AC and CB are each subjected to a temperature increase of 50°F. Disregard the loads given in Problem 1, and take $\alpha_t = 0.000,006,5/°F$.

19. Determine the reactions and member forces for the structure of Problem 4 if, in addition to the specified load, member AD is 0.75 inch too long but is forced to fit.

20 through 23. Use the conventional method of consistent deformations to determine the member forces for each of the statically indeterminate trusses shown for the given loadings.

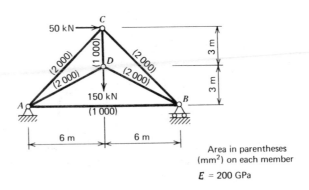

Area in parentheses (mm²) on each member

20.

$E = 200$ GPa

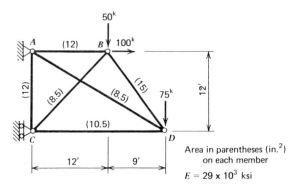

Area in parentheses (in.²)
on each member

$E = 29 \times 10^3$ ksi

21.

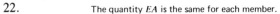

150 kN   200 kN   150 kN
4 @ 5 m = 20 m

22.   The quantity EA is the same for each member.

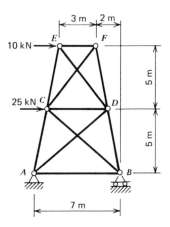

The quantity EA is the same
23.   for each member.

24. Determine the reactions and member forces for the structure of Problem 20 if members AC and CB are subjected to a temperature increase of 30°C. Disregard the loading of Problem 20, and take $\alpha_t = 0.000\ 012/°C$.

25. Determine the reactions and member forces for the structure of Problem 21 if, in addition to the given loads, member BD is 0.50 inch too short but is forced to fit.

26. Determine the reactions and member forces for the structure of Problem 22 if members AB, BC, CD, and DE are subjected to a temperature increase of 25°C. Disregard the loading of Problem 22, and take $\alpha_t = 0.000\ 012/°C$. Express your results in terms of EA.

27 through 33. Use the matrix formulation of Section 10.8 to determine the reactions, the member forces, and the designated free-joint displacements for
each structure shown for the indicated loading. The suggested primary structure and the coordinate system (selected to include the desired free-joint displacements) are shown in each case.

27. The structure and loading of Problem 1.

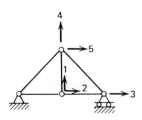

28. The structure and loading of Problem 9.

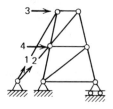

29. The structure and loading of Problem 20.

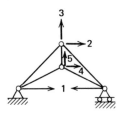

30. The structure and loading of Problem 22.

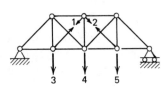

31. The structure and loading shown below.

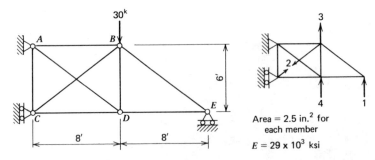

Area = 2.5 in.^{2} for
each member

$E = 29 \times 10^3$ ksi

32. The structure of Problem 20 with member AB 0.5 inch too long but forced to fit. Disregard the prescribed loads of Problem 20, and use the coordinate system of Problem 29.

33. The structure and coordinate system of Problem 31, if member BE is subjected to a temperature increase of 50°F. Disregard the applied loads.

34. Develop a computer program for the solution of statically indeterminate trusses by the method of Section 10.8. The program should be capable of handling either external or internal redundants.

   The truss analysis program of Problem 3.30 and the flexibility matrix program of Problem 5.26 can be incorporated as subprograms.

   Use the program to solve Problems 27, 29, and 31.

35 through 37. Use Castigliano's second theorem to generate the compatibility equations in the form of Eq. 10.29 for each of the designated structures for the specified loadings. The selection of the redundant force components and the free-joint displacements to be included are indicated by the suggested primary structure and coordinate system.

35. The structure and loading of Problem 1 with the primary structure and coordinate system indicated in Problem 27.

36. The structure and loading of Problem 20 with the primary structure and coordinate system indicated in Problem 29.

37. The structure and loading of Problem 31, along with the primary structure and coordinate system shown.

CHAPTER 11

The World Trade Center, New York City (courtesy The Port Authority of New York and New Jersey).

ANALYSIS OF BEAM AND FRAME STRUCTURES BY COMPATIBILITY METHODS

11.1

Compatibility Methods for Beam-Type Structures

Since the underlying principles of compatibility methods are the same regardless of the structure type, much of what will be presented in this chapter has already been introduced in Chapter 10. Thus, it is suggested that the reader consider these two chapters as a single unit of study and that they be taken in order. The emphasis in Chapter 10 was on pin-connected structures, whereas here the emphasis is on beam- and frame-type structures. Thus, the main difference lies in the techniques employed in identifying an appropriate primary structure and in determining the required flexibility coefficients and other displacement quantities.

As was explained for truss analysis, the methods that are characterized as compatibility methods are those in which the fundamental equations that are used in the solution are compatibility equations. The final solution must also satisfy equilibrium; however, the satisfaction of these conditions is woven into the solution as the individual compatibility equations are formulated.

Of the many compatibility methods that are available, only a few of the most common will be treated here. This chapter is also limited to the system approach, in which the displacement quantities are determined from considering the entire structural system. This is in contrast with the element approach, in which the governing equations result from a systematic combination of the compatibility relationships for the individual members.

11.2 Nature of Redundancies

A redundant force was described in Chapter 9 as a force that can be removed from the structure without rendering the structure unstable. Thus, such a force is redundant, or unnecessary, for the stability of the structure. As was true with truss structures, these redundant forces may be external or internal. External redundants are reaction forces, whereas internal redundants are components of internal forces.

This chapter is limited to planar structures, and thus the nature of the redundants and the number of them can be established by considering the provisions of Section 4.12.

11.3 Method of Consistent Deformations for Beam and Frame Structures

The simple statically indeterminate beam problem that was discussed in Sections 9.1 through 9.5 provides an elementary example of the method of consistent deformations for a beam-type structure. The formalization of the method for pin-connected frameworks was accomplished in Chapter 10. These two treatments serve as background material for the development of the method for beam and frame structures as it is presented in this section.

The problem considered in Chapter 9 provides an example of a singly redundant case, and thus we will consider here a case that is statically indeterminate to the second degree. The continuous beam structure of Fig. 11.1a will serve as the example. The criteria of Section 4.12 verify that this structure is twice statically indeterminate. One way to reduce the given structure to a statically determinate primary structure is to remove the two interior reactions, as shown in Fig. 11.1b. These redundant reaction components are identified as R_1 and R_2. The primary structure can now be analyzed by the methods of statics, and the displacements Δ_{10} and Δ_{20}, which correspond to the lines of action of the redundant reactions, can be determined by the methods of Chapter 6. Since Δ_{10} and Δ_{20} are in violation of the boundary conditions of the original structure, it is necessary to modify the solution of the primary structure until the displacements at these points are compatible with the prescribed boundary conditions. The required modification is accomplished by introducing, in turn, unit values of the redundant reactions on the primary structure and determining the effects that these individual loading cases have on the displacements where compatibility is to be restored. These unit load cases are shown in Fig. 11.1c; they can be analyzed in accordance with static considerations and the desired displacements determined by the methods of Chapter 6. Each of these displacements is shown in Fig. 11.1c as f_{ij}, which is the flexibility coefficient that expresses the displacement at the point and in the direction of the redundant reaction R_i that is caused by a unit value of the redundant reaction R_j.

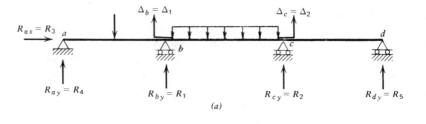

(a)

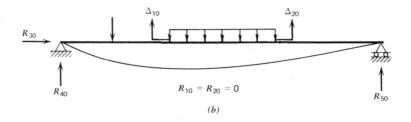

(b)

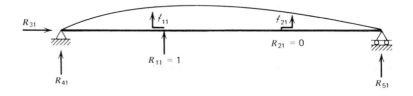

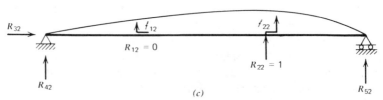

(c)

Fig. 11.1 *Statically indeterminate continuous beam.* (a) *Statically indeterminate beam.* (b) *Statically determinate primary structure.* (c) *Unit values of the redundant reactions.*

The total displacements at the points and in the directions of the redundant reactions that are caused by the combined effects of the redundant reactions are identified as Δ_{1R} and Δ_{2R} and are determined from superposition to be

$$\Delta_{1R} = f_{11}R_1 + f_{12}R_2 \tag{11.1}$$
$$\Delta_{2R} = f_{21}R_1 + f_{22}R_2$$

These displacements must be combined with Δ_{10} and Δ_{20} of Fig. 11.1b to yield the desired displacements of the original structure as defined in Fig. 11.1a. This combination is expressed by compatibility equations in the form

$$\Delta_{10} + \Delta_{1R} = \Delta_1 \tag{11.2}$$
$$\Delta_{20} + \Delta_{2R} = \Delta_2$$

Or, substituting Eq. 11.1 into Eq. 11.2 and rearranging terms, we have

$$\begin{bmatrix} f_{11} & f_{12} \\ f_{21} & f_{22} \end{bmatrix} \begin{Bmatrix} R_1 \\ R_2 \end{Bmatrix} = \begin{Bmatrix} \Delta_1 - \Delta_{10} \\ \Delta_2 - \Delta_{20} \end{Bmatrix} \tag{11.3}$$

This form is identical with Eq. 10.10 and, as was explained there, the f terms form the flexibility matrix that gives the displacement–force relations associated with the actions of the redundant reactions R_1 and R_2.

The solution of Eq. 11.3 gives the magnitudes of the redundant reactions. These reactions can be placed on the original structure, and the remaining reactions can be determined from statics. Or, as a more general procedure, the same superposition pattern that is expressed in Eqs. 11.1 and 11.2 is used for determining any other response quantity of interest, such as reaction, moment, or shear. For instance, if S is taken as such a response quantity, then

$$S = S_0 + S_1 R_1 + S_2 R_2 \tag{11.4}$$

where S_0 is the value of S on the primary structure when the actual loading of the given structure is applied, and S_i is the value of S on the primary structure when a unit value of R_i is applied.

The discussion in Section 10.5 concerning support settlements and elastic supports is applicable here too. For cases of nonyielding supports, $\Delta_1 = \Delta_2 = 0$, whereas if support settlements are permitted, then values of Δ_1 and Δ_2 are prescribed. In either case, Eq. 11.3 is valid, and its solution produces the values of the redundant reactions. The value of any other response quantity of interest can be determined from Eq. 11.4. If there are prescribed settlements for the end support points, then Δ_1 and Δ_2 must be interpreted as the displacements relative to the chord connecting the ends of the beam. This situation is illustrated in Fig. 11.2.

If the redundant reactions of the original structure are provided by elastic supports, then a relationship similar to Eq. 10.15 is valid for each redundant reaction point. Thus, at support i,

$$R_i = -k_{si}\Delta_i \tag{11.5}$$

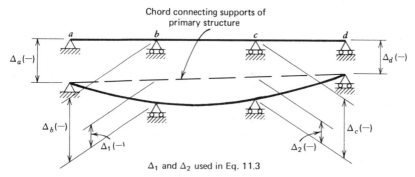

Fig. 11.2 *Settlement of continuous beam. Note that upward displacements are positive.*

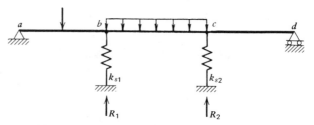

Fig. 11.3 *Continuous beam on elastic supports.*

If the supports at points b and c on Fig. 11.1a are provided by elastic supports, as shown in Fig. 11.3, then Eq. 11.3 takes the form

$$
\begin{bmatrix}
\left(f_{11} + \dfrac{1}{k_{s1}}\right) & f_{12} \\
f_{21} & \left(f_{22} + \dfrac{1}{k_{s2}}\right)
\end{bmatrix}
\begin{Bmatrix} R_1 \\ R_2 \end{Bmatrix}
=
\begin{Bmatrix} -\Delta_{10} \\ -\Delta_{20} \end{Bmatrix}
\tag{11.6}
$$

This arrangement would correspond to the case in which R_1 and R_2 are provided by beams that span in a direction perpendicular to continuous beam $abcd$. Here, the elastic constants, k_{s1} and k_{s2}, would be a function of the properties of the supporting beams.

If points a and d also have elastic supports, Eq. 11.6 is still valid. However, care must be exercised to make certain that the flexibility coefficients f_{ij} and the displacements Δ_{i0}, include the effects of the displacements at points a and d of the primary structure.

In either case, the redundant reactions are provided through the solution of Eq. 11.6, and any other response quantities are determined from Eq. 11.4.

When cases with more than two redundants are considered, the same general procedure is employed. For each redundant removed from the original structure, a compatibility equation is required similar to those given in Eq. 11.1. This leads to a set of simultaneous equations similar to those given as Eq. 11.3, which must be solved simultaneously for the redundant reactions.

It is emphasized that the final solution satisfies all of the requirements of compatibility and equilibrium. The solution of the equations for the redundants ensures that all of the compatibility conditions are satisfied. Also, each solution of the primary structure, for the actual loading and the unit loading conditions, is based on a statically determinate analysis in which equilibrium is ensured. Thus, the superposition indicated by Eq. 11.4 will fully comply with the requirements of compatibility and equilibrium.

11.4

Application of the Method of Consistent Deformations

The application of the method of consistent deformations requires the formulation of compatibility equations of the type given by Eq. 11.3. This requires the determination of the Δ_{i0} displacements and the f_{ij} flexibility coefficients. These are all displacement quantities for the statically determinate primary structure, which are determined by any of the methods presented in Chapter 6.

The example problems given in this section illustrate the application of the method.

Construct the moment diagram for the frame structure given. The quantity EI is **11.4.1 Example** the same for each member.   **problem**

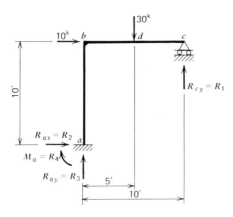

Structure Classification:

$j = 3$ External: $r_a > r;$ $4 > 3$

$m_a = 2$ $\therefore$ Statically indeterminate to first degree

$r_a = 4$

$n = 0$ Overall: $3m_a + r_a > 3j + n;$ $10 > 9$

$r = 3$ $\therefore$ Statically indeterminate to first degree

Primary Structure and Loadings: Select R_{cy} as the redundant reaction R_1.

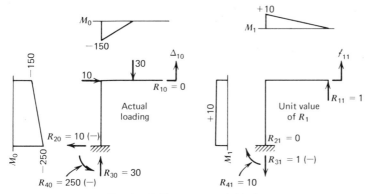

Reactions in kips ; moments in kip-ft

Displacement Calculations: Moment–area method is used.

Δ_{10}

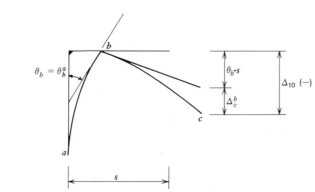

$$\theta_b = \theta_b^a = \left(\frac{250 + 150}{2EI}\right)10 = \frac{2{,}000 \text{ ft}^2\text{-k}}{EI}$$

$$\Delta_c^b = \left(\frac{150}{2EI}\right)5(8.33) = \frac{3{,}124 \text{ ft}^3\text{-k}}{EI}$$

$$\Delta_{10} = \left(\frac{2{,}000}{EI} \times 10\right) + \left(\frac{3{,}124}{EI}\right) = \frac{23{,}124 \text{ ft}^3\text{-k}}{EI}(-)$$

f_{11}

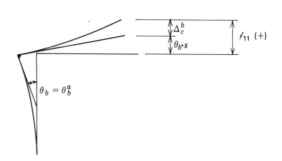

$$\theta_b = \theta_b^a = \left(\frac{10}{EI}\right)10 = \frac{100 \text{ ft}^2\text{-k}}{EI}$$

$$\Delta_c^b = \left(\frac{10}{2EI}\right)10(6.67) = \frac{333.5 \text{ ft}^2\text{-k}}{EI}$$

$$f_{11} = \left(\frac{100}{EI} \times 10\right) + \left(\frac{333.5}{EI}\right) = \frac{1{,}333.5 \text{ ft}^3\text{-k}}{EI}(+)$$

Redundant Reaction:

$$R_1 = \left(\frac{\Delta_1 - \Delta_{10}}{f_{11}}\right)$$

Since $\Delta_1 = 0$,

$$R_1 = -\frac{\Delta_{10}}{f_{11}} = \frac{23{,}124}{1{,}333.5} = 17.34$$

Final Moments:

$$M = M_0 + M_1 \cdot R_1$$

↑ ⌐ Moment diagram for $R_1 = 1$
  on primary structure

└ Moment diagram for actual loading
  on primary structure

In matrix form, for key points on the frame,

$$\begin{Bmatrix} M_a \\ M_b \\ M_c \\ M_d \end{Bmatrix} = \begin{Bmatrix} -250 \\ -150 \\ 0 \\ 0 \end{Bmatrix} + \begin{Bmatrix} +10 \\ +10 \\ 0 \\ +5 \end{Bmatrix} \cdot 17.34 = \begin{Bmatrix} -76.6 \\ +23.4 \\ 0 \\ +86.7 \end{Bmatrix} \quad \text{(kip-ft)}$$

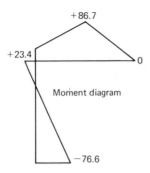

Moment diagram

Construct the final shear and moment diagrams for the structure and loading given. The quantity EI is the same for each span.

11.4.2 Example problem

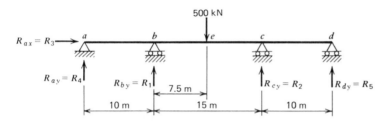

Structure Classification:

$j = 4$ External: $r_a > r$; $5 > 3$
$m_a = 3$ $\therefore$ Statically indeterminate to second degree
$r_a = 5$
$n = 0$ Overall: $3m_a + r_a > 3j + n$; $14 > 12$
$r = 3$ $\therefore$ Statically indeterminate to second degree

Primary Structure and Loadings:

Actual Loading:

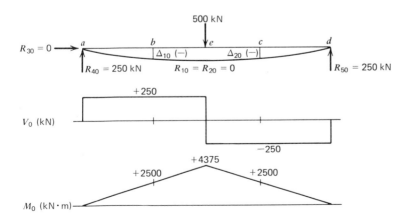

Unit Value of R_1:

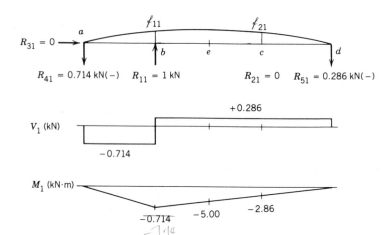

Unit Value of R_2:

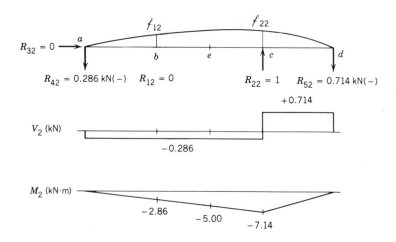

Displacement Calculations: Conjugate beam method is used.

Δ_{10} & Δ_{20}:

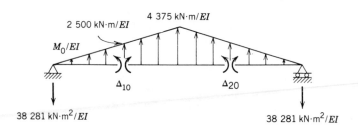

$$\Delta_{10} = -\left(\frac{38\ 281}{EI}\right)10 + \left(\frac{2\ 500}{2EI}\right)(10)(3.33) = \frac{-341\ 185\ \text{kN} \cdot \text{m}^3}{EI}$$

$\Delta_{20} = \Delta_{10}$ by symmetry

Both are negative, since upward displacements are positive.

f_{11} & f_{21}:

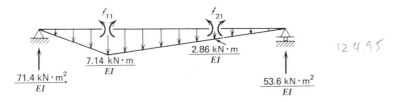

7.14 kN·m
EI

2.86 kN·m
EI

$\dfrac{71.4 \text{ kN} \cdot \text{m}^2}{EI}$

$\dfrac{53.6 \text{ kN} \cdot \text{m}^2}{EI}$

124.95

$$f_{11} = \left(\frac{71.4}{EI}\right)10 - \left(\frac{7.14}{2EI}\right)(10)(3.33) = \frac{+595.1 \text{ kN} \cdot \text{m}^3}{EI}$$

$$f_{21} = \left(\frac{53.6}{EI}\right)10 - \left(\frac{2.86}{2EI}\right)(10)(3.33) = \frac{+488.4 \text{ kN} \cdot \text{m}^3}{EI}$$

f_{12} & f_{22}:

$f_{12} = f_{21}$ by Maxwell's law (Eq. 5.54)

$f_{22} = f_{11}$ by symmetry

Redundant Reactions:

$$\begin{bmatrix} f_{11} & f_{12} \\ f_{21} & f_{22} \end{bmatrix} \begin{Bmatrix} R_1 \\ R_2 \end{Bmatrix} = \begin{Bmatrix} \Delta_1 - \Delta_{10} \\ \Delta_2 - \Delta_{20} \end{Bmatrix} \tag{11.3}$$

Since $\Delta_1 = \Delta_2 = 0$, we have

$$\begin{bmatrix} 595.1 & 588.4 \\ 488.4 & 595.1 \end{bmatrix} \begin{Bmatrix} R_1 \\ R_2 \end{Bmatrix} = \begin{Bmatrix} 341\ 185 \\ 341\ 185 \end{Bmatrix}$$

$$\begin{Bmatrix} R_1 \\ R_2 \end{Bmatrix} = \frac{1}{115\ 609.5} \begin{bmatrix} 595.1 & -488.4 \\ -488.4 & 595.1 \end{bmatrix} \begin{Bmatrix} 341\ 185 \\ 341\ 185 \end{Bmatrix} = \begin{Bmatrix} 314.9 \\ 314.9 \end{Bmatrix}$$

Final Shears and Moments:

$$S = S_0 + S_1 \cdot R_1 + S_2 \cdot R_2 \tag{11.4}$$

where S can be interpreted as the shears or moments. In matrix form, for key points on the structure,

$$\begin{Bmatrix} V_{a-b} \\ V_{b-e} \\ V_{e-c} \\ V_{c-d} \end{Bmatrix} = \begin{bmatrix} +250 \\ +250 \\ -250 \\ -250 \end{bmatrix} + \begin{bmatrix} -0.714 \\ +0.286 \\ +0.286 \\ +0.286 \end{bmatrix} 314.9 + \begin{bmatrix} -0.286 \\ -0.286 \\ -0.286 \\ +0.714 \end{bmatrix} 314.9 = \begin{Bmatrix} -64.9 \\ +250.0 \\ -250.0 \\ +64.9 \end{Bmatrix} \text{(kN)}$$

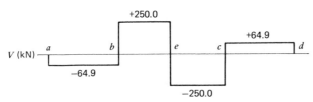

+250.0

+64.9

V (kN)   a   b   e   c   d

−64.9

−250.0

$$\begin{Bmatrix} M_a \\ M_b \\ M_e \\ M_c \\ M_d \end{Bmatrix} = \begin{bmatrix} 0 \\ +2\ 500 \\ +4\ 375 \\ +2\ 500 \\ 0 \end{bmatrix} + \begin{bmatrix} 0 \\ -7.14 \\ -5.00 \\ -2.86 \\ 0 \end{bmatrix} 314.9 + \begin{bmatrix} 0 \\ -2.86 \\ -5.00 \\ -7.14 \\ 0 \end{bmatrix} 314.9 = \begin{Bmatrix} 0 \\ -649 \\ +1\ 226 \\ -649 \\ 0 \end{Bmatrix} \text{(kN} \cdot \text{m)}$$

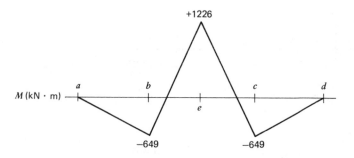

11.4.3 Example problem   Consider the continuous beam of Example 11.4.2, and determine the moment diagram for the following set of support settlements:

$$\Delta_a = -27.5 \text{ mm}; \quad \Delta_b = -47.5 \text{ mm}; \quad \Delta_c = -22 \text{ mm}; \quad \Delta_d = -10 \text{ mm}$$

Note:

Positive displacements are upward; therefore, these settlements are downward.

Settlement Pattern:

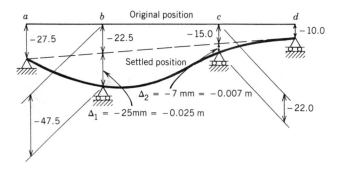

Structure Classification: Same as Example 11.4.2

Determination of Flexibility Coefficients: From Example 11.4.2

$$\begin{bmatrix} f_{11} & f_{12} \\ f_{21} & f_{22} \end{bmatrix} = \frac{1}{EI} \begin{bmatrix} 595.1 & 488.4 \\ 488.4 & 595.1 \end{bmatrix} \quad \left(\frac{\text{kN} \cdot \text{m}^3}{EI} \right)$$

Redundant Reactions:

$$\begin{bmatrix} f_{11} & f_{12} \\ f_{21} & f_{22} \end{bmatrix} \begin{Bmatrix} R_1 \\ R_2 \end{Bmatrix} = \begin{Bmatrix} \Delta_1 - \Delta_{10} \\ \Delta_2 - \Delta_{20} \end{Bmatrix} \tag{11.3}$$

In this case, $\Delta_{10} = \Delta_{20} = 0$, and thus

$$\frac{1}{EI} \begin{bmatrix} 595.1 & 488.4 \\ 488.4 & 595.1 \end{bmatrix} \begin{Bmatrix} R_1 \\ R_2 \end{Bmatrix} = \begin{Bmatrix} -0.025 \\ -0.007 \end{Bmatrix}$$

$$\begin{Bmatrix} R_1 \\ R_2 \end{Bmatrix} = \frac{EI}{115\ 609.5} \begin{bmatrix} 595.1 & -488.4 \\ -488.4 & 595.1 \end{bmatrix} \begin{Bmatrix} -0.025 \\ -0.007 \end{Bmatrix}$$

$$= EI \begin{Bmatrix} -99.1 \times 10^{-6} \\ 69.5 \times 10^{-6} \end{Bmatrix} \quad \left(\frac{EI}{\text{kN} \cdot \text{m}^2} \right)$$

Note:

R_1 and R_2 are unitless; they merely give the magnitudes of the redundant reactions.

Final Moments:

$$M = M_0 + M_1 \cdot R_1 + M_2 \cdot R_2 \qquad (11.4)$$

where, in this case, $M_0 = 0$ and M_1 and M_2 are the moment diagrams given in Example 11.4.2. In matrix form, for key points on the structure,

$$\begin{Bmatrix} M_a \\ M_b \\ M_e \\ M_c \\ M_d \end{Bmatrix} = \begin{Bmatrix} 0 \\ -7.14 \\ -5.00 \\ -2.86 \\ 0 \end{Bmatrix}(-99.1\ EI \times 10^{-6}) + \begin{Bmatrix} 0 \\ -2.86 \\ -5.00 \\ -7.14 \\ 0 \end{Bmatrix}(+69.5\ EI \times 10^{-6})$$

$$= \begin{Bmatrix} 0 \\ 508.8 \\ 148.0 \\ -212.8 \\ 0 \end{Bmatrix} EI \times 10^{-6}\left(\frac{EI}{m}\right)$$

Note:

The reactions and the moments are dependent on the flexural stiffness, EI. For the specific case of $I = 3\,000 \times 10^{-6}\ m^4$ and $E = 200 \times 10^9\ Pa = 200 \times 10^6$ kN/m^2, we obtain

$$\begin{Bmatrix} M_a \\ M_b \\ M_e \\ M_c \\ M_d \end{Bmatrix} = \begin{Bmatrix} 0 \\ 508.8 \\ 148.0 \\ -212.8 \\ 0 \end{Bmatrix}(200 \times 10^6)(3\,000 \times 10^{-6})10^{-6} = \begin{Bmatrix} 0 \\ 305.3 \\ 88.8 \\ -127.7 \\ 0 \end{Bmatrix}(kN \cdot m)$$

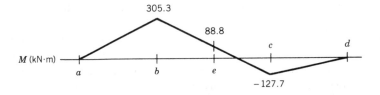

Consider again the beam and loading of Example 11.4.2, but, in this case, assume that points b and c are supported by elastic supports that have stiffnesses of $k_{s1} = k_{s2} = 0.006EI$ (units EI/m^3). Determine the reactions carried by the elastic supports. **11.4.4 Example problem**

Structure Classification: Same as Example 11.4.2

Determination of Displacement Quantities and Flexibility Coefficients: Same as Example 11.4.2

Redundant Reactions:

$$
\begin{bmatrix} \left(f_{11} + \dfrac{1}{k_{s1}}\right) & f_{12} \\[2mm] f_{21} & \left(f_{22} + \dfrac{1}{k_{s2}}\right) \end{bmatrix} \begin{Bmatrix} R_1 \\ R_2 \end{Bmatrix} = \begin{Bmatrix} -\Delta_{10} \\ -\Delta_{20} \end{Bmatrix}
\tag{11.6}
$$

$$
\frac{1}{EI}\begin{bmatrix} (595.1 + 166.7) & 488.4 \\ 488.4 & (595.1 + 166.7) \end{bmatrix} \begin{Bmatrix} R_1 \\ R_2 \end{Bmatrix} = \frac{1}{EI}\begin{Bmatrix} 341\ 185 \\ 341\ 185 \end{Bmatrix}
$$

from which

$$
\begin{Bmatrix} R_1 \\ R_2 \end{Bmatrix} = \begin{Bmatrix} 272.9 \\ 272.9 \end{Bmatrix}
$$

Comparing the results with those of Example 11.4.2, we see that the magnitudes of the interior redundant reactions are reduced by the softening effect of the elastic supports.

11.5

Selection of Internal Force Components as Redundants

Application of the criteria of Section 4.12 will indicate the degree of indeterminacy of beam and frame structures. It is clear from the treatment that a structure may be indeterminate as a result of either external or internal considerations, or a combination of the two.

Consider the m-span continuous beam shown in Fig. 11.4a. Based on the criteria of Section 4.12, this structure is externally indeterminate to the $(m - 1)$th degree. Likewise, the overall classification leads to the conclusion that the structure is indeterminate to the $(m - 1)$th degree. Both of these structural classifications are shown in the figure. Actually, many different combinations can be used in selecting redundants that will render this structure statically determinate. The most obvious approach, which follows the procedure of Section 11.3, is to select the $(m - 1)$ interior reaction components as the redundants, as shown in Fig. 11.4b. The primary structure is then a simply supported beam whose span is equal to the total length of the structure. In this case, the required compatibility for the given structure is violated at each of the interior support points. An alternative approach is to select the internal moments at the $(m - 1)$ interior support points as the redundants. For this selection, the primary structure is a series of m simply supported beams, and in this case compatibility for the given structure is relaxed at each of the interior support points where continuity of slope is violated. Of course, the primary structure could be obtained by releasing any combination of $(m - 1)$ external reaction components and internal force components, as long as the resulting primary structure is statically determinate and stable.

It is noted that the continuous beam of Fig. 11.4 is a special kind of structure; however, the same basic concepts regarding internal and external redundancies can be applied to other structures.

11.6

The Three-Moment Equation

Consider the continuous beam that was introduced in the previous section, and select the internal moments at the $(m - 1)$ support points as the redundants. Thus, the primary structure is composed of the m simply supported beams shown in Fig. 11.4c.

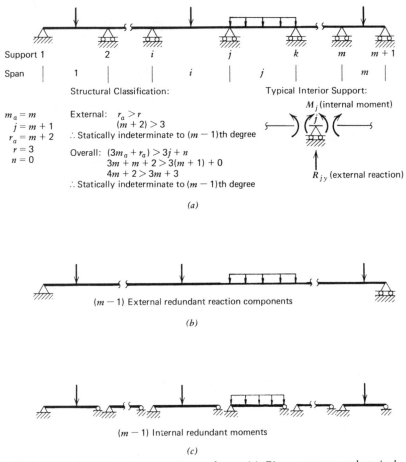

Fig. 11.4 *Statically indeterminate continuous beam.* (a) *Given structure and statical classification.* (b) *Primary structure based on external redundant reactions.* (c) *Primary structure based on internal redundant moments.*

Attention is now focused on the two-span section that reaches over the supports i, j, and k, which is shown in Fig. 11.5a. When the internal moments are removed and the continuous beam is transformed into a series of simply supported beams, there is a slope discontinuity over each of the interior supports. For the actual loading on the primary structure, this discontinuity at support j is θ_{j0}, as indicated on Fig. 11.5b. Since this discontinuity is at variance with the compatibility requirements for the original structure, it must be corrected in order to obtain the correct solution. This correction is accomplished by introducing, through three separate loading conditions, unit values of M_i, M_j, and M_k. The discontinuities in the slope at point j corresponding to each of these unit moment cases are shown as f_{ji}, f_{jj}, and f_{jk}, respectively, which are flexibility coefficients for the primary structure. Applying superposition, we multiply each of these discontinuities by the actual value of the respective redundant moment and combine them with the discontinuity θ_{j0} to obtain the total discontinuity, θ_j. Since θ_j is zero in the given structure, the final compatibility equation at point j becomes

$$f_{ji}M_i + f_{jj}M_j + f_{jk}M_k + \theta_{j0} = \theta_j = 0 \tag{11.7}$$

In the above equation, positive discontinuities are taken to be clockwise, as shown in Fig. 11.5.

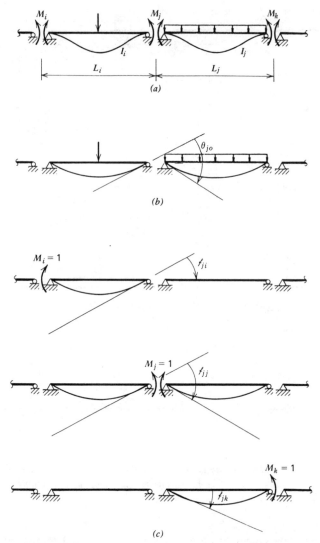

Fig. 11.5 *Development of three-moment equation. (a) Statically indeterminate continuous beam. (b) Statically determinate primary structure. (c) Unit values of redundant moments.*

The required rotation quantities can be determined by the moment–area principle, or any of the other methods presented in Chapter 6. The quantity θ_{jo} depends on the loads that act on spans i and j, and the expressions for the flexibility coefficients are

$$f_{ji} = \frac{L_i}{6EI_i}$$

$$f_{jj} = \frac{L_i}{3EI_i} + \frac{L_j}{3EI_j} \tag{11.8}$$

$$f_{jk} = \frac{L_j}{6EI_j}$$

where L_i and L_j are the lengths for spans i and j, and I_i and I_j are the corresponding moments of inertia. Substitution of Eq. 11.8 into 11.7 and a rearrangement of terms leads to

$$\left(\frac{L_i}{I_i}\right)M_i + \left(\frac{2L_i}{I_i} + \frac{2L_j}{I_j}\right)M_j + \left(\frac{L_j}{I_j}\right)M_k = -6E\theta_{j0} \qquad (11.9)$$

It is possible to develop equations for the quantity θ_{j0} for various loading arrangements on spans i and j. This has been done in some textbooks, but it will not be done here.

For the complete structure, the slope discontinuities that occur at all of the interior supports must be corrected. Thus, for the structure of Fig. 11.4, Eq. 11.9 must be applied with $j = 2, 3, \ldots, m$. However, each of the $(m - 1)$ compatibility equations involves only three moments—the moment at the point where compatibility is being considered and the moments at the far ends of the spans to the left and to the right. For this reason, Eq. 11.9 is commonly referred to as the *three-moment equation*. The original formulation is credited to the French engineer Clapeyron, although he did not develop it as a special case of the method of consistent deformations.

11.7 Application of the Three-Moment Equation

The three-moment equation is particularly useful in determining the internal support moments of a continuous beam. Thus, for the arrangement given in Fig. 11.4a, the equation would be applied at each of the $(m - 1)$ interior support points. This would provide the $(m - 1)$ compatibility equations that are required for the determination of the $(m - 1)$ redundant moments.

For fixed-ended beams, such as the one shown in Fig. 11.6, there is an additional redundant moment at each fixed end. Although a special form of the three-moment equation could be formulated for this case, a convenient artifice is to replace the fixed end by an imaginary end span of zero length. The three-moment equation is then applied at the end support points as well as at the interior points. Thus, an additional compatibility equation is gained for each additional redundant moment.

One of the major advantages of selecting the internal moments as the redundants in the method of consistent deformations, which leads to the three-moment

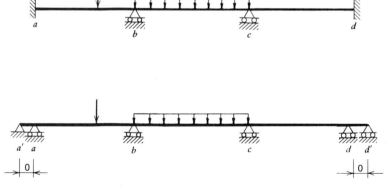

Fig. 11.6 *Treatment of fixed-end beam.*

equation, is that the flexibility matrix used in the solution of the redundant moments is a banded matrix. This greatly simplifies the solution of the simultaneous equations.

It is stressed that strict adherence to the sign convention that was introduced through the derivations of Section 11.6 is necessary. That is, positive moments are those that cause compression on the top fibers of the beam, and support displacements are positive when upward.

Since the solution of the resulting compatibility equations yields member-end moments, the shear and moment diagrams are easily constructed in accordance with the method presented in Section 4.14.

11.7.1 Example problem Determine the support moments for the structure shown by applying the three-moment equation, and construct the shear and moment diagrams.

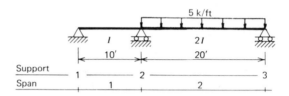

Three-Moment Equation for Support j:

$$\left(\frac{L_i}{I_i}\right)M_i + 2\left(\frac{L_i}{I_i} + \frac{L_j}{I_j}\right)M_j + \left(\frac{L_j}{I_j}\right)M_k = -6E\theta_{j0} \tag{11.9}$$

Support 2:

$i = 1, \quad j = 2, \quad k = 3$
$L_1 = 10', \quad I_1 = I$
$L_2 = 20', \quad I_2 = 2I$

$$(\Delta_3^2)_0 = \frac{250}{2EI} \cdot \frac{40}{3} \cdot 10 = \frac{16,667}{EI}$$

$$\theta_{20} = \frac{(\Delta_3^2)_0}{20} = \frac{833.3}{EI}$$

$$\overset{0}{\underset{\uparrow}{\left(\frac{10}{I}\right)}}M_1 + 2\left(\frac{10}{I} + \frac{20}{2I}\right)M_2 + \overset{0}{\underset{\uparrow}{\left(\frac{20}{2I}\right)}}M_3 = -6E\left(\frac{833.3}{EI}\right)$$

$$\frac{40}{I}M_2 = -\frac{5,000}{I}$$

Solution for Moment:

$$\frac{40}{I}M_2 = -\frac{5,000}{I}$$

$$M_2 = -125'^{-k}$$

Shear and Moment Diagrams:

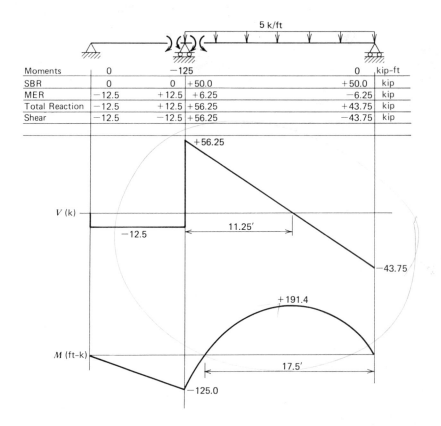

Moments	0		−125			0	kip-ft
SBR	0	0	+50.0			+50.0	kip
MER	−12.5	+12.5	+6.25			−6.25	kip
Total Reaction	−12.5	+12.5	+56.25			+43.75	kip
Shear	−12.5	−12.5	+56.25			−43.75	kip

Determine the support moments for the structure shown by application of the three-moment equation.

11.7.2 Example problem

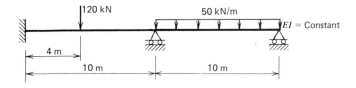

Modified Structure:

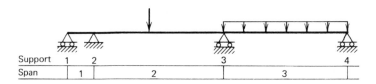

Support	1	2		3		4
Span		1	2		3	

Three-Moment Equation for Support j:

$$L_i M_i + 2(L_i + L_j)M_j + L_j M_k = -6EI\theta_{j0} \qquad (11.9)$$

Support 2:

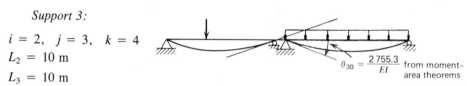

$i = 1, \quad j = 2, \quad k = 3$

$L_1 = 0$

$L_2 = 10$ m

$\theta_{20} = \dfrac{768}{EI}$ from moment-area theorems

$$0 \cdot M_1 + 2(0 + 10)M_2 + 10 \cdot M_3 = -6EI\left(\frac{768}{EI}\right)$$

$$20M_2 + 10M_3 = -4\,608$$

Support 3:

$i = 2, \quad j = 3, \quad k = 4$

$L_2 = 10$ m

$L_3 = 10$ m

$\theta_{30} = \dfrac{2\,755.3}{EI}$ from moment-area theorems

$$10 \cdot M_2 + 2(10 + 10)M_3 + 10 \cdot M_4 = -6EI\left(\frac{2\,755.3}{EI}\right)$$

$$10M_2 + 40M_3 = -16\,532$$

Solution for Moments:

$$20M_2 + 10M_3 = -4\,608$$
$$10M_2 + 40M_3 = -16\,532$$

Solving simultaneously,

$$M_2 = -27.2 \text{ kN} \cdot \text{m}; \quad M_3 = -406.5 \text{ kN} \cdot \text{m}$$

11.8

Support Settlements, Temperature Change, and Fabrication Errors

Support settlements, temperature change, or fabrication errors induce forces within the system without the application of external forces. For this reason, problems of this type are referred to as *self-straining problems.*

For the case of support settlements, consider again the two-span section shown in Fig. 11.5. If the supports displace as shown in Fig. 11.7, where upward displacements are taken as positive, then there is a chord discontinuity of θ_{js} at support j that is given by the expression

$$\theta_{js} = \left[-\frac{\Delta_i}{L_i} + \Delta_j\left(\frac{1}{L_i} + \frac{1}{L_j}\right) - \frac{\Delta_k}{L_j} \right] \qquad (11.10)$$

where Δ_i, Δ_j, and Δ_k are the displacements at the support points. This rotation must be included in the compatibility equation at point j, and thus Eq. 11.7 takes the form

$$f_{ji}M_i + f_{jj}M_j + f_{jk}M_k + \theta_{js} + \theta_{j0} = \theta_j = 0 \qquad (11.11)$$

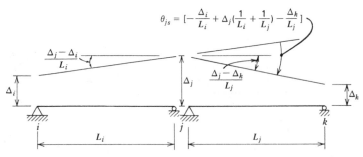

$$\theta_{js} = \left[-\frac{\Delta_i}{L_i} + \Delta_j\left(\frac{1}{L_i} + \frac{1}{L_j}\right) - \frac{\Delta_k}{L_j}\right]$$

Fig. 11.7 *Chord discontinuity from support displacements.*

Substitution of Eqs. 11.8 and 11.10 into Eq. 11.11 gives

$$\left(\frac{L_i}{I_i}\right)M_i + \left(\frac{2L_i}{I_i} + \frac{2L_j}{I_j}\right)M_j + \left(\frac{L_j}{I_j}\right)M_k$$
$$= 6E\left[\frac{\Delta_i}{L_i} - \Delta_j\left(\frac{1}{L_i} + \frac{1}{L_j}\right) + \frac{\Delta_k}{L_j}\right] - 6E\theta_{j0} \quad (11.12)$$

For the special case of $I_i = I_j = I$, Eq. 11.12 reduces to

$$L_iM_i + 2(L_i + L_j)M_j + L_jM_k$$
$$= 6EI\left[\frac{\Delta_i}{L_i} - \Delta_j\left(\frac{1}{L_i} + \frac{1}{L_j}\right) + \frac{\Delta_k}{L_j}\right] - 6EI\theta_{j0} \quad (11.13)$$

Equation 11.12 or 11.13 replaces Eq. 11.9 as the basic three-moment equation when support settlements are present.

For the effects of temperature change, consider the single-span structure shown in Fig. 11.8a. The temperature gradient ΔT is given by

$$\Delta T = T_b - T_t \quad (11.14)$$

where T_b and T_t are the temperatures at the bottom and top fibers of the beam, respectively. For the element shown in Fig. 11.8b, it is clear that

$$d\theta = \frac{\alpha(T_b - T_t)\,dx}{h} = \frac{\alpha\,\Delta T\,dx}{h} \quad (11.15)$$

where α is the coefficient of thermal expansion and h is the member depth. Therefore, the curvature is

$$\frac{d\theta}{dx} = \frac{\alpha\,\Delta T}{h} \quad (11.16)$$

This curvature, of course, plays the same role as M/EI for a beam under loading and, therefore, for constant temperature gradient along the member, the curvature diagram of Fig. 11.8c results. Application of the moment–area method leads to end slopes of

$$\theta = \frac{\alpha\,\Delta T\,L}{2h} \quad (11.17)$$

which are shown in Fig. 11.8d. In the above, positive ΔT produces positive curvature.

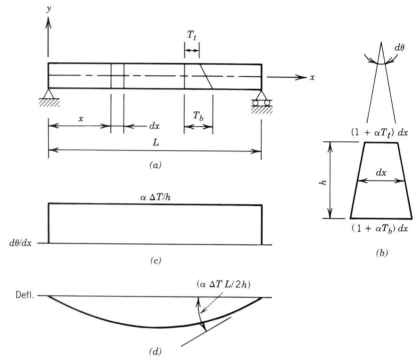

Fig. 11.8 *Temperature gradient on beam.* (a) *Gradient.* (b) *Stained element.* (c) *Induced curvatures.* (d) *Deflected beam with end rotation.*

Returning to Fig. 11.5, and replacing the loading of Fig. 11.5a with separate temperature gradients for each span, we see that

$$\theta_{j0} = \left(\frac{\alpha \, \Delta T \, L}{2h}\right)_i + \left(\frac{\alpha \, \Delta T \, L}{2h}\right)_j \qquad (11.18)$$

where the subscripts refer to the two separate spans. For constant α, ΔT, and h, Eq. 11.18 becomes

$$\theta_{j0} = \left(\frac{\alpha \, \Delta T}{2h}\right)(L_i + L_j) \qquad (11.19)$$

Equation 11.9 remains valid for the temperature problem, with θ_{j0} of Eq. 11.19 being used. Of course, in the case of combined loading and temperature change, θ_{j0} must include the superposition of the two effects, and if settlement is involved, Eq. 11.13 must be used.

For beam structures, fabrication errors include initial crookedness of the members and the associated end slopes. These end slopes must be included in the determination of θ_{j0}.

11.8.1 Example problem   Determine the support moments for the structure shown for the given temperature gradient by applying the three-moment equation.

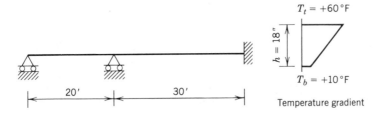

$$E = 30 \times 10^3 \text{ ksi}; \quad I = 1{,}500 \text{ in.}^4; \quad EI = 45 \times 10^6 \text{ k-in.}^2 = 312.5 \times 10^3 \text{ k-ft}^2$$
$$\alpha = 0.000{,}006{,}5/°\text{F}; \quad \Delta T = T_b - T_t = 10 - 60 = -50°\text{F}$$

Modified Structure:

Support   1   2   3   4

Span   1   2   3

Three-Moment Equation for Support **j:**

$$L_i M_i + 2(L_i + L_j)M_j + L_j M_k = -6EI\theta_{j0} \qquad (11.13)$$

where

$$\theta_{j0} = \left(\frac{\alpha\,\Delta T}{2h}\right)(L_i + L_j) \qquad (11.19)$$

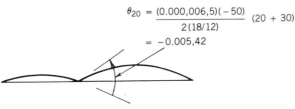

Support 2:

$i = 1, \quad j = 2, \quad k = 3$
$L_1 = 20'$
$L_2 = 30'$

$$\theta_{20} = \frac{(0.000,006,5)(-50)}{2(18/12)}(20 + 30)$$
$$= -0.005,42$$

$$\overset{0}{\underset{\uparrow}{}}$$
$$20 \cdot M_1 + 2(20 + 30)M_2 + 30 \cdot M_3 = -6EI(-0.005,42)$$
$$100M_2 + 30M_3 = 10,163$$

Support 3:

$i = 2, \quad j = 3, \quad k = 4$
$L_2 = 30'$
$L_3 = 0'$

$$\theta_{30} = \frac{(0.000,006,5)(-50)}{2(18/12)}(30 + 0)$$
$$= -0.003,25$$

$$30 \cdot M_2 + 2(30 + 0)M_3 + 0 \cdot M_4 = -6EI(-0.003,25)$$
$$30M_2 + 60M_3 = 6,094$$

Solution for Moments:

$$100M_2 + 30M_3 = 10,163$$
$$30M_2 + 60M_3 = 6,094$$

Solving simultaneously,

$$M_2 = 83.7 \text{ ft-kips}; \qquad M_3 = 59.8 \text{ ft-kips}$$

Note:

This problem is resolved in Section 13.7.1, and the moment diagram and deflected structure are discussed in detail.

11.8.2 Example problem Determine the support moments for the structure of Section 11.8.1 for the lack-of-fit situation described. Use the three-moment equation, and disregard the temperature gradient previously given.

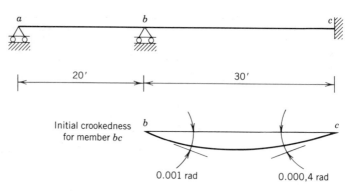

Three-Moment Equation for Support j:

Equation 11.13 is again used, as it was in the previous example, with θ_{j0} given directly from the initial crookedness for member bc.

Support 2:

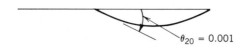

$$100M_2 + 30M_3 = -6EI(0.001) = -1,875$$

Support 3:

$$30M_2 + 60M_3 = -6EI(0.000,4) = -750$$

Solution for Moments:

$$100M_2 + 30M_3 = -1,875$$
$$30M_2 + 60M_3 = -750$$

Solving simultaneously,

$$M_2 = -17.65 \text{ ft-kips}; \qquad M_3 = -3.67 \text{ ft-kips}$$

11.9

Matrix
Formulation
of Method of
Consistent
Deformations

In general, a statically indeterminate beam- or frame-type structure may have internal, external, or a combination of both internal and external redundancies. Thus, Eq. 10.29 is valid here as a general representation of the displacement–force equations for the resulting primary structure. The $\{P\}$ vector includes a total of $(k + m + n)$ components, in which the components P_1 through P_k represent the k unknown external redundant forces, P_{k+1} through P_q represent the m unknown internal redundant force components, and P_{q+1} through P_s correspond to the n known forces that are applied to the structure.

Corresponding to the $\{P\}$ vector is the $\{\Delta\}$ vector that includes the displacements on the primary structure. The first k elements give the displacements that are associated with the external redundants, and the next m elements are associated with the internal member releases. These are zero unless support displacements or member discontinuities are prescribed. The final n elements of the $\{\Delta\}$ vector are the unknown displacements that correspond to the degrees of freedom associated with the applied forces.

The $/$ terms of Eq. 10.29 are the flexibility coefficients that give the interactive relationships among the full arrays of displacements and forces on the primary structure.

Equation 10.29 is again partitioned as shown in Eq. 10.30, where $[/]_{ij}$ relates the $\{P\}_j$ forces to the $\{\Delta\}_i$ displacements. The partitioning separates the unknown redundant forces from the known applied forces. This, in turn, separates the known displacements, which are associated with the redundant forces, from the unknown displacements, which correspond to the applied forces.

The expanded forms of the partitioned equations are given as Eqs. 10.31 and 10.32. The first of these equations can be written as

$$[/]_{11}\{P\}_1 = \{\Delta\}_1 - [/]_{12}\{P\}_2 \tag{11.20}$$

or

$$[/]_{11}\{P\}_1 = \{\Delta\}_1 - \{\Delta_0\}_1 \tag{11.21}$$

This is the form that was given earlier as Eq. 10.33, where $\{\Delta_0\}_1$ represents the displacements associated with the redundant forces that are caused by the action of applied loads on the primary structure. In the present context, these terms must be broadened to include the contributions from temperature change and fabrication errors.

For the special case of a continuous beam, Eq. 11.21 represents a matrix generalization of Eq. 11.11, which is the three-moment equation. Here, $\{P\}_1$ is composed entirely of the redundant support moments, and $(\{\Delta\}_1 - \{\Delta_0\}_1)$ includes terms given by $[\theta_j - (\theta_{js} + \theta_{j0})]$, in which j ranges over the interior support points corresponding to the redundant moments.

The solution of Eq. 11.21 gives the redundant force components $\{P\}_1$ in the form

$$\{P\}_1 = [/]_{11}^{-1}(\{\Delta\}_1 - \{\Delta_0\}_1) \tag{11.22}$$

If the unknown displacements are desired, Eq. 11.22 is substituted into Eq. 10.32 to obtain

$$\{\Delta\}_2 = [/]_{21}[/]_{11}^{-1}(\{\Delta\}_1 - \{\Delta_0\}_1) + [/]_{22}\{P\}_2 \tag{11.23}$$

With the redundant reactions and member forces having been established from Eq. 11.22, the remaining member forces and reactions can be determined by applying statics to the primary structure. Here, Eq. 4.20 can be rewritten in the form

$$[C]_P \begin{Bmatrix} \{V\}_P \\ \{M\}_P \\ \{R\}_P \end{Bmatrix} = \begin{Bmatrix} \{P\}_1 \\ \{P\}_2 \end{Bmatrix} \tag{11.24}$$

where $[C]_P$ is the overall statics matrix for the primary structure and $\{V\}_P$, $\{M\}_P$, and $\{R\}_P$ are the unknown shears, moments, and reactions, respectively, for the primary structure. The vector $\{P\}_1$, of course, includes the redundant force components as determined from Eq. 11.22 and $\{P\}_2$ contains the designated applied forces. The combined vector of $\{P\}_1$ and $\{P\}_2$ forms the overall load vector for the primary structure.

The solution of Eq. 11.24 yields

$$\begin{Bmatrix} \{V\}_P \\ \{M\}_P \\ \{R\}_P \end{Bmatrix} = [C]_P^{-1} \begin{Bmatrix} \{P\}_1 \\ \{P\}_2 \end{Bmatrix} \tag{11.25}$$

The statics matrix, $[C]_P$, could be generated as explained in Section 4.13 and its inverse determined as indicated in Eq. 11.25. However, it is clear that each column of $[C]_P^{-1}$ is composed of elements of $\{V\}_P$, $\{M\}_P$, or $\{R\}_P$ that are associated with a unit value of the element of $\{P\}_1$ or $\{P\}_2$ that multiplies that column. Since the shears, moments, and reactions for these individual unit load cases are needed to generate the flexibility coefficients of Eq. 10.29, the explicit development of $[C]_P$, and the determination of its inverse, is not necessary.

Therefore, Eq. 11.25 gives the final member forces and reactions for the primary structure through a superposition of the effects of unit load components being applied to the primary structure. This parallels the format of Eq. 10.29, in which the final displacements are expressed as the superposition of the contributions from individual unit load cases.

11.9.1 Example problem

Rework Example 11.7.2 by the general matrix formulation. In this case, determine the reactions and the support moments, and indicate how the displacements corresponding to the applied loads would be determined.

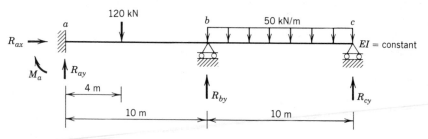

Structure Classification:

$m_a = 2$	External:	$r_a > r;\ \ 5 > 3$
$r_a = 5$		Statically indeterminate to second degree
$j = 3$		
$n = 0$	Overall:	$3m_a + r_a > 3j + n;\ \ 11 > 9$
$r = 3$		Statically indeterminate to second degree

Primary Structure: Select M_a as a redundant reaction component and M_{bc} as a redundant internal force component.

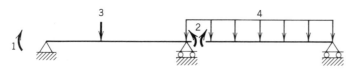

$$\{P\} = \begin{Bmatrix} P_1 \\ P_2 \\ \overline{P_3} \\ P_4 \end{Bmatrix} = \begin{Bmatrix} M_a \\ M_{bc} \\ \overline{120} \\ 50 \end{Bmatrix} = \begin{Bmatrix} \{P\}_1 \\ \overline{\{P\}_2} \end{Bmatrix}$$

Note:

• P_4 is intensity of distributed load.

$$\{\Delta\} = \begin{Bmatrix} \Delta_1 \\ \Delta_2 \\ \overline{\Delta_3} \\ \Delta_4 \end{Bmatrix} \quad \begin{Bmatrix} 0 \\ 0 \\ \overline{\Delta_3} \\ \Delta_4 \end{Bmatrix} = \begin{Bmatrix} \{\Delta\}_1 \\ \overline{\{\Delta\}_2} \end{Bmatrix}$$

Note:

• Δ_2 is the slope discontinuity at the cut section.

• Δ_4 is taken as the midspan deflection for span bc.

Determination of Flexibility Coefficients:

$P_1 = 1.$

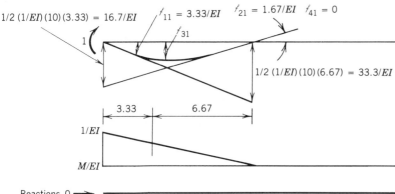

$P_2 = 1$

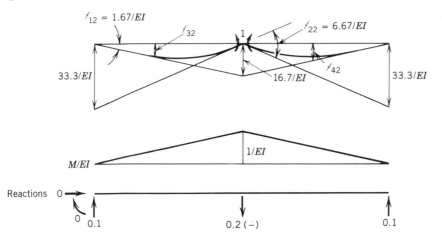

$P_3 = 1$

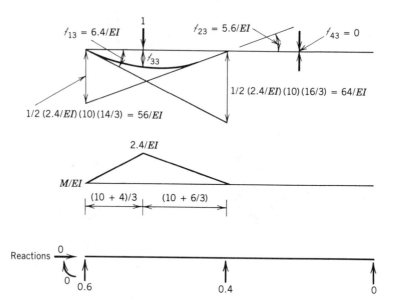

$P_4 = 1$

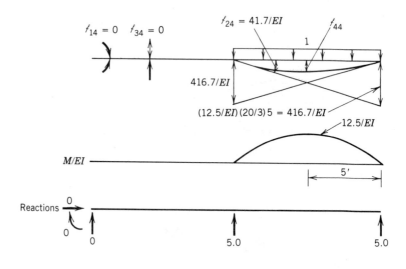

$$[f] = \frac{1}{EI} \left[\begin{array}{cc|cc} 3.33 & 1.67 & 6.40 & 0 \\ 1.67 & 6.67 & 5.60 & 41.7 \\ \hline f_{31} & f_{32} & f_{33} & 0 \\ 0 & f_{42} & 0 & f_{44} \end{array} \right] = \left[\begin{array}{c|c} [f]_{11} & [f]_{12} \\ \hline [f]_{21} & [f]_{22} \end{array} \right]$$

Note:

$f_{31}, f_{32}, f_{33}, f_{42}$, and f_{44} could readily be evaluated by moment–area theorems.

Redundant Forces and Free Displacements:

$$[/]_{11}\{P\}_1 = \{\Delta\}_1 - [/]_{12}\{P\}_2 \tag{11.20}$$

$$\{P\}_1 = [/]_{11}^{-1}(\{\Delta\}_1 - [/]_{12}\{P\}_2)$$

$$\{P\}_1 = \begin{Bmatrix} M_a \\ M_{bc} \end{Bmatrix} = \frac{EI}{19.42}\begin{bmatrix} 6.67 & -1.67 \\ -1.67 & 3.33 \end{bmatrix}\left(\begin{Bmatrix} 0 \\ 0 \end{Bmatrix} - \frac{1}{EI}\begin{bmatrix} 6.40 & 0 \\ 5.60 & 41.7 \end{bmatrix}\begin{Bmatrix} 120 \\ 50 \end{Bmatrix}\right)$$

$$\{P\}_1 = \begin{Bmatrix} M_a \\ M_{bc} \end{Bmatrix} = \begin{Bmatrix} -26.9 \\ -406.4 \end{Bmatrix} kN \cdot m$$

$$\{\Delta\}_2 = [/]_{21}\{P\}_1 + [/]_{22}\{P\}_2 \tag{11.23}$$

$$\{\Delta\}_2 = \frac{1}{EI}\begin{bmatrix} /_{31} & /_{32} \\ 0 & /_{42} \end{bmatrix}\begin{Bmatrix} -26.9 \\ -406.4 \end{Bmatrix} + \begin{bmatrix} /_{33} & 0 \\ 0 & /_{44} \end{bmatrix}\begin{Bmatrix} 120 \\ 50 \end{Bmatrix}$$

$$\{\Delta\}_2 = \begin{Bmatrix} \Delta_3 \\ \Delta_4 \end{Bmatrix} = \frac{1}{EI}\begin{Bmatrix} -26.9/_{31} - 406.4/_{32} + 120/_{33} \\ -406.4/_{42} + 50/_{44} \end{Bmatrix}$$

Reactions for Primary Structure: Since only the reactions for the primary structure are desired, Eqs. 11.24 and 11.25 are simplified.

$$\{R\}_P = [C]_P^{-1}\begin{Bmatrix} \{P\}_1 \\ \{P\}_2 \end{Bmatrix} \tag{11.25}$$

$$\{R\}_P = \begin{Bmatrix} R_{ax} \\ R_{ay} \\ R_{by} \\ R_{cy} \end{Bmatrix} = \begin{bmatrix} 0 & 0 & 0 & 0 \\ -0.1 & 0.1 & 0.6 & 0 \\ 0.1 & -0.2 & 0.4 & 5.0 \\ 0 & 0.1 & 0 & 5.0 \end{bmatrix}\begin{Bmatrix} -26.9 \\ -406.4 \\ 120.0 \\ 50.0 \end{Bmatrix} = \begin{Bmatrix} 0 \\ 34.1 \\ 376.3 \\ 209.4 \end{Bmatrix} kN$$

11.10 Compatibility Equations by Energy Methods

The procedure developed in Section 10.10 for generating the compatibility equations for statically indeterminate truss analysis can be similarly applied to indeterminate beam- and frame-type structures. Castigliano's second theorem is again employed, and its form is recalled from Section 5.9.2 to be

$$\frac{\partial U}{\partial P_i} = \Delta_i \tag{11.26}$$

In this equation, U is the total strain energy for the structure, and P_i and Δ_i represent the ith applied load and its companion displacement, respectively. In this case, the strain energy results from flexure, and for a structure that is composed of m members, it was shown in Section 6.11 that

$$U = \sum_{i=1}^{m} \frac{1}{2E}\int_{x=0}^{x=l_i} \frac{M_i^2}{I_i} dx \tag{11.27}$$

where M_i is the bending moment as a function of x, l_i is the member length, I_i is the moment of inertia, and E is the modulus of elasticity. The moment M_i must be expressed in terms of the applied P loads so that the differentiation of Eq. 11.26 can be carried out.

Equation 11.26 must be applied for each displacement degree of freedom for which a compatibility relation is desired on the primary structure. This leads to a set of displacement–force equations that are equivalent to Eq. 10.29. These equations are then partitioned as indicated in Eq. 10.30, and the solution proceeds according to the method outlined in Section 11.9. It should be noted that the strain energy expressed as given by Eq. 11.27 becomes very complicated when it is expressed in terms of the P loads, and the required differentiation is tedious.

The treatment of this section assumes that each member is subjected to flexure without any axial load. In cases where both flexure and axial load are present, the strain energy for each member results from summing the individual strain energies given Eqs. 10.39 and 11.27.

**11.10.1
Example
problem**

Determine the reactions and the vertical displacements at points b and d for the given structure.

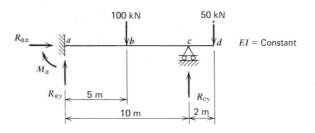

Structure Classification:

$r_a = 4$ External: $r_a > r$; $4 > 3$
$r = 3$ Statically indeterminate to first degree
$n = 0$
$m_a = 2$ Overall: $3m_a + r_a > 3j + n$; $6 + 4 > 9$
$j = 3$ Statically indeterminate to first degree

Primary Structure: Select M_a as a redundant reaction component.

$$\{P\} = \begin{Bmatrix} P_1 \\ P_2 \\ P_3 \end{Bmatrix} = \begin{Bmatrix} M_a \\ \overline{100} \\ 50 \end{Bmatrix} = \begin{Bmatrix} \{P\}_1 \\ \overline{\{P\}_2} \end{Bmatrix}; \qquad \{\Delta\} = \begin{Bmatrix} \Delta_1 \\ \overline{\Delta_2} \\ \Delta_3 \end{Bmatrix} = \begin{Bmatrix} 0 \\ \overline{v_b} \\ v_d \end{Bmatrix} = \begin{Bmatrix} \{\Delta\}_1 \\ \overline{\{\Delta\}_2} \end{Bmatrix}$$

Region	$a - b$	$d - c$	$c - b$
Function	$0 < x < 5$	$0 < x < 2$	$0 < x < 5$
M	$P_1 + (-0.1P_1 + 0.5P_2 - 0.2P_3)x$	$-P_3x$	$-P_3(2 + x)$ $+ (0.1P_1 + 0.5P_2 + 1.2P_3)x$

Strain Energy:

$$U = \sum_{i=1}^{m} \frac{1}{2EI_i} \int_0^{l_i} M_i^2 \, dx \tag{11.27}$$

$$U = \frac{1}{2EI} \int_0^5 [P_1 + (-0.1P_1 + 0.5P_2 - 0.2P_3)x]^2 \, dx + \frac{1}{2EI} \int_0^2 [-P_3 x]^2 \, dx$$

$$+ \frac{1}{2EI} \int_0^5 [-P_3(2 + x) + (0.1P_1 + 0.5P_2 + 1.2P_3)x]^2 \, dx$$

Compatibility Equations from Castigliano's Second Theorem:

$$\frac{\partial U}{\partial P_1} = \Delta_1 \Rightarrow \frac{1}{EI}(3.333P_1 + 6.250P_2 - 3.333P_3) = \Delta_1$$

$$\frac{\partial U}{\partial P^2} = \Delta_2 \Rightarrow \frac{1}{EI}(6.250P_1 + 20.833P_2 - 12.500P_3) = \Delta_2$$

$$\frac{\partial U}{\partial P_3} = \Delta_3 \Rightarrow \frac{1}{EI}(-3.333P_1 - 12.500P_2 + 11.000P_3) = \Delta_3$$

Determination of Reactions and Displacements:

$$[/] = \frac{1}{EI} \begin{bmatrix} 3.333 & 6.250 & -3.333 \\ \hline 6.250 & 20.833 & -12.500 \\ -3.333 & -12.500 & 11.000 \end{bmatrix} = \begin{bmatrix} [/]_{11} & [/]_{12} \\ \hline [/]_{21} & [/]_{22} \end{bmatrix}$$

$$\{P\}_1 = [/]_{11}^{-1}(\{\Delta\}_1 - [/]_{12}\{P\}_2) \tag{11.20}$$

$$\{P\}_1 = \{M_a\} = \frac{EI}{3.33}\left(\{0\} - \frac{1}{EI}[6.250 \quad -3.333]\begin{Bmatrix} 100 \\ 50 \end{Bmatrix}\right) = -137.5 \text{ kN}$$

From statics: $R_{ax} = 0$; $R_{ay} = +53.75$ kN; $R_{cy} = +96.25$ kN

$$\{\Delta\}_2 = [/]_{21}\{P\}_1 + [/]_{22}\{P\}_2 \tag{11.23}$$

$$\{\Delta\}_2 = \begin{Bmatrix} v_b \\ v_d \end{Bmatrix} = \frac{1}{EI}\begin{bmatrix} 6.250 \\ -3.333 \end{bmatrix}\{-137.5\}$$

$$+ \frac{1}{EI}\begin{bmatrix} 20.833 & -12.500 \\ -12.500 & 11.000 \end{bmatrix}\begin{Bmatrix} 100 \\ 50 \end{Bmatrix} = \frac{1}{EI}\begin{Bmatrix} 598.9 \\ -241.7 \end{Bmatrix}$$

11.11 Multiple Loadings

As was pointed out earlier in Section 10.11, it is frequently necessary to analyze a structure for a number of separate loading conditions. Although the presentation given there was in the context of pin-connected structures, it is equally applicable to the beam- and frame-type structures presently under consideration. Equation 10.40 is directly applicable, where the redundant forces for the *i*th loading case $\{P\}_1^i$ are associated with the displacements $\{\Delta\}_1^i = (\{\Delta\}_1 - \{\Delta_0\}_1)^i$ for the *i*th loading case. The flexibility matrix $[/]_{11}$ is a function of the primary structure and is independent of the loading on the structure.

11.12 Selection of Redundants

The method of consistent deformations requires that a primary structure be iden-tified, and therefore, redundant force components must be selected. This topic is treated in greater detail in Chapter 18, but a few remarks are in order here.

In general, it is best that the primary structure behavior be, as nearly as possible, similar to that of the actual structure. In this way, the redundants merely induce corrections to the primary structure to restore the compatibility requirements of the actual structure. The smaller the corrections are, the better conditioned will be the $[/]_{11}$ matrix, which must be inverted according to Eq. 11.22. This will minimize the effect of round-off errors in the solution process.

11.13

Additional Reading

Beaufait, F. W., *Basic Concepts of Structural Analysis,* Sections 8.1 through 8.3, Prentice–Hall, Englewood Cliffs, N.J., 1977.

Gerstle, K. H., *Basic Structural Analysis,* Chapters 10, 11, and 12, Prentice–Hall, Englewood Cliffs, N.J., 1974.

Ghali, A., and Neville, A. M., *Structural Analysis,* Chapter 2, Intext Educational Publishers, Scranton, Pa., 1972.

Hsieh, Yuan-Yu, *Elementary Theory of Structures,* Chapter 9, Prentice–Hall, Englewood Cliffs, N.J., 1970.

Wang, Chu-Kia, *Matrix Methods of Structural Analysis,* 2nd Ed., Chapter 9, International Textbook, Scranton, Pa., 1970.

11.14

Suggested Problems

1 through 8. Analyze each of the statically indeterminate structures given for the specified loading. Use the method of consistent deformations, selecting external reaction components as the redundant forces. Construct the shear and moment diagrams for each member.

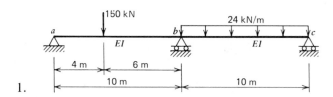

1.

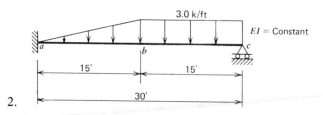

2.

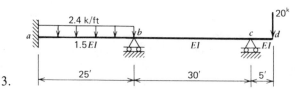

3.

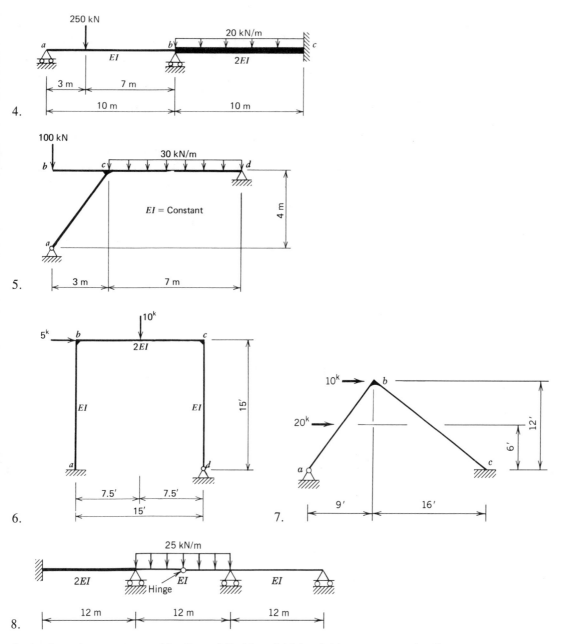

4.

5.

6.

7.

8.

9. Analyze the structure and loading of Problem 1 if the rigid support at point C is replaced by the elastic support shown below.

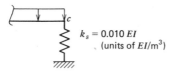

$k_s = 0.010\,EI$
(units of EI/m^3)

10. Analyze the structure and loading of Problem 4 if the rigid supports at points a and b are replaced by elastic supports with stiffnesses of $k_s = 0.005EI$ (units EI/m^3).

11. Determine the reactions and construct the shear and moment diagrams for the structure of Problem 1 if, in addition to the given loading, point C settles 0.012 m downward. Take $EI = 700\ 000$ kN · m².

12. Determine the reactions, shears, and moments for the structure of Problem 2 if point a experiences a rotation of 0.001 radian clockwise. Do not include the loads of Problem 2, and take the value of EI to be 30×10^6 k-in².

13. Consider the structure of Problem 4 in the absence of any applied loads. Determine the shears and moments throughout the structure for the following pattern of support settlements: Point a, 15 mm downward; point b, 25 mm downward; point c, 0.000 5 radian counterclockwise. Express your results in terms of EI.

14 through 23. Use the three-moment equation formulation to determine the support moments for each of the given structures for the indicated loading.

14. The structure and loading of Problem 1.

15. The structure and loading of Problem 3.

16. The structure and loading of Problem 4.

17. The structure and loading of Problem 1 with the settlement prescribed in Problem 11.

18. The structure and loading of Problem 2 with the support movement described in Problem 12.

19. The structure of Problem 4 with the support settlements designated in Problem 13. Disregard the loads of Problem 4, and express the results in terms of EI.

20. The structure of Problem 1 with a temperature gradient of $\Delta T = T_b - T_t = 30°C$ applied to both spans. Disregard the given loads of Problem 1, and take $\alpha_t = 0.000\ 012/°C$, $h = 40$ mm, and $EI = 500\ 000$ kN · m².

21. The structure of Problem 3 with a temperature gradient of $\Delta T = T_b - T_t = -50°F$ acting on span ab. Assume that the loads of Problem 3 remain on the structure, and take $\alpha_t = 0.000,006,5/°F$, $EI = 25 \times 10^6$ k-in.², and $h = 15$ inches.

22. The structure of Problem 4 in which member ab has an initial crookedness as noted below, but is forced into place. Disregard the prescribed loads of Problem 4, and take $EI = 400\ 000$ kN · m².

0.000 5 rad   0.001 rad

23. The structure of Problem 3 in which member bc has the following initial crookedness but is forced to fit. Do not include the loads prescribed in Problem 3, and take $EI = 30 \times 10^6$ k-in². *Hint:* Convert initial crookedness to end slopes only.

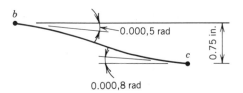

0.000,5 rad

0.75 in.

0.000,8 rad

24 through 28. Use the matrix formulation of Section 11.9 to determine the reactions, the complete shear and moment diagrams, and the designated free-
joint displacements for each structure shown for the indicated loading. The
suggested primary structure and the coordinate system (selected to include the
desired free-joint displacements) are given for each structure.

24. The structure and loading of Problem 1.

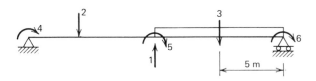

5 m

25. The structure and loading of Problem 4.

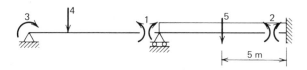

5 m

26. The structure and loading of Problem 6.

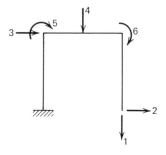

27. The structure of Problem 1 if, in addition to the given loads, the support of
point b settles 15 mm downward. Use the coordinate system given in Problem
24 and take $EI = 700\ 000\ kN \cdot m^2$.

28. The structure of Problem 4 with the initial crookedness for member ab that is
prescribed in Problem 22. Use the coordinate system given in Problem 25
with $EI = 400\ 000\ kN \cdot m^2$.

29. Develop a computer program for the solution of continuous beams by the
three-moment equation formulation of Section 11.7. Use the program to solve
Problems 1, 2, 3, and 4.

30 and 31. Use Castigliano's second theorem to determine the complete set of reactions and the indicated displacements for each of the structures and loadings indicated.

30. The structure and loading of Problem 2, with the desired displacements being the vertical displacement at point b and the rotation at point c.

31. The structure and loading of Problem 5, with the desired displacements being the vertical displacements at point b and at the middle of span cd.

The Eads Bridge, St. Louis, Mo. (reprinted from Modern Steel Construction, Third Quarter, 1974, courtesy of the American Institute of Steel Construction, New York, N.Y.).

ANALYSIS OF PIN-CONNECTED FRAMEWORKS BY EQUILIBRIUM METHODS

12.1

For pedagogic reasons, the analysis of statically determinate structures and statically indeterminate structures is presented separately in Parts I and II, respectively, of the book. This separation follows the classical presentations for teaching the fundamentals of structural analysis. This appcoach is indeed logical for the compatibility methods of Chapters 10 and 11, where redundant forces are identified and determined. However, as will be seen, *equilibrium methods* are not uniquely identified with statically indeterminate analysis, but are equally applicable to statically determinate systems. When the conventional methods of static analysis are applied to statically determinate structures, only the member forces are determined. The desired displacements must then be calculated separately by the methods that are appropriate for the structure type. However, the equilibrium

Equilibrium Method for Truss-Type Structures

methods that are presented here are more comprehensive in that they readily produce the complete set of structure displacements and member forces.

To reiterate, all methods of structural analysis must ultimately produce a solution that satisfies both equilibrium and compatibility. Equilibrium methods derive their name from the fact that the fundamental equations used in the formation are equations of equilibrium. The conditions of compatibility, which include the appropriate displacement boundary conditions, are taken into account as the equations of equilibrium are developed. Since the resulting equilibrium equations are composed of stiffness quantities, and since the solution of these equations yields displacements, the equilibrium method is alternatively referred to as the *stiffness method* or the *displacement method*.

There are many specific techniques that fall under the broad category of equilibrium methods, but each possesses the central characteristics described above. This chapter presents the fundamental procedures that must be followed in formulating a solution by the equilibrium method for pin-connected frameworks. It will be seen that an element approach is quite natural here and that the final equations result from a synthesis of the individual member characteristics.

12.2

Kinematic Indeterminacy

For the compatibility methods presented in Chapters 10 and 11, the concept of redundant forces was central to the discussion. Each redundant force was set equal to zero to form a statically determinate primary structure, and the degree of statical indeterminacy was equal to the number of redundant forces. The solution of the primary structure violated the compatibility requirements with regard to the displacement quantity associated with each redundant force, and the correct solution for the original indeterminate structure required a restoration of compatibility. In general, for a structure that is statically indeterminate to the nth degree, there are n redundant forces, which are removed to form the statically determinate primary structure. To restore the structure to its original form, it is necessary to solve n compatibility equations for the n redundant forces.

The concept of statical indeterminacy is not crucial in formulating a problem by the equilibrium method. Instead, an analogous concept of kinematic indeterminacy is useful. Here, attention is focused on the number of independent displacement degrees of freedom that the structure possesses, and the degree of kinematic indeterminacy is equal to the number of these displacement components. When these displacements are set equal to zero, the structure is reduced to a kinematically determinate primary structure; however, the solution of this primary structure violates the equilibrium conditions. To obtain the correct solution, equilibrium must be restored throughout the structure. In general, if a structure is kinematically indeterminate to the nth degree, there are n independent displacement quantities that are required to completely define the response of the structure for any loading condition. These displacements are restrained to form the kinematically determinate primary structure. In relaxing these artificially imposed restraints, it is necessary to solve n equilibrium equations for the n displacements. The significance of the solution for the kinematically determinate primary structure is more effectively illustrated for beam- and frame-type structures than it is for pin-connected frameworks, as will be seen in Chapter 13.

The differences between static and kinematic indeterminacy were discussed in Section 9.6, but to review the concepts, consider the simple truss shown in Fig. 12.1a. This is the same structure that was shown in Fig. 10.2, where R_1 and R_2

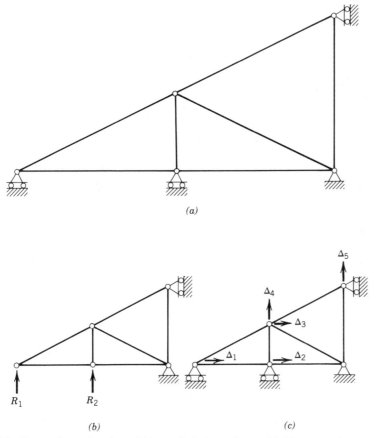

Fig. 12.1 *Comparison of static and kinematic indeterminacy.* (a) *Given truss structure.* (b) *Redundant reaction components for compatibility analysis.* (c) *Unknown displacement quantities for equilibrium analysis.*

were taken as the redundant reactions, as is shown in Fig. 12.1b. Two compatibility equations were necessary (Eq. 10.10), and the solution to these equations would lead to the determination of the redundant reactions. On the other hand, Fig. 12.1c indicates that there are five independent displacement components, and, thus, the structure is kinematically indeterminate to the fifth degree. One equilibrium equation can be written for each displacement degree of freedom, and the solution of these equations yields these five displacements.

From the above discussion, two observations are of importance. First, the equilibrium approach is more straightforward because all of the unknown displacements are restrained to produce the kinematically determinate structure. On the other hand, in the compatibility method, some selectivity is required in designating the redundant forces. Second, since computational effort is largely dependent on the number of unknowns, a comparison of the degrees of statical indeterminacy and kinematic indeterminacy is indicative of the labor involved in completing the solution. This criterion can be misleading, however, since it fails to account for the differences in effort required to generate the governing equations for the two approaches. It should also be noted that the pattern indicated by the simple problem of Fig. 12.1 does not always hold. In some structures the degree of statical indeterminacy is greater than the degree of kinematic indeterminacy.

12.3

Member Force–Displacement Relations

A crucial first step in the analysis of a structure by the equilibrium method is to have a clear understanding of the force–displacement relations for the individual members of the structure. These relationships are developed in Chapter 5 and are given by Eq. 5.11 in the form

$$\left\{ \begin{array}{c} F_1 \\ F_2 \end{array} \right\}_i = \begin{bmatrix} \dfrac{EA}{l} & \dfrac{EA}{l} \\[2mm] \dfrac{EA}{l} & \dfrac{EA}{l} \end{bmatrix}_i \left\{ \begin{array}{c} \delta_1 \\ \delta_2 \end{array} \right\}_i \tag{12.1}$$

where A is the member cross-sectional area, E is the modulus of elasticity, l is the member length, F_1 and F_2 are the member-end forces, and δ_1 and δ_2 are the member-end displacements. The i subscript indicates that all of these quantities are associated with the ith member, as shown in Fig. 12.2.

Equation 12.1 can be written in an abbreviated form similar to Eq. 5.12. This takes the form

$$\{F\}_i = [k]_{ai}\{\delta\}_i \tag{12.2}$$

where $\{F\}_i$ is the member-end force vector, $\{\delta\}_i$ is the member-end displacement vector, and $[k]_{ai}$ is the member stiffness matrix.

In an expanded form, Eq. 12.1 becomes

$$(F_1)_i = (F_2)_i = \left(\frac{EA}{l}\right)_i [(\delta_1)_i + (\delta_2)_i] \tag{12.3}$$

which verifies the obvious requirement of longitudinal equilibrium for the ith member.

It should be noted that it is sometimes convenient to alter the numbering scheme for member-end forces and displacements. Instead of repeating the numbers 1 and 2 for each member, it is sometimes easier simply to use consecutive numbers. That is, 1 and 2 are used for the first member, 3 and 4 for the second member, and so on. This technique is illustrated in the matrix formulation of Section 12.5.

12.4

Formulation of Equilibrium Method of Analysis for Trusses

The fundamental steps in analyzing a structure according to the equilibrium method will be illustrated by considering the simple three-bar truss of Fig. 12.3. This structure is statically indeterminate to the first degree, although this does not affect the formulation of the solution. The solution procedure would be essentially the same for a statically determinate two-bar truss. Since there are two displace-

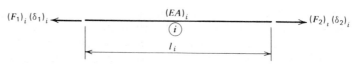

Fig. 12.2 *Member-end forces and displacements for pin-ended member.*

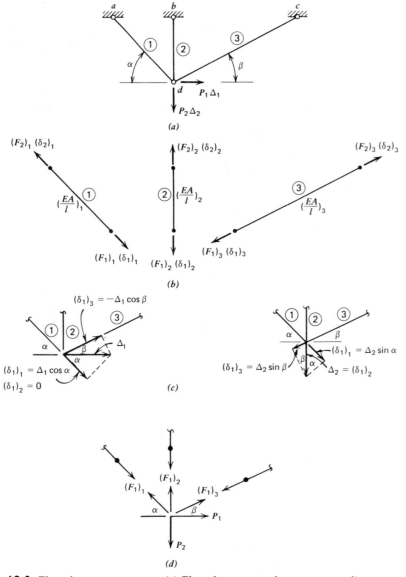

Fig. 12.3 *Three-bar truss structure. (a) Three-bar truss and structure coordinate system. (b) Member coordinate systems; member-end forces and displacements. (c) Relations between structure displacements and member-end displacements. (d) Member-end forces and structure forces at joint.*

ment degrees of freedom, the structure is kinematically indeterminate to the second degree. Thus, we anticipate the need for two equilibrium equations, whose solution will provide the two unknown displacements, Δ_1 and Δ_2.

The member arrangement is shown in Fig. 12.3a along with the structure coordinate system, which shows the positive directions for the P forces and the Δ displacements. The individual members are isolated in Fig. 12.3b, and the member coordinate systems are shown. These systems indicate the positive directions for the F member forces and the δ member displacements.

To ensure that the structure remains compatibly interconnected as it deforms, the structure displacements at joint d must conform to the member-end displacements for each of the members that terminate at that joint. This is done by expressing the member-end displacements in terms of the structure displacement in the form of *compatibility equations*. These are sometimes called *connectivity equations*. These equations are generated with the aid of Fig. 12.3c. In this figure, positive structure displacements of Δ_1 and Δ_2 are induced, and their projections along each of the members framing into joint d become the δ_1 displacement for the respective member. Careful attention must be given to the sign conventions that are established by the structure and member coordinate systems. For instance, a positive Δ_1 induces a negative $(\delta_1)_3$. Combining the effects of Δ_1 and Δ_2, we determine the following compatibility equations:

$$(\delta_1)_1 = \Delta_1 \cos \alpha + \Delta_2 \sin \alpha$$
$$(\delta_1)_2 = \Delta_2 \tag{12.4}$$
$$(\delta_1)_3 = -\Delta_1 \cos \beta + \Delta_2 \sin \beta$$

Note that these equations implicitly limit the analysis to small displacements since the member displacements are assumed to act along the original member directions. In addition to the compatibility equations expressed by Eq. 12.4, the member displacements are also constrained by the boundary conditions. In this case, since joints a, b, and c are nonyielding support points, the boundary conditions require that

$$(\delta_2)_1 = (\delta_2)_2 = (\delta_2)_3 = 0 \tag{12.5}$$

These compatibility and boundary conditions can now be used to express the member forces in terms of the structure displacements. This is done by substituting Eqs. 12.4 and 12.5 into Eq. 12.3 to obtain

$$(F_1)_1 = (F_2)_1 = \left(\frac{EA}{l}\right)_1 (\delta_1)_1 = \left(\frac{EA}{l}\right)_1 (\Delta_1 \cos \alpha + \Delta_2 \sin \alpha)$$
$$(F_1)_2 = (F_2)_2 = \left(\frac{EA}{l}\right)_2 (\delta_1)_2 = \left(\frac{EI}{l}\right)_2 (\Delta_2) \tag{12.6}$$
$$(F_1)_3 = (F_2)_3 = \left(\frac{EA}{l}\right)_3 (\delta_1)_3 = \left(\frac{EA}{l}\right)_3 (-\Delta_1 \cos \beta + \Delta_2 \sin \beta)$$

The kinematically determinate primary structure, which has zero displacements at point d, will have zero bar forces in the three bars framing into that point. Thus, there is no way that the forces P_1 and P_2 can be equilibrated. Only by relaxing the artificially imposed restraints is it possible to satisfy equilibrium. When joint d is released, the structure deforms, and forces develop in the members. Deformation continues until joint d comes into equilibrium. The forces acting at joint d are shown in Fig. 12.3d, which assumes both positive structure and member forces. Applying equilibrium, we have

$$-(F_1)_1 \cos \alpha + (F_1)_3 \cos \beta + P_1 = 0 \tag{12.7}$$
$$-(F_1)_1 \sin \alpha - (F_1)_2 - (F_1)_3 \sin \beta + P_2 = 0$$

Again, small displacements are implied by showing unaltered member directions for the final equilibrium condition.

Substitution of Eq. 12.6 into Eq. 12.7 yields

$$
\left[\left(\frac{EA}{l} \right)_1 \cos^2 \alpha + \left(\frac{EA}{l} \right)_3 \cos^2 \beta \right] \Delta_1
$$

$$
+ \left[\left(\frac{EA}{l} \right)_1 \sin \alpha \cos \alpha - \left(\frac{EA}{l} \right)_3 \sin \beta \cos \beta \right] \Delta_2 = P_1
$$

$$
(12.8)
$$

$$
\left[\left(\frac{EA}{l} \right)_1 \sin \alpha \cos \alpha - \left(\frac{EA}{l} \right)_3 \sin \beta \cos \beta \right] \Delta_1
$$

$$
+ \left[\left(\frac{EA}{l} \right)_1 \sin^2 \alpha + \left(\frac{EA}{l} \right)_2 + \left(\frac{EA}{l} \right)_3 \sin^2 \beta \right] \Delta_2 = P_2
$$

These are the equilibrium equations for the structure, and they are expressed in terms of the unknown structure displacements. In matrix form,

$$
\begin{bmatrix} K_{11} & K_{12} \\ K_{21} & K_{22} \end{bmatrix} \begin{Bmatrix} \Delta_1 \\ \Delta_2 \end{Bmatrix} = \begin{Bmatrix} P_1 \\ P_2 \end{Bmatrix}
$$

$$
(12.9)
$$

where K_{ij} represents the structure stiffness coefficients. Specifically, these coefficients are

$$
K_{11} = \left[\left(\frac{EA}{l} \right)_1 \cos^2 \alpha + \left(\frac{EA}{l} \right)_3 \cos^2 \beta \right]
$$

$$
K_{12} = K_{21} = \left[\left(\frac{EA}{l} \right)_1 \sin \alpha \cos \alpha - \left(\frac{EA}{l} \right)_3 \sin \beta \cos \beta \right] \quad (12.10)
$$

$$
K_{22} = \left[\left(\frac{EA}{l} \right)_1 \sin^2 \alpha + \left(\frac{EA}{l} \right)_2 + \left(\frac{EA}{l} \right)_3 \sin^2 \beta \right]
$$

It should be noted that Eq. 12.9 is the inverse relationship of Eq. 5.30, where K_{ij} gives the value of the ith force that accompanies a unit value of the jth displacement.

In a typical situation, the P loads are given, and the elements of the stiffness matrix are determined from Eq. 12.10. Equation 12.9 is then solved for the Δ displacements in the form

$$
\begin{Bmatrix} \Delta_1 \\ \Delta_2 \end{Bmatrix} = \begin{bmatrix} K_{11} & K_{12} \\ K_{21} & K_{22} \end{bmatrix}^{-1} \begin{Bmatrix} P_1 \\ P_2 \end{Bmatrix}
$$

$$
(12.11)
$$

These displacements can then be substituted into Eqs. 12.4 and 12.6 to give the member displacements and member forces, respectively.

For a more complicated structure, the same basic procedure would be employed. In general, the individual member forces will depend on the displacements at each end of the member, and there will be two equations of equilibrium for each displaceable joint.

12.4.1 Example problem   Determine the displacements at joint d and the individual member forces for the three-bar truss shown under the designated loading condition. The cross-sectional area of each member is 2 000 mm², and E is 10 GN/m².

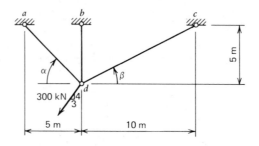

Coordinate Systems:

Structure:

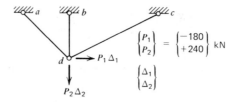

$$\begin{Bmatrix} P_1 \\ P_2 \end{Bmatrix} = \begin{Bmatrix} -180 \\ +240 \end{Bmatrix} \text{ kN}$$

$$\begin{Bmatrix} \Delta_1 \\ \Delta_2 \end{Bmatrix}$$

Members:

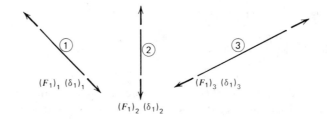

Stiffness Matrix: (Eqs. 12.10)

$$K_{11} = \left[\left(\frac{EA}{l} \right)_1 \cos^2 \alpha + \left(\frac{EA}{l} \right)_3 \cos^2 \beta \right]$$

$$= \left[\frac{EA}{7.07} \left(\frac{1}{\sqrt{2}} \right)^2 + \frac{EA}{11.18} \left(\frac{2}{\sqrt{5}} \right)^2 \right] = 0.142\ 3EA$$

$$K_{12} = K_{21}$$

$$= \left[\left(\frac{EA}{l} \right)_1 \sin \alpha \cos \alpha - \left(\frac{EA}{l} \right)_3 \sin \beta \cos \beta \right]$$

$$= \left[\frac{EA}{7.07} \left(\frac{1}{\sqrt{2}} \right)^2 - \frac{EA}{11.18} \left(\frac{1}{\sqrt{5}} \cdot \frac{2}{\sqrt{5}} \right) \right] = 0.034\ 9EA$$

$$K_{22} = \left[\left(\frac{EA}{l}\right)_1 \sin^2\alpha + \left(\frac{EA}{l}\right)_2 + \left(\frac{EA}{l}\right)_3 \sin^2\beta\right]$$

$$= \left[\frac{EA}{7.07}\left(\frac{1}{\sqrt{2}}\right)^2 + \left(\frac{EA}{5}\right) + \frac{EA}{11.18}\left(\frac{1}{\sqrt{5}}\right)^2\right] = 0.288\ 6EA$$

$$\begin{bmatrix} K_{11} & K_{12} \\ K_{21} & K_{22} \end{bmatrix} = EA\begin{bmatrix} 0.142\ 3 & 0.034\ 9 \\ 0.034\ 9 & 0.288\ 6 \end{bmatrix} \left(\frac{EA}{m}\right)$$

Structure Displacements:

$$\begin{Bmatrix} \Delta_1 \\ \Delta_2 \end{Bmatrix} = \begin{bmatrix} K_{11} & K_{12} \\ K_{21} & K_{22} \end{bmatrix}^{-1} \begin{Bmatrix} P_1 \\ P_2 \end{Bmatrix} \tag{12.11}$$

$$\begin{Bmatrix} \Delta_1 \\ \Delta_2 \end{Bmatrix} = \frac{1}{0.039\ 85EA}\begin{bmatrix} 0.288\ 6 & -0.034\ 9 \\ -0.034\ 9 & 0.142\ 3 \end{bmatrix}\begin{Bmatrix} -180 \\ +240 \end{Bmatrix} = \begin{Bmatrix} -0.075\ 7 \\ 0.050\ 7 \end{Bmatrix} \text{(m)}$$

where $EA = (10 \times 10^6 \text{ kN/m}^2)(0.002 \text{ m}^2) = 20 \times 10^3$ kN

Member Displacements: (Eqs. 12.4)

$$(\delta_1)_1 = \Delta_1 \cos\alpha + \Delta_2 \sin\alpha$$

$$= (-0.075\ 7)\left(\frac{1}{\sqrt{2}}\right) + (0.050\ 7)\left(\frac{1}{\sqrt{2}}\right) = -0.017\ 7 \text{ m}$$

$$(\delta_1)_2 = \Delta_2 = 0.050\ 7\text{m}$$
$$(\delta_1)_3 = -\Delta_1 \cos\beta + \Delta_2 \sin\beta$$

$$= -(-0.075\ 7)\left(\frac{2}{\sqrt{5}}\right) + (0.050\ 7)\left(\frac{1}{\sqrt{5}}\right) = 0.090\ 4 \text{ m}$$

Member Forces: (Eqs. 12.6)

$$(F_1)_1 = \left(\frac{EA}{l}\right)_1 (\delta_1)_1 = \frac{20 \times 10^3}{7.07}(-0.017\ 7) = -50.1 \text{ kN}$$

$$(F_1)_2 = \left(\frac{EA}{l}\right)_2 (\delta_1)_2 = \frac{20 \times 10^3}{5}(0.050\ 7) = 202.8 \text{ kN}$$

$$F_1)_3 = \left(\frac{EA}{l}\right)_3 (\delta_1)_3 = \frac{20 \times 10^3}{11.18}(0.090\ 4) = 161.7 \text{ kN}$$

Equilibrium Check: (Eqs. 12.7)

$$-(F_1)_1 \cos\alpha + (F_1)_3 \cos\beta + P_1 \overset{?}{=} 0$$

$$-(-50.1)\left(\frac{1}{\sqrt{2}}\right) + (161.7)\left(\frac{2}{\sqrt{5}}\right) - 180 = 0.05 \quad \therefore \text{ ok}$$

$$-(F_1)_1 \sin\alpha - (F_1)_2 - (F_1)_3 \sin\beta + P_2 \overset{?}{=} 0$$

$$-(-50.1)\left(\frac{1}{\sqrt{2}}\right) - 202.8 - (161.7)\left(\frac{1}{\sqrt{5}}\right) + 240 = 0.31 \quad \therefore \text{ ok}$$

12.5

Matrix Formulation of the Equilibrium Method of Truss Analysis

The fundamental procedures outlined in Section 12.4 for the analysis of pin-connected frameworks by the equilibrium method are very systematic. By including additional member-end forces and displacements, the same approach can readily be extended to cover beam- and frame-type structures, as will be seen in Chapter 13. The orderliness of the method makes it particularly well suited for a matrix formulation, and this section is devoted to the development of the equilibrium method of truss analysis in matrix format. In Chapter 13, a similar matrix approach is given for beam and frame structures, and Chapter 17 presents a completely general matrix formulation of the equilibrium method.

Consider the general truss structure shown in Fig. 12.4a, in which the force vector $\{P\}$ and the displacement vector $\{\Delta\}$ are defined for the overall structure. The elements of the force and displacement systems are limited to those points where displacements are admissible and are related by

$$\{P\} = [K]\{\Delta\} \tag{12.12}$$

in which $[K]$ is the structure stiffness matrix, which was shown in an expanded form in Eq. 12.9.

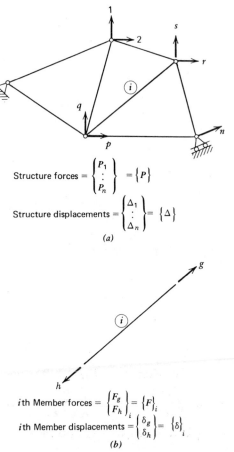

Structure forces $= \begin{Bmatrix} P_1 \\ \vdots \\ P_n \end{Bmatrix} = \{P\}$

Structure displacements $= \begin{Bmatrix} \Delta_1 \\ \vdots \\ \Delta_n \end{Bmatrix} = \{\Delta\}$

(a)

ith Member forces $= \begin{Bmatrix} F_g \\ F_h \end{Bmatrix}_i = \{F\}_i$

ith Member displacements $= \begin{Bmatrix} \delta_g \\ \delta_h \end{Bmatrix}_i = \{\delta\}_i$

(b)

Fig. 12.4 *Structure and member coordinate systems. (a) Structure coordinate system. (b) Member coordinate system for ith member.*

In a similar fashion, Fig. 12.4b gives the ith truss member with the vector of member-end forces, $\{F\}_i$, and the corresponding vector of member-end displacements, $\{\delta\}_i$. These forces and displacements are related by

$$\{F\}_i = [k]_{ai}\{\delta\}_i \qquad (12.13)$$

in which $[k]_{ai}$ is the stiffness matrix of the ith member. This corresponds to the stiffness matrix introduced in Section 5.3.

So that the individual members of the truss fit properly into the structure as a whole, the member displacements must be compatible with the structure displacements. This is ensured by the relation

$$\{\delta\}_i = [\beta]_i\{\Delta\} \qquad (12.14)$$

where $[\beta]_i$ is referred to as the *connectivity or compatibility matrix* for the ith member. This equation is a generalization of Eqs. 12.4 and 12.5 in which the individual elements of the $[\beta]_i$ matrix are controlled by the inclination of the members relative to the structural coordinates. For instance, consider the ith member of Fig. 12.4a. As shown in Fig. 12.5, the total member-end displacement results from the summation of the contributions of each structural displacement that occurs at the member end. Thus, using the direction cosines of member i, as given in Fig. 12.5, we obtain

$$\delta_g = -(\Delta_r \cos \theta + \Delta_s \cos \phi) \qquad (12.15)$$
$$\delta_h = -(\Delta_p \cos \eta + \Delta_q \cos \xi)$$

Substitution of Eq. 12.14 into Eq. 12.13 yields

$$\{F\}_i = [k]_{ai}[\beta]_i\{\Delta\} \qquad (12.16)$$

which gives the member-end forces in .erms of the structure displacements and is the general form of Eq. 12.6.

When the kinematically determinate primary structure is released, the work done as measured in the structure coordinate system must be equal to the work done as measured in the member coordinate system. Thus for the entire truss of m members, we have, according to the matrix form of Eq. 2.27,

$$\tfrac{1}{2}\{P\}^T\{\Delta\} = \sum_{i=1}^{m} \tfrac{1}{2}\{F\}_i^T\{\delta\}_i \qquad (12.17)$$

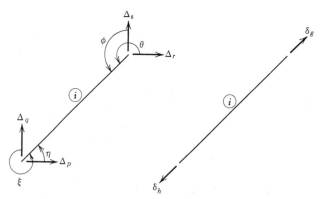

Fig. 12.5 *Translation from structure displacements to member displacements.*

Substitution of Eq. 12.14 into Eq. 12.17 gives

$$\frac{1}{2}\{P\}^T\{\Delta\} = \sum_{i=1}^{m} \frac{1}{2}\{F\}_i^T[\beta]_i\{\Delta\} \tag{12.18}$$

Transposition of each side and a subsequent rearrangement of terms gives

$$\{\Delta\}^T\left(\{P\} - \sum_{i=1}^{m} [\beta]_i^T\{F\}_i\right) = 0$$

Since $\{\Delta\}^T$ can be controlled arbitrarily through various patterns of release, the bracketed term must vanish. Thus, we have

$$\{P\} = \sum_{i=1}^{m} [\beta]_i^T\{F\}_i \tag{12.19}$$

This expression gives the equilibrium conditions between the member forces and the structure forces, and it is the equation for which Eq. 12.7 is a specific case.

Substitution of Eq. 12.16 into Eq. 12.19 gives

$$\{P\} = \sum_{i=1}^{m} [\beta]_i^T[k]_{ai}[\beta]_i\{\Delta\} \tag{12.20}$$

which represents a generalization of Eq. 12.8. Comparing Eqs. 12.12 and 12.20, we see that the structure stiffness matrix $[K]$ is given by

$$[K] = \sum_{i=1}^{m} [\beta]_i^T[k]_{ai}[\beta]_i \tag{12.21}$$

This equation expresses the manner in which the individual member stiffness characteristics are synthesized to obtain the overall stiffness characteristics of the structure.

In a typical problem, the structure load vector $\{P\}$ is given, and the structure stiffness matrix can be determined from Eq. 12.21. The truss displacements are then determined from solving Eq. 12.12 in the form

$$\{\Delta\} = [K]^{-1}\{P\} \tag{12.22}$$

Once the structure displacements are obtained, the member-end displacements $\{\delta\}_i$ can be computed from Eq. 12.14, and the member-end forces $\{F\}_i$ can be determined from Eq. 12.16 for each member.

12.6

Application of Matrix Formulation

As has already been noted, the equilibrium method is equally applicable for the analysis of statically determinate and statically indeterminate structures. For statically determinate systems, it is frequently easier to determine the member forces for the methods of static analysis that were presented in Chapter 3. The desired displacement quantities are then determined by the methods given in Chapter 5. However, the general matrix format that is presented here will yield the complete array of structure displacements from which all of the member forces can be determined. The first example problem of this section is a reanalysis of a statically determinate problem that was considered earlier in the book.

In analyzing statically indeterminate structures, the dual requirements of compatibility and equilibrium must be satisfied. The equilibrium approach inherently

satisfies both of these in that the governing equations explicitly state the equilibrium conditions, and the compatibility conditions are included in the generation
of these equations.

Determine the member forces and the joint displacements for the statically deter- **12.6.1 Example**
minate truss shown below with the indicated loading. This structure was previ- **problem**
ously considered in Chapter 5.

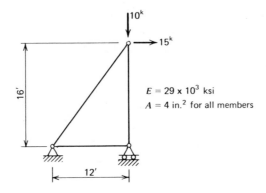

$E = 29 \times 10^3$ ksi
$A = 4$ in.^{2} for all members

Coordinate Systems:

Structure:

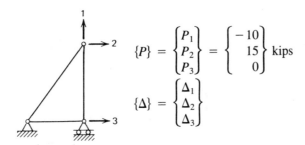

$$\{P\} = \begin{Bmatrix} P_1 \\ P_2 \\ P_3 \end{Bmatrix} = \begin{Bmatrix} -10 \\ 15 \\ 0 \end{Bmatrix} \text{ kips}$$

$$\{\Delta\} = \begin{Bmatrix} \Delta_1 \\ \Delta_2 \\ \Delta_3 \end{Bmatrix}$$

Members:

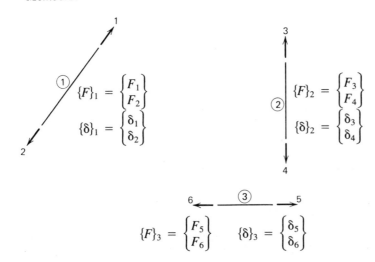

$$\{F\}_1 = \begin{Bmatrix} F_1 \\ F_2 \end{Bmatrix} \qquad \{F\}_2 = \begin{Bmatrix} F_3 \\ F_4 \end{Bmatrix}$$

$$\{\delta\}_1 = \begin{Bmatrix} \delta_1 \\ \delta_2 \end{Bmatrix} \qquad \{\delta\}_2 = \begin{Bmatrix} \delta_3 \\ \delta_4 \end{Bmatrix}$$

$$\{F\}_3 = \begin{Bmatrix} F_5 \\ F_6 \end{Bmatrix} \qquad \{\delta\}_3 = \begin{Bmatrix} \delta_5 \\ \delta_6 \end{Bmatrix}$$

Member Forces:

$$\{F\}_i = [k]_{ai}\{\delta\}_i \tag{12.13}$$

$$\{F\}_i = \begin{bmatrix} \dfrac{EA}{l} & \dfrac{EA}{l} \\ \dfrac{EA}{l} & \dfrac{EA}{l} \end{bmatrix}_i \{\delta\}_i \tag{12.1}$$

Structure Connectivity:

$$\{\delta\}_i = [\beta]_i\{\Delta\} \tag{12.14}$$

Member 1 (i = 1)

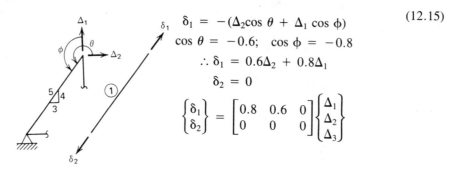

$$\delta_1 = -(\Delta_2\cos\theta + \Delta_1\cos\phi) \tag{12.15}$$

$$\cos\theta = -0.6; \quad \cos\phi = -0.8$$

$$\therefore \delta_1 = 0.6\Delta_2 + 0.8\Delta_1$$

$$\delta_2 = 0$$

$$\begin{Bmatrix} \delta_1 \\ \delta_2 \end{Bmatrix} = \begin{bmatrix} 0.8 & 0.6 & 0 \\ 0 & 0 & 0 \end{bmatrix} \begin{Bmatrix} \Delta_1 \\ \Delta_2 \\ \Delta_3 \end{Bmatrix}$$

Member 2 (i = 2)

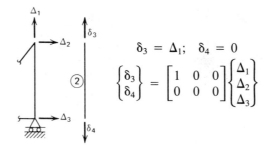

$$\delta_3 = \Delta_1; \quad \delta_4 = 0$$

$$\begin{Bmatrix} \delta_3 \\ \delta_4 \end{Bmatrix} = \begin{bmatrix} 1 & 0 & 0 \\ 0 & 0 & 0 \end{bmatrix} \begin{Bmatrix} \Delta_1 \\ \Delta_2 \\ \Delta_3 \end{Bmatrix}$$

Member 3 (i = 3)

$$\delta_5 = \Delta_3; \quad \delta_6 = 0$$

$$\begin{Bmatrix} \delta_5 \\ \delta_6 \end{Bmatrix} = \begin{bmatrix} 0 & 0 & 1 \\ 0 & 0 & 0 \end{bmatrix} \begin{Bmatrix} \Delta_1 \\ \Delta_2 \\ \Delta_3 \end{Bmatrix}$$

Structure Stiffness:

$$[K] = \sum_{i=1}^{m} [\beta]_i^T [k]_{ai} [\beta]_i \tag{12.21}$$

$$[K] = \begin{bmatrix} 0.8 & 0 \\ 0.6 & 0 \\ 0 & 0 \end{bmatrix} \begin{bmatrix} \dfrac{EA}{l} & \dfrac{EA}{l} \\ \dfrac{EA}{l} & \dfrac{EA}{l} \end{bmatrix}_1 \begin{bmatrix} 0.8 & 0.6 & 0 \\ 0 & 0 & 0 \end{bmatrix} + \begin{bmatrix} 1 & 0 \\ 0 & 0 \\ 0 & 0 \end{bmatrix} \begin{bmatrix} \dfrac{EA}{l} & \dfrac{EA}{l} \\ \dfrac{EA}{l} & \dfrac{EA}{l} \end{bmatrix}_2 \begin{bmatrix} 1 & 0 & 0 \\ 0 & 0 & 0 \end{bmatrix}$$

$$+ \begin{bmatrix} 0 & 0 \\ 0 & 0 \\ 1 & 0 \end{bmatrix} \begin{bmatrix} \dfrac{EA}{l} & \dfrac{EA}{l} \\ \dfrac{EA}{l} & \dfrac{EA}{l} \end{bmatrix}_3 \begin{bmatrix} 0 & 0 & 1 \\ 0 & 0 & 0 \end{bmatrix}$$

$$[K] = EA \begin{bmatrix} \left(\dfrac{0.64}{l_1} + \dfrac{1}{l_2}\right) & \left(\dfrac{0.48}{l_1}\right) & 0 \\ \left(\dfrac{0.48}{l_1}\right) & \left(\dfrac{0.36}{l_1}\right) & 0 \\ 0 & 0 & \left(\dfrac{1}{l_3}\right) \end{bmatrix}$$

For $l_1 = 240''$, $l_2 = 192''$, $l_3 = 144''$

$$[K] = EA \begin{bmatrix} 0.007{,}88 & 0.002 & 0 \\ 0.002 & 0.001{,}65 & 0 \\ 0 & 0 & 0.006{,}94 \end{bmatrix}$$

Equations of Equilibrium and Solution for Displacements:

$$\{P\} = [K]\{\Delta\} \tag{12.12}$$

$$\{\Delta\} = [K]^{-1}\{P\} \tag{12.22}$$

$$\begin{Bmatrix} \Delta_1 \\ \Delta_2 \\ \Delta_3 \end{Bmatrix} = \frac{1}{EA} \begin{bmatrix} 191.8 & -255.7 & 0 \\ -255.7 & 1{,}007.5 & 0 \\ 0 & 0 & 144.1 \end{bmatrix} \begin{Bmatrix} P_1 \\ P_2 \\ P_3 \end{Bmatrix}$$

Note that $[K]^{-1}$ equals the structure flexibility matrix of Section 5.8.1. For the given $\{P\}$ and $EA = 4 \times 29 \times 10^3 = 116{,}000$ kips, we have

$$\begin{Bmatrix} \Delta_1 \\ \Delta_2 \\ \Delta_3 \end{Bmatrix} = \frac{1}{EA} \begin{Bmatrix} -5{,}753.5 \\ +17{,}669.5 \\ 0 \end{Bmatrix} = \begin{Bmatrix} -0.049{,}6 \\ +0.152{,}3 \\ 0 \end{Bmatrix} \text{ inch}$$

These results agree with those of Example 5.8.1.

Member Forces:

$$\{F\}_i = [k]_{ai}[\beta]_i\{\Delta\} \tag{12.16}$$

Member 1 $(i = 1)$

$$\begin{Bmatrix} F_1 \\ F_2 \end{Bmatrix} = \begin{bmatrix} \dfrac{EA}{l} & \dfrac{EA}{l} \\ \dfrac{EA}{l} & \dfrac{EA}{l} \end{bmatrix}_1 \begin{bmatrix} 0.8 & 0.6 & 0 \\ 0 & 0 & 0 \end{bmatrix} \frac{1}{EA} \begin{Bmatrix} -5{,}753.5 \\ +17{,}669.5 \\ 0 \end{Bmatrix} = \begin{Bmatrix} +25.0 \\ +25.0 \end{Bmatrix} \text{ kips}$$

Member 2 (i = 2)

$$\begin{Bmatrix} F_3 \\ F_4 \end{Bmatrix} = \begin{bmatrix} \dfrac{EA}{l} & \dfrac{EA}{l} \\ \dfrac{EA}{l} & \dfrac{EA}{l} \end{bmatrix}_2 \begin{bmatrix} 1 & 0 & 0 \\ 0 & 0 & 0 \end{bmatrix} \dfrac{1}{EA} \begin{Bmatrix} -5,753.5 \\ +17,669.5 \\ 0 \end{Bmatrix} = \begin{Bmatrix} -30.0 \\ -30.0 \end{Bmatrix} \text{ kips}$$

Member 3 (i = 3)

$$\begin{Bmatrix} F_5 \\ F_6 \end{Bmatrix} = \begin{bmatrix} \dfrac{EA}{l} & \dfrac{EA}{l} \\ \dfrac{EA}{l} & \dfrac{EA}{l} \end{bmatrix}_3 \begin{bmatrix} 0 & 0 & 1 \\ 0 & 0 & 0 \end{bmatrix} \dfrac{1}{EA} \begin{Bmatrix} -5,753.5 \\ +17,669.5 \\ 0 \end{Bmatrix} = \begin{Bmatrix} 0 \\ 0 \end{Bmatrix} \text{ kips}$$

These results confirm the obvious requirement that the forces are equal at the ends of a pin-ended member. Also, the member forces agree with those determined from statics, as shown below.

Summary:

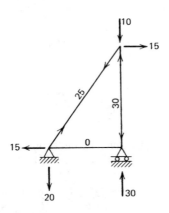

12.6.2 Example problem   Determine the displacements at joint *d* and the individual member forces for the structure and loading of Example 12.4.1. The structure is redrawn below, and $EA = 20 \times 10^3$ kN.

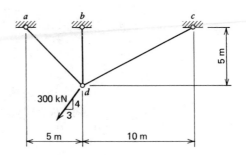

Coordinate Systems:

Structure:

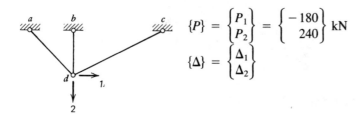

$$\{P\} = \begin{Bmatrix} P_1 \\ P_2 \end{Bmatrix} = \begin{Bmatrix} -180 \\ 240 \end{Bmatrix} \text{ kN}$$

$$\{\Delta\} = \begin{Bmatrix} \Delta_1 \\ \Delta_2 \end{Bmatrix}$$

Members:

$$\{F\}_1 = \begin{Bmatrix} F_1 \\ F_2 \end{Bmatrix}$$

$$\{\delta\}_1 = \begin{Bmatrix} \delta_1 \\ \delta_2 \end{Bmatrix}$$

$$\{F\}_2 = \begin{Bmatrix} F_3 \\ F_4 \end{Bmatrix}$$

$$\{\delta\}_2 = \begin{Bmatrix} \delta_3 \\ \delta_4 \end{Bmatrix}$$

$$\{F\}_3 = \begin{Bmatrix} F_5 \\ F_6 \end{Bmatrix}$$

$$\{\delta\}_3 = \begin{Bmatrix} \delta_5 \\ \delta_6 \end{Bmatrix}$$

Member Forces:

$$\{F\}_i = [k]_{ai}\{\delta\}_i \tag{12.13}$$

$$\{F\}_i = \begin{bmatrix} \dfrac{EA}{l} & \dfrac{EA}{l} \\ \dfrac{EA}{l} & \dfrac{EA}{l} \end{bmatrix}_i \{\delta\}_i \tag{12.1}$$

Structure Connectivity:

$$\{\delta\}_i = [\beta]_i\{\Delta\} \tag{12.14}$$

Member 1 ($i = 1$)

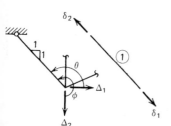

$$\delta_1 = -(\Delta_1 \cos \theta + \Delta_2 \cos \phi \tag{12.15}$$

$$\cos \theta = -0.707; \quad \cos \phi = -0.707$$

$$\delta_1 = 0.707\Delta_1 + 0.707\Delta_2$$

$$\delta_2 = 0$$

$$\begin{Bmatrix} \delta_1 \\ \delta_2 \end{Bmatrix} = \begin{bmatrix} 0.707 & 0.707 \\ 0 & 0 \end{bmatrix} \begin{Bmatrix} \Delta_1 \\ \Delta_2 \end{Bmatrix}$$

Member 2 (i = 2)

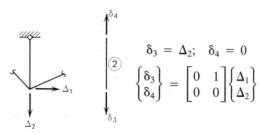

$$\delta_3 = \Delta_2; \quad \delta_4 = 0$$

$$\begin{Bmatrix} \delta_3 \\ \delta_4 \end{Bmatrix} = \begin{bmatrix} 0 & 1 \\ 0 & 0 \end{bmatrix} \begin{Bmatrix} \Delta_1 \\ \Delta_2 \end{Bmatrix}$$

Member 3 (i = 3)

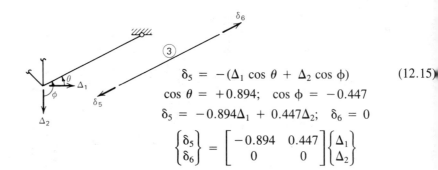

$$\delta_5 = -(\Delta_1 \cos \theta + \Delta_2 \cos \phi) \quad (12.15)$$

$$\cos \theta = +0.894; \quad \cos \phi = -0.447$$

$$\delta_5 = -0.894\Delta_1 + 0.447\Delta_2; \quad \delta_6 = 0$$

$$\begin{Bmatrix} \delta_5 \\ \delta_6 \end{Bmatrix} = \begin{bmatrix} -0.894 & 0.447 \\ 0 & 0 \end{bmatrix} \begin{Bmatrix} \Delta_1 \\ \Delta_2 \end{Bmatrix}$$

Structure Stiffness:

$$[K] = \sum_{i=1}^{m} [\beta]_i^T [k]_{ai} [\beta]_i \quad (12.21)$$

$$[K] = \begin{bmatrix} 0.707 & 0 \\ 0.707 & 0 \end{bmatrix} \begin{bmatrix} \dfrac{EA}{l} & \dfrac{EA}{l} \\ \dfrac{EA}{l} & \dfrac{EA}{l} \end{bmatrix}_1 \begin{bmatrix} 0.707 & 0.707 \\ 0 & 0 \end{bmatrix}$$

$$+ \begin{bmatrix} 0 & 0 \\ 1 & 0 \end{bmatrix} \begin{bmatrix} \dfrac{EA}{l} & \dfrac{EA}{l} \\ \dfrac{EA}{l} & \dfrac{EA}{l} \end{bmatrix}_2 \begin{bmatrix} 0 & 1 \\ 0 & 0 \end{bmatrix}$$

$$+ \begin{bmatrix} -0.894 & 0 \\ 0.447 & 0 \end{bmatrix} \begin{bmatrix} \dfrac{EA}{l} & \dfrac{EA}{l} \\ \dfrac{EA}{l} & \dfrac{EA}{l} \end{bmatrix}_3 \begin{bmatrix} -0.894 & 0.447 \\ 0 & 0 \end{bmatrix}$$

$$[K] = EA \begin{bmatrix} \left\{ \dfrac{(0.707)^2}{l_1} + \dfrac{(0.894)^2}{l_3} \right\} & \left\{ \dfrac{(0.707)^2}{l_1} + \dfrac{(-0.894)(0.447)}{l_3} \right\} \\ \left\{ \dfrac{(0.707)^2}{l_1} + \dfrac{(0.447)(-0.894)}{l_3} \right\} & \left\{ \dfrac{(0.707)^2}{l_1} + \dfrac{1}{l_2} + \dfrac{(0.447)^2}{l_3} \right\} \end{bmatrix}$$

for $l_1 = 7.07$ m, $l_2 = 5$ m, $l_3 = 11.18$ m

$$[K] = EA \begin{bmatrix} 0.142\ 2 & 0.035\ 0 \\ 0.035\ 0 & 0.288\ 6 \end{bmatrix} \left(\dfrac{EA}{m} \right)$$

Structure Displacements:

$$\{\Delta\} = [K]^{-1}\{P\} \tag{12.22}$$

This inversion was accomplished in Example 12.4.1, from which

$$\{\Delta\} = \frac{1}{EA}\left\{\begin{matrix} -1\,513.8 \\ 1\,014.7 \end{matrix}\right\} = \left\{\begin{matrix} -0.075\,7 \\ 0.050\,7 \end{matrix}\right\} \quad (m)$$

Member Forces:

$$\{F\}_i = [k]_{ai}[\beta]_i\{\Delta\} \tag{12.16}$$

Member 1 ($i = 1$)

$$\left\{\begin{matrix} F_1 \\ F_2 \end{matrix}\right\} = \frac{EA}{l_1}\begin{bmatrix} 1 & 1 \\ 1 & 1 \end{bmatrix}\begin{bmatrix} 0.707 & 0.707 \\ 0 & 0 \end{bmatrix}\frac{1}{EA}\left\{\begin{matrix} -1\,513.8 \\ 1\,014.7 \end{matrix}\right\} = \left\{\begin{matrix} -49.9 \\ -49.9 \end{matrix}\right\} \text{ kN}$$

Member 2 ($i = 2$)

$$\left\{\begin{matrix} F_3 \\ F_4 \end{matrix}\right\} = \frac{EA}{l_2}\begin{bmatrix} 1 & 1 \\ 1 & 1 \end{bmatrix}\begin{bmatrix} 0 & 1 \\ 0 & 0 \end{bmatrix}\frac{1}{EA}\left\{\begin{matrix} -1\,513.8 \\ 1\,014.7 \end{matrix}\right\} = \left\{\begin{matrix} 202.9 \\ 202.9 \end{matrix}\right\} \text{ kN}$$

Member 3 ($i = 3$)

$$\left\{\begin{matrix} F_5 \\ F_6 \end{matrix}\right\} = \frac{EA}{l_3}\begin{bmatrix} 1 & 1 \\ 1 & 1 \end{bmatrix}\begin{bmatrix} -0.894 & 0.447 \\ 0 & 0 \end{bmatrix}\frac{1}{EA}\left\{\begin{matrix} -1\,513.8 \\ 1\,014.7 \end{matrix}\right\} = \left\{\begin{matrix} 161.6 \\ 161.6 \end{matrix}\right\} \text{ kN}$$

These results verify those of Example 12.4.1.

Determine the displacements at points b and c and the forces in members ab and ac for the structure shown. The quantity EA is the same for each member.

12.6.3 Example problem

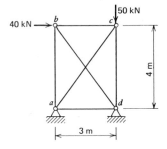

Coordinate Systems:

Structure:

$$\{P\} = \left\{\begin{matrix} P_1 \\ P_2 \\ P_3 \\ P_4 \end{matrix}\right\} = \left\{\begin{matrix} 40 \\ 0 \\ 0 \\ -50 \end{matrix}\right\}$$

$$\{\Delta\} = \left\{\begin{matrix} \Delta_1 \\ \Delta_2 \\ \Delta_3 \\ \Delta_4 \end{matrix}\right\}$$

Members:

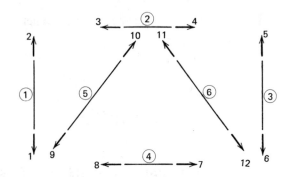

Member Forces:

$$\{F\}_i = [k]_{ai}\{\delta\}_i \qquad (12.13)$$

Structure Connectivity:

$$\{\delta\}_i = [\beta]_i\{\Delta\} \qquad (12.14)$$

Member 1 (i = 1): $\quad \delta_1 = 0, \quad \delta_2 = \Delta_2, \quad [\beta]_1 = \begin{bmatrix} 0 & 0 & 0 & 0 \\ 0 & 1 & 0 & 0 \end{bmatrix}$

Member 2 (i = 2): $\quad \delta_3 = -\Delta_1, \quad \delta_4 = \Delta_3, \quad [\beta]_2 = \begin{bmatrix} -1 & 0 & 0 & 0 \\ 0 & 0 & 1 & 0 \end{bmatrix}$

Member 3 (i = 3): $\quad \delta_5 = \Delta_4, \quad \delta_6 = 0, \quad [\beta]_3 = \begin{bmatrix} 0 & 0 & 0 & 1 \\ 0 & 0 & 0 & 0 \end{bmatrix}$

Member 4 (i = 4): $\quad \delta_7 = 0, \quad \delta_8 = 0, \quad [\beta]_4 = \begin{bmatrix} 0 & 0 & 0 & 0 \\ 0 & 0 & 0 & 0 \end{bmatrix}$

Member 5 (i = 5): $\quad \delta_9 = 0,$

$$\delta_{10} = 0.6\Delta_3 + 0.8\Delta_4$$

$$[\beta]_5 = \begin{bmatrix} 0 & 0 & 0 & 0 \\ 0 & 0 & 0.6 & 0.8 \end{bmatrix}$$

Member 6 (i = 6): $\quad \delta_{11} = -0.6\Delta_1 + 0.8\Delta_2, \quad [\beta]_6 = \begin{bmatrix} -0.6 & 0.8 & 0 & 0 \\ 0 & 0 & 0 & 0 \end{bmatrix}$

$$\delta_{12} = 0$$

Structure Stiffness:

$$[K] = \sum_{i=1}^{m} [\beta]_i^T [k]_{ai} [\beta]_i \tag{12.21}$$

$$[K] = EA \begin{bmatrix} \dfrac{1}{l_2} + \dfrac{0.36}{l_6} & \dfrac{-0.48}{l_6} & \dfrac{-1}{l_2} & 0 \\[2mm] \dfrac{0.48}{l_6} & \dfrac{1}{l_1} + \dfrac{0.64}{l_6} & 0 & 0 \\[2mm] -\dfrac{1}{l_2} & 0 & \dfrac{1}{l_2} + \dfrac{0.36}{l_5} & \dfrac{0.48}{l_5} \\[2mm] 0 & 0 & \dfrac{0.48}{l_5} & \dfrac{1}{l_3} + \dfrac{0.64}{l_5} \end{bmatrix}$$

$l_1 = l_3 = 4$ m; $l_2 = l_4 = 3$ m; $l_5 = l_6 = 5$ m

$$[K] = EA \begin{bmatrix} 0.405\ 3 & -0.096\ 0 & -0.333\ 3 & 0 \\ -0.096\ 0 & 0.378\ 0 & 0 & 0 \\ -0.333\ 3 & 0 & 0.405\ 3 & 0.096\ 0 \\ 0 & 0 & 0.096\ 0 & 0.378\ 0 \end{bmatrix} \left(\dfrac{EA}{m} \right)$$

Equations of Equilibrium and Solution for Displacements:

$$\{P\} = [K]\{\Delta\} \tag{12.12}$$
$$\{\Delta\} = [K]^{-1}\{P\} \tag{12.22}$$

$$\begin{Bmatrix} \Delta_1 \\ \Delta_2 \\ \Delta_3 \\ \Delta_4 \end{Bmatrix} = \dfrac{1}{0.004\ 861\ EA} \begin{bmatrix} 0.054\ 43 & 0.013\ 82 & 0.047\ 62 & -0.012\ 09 \\ 0.013\ 82 & 0.016\ 37 & 0.012\ 09 & -0.003\ 072 \\ 0.047\ 62 & 0.012\ 90 & 0.054\ 43 & -0.013\ 82 \\ -0.012\ 09 & -0.003\ 072 & -0.013\ 82 & 0.016\ 37 \end{bmatrix} \begin{Bmatrix} 40 \\ 0 \\ 0 \\ -50 \end{Bmatrix}$$

$$= \dfrac{1}{EA} \begin{Bmatrix} 572.25 \\ 145.32 \\ 534.01 \\ -267.87 \end{Bmatrix} \left(\dfrac{kN \cdot m}{EA} \right)$$

These results compare favorably with those of Example 10.8.1.

Member Forces:

$$\{F\}_i = [k]_{ai} [\beta]_i \{\Delta\} \tag{12.16}$$

Member ab ($i = 1$)

$$\begin{Bmatrix} F_1 \\ F_2 \end{Bmatrix} = \dfrac{EA}{4} \begin{bmatrix} 1 & 1 \\ 1 & 1 \end{bmatrix} \begin{bmatrix} 0 & 0 & 0 & 0 \\ 0 & 1 & 0 & 0 \end{bmatrix} \dfrac{1}{EA} \begin{Bmatrix} 572.25 \\ 145.32 \\ 534.01 \\ -267.87 \end{Bmatrix} = \begin{Bmatrix} 36.3 \\ 36.3 \end{Bmatrix} kN$$

Member ac ($i = 5$)

$$\begin{Bmatrix} F_9 \\ F_{10} \end{Bmatrix} = \dfrac{EA}{5} \begin{bmatrix} 1 & 1 \\ 1 & 1 \end{bmatrix} \begin{bmatrix} 0 & 0 & 0 & 0 \\ 0 & 0 & 0.6 & 0.8 \end{bmatrix} \dfrac{1}{EA} \begin{Bmatrix} 572.25 \\ 145.32 \\ 534.01 \\ -267.87 \end{Bmatrix} = \begin{Bmatrix} 21.2 \\ 21.2 \end{Bmatrix} kN$$

These results also compare favorably with those of Example 10.8.1.

12.7

Determination of Reactions and Effects of Support Settlements

In the problems considered in Sections 12.4 and 12.6, the analysis leads to the structure displacements, from which the member forces are then determined. From these member forces, the reactions can be determined using statics; however, there is a direct matrix approach that will yield the reactions. As a by-product of this approach, the problem of support settlements is easily handled.

Consider the structure shown in Fig. 12.6a, in which there are p independent admissible displacements. If in addition, the restraints associated with the reactions are removed, there is an additional set of k displacements, as shown in Fig. 12.6b. Assuming initially that the full array of $p + k = s$ displacements can occur, we may generate a set of s force–displacement equations that represent the equations of equilibrium for the entire structure. These equations take the form

$$
\begin{bmatrix}
K_{11} & \cdots & K_{1p} & K_{1,p+1} & \cdots & K_{1s} \\
\vdots & & \vdots & \vdots & & \vdots \\
K_{p1} & \cdots & K_{pp} & K_{p,p+1} & \cdots & K_{ps} \\
\hline
K_{p+1,1} & \cdots & K_{p+1,p} & K_{p+1,p+1} & \cdots & K_{p+1,s} \\
\vdots & & \vdots & \vdots & & \vdots \\
K_{s1} & \cdots & K_{sp} & K_{s,p+1} & \cdots & K_{ss}
\end{bmatrix}
\begin{Bmatrix}
\Delta_1 \\ \vdots \\ \Delta_p \\ \hline \Delta_{p+1} \\ \vdots \\ \Delta_s
\end{Bmatrix}
=
\begin{Bmatrix}
P_1 \\ \vdots \\ P_p \\ \hline P_{p+1} \\ \vdots \\ P_s
\end{Bmatrix}
\qquad (12.23)
$$

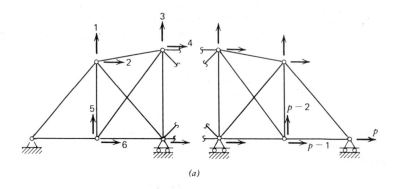

(a)

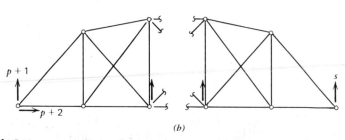

(b)

Fig. 12.6 *Structure coordinates for entire structure.* (a) *Structure coordinates corresponding to admissible displacements.* (b) *Structure coordinates corresponding to support restraints.*

If this equation is partitioned along the dashed lines of Eq. 12.23, we obtain the form

$$
\begin{bmatrix} [K]_{11} & \vdots & [K]_{12} \\ \text{---} & \text{---} & \text{---} \\ [K]_{21} & \vdots & [K]_{22} \end{bmatrix}
\begin{Bmatrix} \{\Delta\}_1 \\ \text{---} \\ \{\Delta\}_2 \end{Bmatrix}
= \begin{Bmatrix} \{P\}_1 \\ \text{---} \\ \{P\}_2 \end{Bmatrix}
\tag{12.24}
$$

where $\{\Delta\}_1$ represents the unknown structure displacements that correspond to the known applied $\{P\}_1$ forces, and $\{P\}_2$ includes the unknown reaction components associated with the designated support displacements $\{\Delta\}_2$. Each submatrix of the stiffness matrix has a unique meaning; $[K]_{ij}$ gives the interaction between the $\{P\}_i$ forces and the $\{\Delta\}_j$ displacements.

Equation 12.24 can be expanded to give

$$
[K]_{11}\{\Delta\}_1 + [K]_{12}\{\Delta\}_2 = \{P\}_1
\tag{12.25}
$$

$$
[K]_{21}\{\Delta\}_1 + [K]_{22}\{\Delta\}_2 = \{P\}_2
\tag{12.26}
$$

The elements of $\{\Delta\}_2$ are known: each element corresponding to a fully restrained component is zero, while the elements associated with settlements have designated nonzero values. Thus, since $\{P\}_1$ is also known, Eq. 12.25 can be solved for $\{\Delta\}_1$ in the form

$$
\{\Delta\}_1 = [K]_{11}^{-1}(\{P\}_1 - [K]_{12}\{\Delta\}_2)
\tag{12.27}
$$

Here, it is clear that $[K]_{12}\{\Delta\}_2$ establishes a set of equivalent forces, which are nonzero whenever support settlements are present. Knowing $\{\Delta\}_1$, we can determine the reaction forces $\{P\}_2$ directly from Eq. 12.26. Or, substituting Eq. 12.27 into Eq. 12.26, we obtain

$$
\{P\}_2 = [K]_{21}[K]_{11}^{-1}(\{P\}_1 - [K]_{12}\{\Delta\}_2) + [K]_{22}\{\Delta\}_2
\tag{12.28}
$$

This gives the reaction forces in terms of the known forces $\{P\}_1$ and the designated support displacements $\{\Delta\}_2$.

It should be noted that $[K]_{11}$ is the usual stiffness matrix that evolves when the support displacements are excluded from the formulation. Also, it should be emphasized that $\{P\}_2$ and $\{\Delta\}_2$ need only include those elements corresponding to a desired reaction component or a designated settlement element. The problem that follows is an example where $\{\Delta\}_2$ is restricted to one element.

Consider again the three-bar truss of Exmaple 12.6.2. Determine the displacements at joint d and the member forces for the loading given, assuming that joint b settles vertically 10 mm downward.

12.7.1 Example problem

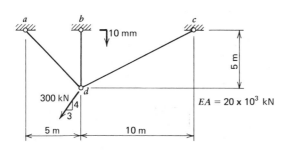

$EA = 20 \times 10^3$ kN

Coordinate Systems:

Structure:

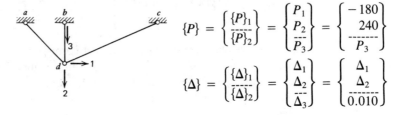

$$\{P\} = \left\{\frac{\{P\}_1}{\{P\}_2}\right\} = \left\{\begin{array}{c} P_1 \\ P_2 \\ \hline P_3 \end{array}\right\} = \left\{\begin{array}{c} -180 \\ 240 \\ \hline P_3 \end{array}\right\}$$

$$\{\Delta\} = \left\{\frac{\{\Delta\}_1}{\{\Delta\}_2}\right\} = \left\{\begin{array}{c} \Delta_1 \\ \Delta_2 \\ \hline \Delta_3 \end{array}\right\} = \left\{\begin{array}{c} \Delta_1 \\ \Delta_2 \\ \hline 0.010 \end{array}\right\}$$

Members:

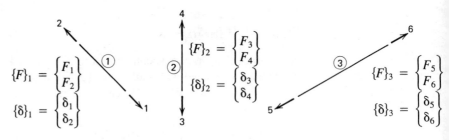

$$\{F\}_1 = \left\{\begin{array}{c} F_1 \\ F_2 \end{array}\right\}$$

$$\{\delta\}_1 = \left\{\begin{array}{c} \delta_1 \\ \delta_2 \end{array}\right\}$$

$$\{F\}_2 = \left\{\begin{array}{c} F_3 \\ F_4 \end{array}\right\}$$

$$\{\delta\}_2 = \left\{\begin{array}{c} \delta_3 \\ \delta_4 \end{array}\right\}$$

$$\{F\}_3 = \left\{\begin{array}{c} F_5 \\ F_6 \end{array}\right\}$$

$$\{\delta\}_3 = \left\{\begin{array}{c} \delta_5 \\ \delta_6 \end{array}\right\}$$

Member Forces:

$$\{F\}_i = \left(\frac{EA}{l}\right)_i \begin{bmatrix} 1 & 1 \\ 1 & 1 \end{bmatrix} \{\delta\}_i \qquad (12.1)$$

Structure Connectivity:

$$\{\delta\}_i = [\beta]_i \{\Delta\} \qquad (12.14)$$

Member 1 ($i = 1$):   see Example 12.6.2

$$\left\{\begin{array}{c} \delta_1 \\ \delta_2 \end{array}\right\} = \begin{bmatrix} 0.707 & 0.707 & 0 \\ 0 & 0 & 0 \end{bmatrix} \left\{\begin{array}{c} \Delta_1 \\ \Delta_2 \\ \hline \Delta_3 \end{array}\right\}$$

Member 2 ($i = 2$)

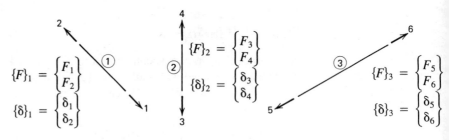

$$\delta_3 = \Delta_2, \quad \delta_4 = -\Delta_3$$

$$\left\{\begin{array}{c} \delta_3 \\ \delta_4 \end{array}\right\} = \begin{bmatrix} 0 & 0 & 0 \\ 0 & 0 & -1 \end{bmatrix} \left\{\begin{array}{c} \Delta_1 \\ \Delta_2 \\ \Delta_3 \end{array}\right\}$$

Member 3 ($i = 3$):   see Example 12.6.2

$$\left\{\begin{array}{c} \delta_5 \\ \delta_6 \end{array}\right\} = \begin{bmatrix} -0.894 & 0.447 & 0 \\ 0 & 0 & 0 \end{bmatrix} \left\{\begin{array}{c} \Delta_1 \\ \Delta_2 \\ \Delta_3 \end{array}\right\}$$

Structure Stiffness:

$$[K] = \sum_{i=1}^{m} [\beta]_i^T [k]_{ai} [\beta]_i \qquad (12.21)$$

$$[K] = EA \begin{bmatrix} \left\{ \dfrac{(0.707)^2}{l_1} + \dfrac{(0.894)^2}{l_3} \right\} & \left\{ \dfrac{(0.707)^2}{l_1} + \dfrac{(-0.894)(0.447)}{l_3} \right\} & 0 \\[2ex] \left\{ \dfrac{(0.707)^2}{l_1} + \dfrac{(0.447)(0.894)}{l_3} \right\} & \left\{ \dfrac{(0.707)^2}{l_1} + \dfrac{1}{l_2} + \dfrac{(0.447)^2}{l_3} \right\} & \left\{ -\dfrac{1}{l_2} \right\} \\[2ex] 0 & \left\{ -\dfrac{1}{l_2} \right\} & \left\{ \dfrac{1}{l_2} \right\} \end{bmatrix}$$

$$[K] = EA \left[\begin{array}{cc|c} 0.142\ 2 & 0.035\ 0 & 0 \\ 0.035\ 0 & 0.288\ 6 & -0.200\ 0 \\ \hline 0 & -0.200\ 0 & 0.200\ 0 \end{array} \right] = \left[\begin{array}{c|c} [K]_{11} & [K]_{12} \\ \hline [K]_{21} & [K]_{22} \end{array} \right]$$

Note:

$[K]_{11}$ is identical to $[K]$ of Examples 12.4.1 and 12.6.2

Structure Displacements:

$$\{\Delta\}_1 = [K]_{11}^{-1} (\{P\}_1 - [K]_{12} \{\Delta\}_2) \qquad (12.27)$$

Taking $[K]_{11}^{-1}$ from Example 12.4.1, we have

$$\begin{Bmatrix} \Delta_1 \\ \Delta_2 \end{Bmatrix} = \{\Delta\}_1 = \frac{1}{0.039\ 85\ EA} \begin{bmatrix} 0.288\ 6 & -0.034\ 9 \\ -0.034\ 9 & 0.142\ 3 \end{bmatrix}$$

$$\times \left(\begin{Bmatrix} -180 \\ 240 \end{Bmatrix} - EA \begin{Bmatrix} 0 \\ -0.200\ 0 \end{Bmatrix} (0.010) \right)$$

$$\{\Delta\}_1 = \left(\begin{Bmatrix} -0.075\ 7 \\ 0.050\ 7 \end{Bmatrix} + \begin{Bmatrix} -0.001\ 8 \\ 0.007\ 1 \end{Bmatrix} \right) = \begin{Bmatrix} -0.077\ 5 \\ 0.057\ 8 \end{Bmatrix}$$

From $\{P\}_1$—same
as in Example 12.6.2

From $\{\Delta\}_2$—induced
through support settlement

Thus,

$$\{\Delta\} = \begin{Bmatrix} \Delta_1 \\ \Delta_2 \\ \Delta_3 \end{Bmatrix} = \begin{Bmatrix} -0.077\ 5 \\ 0.057\ 8 \\ 0.010\ 0 \end{Bmatrix} \text{ m}$$

Member Forces:

$$\{F\}_1 = [k]_{ai} [\beta]_i \{\Delta\} \qquad (12.16)$$

Member 1 ($i = 1$)

$$\begin{Bmatrix} F_1 \\ F_2 \end{Bmatrix} = \frac{20 \times 10^3}{7.07} \begin{bmatrix} 1 & 1 \\ 1 & 1 \end{bmatrix} \begin{bmatrix} 0.707 & 0.707 & 0 \\ 0 & 0 & 0 \end{bmatrix} \begin{Bmatrix} -0.077\ 5 \\ 0.057\ 8 \\ 0.010\ 0 \end{Bmatrix} = \begin{Bmatrix} -39.4 \\ -39.4 \end{Bmatrix} \text{ kN}$$

Member 2 (i = 2)

$$\begin{Bmatrix} F_3 \\ F_4 \end{Bmatrix} = \frac{20 \times 10^3}{5.00} \begin{bmatrix} 1 & 1 \\ 1 & 1 \end{bmatrix} \begin{bmatrix} 0 & 1 & 0 \\ 0 & 0 & -1 \end{bmatrix} \begin{Bmatrix} -0.077\ 5 \\ 0.057\ 8 \\ 0.010\ 0 \end{Bmatrix} = \begin{Bmatrix} 191.2 \\ 191.2 \end{Bmatrix} \text{kN}$$

Member 3 (i = 3)

$$\begin{Bmatrix} F_5 \\ F_6 \end{Bmatrix} = \frac{20 \times 10^3}{11.18} \begin{bmatrix} 1 & 1 \\ 1 & 1 \end{bmatrix} \begin{bmatrix} -0.894 & 0.447 & 0 \\ 0 & 0 & 0 \end{bmatrix} \begin{Bmatrix} -0.077\ 5 \\ 0.057\ 8 \\ 0.010\ 0 \end{Bmatrix} = \begin{Bmatrix} 170.2 \\ 170.2 \end{Bmatrix} \text{kN}$$

12.8

Elastic Supports

Structures that have elastic supports are easily handled by the equilibrium method. For instance, consider the structure shown in Fig. 12.7. The vectors at the joints indicate the admissible structure displacements, and it is noted that two of these displacements are at point *a*. This point is elastically restrained by a support spring, which has a spring constant of k_s.

As far as joint *a* is concerned, the spring simply acts as a member, which can carry only axial load and is thus similar to any other pin-ended truss member that frames into joint *a*. Whereas the axial stiffness of a typical truss member is (EA/l), the axial stiffness of the spring is k_s; therefore, the member force–displacement relations for the spring are given by modifying Eq. 12.1 to the form

$$\begin{Bmatrix} F_1 \\ F_2 \end{Bmatrix}_i = \begin{bmatrix} k_s & k_s \\ k_s & k_s \end{bmatrix}_i \begin{Bmatrix} \delta_1 \\ \delta_2 \end{Bmatrix}_i \tag{12.29}$$

Of course, the member-end spring displacements must be related to the structure displacements by a compatibility relationship in the form of Eq. 12.14.

The analysis proceeds in the normal fashion. The structure stiffness matrix is determined from Eq. 12.21, in which the spring is included as one of the *m* members. The structure displacements are obtained from Eq. 12.22, and the member-end displacements are given by Eq. 12.14. The member forces are determined from Eq. 12.1 for regular truss members and from Eq. 12.29 for the support springs.

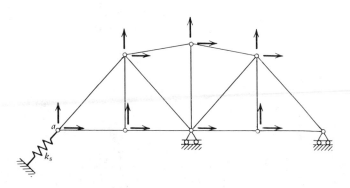

Fig. 12.7 *Structure with elastic support.*

Consider again the structure of Fig. 12.4a. The kinematically determinate primary **Temperature** structure would have zero displacements at all joints. Now, assume that if member **Effects and** i were free to change length, it would tend to shorten by an amount $\bar{\delta}_i$. This could **Fabrication** result from a fabrication error in which member i was too short by an amount $\bar{\delta}_i$ **Errors** or from a decrease in temperature that would cause a free shortening of $\bar{\delta}_i$. However, if the member is to fit compatibly within the kinematically determinate primary structure, then it must be elongated so as to negate the $\bar{\delta}_i$ shortening. Under this condition, member i must be subjected to the member forces F_g and F_h, as shown in Fig. 12.8a. Figure 12.8b shows the forces acting at joints g and h, and equilibrium considerations lead to

$$P_r = -F_g \cos \theta$$
$$P_s = -F_g \cos \phi$$
$$P_p = -F_h \cos \eta \qquad (12.30)$$
$$P_q = -F_h \cos \xi$$

where the cosine terms are the direction cosines of the member. The joint forces given by Eq. 12.30 represent the *restraining forces* that are needed to sustain the structure in its kinematically determinate position. It is further recognized that the member forces associated with this kinematically determinate state are given by

$$F_g = F_h = -\left(\frac{EA}{l}\right)_i \bar{\delta}_i \qquad (12.31)$$

This expression follows from Eq. 5.8, in which $\bar{\delta}_i$ is taken as positive for member elongations—that is, for members that are fabricated too long or for elongations that evolve from temperature increases. The negative sign in Eq. 12.31 recognizes that a positive $\bar{\delta}_i$ induces a compressive force, and vice versa.

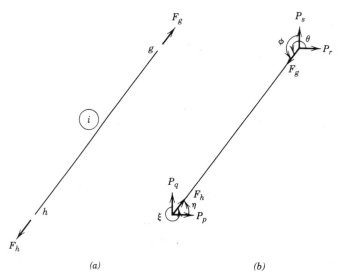

(a)   (b)

Fig. 12.8 *Temperature effects.* (a) *Positive member forces on restrained element.* (b) *Joint forces for kinematically determinate structure.*

Substitution of Eq. 12.31 into Eq. 12.30 yields

$$P_r = \left(\frac{EA}{l}\right)_i \bar{\delta}_i \cos\theta$$

$$P_s = \left(\frac{EA}{l}\right)_i \bar{\delta}_i \cos\phi$$

$$P_p = \left(\frac{EA}{l}\right)_i \bar{\delta}_i \cos\eta \qquad (12.32)$$

$$P_q = \left(\frac{EA}{l}\right)_i \bar{\delta}_i \cos\xi$$

These restraining forces can be accommodated only at support points. At joints that are kinematically free, these forces must be systematically removed through the application of equal and opposite *applied forces* or *equivalent temperature loads*. These applied force components become part of the force vector of Eq. 12.22, and these must, of course, be in addition to any regularly applied forces.

The final reactions and member forces are determined by superimposing the forces associated with the kinematically determinate system and the forces that result from the application of the equivalent temperature loads corresponding to the kinematic degrees of freedom at all displacement joints.

12.9.1 Example problem   Consider again the structure of Example 12.6.3. In this instance, determine the displacements at points *b* and *c* and the forces in members *ab* and *bd* that occur as a result of a temperature decrease of 25°C in member *bd*.

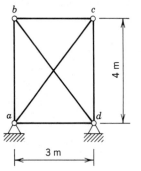

$$\Delta T_{bd} = -25°C$$
$$EA = 200 \times 10^3 \text{ kN} \quad \text{(all members)}$$
$$\alpha_t = 0.000\ 012/°C$$

Restraining Forces: Consider member *bd*. In applying Eqs. 12.32 and 12.31, we have

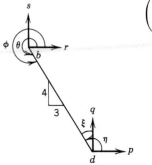

$$\left(\frac{EA}{l}\right)_{bd} = \frac{200 \times 10^3}{5\ 000} = 40 \text{ kN/m}$$

$$\bar{\delta}_{bd} = -\alpha_t \cdot \Delta T \cdot l$$
$$= -(0.000\ 012)(25)(5\ 000) = -1.5 \text{ mm}$$

$$\cos\theta = 0.6 \qquad \cos\phi = -0.8$$
$$\cos\eta = -0.6 \qquad \cos\xi = 0.8$$

from which,

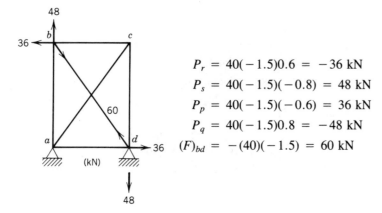

$$P_r = 40(-1.5)0.6 = -36 \text{ kN}$$
$$P_s = 40(-1.5)(-0.8) = 48 \text{ kN}$$
$$P_p = 40(-1.5)(-0.6) = 36 \text{ kN}$$
$$P_q = 40(-1.5)0.8 = -48 \text{ kN}$$
$$(F)_{bd} = -(40)(-1.5) = 60 \text{ kN}$$

Equivalent Temperature Loads: Since joint b is free to displace, it cannot sustain the restraining forces; therefore, applied forces that are equal and opposite to P_r and P_s must be applied.

For the structure coordinate system of Example 12.6.3, we thus have $P_1 = -P_r = +36$ kN and $P_2 = -P_s = -48$ kN.

$$\{P\} = \begin{Bmatrix} P_1 \\ P_2 \\ P_3 \\ P_4 \end{Bmatrix} = \begin{Bmatrix} 36 \\ -48 \\ 0 \\ 0 \end{Bmatrix} \qquad \{\Delta\} = \begin{Bmatrix} \Delta_1 \\ \Delta_2 \\ \Delta_3 \\ \Delta_4 \end{Bmatrix}$$

Equations at Equilibrium and Solution for Displacement:

$$\{P\} = [K]\{\Delta\} \tag{12.12}$$
$$\{\Delta\} = [K]^{-1}\{P\} \tag{12.22}$$

In the above, $[K]$ is the same as for Example 12.6.3, and the load vector is based on the equivalent temperature loads.

$$\begin{Bmatrix} \Delta_1 \\ \Delta_2 \\ \Delta_3 \\ \Delta_4 \end{Bmatrix} = [K]^{-1} \begin{Bmatrix} 36 \\ -48 \\ 0 \\ 0 \end{Bmatrix} = \frac{1}{EA} \begin{Bmatrix} 266.63 \\ -59.26 \\ 233.33 \\ -59.26 \end{Bmatrix} \left(\frac{\text{kN} \cdot \text{m}}{EA} \right)$$

For $EA = 200 \times 10^3$ kN, these displacements agree with those of Example 10.9.2.

Member Forces:

$$\{F\}_i = [k]_{ai}[\beta]_i\{\Delta\} + \{F\}_{ir}$$

where

$\{F\}_{ir}$ = the member forces associated with the restrained case.
$[k]_{ai}$ and $[\beta]_i$ are taken from Example 12.6.3.
$\{\Delta\}$ = displacements caused by equivalent temperature forces

Member ab (i = *1*)

$$\begin{Bmatrix} F_1 \\ F_2 \end{Bmatrix} = \frac{EA}{4}\begin{bmatrix} 1 & 1 \\ 1 & 1 \end{bmatrix}\begin{bmatrix} 0 & 0 & 0 & 0 \\ 0 & 1 & 0 & 0 \end{bmatrix}\frac{1}{EA}\begin{Bmatrix} 266.63 \\ -59.26 \\ 233.33 \\ -59.26 \end{Bmatrix} + \begin{Bmatrix} 0 \\ 0 \end{Bmatrix} = \begin{Bmatrix} -14.8 \\ -14.8 \end{Bmatrix} \text{ kN}$$

Member bd (i = *6*)

$$\begin{Bmatrix} F_{11} \\ F_{12} \end{Bmatrix} = \frac{EA}{5}\begin{bmatrix} 1 & 1 \\ 1 & 1 \end{bmatrix}\begin{bmatrix} -0.6 & 0.8 & 0 & 0 \\ 0 & 0 & 0 & 0 \end{bmatrix}\frac{1}{EA}\begin{Bmatrix} 266.63 \\ -59.26 \\ 233.33 \\ -59.26 \end{Bmatrix}$$

$$+ \begin{Bmatrix} 60.0 \\ 60.0 \end{Bmatrix} = \begin{Bmatrix} 18.5 \\ 18.5 \end{Bmatrix} \text{ kN}$$

These forces agree with those obtained in Example 10.9.2.

12.10

Equilibrium Equations by Energy Methods

Castigliano's first theorem provides an alternative procedure for generating the equations of equilibrium. This theorem was given in mathematical form as Eq. 5.37, which has the form

$$\frac{\partial U}{\partial \Delta_i} = P_i \tag{12.33}$$

where U is the total strain energy stored in the system, P_i is the ith externally applied load, and Δ_i is the corresponding displacement.

In the present context, since we are dealing with pin-ended members, the strain energy expression developed in Section 5.10 is valid. Thus, for a truss of m members

$$U = \sum_{j=1}^{m} \frac{F_j^2 l_j}{2E_j A_j} \tag{12.34}$$

where F is the axial force on the member, l is the member length, A is the cross-sectional area, and E is the modulus of elasticity. The j subscript is merely a dummy index, which indicates that summation must take place over all m members. Each member force can be expressed in terms of the member elongation in accordance with the axial force–displacement relation given by Eq. 5.5. In this case, representing the member elongation as e_j, we have

$$e_j = \frac{F_j l_j}{E_j A_j} \tag{12.35}$$

Solving this expression for F_j, and substituting it into Eq. 12.34, we obtain

$$U = \sum_{j=1}^{m} \frac{E_j A_j e_j^2}{2l_j} \tag{12.36}$$

The differentiation indicated in Eq. 12.33 can now be performed if it is understood that the member elongation e_j can be expressed in terms of the Δ_i displacements for the structure. Thus, application of Eq. 12.33 leads to

$$\sum_{j=1}^{m} \frac{E_j A_j e_j}{l_j} \left(\frac{\partial e_j}{\partial \Delta_i} \right) = P_i \tag{12.37}$$

For the jth member, the member elongation can be expressed in the form

$$e_j = (\delta_1)_j + (\delta_2)_j \tag{12.38}$$

where $(\delta_1)_j$ and $(\delta_2)_j$ are the member-end axial displacements, as are shown in Fig. 12.2. Compatibility equations, which are similar to those given as Eq. 12.4 for the three-bar truss example, are required in order to transform Eq. 12.38 into the differentiable form required in Eq. 12.37.

Equation 12.37 must be applied for each kinematic degree of freedom Δ_i, and the resulting expression is an equation of equilibrium corresponding to the force P_i. This leads to a set of force–displacement relations that are equivalent to Eq. 12.12, and the solution for the structure displacements proceeds according to Eq. 12.22.

Use Castigliano's first theorem to generate the equations of equilibrium for the structure shown below. This is the same problem that was considered in Example 12.6.3, and, as noted there, EA is the same for each member. **12.10.1 Example problem**

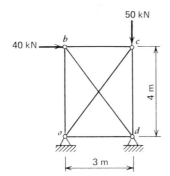

Coordinate Systems: See Example 12.6.3

Structure Connectivity: From Example 12.6.3, we have

Member 1:	$\delta_1 = 0;\quad \delta_2 = \Delta_2$
Member 2:	$\delta_3 = -\Delta_1;\quad \delta_4 = \Delta_3$
Member 3:	$\delta_5 = \Delta_4;\quad \delta_6 = 0$
Member 4:	$\delta_7 = 0;\quad \delta_8 = 0$
Member 5:	$\delta_9 = 0;\quad \delta_{10} = 0.6\Delta_3 + 0.8\Delta_4$
Member 6:	$\delta_{11} = -0.6\Delta_1 + 0.8\Delta_2;\quad \delta_{12} = 0$

Member Elongations:

$$e_i = (\delta_1)_j + (\delta_2)_i \tag{12.38}$$

Member 1: $e_1 = \delta_1 + \delta_2 = \Delta_2$

Member 2: $e_2 = \delta_3 + \delta_4 = -\Delta_1 + \Delta_3$

Member 3: $e_3 = \delta_5 + \delta_6 = \Delta_4$

Member 4: $e_4 = \delta_7 + \delta_8 = 0$

Member 5: $e_5 = \delta_9 + \delta_{10} = 0.6\Delta_3 + 0.8\Delta_4$

Member 6: $e_6 = \delta_{11} + \delta_{12} = -0.6\Delta_1 + 0.8\Delta_2$

Equations of Equilibrium:

$$\sum_{j=1}^{6} \frac{E_j A_j e_j}{l_j}\left(\frac{\partial e_j}{\partial \Delta_i}\right) = P_i \qquad (i = 1, 2, 3, 4) \tag{12.37}$$

$(i = 1)$:

$$\left(\frac{EA}{l}\right)_2 e_2 \left(\frac{\partial e_2}{\partial \Delta_1}\right) + \left(\frac{EA}{l}\right)_6 e_6 \left(\frac{\partial e_6}{\partial \Delta_1}\right) = P_1$$

$$EA\left[\frac{1}{l_2}(-\Delta_1 + \Delta_3)(-1) + \frac{1}{l_6}(-0.6\Delta_1 + 0.8\Delta_2)(-0.6)\right] = P_1$$

$$EA[0.405\ 3\Delta_1 - 0.096\ 0\Delta_2 - 0.333\ 3\Delta_3] = 40$$

$(i = 2)$:

$$\left(\frac{EA}{l}\right)_1 e_1 \left(\frac{\partial e_1}{\partial \Delta_2}\right) + \left(\frac{EA}{l}\right)_6 e_6 \left(\frac{\partial e_6}{\partial \Delta_2}\right) = P_2$$

$$EA\left[\frac{1}{l_1}(\Delta_2)(1) + \frac{1}{l_6}(-0.6\Delta_1 + 0.8\Delta_2)(0.8)\right] = P_2$$

$$EA[-0.096\ 0\Delta_1 + 0.378\ 0\Delta_2] = 0$$

Equation 12.37 is similarly applied for $i = 3, 4$. The final equilibrium equations are

$$EA\begin{bmatrix} 0.405\ 3 & -0.096\ 0 & -0.333\ 3 & 0 \\ -0.096\ 0 & 0.378\ 0 & 0 & 0 \\ -0.333\ 3 & 0 & 0.405\ 3 & 0.096\ 0 \\ 0 & 0 & 0.096\ 0 & 0.378\ 0 \end{bmatrix}\begin{Bmatrix} \Delta_1 \\ \Delta_2 \\ \Delta_3 \\ \Delta_4 \end{Bmatrix} = \begin{Bmatrix} 40 \\ 0 \\ 0 \\ -50 \end{Bmatrix}$$

This agrees with the equation generated in Example 12.6.3, and the solution for the structure displacements and member forces would proceed as given earlier.

12.11

Multiple Loading Cases

As has been noted previously, most design situations require that the structure be analyzed for a number of alternative loading cases. Examination of Eq. 12.22 shows that each load vector $\{P\}$ causes a response that is registered through a unique set of displacements $\{\Delta\}$. The stiffness matrix $[K]$, however, and thus its inverse $[K]^{-1}$, is entirely a function of the properties of the structure. This is clearly indicated by Eq. 12.21.

Thus, it is possible to represent the analysis of a structure for r separate loading conditions by extending Eq. 12.22 into the single relationship

$$[\{\Delta\}_1\{\Delta\}_2 \cdots \{\Delta\}_i \cdots \{\Delta\}_r] = [K]^{-1}[\{P\}_1\{P\}_2 \cdots \{P\}_i \cdots \{P\}_r] \quad (12.39)$$

Here, each column of the rectangular matrix of displacements on the left-hand side of the equation gives the response to the loads that are given in the corresponding column of the rectangular array of loads on the right-hand side of the equation.

12.12

Additional Reading

Gere, J. M., and Weaver, W., Jr., *Analysis of Framed Structures,* Sections 2.8 through 2.12, Van Nostrand, Princeton, N.J., 1965.

Wang, Chu-Kia, *Matrix Methods of Structural Analysis,* 2nd Ed., Chapter 3, International Textbook, Scranton, Pa., 1970.

12.13

Suggested Problems

1 through 4. Use the conventional equilibrium method of Section 12.4 to determine the joint displacements and member forces for the truss structures and loading conditions specified.

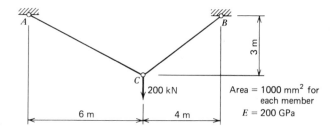

1.

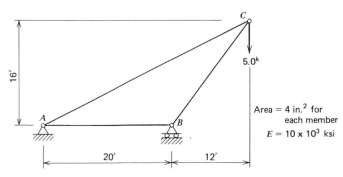

2.

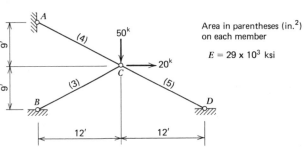

3.

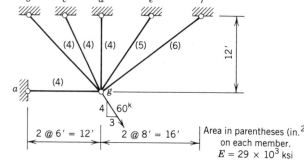

4.

5 through 8. Use the conventional equilibrium method of Section 12.4 to formulate the equations of joint equilibrium for the truss structures and loading conditions given.

5. The structure and loading of Problem 1 in Section 10.14.

6. The structure and loading of Problem 2 in Section 10.14.

7. The structure and loading of Problem 20 in Section 10.14.

8. The structure and loading of Problem 21 in Section 10.14.

9 through 12. Use the matrix formulation of Section 12.5 to determine the joint displacements and member forces for the truss structures and loading conditions specified.

9. The structure and loading of Problem 1.

10. The structure and loading of Problem 2.

11. The structure and loading of Problem 3.

12. The structure and loading of Problem 4.

13 through 16. Use the matrix formulation of Section 12.5 to generate the equations of joint equilibrium for the truss structures and loading conditions specified.

13. The structure and loading of Problem 1 in Section 10.14.

14. The structure and loading of Problem 2 in Section 10.14.

15. The structure and loading of Problem 20 in Section 10.14.

16. The structure and loading of Problem 21 in Section 10.14.

17 and 18. Use the matrix formulation to determine the joint displacements, the member forces, and the reactions for the structures and settlement patterns indicated:

17. Structure of Problem 3 with the following support displacements:
    Joint B: 0.5 in. downward; 0.2 in. right
    Joint D: 0.3 in. downward
    Disregard the loading given in Problem 3.

18. Structure of Problem 4 if, in addition to the given load, there are the following support displacements:
    Joint b: 0.4 in. downward; 0.3 in right
    Joint e: 0.5 in. downward; 0.2 in. left

19. Use the matrix formulation to establish the equations of joint equilibrium for the structure and loading of Problem 1 in Section 10.14 if the rigid support at point D is replaced by the elastic support shown below.

$k_s = 2{,}000$ k/in.

20. Use the matrix formulation to determine the joint displacements and member forces for the structure and loading of Problem 4 if the rigid support and point d is replaced by the elastic support shown below.

$k_s = 1{,}500$ k/in.

21. For the structure of Problem 3, use the matrix formulation to determine the joint displacements and member forces if members AC and CD experience a 40°F increase in temperature. Disregard the loads that are specified in Problem 3, and take $\alpha_t = 0.000{,}006{,}5/°F$.

22. Use the matrix approach to determine the joint displacements and member forces for the structure of Problem 21 in Section 10.14 if, in addition to the given loads, members AB and BD undergo a 50°F decrease in temperature. Take $\alpha_t = 0.000{,}006{,}5/°F$.

23. Use the matrix approach to determine the joint displacements and member forces for the structure of Problem 4 if, in addition to the applied load, members bg and eg are each 0.5 in. too short but are forced to fit.

24. For the structure of Problem 21 in Section 10.14, use the matrix procedure to determine the joint displacements and member forces if member AD is 0.5 in. too long and member BC is 0.35 in. too short but both are forced into place.

25. Develop a computer program for the solution of statically indeterminate trusses by the method of Section 12.5. Use the program to solve Problem 2, 3, and 4, and Problems 5 and 20 of Section 10.14.

26 through 29. Use Castigliano's first theorem to establish the equations of joint equilibrium in the form of Eq. 12.23 for the structures and loadings specified. Partition these equations in the form of Eq. 12.24 and extract $[K]_{11}$, the stiffness matrix that relates the forces to the displacements at the displaceable joints. A coordinate system is specified in each case.

26. The structure and loading of Problem 1.

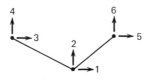

27. The structure and loading of Problem 3.

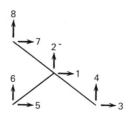

28. The structure and loading of Problem 4.

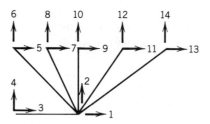

29. The structure and loading of Problem 1 in Section 10.14.

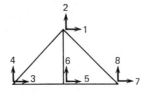

CHAPTER 13

Lake Point Tower, Chicago, Ill. (courtesy Portland Cement Association).

ANALYSIS OF BEAM AND FRAME STRUCTURES BY EQUILIBRIUM METHODS— DIRECT SOLUTIONS

13.1

As was stated at the beginning of Chapter 12, equilibrium methods are based on the solution of the equilibrium equations for the entire structural system. One equilibrium equation is written for each kinematic degree of freedom, while the conditions of compatibility and the boundary conditions are satisfied as the equations are formulated. The solution of these equations leads to a set of displacements, from which the forces throughout the structure can be determined.

Equilibrium Methods for Beam- and Frame-Type Structures

Many methods fall into the general category of equilibrium methods. In this chapter, the emphasis is on the *slope deflection method*. This is one of the classical methods of analysis for statically indeterminate beam- and frame-type structures. Although the conventional form of the slope deflection method is considered to be obsolete by today's standards, it is still useful for hand solutions to small problems. It also serves as an effective introduction to a matrix formulation of the equilibrium method for beam and frame problems. Later, in Chapter 17, a completely general matrix formulation of the equilibrium method, applicable to all classes of structures, is presented.

The concept of kinematic indeterminacy, which was introduced in Section 12.2, is of particular importance for beam and frame structures. As was explained earlier, a kinematically determinate primary structure is formed by restraining each structure displacement component, and the number of restraints gives the degree of kinematic indeterminacy. In this case, however, since forces are applied between the member ends, the individual members are under load in the restrained position. Since the primary structure does not conform to the structure originally given, the artificially imposed restraints must be relaxed. As this relaxation takes place, equilibrium must be satisfied for each displacement degree of freedom. This requires the inclusion of the loads applied directly at the joints as well as the effects of the loads applied along the member lengths.

The methods of this chapter require direct solutions to the equilibrium equations. This requirement makes it difficult to use these methods for hand solutions, except for small problems. Hand calculations are more easily handled by using iterative solutions for the equilibrium equations. One such iterative approach is considered in Chapter 14.

13.2

The Slope Deflection Equation

The slope deflection method was presented by G. A. Maney in 1915 as a method of analysis for rigid-jointed beam and frame structures. The method is an equilibrium method that accounts for flexural deformation but ignores axial and shear deformations.

The member force–displacement equations that are needed for the slope deflection method will be developed by considering member AB of Fig. 13.1. This member, which has its undeformed position along the x axis, is deformed into the configuration shown. The positive axes, along with the positive member-end force

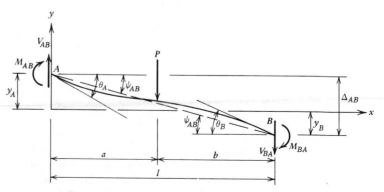

Fig. 13.1 *Deviation of slope deflection equation.*

components and displacement components, are shown in the figure. The axis convention corresponds to that which was employed in Chapter 6; however, the focus on member-end quantities forces a departure from the customary convention used in plotting shear and moment diagrams. These differences will be reconciled later in the example problems. In the deformed position, and for the adopted sign conventions, the boundary conditions require that at $x = 0$

$$y(x = 0) = y_A; \qquad \frac{dy}{dx}(x = 0) = -\theta_A \qquad (13.1)$$

and at $x = l$

$$y(x = l) = -y_B; \qquad \frac{dy}{dx}(x = l) = -\theta_B \qquad (13.2)$$

As the beam deforms, the moment–curvature relationships developed in Chapter 6 require that

$$M = EI\frac{d^2y}{dx^2} \qquad (13.3)$$

For the member shown in Fig. 13.1, the moment as a function of x is given by

$$M = M_{AB} + V_{AB}x - P\{x - a\} \qquad (13.4)$$

where the quantity $\{x - a\}$ is taken as zero if negative. Substitution of Eq. 13.4 into Eq. 13.3 and subsequent realignment of terms yields

$$\frac{d^2y}{dx^2} = \frac{M_{AB}}{EI} + \frac{V_{AB}x}{EI} - \frac{P}{EI}\{x - a\} \qquad (13.5)$$

Integration of Eq. 13.5 and evaluation of the constants of integration by use of Eq. 13.1 leads to

$$\frac{dy}{dx} = \frac{M_{AB}x}{EI} + \frac{V_{AB}x^2}{2EI} - \frac{P}{2EI}\{x - a\}^2 - \theta_A \qquad (13.6)$$

$$y = \frac{M_{AB}x^2}{2EI} + \frac{V_{AB}x^3}{6EI} - \frac{P}{6EI}\{x - a\}^3 - \theta_A x + y_A \qquad (13.7)$$

Application of Eqs. 13.6 and 13.7 at $x = l$ along with the boundary conditions specified by Eq. 13.2 gives

$$-\theta_B = \frac{M_{AB}l}{EI} + \frac{V_{AB}l^2}{2EI} - \frac{P}{2EI}\{l - a\}^2 - \theta_A. \qquad (13.8)$$

$$-y_B = \frac{M_{AB}l^2}{2EI} + \frac{V_{AB}l^3}{6EI} - \frac{P}{6EI}\{l - a\}^3 - \theta_A l + y_A \qquad (13.9)$$

Simultaneously solving these two equations for M_{AB} and V_{AB}, we obtain

$$M_{AB} = \frac{2EI}{l}\left(2\theta_A + \theta_B - \frac{3y_A}{l} - \frac{3y_B}{l}\right) - \frac{Pab^2}{l^2} \qquad (13.10)$$

$$V_{AB} = \frac{6EI}{l^2}\left(-\theta_A - \theta_B + \frac{2y_A}{l} + \frac{2y_B}{l}\right) + \frac{Pb^2(l + 2a)}{l^3} \qquad (13.11)$$

Applying statics to member AB, we then find

$$M_{BA} = \frac{2EI}{l}\left(2\theta_B + \theta_A - \frac{3y_A}{l} - \frac{3y_B}{l}\right) + \frac{Pa^2b}{l^2} \tag{13.12}$$

$$V_{BA} = \frac{-6EI}{l^2}\left(\theta_A + \theta_B - \frac{2y_A}{l} - \frac{2y_B}{l}\right) - \frac{Pa^2(l + 2b)}{l_3} \tag{13.13}$$

In matrix form, Eqs. 13.10, 13.11, 13.12, and 13.13 become

$$\begin{Bmatrix} M_{AB} \\ V_{AB} \\ M_{BA} \\ V_{BA} \end{Bmatrix} = \frac{2EI}{l} \begin{bmatrix} 2 & -3/l & 1 & -3/l \\ -3/l & 6/l^2 & -3/l & 6/l^2 \\ 1 & -3/l & 2 & -3/l \\ -3/l & 6/l^2 & -3/l & 6/l^2 \end{bmatrix} \begin{Bmatrix} \theta_A \\ y_A \\ \theta_B \\ y_B \end{Bmatrix}$$

$$+ \begin{Bmatrix} -Pab^2/l^2 \\ +Pb^2(l + 2a)/l^3 \\ +Pa^2b/l^2 \\ -Pa^2(l + 2b)/l^3 \end{Bmatrix} \tag{13.14}$$

Equation 13.14 can be written as

$$\{F\} = [k]\{\delta\} + \{F\}^f \tag{13.15}$$

where $\{F\}$ is the member-end force vector, $[k]$ is the member stiffness matrix, and $\{\delta\}$ is the member-end displacement vector. In this equation, the reader is reminded that the terms "end forces" and "end displacements" are used in the general sense. That is, an end force may be end shear or end moment and an end displacement may be end translation or end rotation. If the end displacements are all zero (i.e., member ends are fixed), then $\{F\} = \{F\}^f$ and thus $\{F\}^f$ is referred to as the vector of *fixed-end forces*. These forces are the member-end forces for the kinematically determinate primary structure. In Eq. 13.14, the fixed-end forces correspond to the case of the concentrated load P as shown in Fig. 13.1. For other loading conditions, the fixed-end forces would have different forms as shown in Table 13.1. For general loading, Eqs. 13.10 and 13.12 may be written as

$$M_{AB} = \frac{2EI}{l}(2\theta_A + \theta_B - 3\psi_{AB}) + FEM_{AB} \tag{13.16}$$

and

$$M_{BA} = \frac{2EI}{l}(2\theta_B + \theta_A - 3\psi_{AB}) + FEM_{BA} \tag{13.17}$$

where FEM_{AB} and FEM_{BA} are the fixed-end moments at ends A and B, respectively, and ψ_{AB} is the rotation of the chord of member AB as shown in Fig. 13.1. Taking Δ_{AB} as the relative displacement between the member ends, we have

$$\psi_{AB} = \frac{\Delta_{AB}}{l} = \frac{y_B + y_A}{l} \tag{13.18}$$

TABLE 13.1 Fixed-End Forces

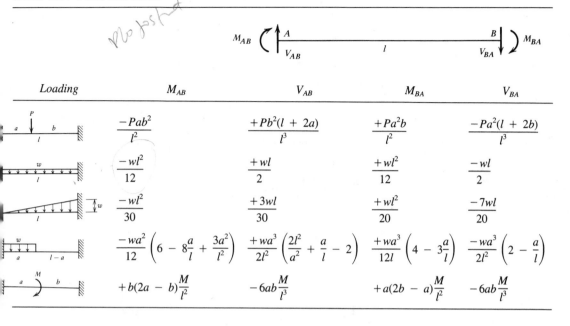

Loading	M_{AB}	V_{AB}	M_{BA}	V_{BA}
P, a, b, l	$\dfrac{-Pab^2}{l^2}$	$\dfrac{+Pb^2(l + 2a)}{l^3}$	$\dfrac{+Pa^2b}{l^2}$	$\dfrac{-Pa^2(l + 2b)}{l^3}$
w, l	$\dfrac{-wl^2}{12}$	$\dfrac{+wl}{2}$	$\dfrac{+wl^2}{12}$	$\dfrac{-wl}{2}$
w, l	$\dfrac{-wl^2}{30}$	$\dfrac{+3wl}{30}$	$\dfrac{+wl^2}{20}$	$\dfrac{-7wl}{20}$
w, a, $l-a$	$\dfrac{-wa^2}{12}\left(6 - 8\dfrac{a}{l} + \dfrac{3a^2}{l^2}\right)$	$\dfrac{+wa^3}{2l^2}\left(\dfrac{2l^2}{a^2} + \dfrac{a}{l} - 2\right)$	$\dfrac{+wa^3}{12l}\left(4 - 3\dfrac{a}{l}\right)$	$\dfrac{-wa^3}{2l^2}\left(2 - \dfrac{a}{l}\right)$
M, a, b	$+b(2a - b)\dfrac{M}{l^2}$	$-6ab\dfrac{M}{l^3}$	$+a(2b - a)\dfrac{M}{l^2}$	$-6ab\dfrac{M}{l^3}$

Equations 13.16 and 13.17 are called the slope deflection equations, and they can be further generalized as a single equation in the form

$$M_{nf} = 2EK_{nf}(2\theta_n + \theta_f - 3\psi_{nf}) + FEM_{nf} \qquad (13.19)$$

where the subscripts n and f refer to the near end and far end, respectively, of member nf. The quantity K_{nf} is referred to as the stiffness factor for member nf and is equal to I/l for the member.

13.3 Interpretation of the Slope Deflection Equation

The slope deflection equation shows that the moment at the end of a member is dependent on a number of quantities. It is easier to see these dependencies if Eq. 13.19 is written in the form

$$M_{nf} = \frac{4EI}{l}\theta_n + \frac{2EI}{l}\theta_f - \frac{6EI}{l^2}\Delta_{nf} + FEM_{nf} \qquad (13.20)$$

In this form, it is clear that each of the first three terms involves a coupling of a stiffness term, which reflects the member characteristics E, I, and l, and a displacement equantity. The fourth term reflects the transverse loading on the member.

Figure 13.2 illustrates the physical meaning of each of the four terms in the expanded slope deflection equation. For example, Fig. 13.2b shows the moment at the near end associated with the rotation θ_n, while $\theta_f = \Delta_{nf} = FEM_{nf} = 0$. Similarly, Figs. 13.2c, d, and e show the near-end moment associated with separate inducements of θ_f, Δ_{nf}, and FEM_{nf}, respectively. The total moment on the member end is thus seen to be the superposition of these individual effects.

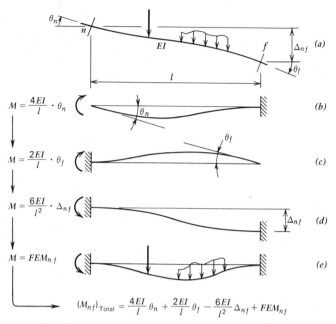

$$(M_{nf})_{\text{Total}} = \frac{4EI}{l}\theta_n + \frac{2EI}{l}\theta_f - \frac{6EI}{l^2}\Delta_{nf} + FEM_{nf}$$

Fig. 13.2 *Interpretation of slope deflection equation.*

13.4

Application of the Slope Deflection Method to Beam Problems

The slope deflection method is used to determine the end moments for statically indeterminate beams and frames. The general procedure used in solving a problem by this method will be explained by using several example problems.

The first example is the two-span beam shown in Fig. 13.3a, which is statically indeterminate to the first degree. A qualitative sketch of the deflected structure is shown in Fig. 13.3b. This sketch aids in establishing the controlling displacement boundary conditions and compatibility conditions. From this sketch, it is clear that the boundary conditions require that points a, b, and c are nondeflecting supports, and thus $\Delta_{ab} = \Delta_{bc} = 0$. Also, since none of the supports provides rotational restraint, there will be rotations θ_a, θ_b, and θ_c at the three supports, respectively. Thus, the structure is kinematically indeterminate to the third degree.

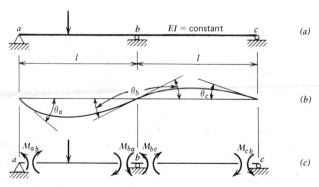

Fig. 13.3 *Beam problem by slope deflection method.*

At point b, there is a compatibility condition that requires that θ_b be the common rotation for the b end of both members framing into joint b.

Based on the compatibility conditions and the boundary conditions expressed earlier, the slope deflection equation (Eq. 13.19) can be written for each member end throughout the structure. Noting that $K_{ab} = K_{ba}$ and $K_{bc} = K_{cb}$, we have

$$
\begin{aligned}
M_{ab} &= 2EK_{ab}(2\theta_a + \theta_b) + FEM_{ab} \\
M_{ba} &= 2EK_{ab}(2\theta_b + \theta_a) + FEM_{ba} \\
M_{bc} &= 2EK_{bc}(2\theta_b + \theta_c) \\
M_{cb} &= 2EK_{bc}(2\theta_c + \theta_b)
\end{aligned}
\tag{13.21}
$$

The fixed-end moments, FEM_{ab} and FEM_{ba}, depend on the loading on span ab. Since there is no loading on span bc, $FEM_{bc} = FEM_{cb} = 0$.

A moment equilibrium equation can be written for each joint using the free-body diagrams of the joints that are shown in Fig. 13.3c. These equations must reflect the force boundary conditions, such as the moment-free supports at points a and c, as well as the internal equilibrium condition at point b. Thus, these equations take the form

$$
\begin{aligned}
\sum M_a &= M_{ab} = 0 \\
\sum M_b &= M_{ba} + M_{bc} = 0 \\
\sum M_c &= M_{cb} = 0
\end{aligned}
\tag{13.22}
$$

The equations of equilibrium can be written in terms of displacement quantities by substituting Eqs. 13.21 into Eqs. 13.22. This leads to

$$
\begin{aligned}
4EK_{ab}\theta_a + 2EK_{ab}\theta_b &= -FEM_{ab} \\
2EK_{ab}\theta_a + (4EK_{ab} + 4EK_{bc})\theta_b + 2EK_{bc}\theta_c &= -FEM_{ba} \\
2EK_{bc}\theta_b + 4EK_{bc}\theta_c &= 0
\end{aligned}
\tag{13.23}
$$

In matrix form, these equations take the form

$$
\begin{bmatrix}
4EK_{ab} & 2EK_{ab} & 0 \\
2EK_{ab} & (4EK_{ab} + 4EK_{bc}) & 2EK_{bc} \\
0 & 2EK_{bc} & 4EK_{bc}
\end{bmatrix}
\begin{Bmatrix} \theta_a \\ \theta_b \\ \theta_c \end{Bmatrix}
=
\begin{Bmatrix} -FEM_{ab} \\ -FEM_{ba} \\ 0 \end{Bmatrix}
\tag{13.24}
$$

The elements of the square coefficient matrix are stiffness quantities that express joint moment per unit of joint rotation. These elements depend entirely on quantities that are related to the structure, and they collectively form the so-called *structure stiffness matrix*. The vector on the right-hand side of Eq. 13.24 contains the fixed-end moment terms, and these are easily determined once the loads are specified. A few typical cases are given in Table 13.1. Thus, Eq. 13.24 can be solved for the displacements enumerated in the displacement vector. These displacement quantities can now be substituted into Eq. 13.21 to determine the member-end moments.

If there are any support settlements, the Δ_{nf} quantity for each member must reflect these conditions. The slope deflection equations are altered accordingly, and the equations of equilibrium have additional terms in the vector on the right-hand side of Eq. 13.24. As before, Eq. 13.24 is solved for the displacements, and the member-end moments are determined from Eq. 13.19.

The problem discussed here has only three unknown displacement quantities. For large beam problems, there would be additional displacements to consider, but the general approach would be unaltered.

13.5

Application of the Slope Deflection Method to Frame Problems

The general procedure used to solve frame problems by the slope deflection method is essentially the same as with beam problems. For example, consider the frame shown in Fig. 13.4a, which is statically indeterminate to the first degree and kinematically indeterminate to the third degree.

The qualitative deflected share of Fig. 13.4b shows that consideration of the boundary conditions and the assumption that the members are inextensible leads to $\Delta_{ab} = \Delta_{bc} = 0$. The rotations θ_a and θ_c develop at the pinned supports, and compatibility requirements that the rotation θ_b be common to the top of member ab and the left end of member bc.

The remainder of this frame problem is the same as that which was presented for the beam problem in the previous section. The slope deflection equations and the equations of equilibrium, Eq. 13.21 and 13.24, respectively, are the same as in the beam problem. Equation 13.24 is solved for the displacements, and Eqs. 13.21 yield the final end moments.

If the boundary conditions of the previous frame are changed by fixing point a against rotation and placing a roller at point c, as shown in Fig. 13.5a, the response of the structure is quite different. Again, the first step is to construct a qualitative sketch of the deflected structure that will reflect that governing boundary conditions and compatibility requirements. As seen in Fig. 13.5b, the boundary condition at point a requires that $\theta_a = 0$. In addition to the rotations θ_b and θ_c, the frame sways through the displacement Δ. Since member extensions are ignored, Δ is the common horizontal displacement at points b and c, while these

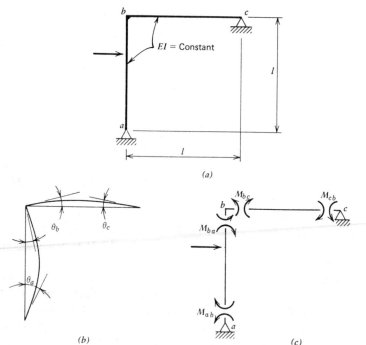

(a)

(b)   (c)

Fig. 13.4 *Frame problem by slope deflection method.*

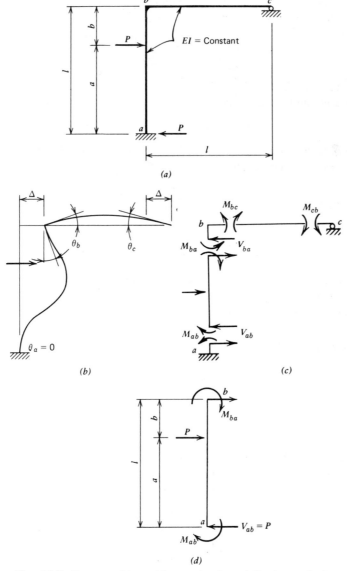

Fig. 13.5 *Frame problem with sway by slope deflection method.*

points do not displace vertically. Thus, the individual member distortions required in the slope deflection equations are $\Delta_{ab} = \Delta$ and $\Delta_{bc} = 0$. At point b, compatibility requires that θ_b be common to both members framing into joint b.

The slope deflection equation can now be written for each member end. These are

$$M_{ab} = 2EK_{ab}\left(\theta_b - 3\frac{\Delta}{l}\right) + FEM_{ab}$$

$$M_{ba} = 2EK_{ab}\left(2\theta_b - 3\frac{\Delta}{l}\right) + FEM_{ba} \qquad (13.25)$$

$$M_{bc} = 2EK_{bc}(2\theta_b + \theta_c)$$

$$M_{cb} = 2EK_{bc}(2\theta_c + \theta_b)$$

As in the previous frame problem, equilibrium equations can be written for moment equilibrium at joints b and c. These are

$$\sum M_b = M_{ba} + M_{bc} = 0$$
$$\sum M_c = M_{cb} = 0$$
(13.26)

Consideration of the free-body diagram of member ab shown in Fig. 13.5d shows that moment equilibrium about point b requires that

$$M_{ab} + M_{ba} + V_{ab}l - Pb = 0 \qquad (13.27)$$

where V_{ab} is the column shear at the base of member ab. However, equilibrium for the entire structure requires that $V_{ab} = P$. Thus, Eq. 13.27 can be written as

$$P = V_{ab} = \frac{Pb - M_{ab} - M_{ba}}{l}$$

or

$$M_{ab} + M_{ba} = Pb - Pl \qquad (13.28)$$

Equations 13.26 and 13.28 are the governing equations of equilibrium.

Substitution of Eqs. 13.25 into Eqs. 13.26 and 13.28 yields the equilibrium equations in terms of displacements.

$$(4EK_{ab} + 4EK_{bc})\theta_b + 2EK_{bc}\theta_c - 6EK_{ab}\frac{\Delta}{l} = -FEM_{ba}$$

$$2EK_{bc}\theta_b + 4EK_{bc}\theta_c = 0 \qquad (13.29)$$

$$6EK_{ab}\theta_b - 12EK_{ab}\frac{\Delta}{l} = -FEM_{ab} - FEM_{ba}$$

$$+ Pb - Pl$$

Multiplication of the third of the above equations by $(-1/l)$ and substitution of the expressions for the fixed-end moments from Table 13.1 lead to the following matrix form of the equations of equilibrium:

$$\begin{bmatrix} (4EK_{ab} + 4EK_{bc}) & 2EK_{bc} & \dfrac{-6EK_{ab}}{l} \\ 2EK_{bc} & 4EK_{bc} & 0 \\ \dfrac{-6EK_{ab}}{l} & 0 & \dfrac{12EK_{ab}}{l^2} \end{bmatrix} \begin{Bmatrix} \theta_b \\ \theta_c \\ \Delta \end{Bmatrix} = \begin{Bmatrix} \dfrac{-Pa^2b}{l^2} \\ 0 \\ \dfrac{Pa^2(l + 2b)}{l^3} \end{Bmatrix} \quad (13.30)$$

The solution of Eq. 13.30 provides the displacement quantities, which, when substituted into Eqs. 13.25, yield the member-end moments.

13.6

Numerical Slope Deflection Problems

Before discussing numerical problems, some reflection on Sections 13.4 and 13.5 is in order. In each example, a qualitative sketch of the deflected structure revealed the displacement boundary conditions, the compatibility requirements, and the unknown displacement quantities. The slope deflection equation was written for each member end in terms of these unknown displacements. Also, corresponding to each structure displacement, or kinematic degree of freedom, an equilibrium equation was written. That is, each joint where rotational freedom existed required an equilibrium equation involving the moments acting on that joint. Similarly, a sway freedom required an equilibrium equation involving column shears and moments. Thus, in all cases, there are as many equilibrium equations as there are unknown displacements. This ensures the determination of the unknown displacements through the simultaneous solution of the equilibrium equations after they have been expressed in terms of these displacements. The displacements then led to the end moments through the application of the slope deflection equation.

Equations 13.24 and 13.30 are equations of equilibrium. The elements of the coefficient matrices are stiffness quantities, while the solutions result in displacements. For these reasons, this formulation is alternatively referred to as the equilibrium method, stiffness method, or displacement method. It is important to note, however, that even though the solution centers around the solution of equilibrium equations, compatibility has been systematically satisfied. All methods of solving statically indeterminate problems must satisfy both equilibrium and compatibility.

Determine the end moments and construct the shear and moment diagrams for the structure shown below. This structure was previously considered in Section 11.7.2.   **13.6.1 Example problem**

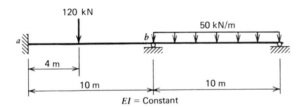

EI = Constant

Compatibility and Boundary Conditions:

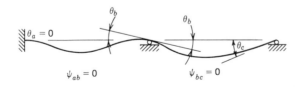

Moment Equations:

$$M_{nf} = 2EK_{nf}(2\theta_n + \theta_f - 3\psi_{nf}) + FEM_{nf} \qquad (13.19)$$

$$K_{ab} = K_{bc} = I/10 \text{ m} = K$$

$$FEM_{ab} = \frac{-120 \times 4 \times 6^2}{10^2} = -172.8 \text{ kN} \cdot \text{m}$$

$$FEM_{ba} = \frac{120 \times 4^2 \times 6}{10^2} = +115.2 \text{ kN} \cdot \text{m}$$

$$FEM_{bc} = \frac{-50 \times 10^2}{12} = -416.7 \text{ kN} \cdot \text{m}; \qquad FEM_{cb} = +416.7 \text{ kN} \cdot \text{m}$$

$$M_{ab} = 2EK(\theta_b) - 172.8$$
$$M_{ba} = 2EK(2\theta_b) + 115.2$$
$$M_{bc} = 2EK(2\theta_b + \theta_c) - 416.7$$
$$M_{cb} = 2EK(2\theta_c + \theta_b) + 416.7$$

Equilibrium Equations: Unknown moments are assumed to act positively (clockwise) on member ends.

At joint b: M_{ba} M_{bc} $M_{ba} + M_{bc} = 0$

At joint c: M_{cb} $M_{cb} = 0$

Substitution of moment equations into equilibrium equations yields

$$8EK\theta_b + 2EK\theta_c = 301.5$$
$$2EK\theta_b + 4EK\theta_c = -416.7$$

In matrix form,

$$\begin{bmatrix} 8 & 2 \\ 2 & 4 \end{bmatrix} \begin{Bmatrix} EK\theta_b \\ EK\theta_c \end{Bmatrix} = \begin{Bmatrix} 301.5 \\ -416.7 \end{Bmatrix}$$

Solution for Displacements: Solution of the above equation yields

$$\begin{Bmatrix} EK\theta_b \\ EK\theta_c \end{Bmatrix} = \begin{Bmatrix} 72.8 \\ -140.6 \end{Bmatrix} \text{ kN} \cdot \text{m}$$

Final Moments: Substitution of $EK\theta$ values into the moment equations yields the final moments

$$M_{ab} = 2(72.8) - 172.8 = -27.2 \text{ kN} \cdot \text{m}$$
$$M_{ba} = 4(72.8) + 115.2 = +406.5 \text{ kN} \cdot \text{m}$$
$$M_{bc} = 4(72.8) + 2(-140.6) - 416.7 = -406.5 \text{ kN} \cdot \text{m}$$
$$M_{cb} = 4(-140.6) + 2(72.8) + 416.7 = 0 \text{ kN} \cdot \text{m}$$

Shear and Moment Diagrams: The moments are shown to act on the member ends according to the slope deflection sign convention introduced at the beginning of Section 13.2. The shear and moment diagrams are then constructed according to the method of Section 4.14 using the sign convention of Section 4.3.

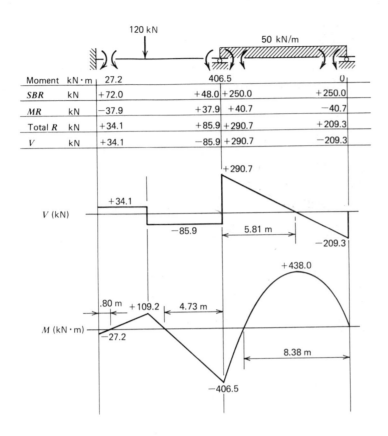

Moment	kN·m	27.2		406.5		0
SBR	kN	+72.0		+48.0	+250.0	+250.0
MR	kN	−37.9		+37.9	+40.7	−40.7
Total R	kN	+34.1		+85.9	+290.7	+209.3
V	kN	+34.1		−85.9	+290.7	−209.3

Determine the end moments and the maximum bending stress for the structure of Problem 13.6.1 if point b settles 0.03 m.

13.6.2 Example problem

$$E = 200 \times 10^9 \text{ Pa} = 200 \text{ GN/m}^2; \quad I = 2\,000 \times 10^{-6} \text{ m}^4; \quad d = 0.3 \text{ m (beam depth)}$$

Compatibility and Boundary Conditions:

$$\psi_{ab} = 0.03/10 = 0.003 \qquad \psi_{bc} = -0.003$$

Moment Equations:

$$M_{nf} = 2EK_{nf}(2\theta_n + \theta_f - 3\psi_{nf}) + FEM_{nf} \qquad (13.19)$$

All FEMs are zero

$$M_{nf} = 2EK_{nf}(2\theta_n + \theta_f) - 6EK_{nf}\psi_{nf}$$

$$K_{ab} = K_{bc} = K = \frac{2\,000 \times 10^{-6} \text{ m}^4}{10 \text{ m}} = 2\,000 \times 10^{-7} \text{ m}^3$$

$$EK = 200\ \frac{GN}{m^2} \times 2\ 000 \times 10^{-7}\ m^3 = 40\ MN \cdot m$$

$$6EK\psi_{ab} = 6 \times 40\ MN \cdot m \times 0.003 = 720\ kN \cdot m$$

$$6EK\psi_{bc} = -720\ kN \cdot m$$

$$M_{ab} = 2EK(\theta_b) - 720$$

$$M_{ba} = 2EK(2\theta_b) - 720$$

$$M_{bc} = 2EK(2\theta_b + \theta_c) + 720$$

$$M_{cb} = 2EK(2\theta_c + \theta_b) + 720$$

Equilibrium Equations: Same equilibrium equations as for Problem 13.6.1.

$$M_{ba} + M_{bc} = 0$$

$$M_{cb} = 0$$

Substitution of moment equations into equilibrium equations yields

$$8EK\theta_b + 2EK\theta_c = 0$$

$$2EK\theta_b + 4EK\theta_c = -720$$

Solution for Displacements:

$$EK\theta_b = +51.4; \qquad EK\theta_c = -205.7 \quad (kN \cdot m)$$

Final Moments:

$$M_{ab} = 2(51.4) - 720 = -617.2\ kN \cdot m$$

$$M_{ba} = 4(51.4) - 720 = -514.4\ kN \cdot m$$

$$M_{bc} = 4(51.4) + 2(-205.7) + 720 = +514.2\ kN \cdot m$$

$$M_{cb} = 4(-205.7) + 2(51.4) + 720 = 0\ kN \cdot m$$

Maximum Bending Stress:

$$f = \frac{Mc}{I} = \frac{617.2\ kN \cdot m \times 0.15\ m}{2\ 000 \times 10^{-6}m^4} = 46.29\ \frac{MN}{m^2}$$

13.6.3 Example problem   Determine the end moments for all members of the frame shown below and construct the moment diagrams.

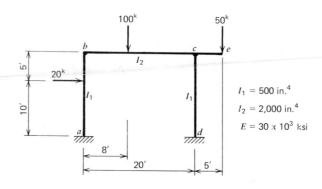

$I_1 = 500\ in.^4$

$I_2 = 2,000\ in.^4$

$E = 30 \times 10^3\ ksi$

Compatibility and Boundary Conditions:

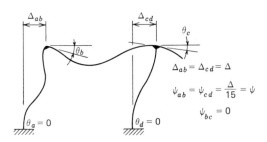

$$\Delta_{ab} = \Delta_{cd} = \Delta$$

$$\psi_{ab} = \psi_{cd} = \frac{\Delta}{15} = \psi$$

$$\psi_{bc} = 0$$

Moment Equations:

$$M_{nf} = 2EK_{nf}(2\theta_n + \theta_f - 3\psi_{nf}) + FEM_{nf} \qquad (13.19)$$

$$K_{ab} = K_{cd} = I_1/15 = K; \quad K_{bc} = I_2/20 = 4I_1/20 = 3K$$

$$FEM_{ab} = \frac{-20 \times 10 \times 5^2}{15^2} = -22.2'^{-k}$$

$$FEM_{ba} = \frac{+20 \times 10^2 \times 5}{15^2} = +44.4'^{-k}$$

$$FEM_{bc} = \frac{-100 \times 8 \times 12^2}{20^2} = -288'^{-k}$$

$$FEM_{cb} = \frac{+100 \times 8^2 \times 12}{20^2} = +192'^{-k}$$

$$FEM_{cd} = FEM_{dc} = 0$$

$$M_{ce} = -(50 \times 5) = -250'^{-k} \quad \text{(from statics)}$$

$$M_{ab} = 2EK(\theta_b - 3\psi) - 22.2$$

$$M_{ba} = 2EK(2\theta_b - 3\psi) + 44.4$$

$$M_{bc} = 2E \cdot 3K(2\theta_b + \theta_c) - 288$$

$$M_{cb} = 2E \cdot 3K(2\theta_c + \theta_b) + 192$$

$$M_{cd} = 2EK(2\theta_c - 3\psi)$$

$$M_{dc} = 2EK(\theta_c - 3\psi)$$

Equilibrium Equations:

At joint *b*: M_{bc} M_{ba} $M_{ba} + M_{bc} = 0$

At joint *c*: M_{cb} 250^{ft-k} M_{cd} $M_{cb} + M_{cd} - 250 = 0$

Columns *ab* and *cd*:

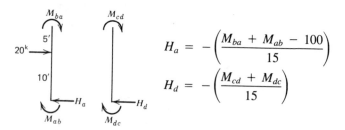

$$H_a = -\left(\frac{M_{ba} + M_{ab} - 100}{15}\right)$$

$$H_d = -\left(\frac{M_{cd} + M_{dc}}{15}\right)$$

but from consideration of the entire structure,

$$H_a + H_d = 20$$

$$\therefore M_{ab} + M_{ba} + M_{cd} + M_{dc} = -200$$

Substitution of moment equations into the equilibrium equations yields

$$16EK\theta_b + 6EK\theta_c - 6EK\psi = 243.6$$
$$6EK\theta_b + 16EK\theta_c - 6EK\psi = 58.0$$
$$6EK\theta_b + 6EK\theta_c - 24EK\psi = -222.2$$

In matrix form,

$$\begin{bmatrix} 16 & 6 & -6 \\ 6 & 16 & -6 \\ -6 & -6 & 24 \end{bmatrix} \begin{Bmatrix} EK\theta_b \\ EK\theta_c \\ EK\psi \end{Bmatrix} = \begin{Bmatrix} 243.6 \\ 58.0 \\ 222.2 \end{Bmatrix}$$

Solution for Displacements: Solution of the above equation yields

$$\begin{Bmatrix} EK\theta_b \\ EK\theta_c \\ EK\psi \end{Bmatrix} = \begin{Bmatrix} +20.14 \\ +1.58 \\ +14.69 \end{Bmatrix} \text{ ft-k}$$

Final Moments: Substitution of $EK\theta_b$, $EK\theta_c$, and $EK\psi$ values into the moment equations gives the final moments.

$$M_{ab} = 2(20.14) - 6(14.69) - 22.2 = -70.06'^{-k}$$
$$M_{ba} = 4(20.14) - 6(14.69) + 44.4 = +36.82$$
$$M_{bc} = 12(20.14) + 6(1.58) - 288 = -36.84$$
$$M_{cb} = 12(1.58) + 6(20.14) + 192 = +331.80$$
$$M_{cd} = 4(1.58) - 6(14.69) = -81.82$$
$$M_{dc} = 2(1.58) - 6(14.69) = -84.98$$
$$M_{ce} = -250 \quad \text{(from statics)}$$

Moment Diagrams: (Columns viewed from right side.)

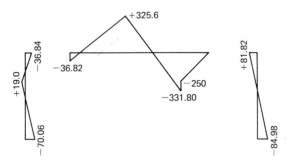

Moments in ft-k—detailed construction of M diagrams is not shown here. Follow method of Example 4.14.1.

13.7
Temperature Change

In defining the fixed-end forces of Eq. 13.15, it became clear that these are member-end forces that are induced through the application of transverse forces on a member that is fixed at its ends. However, the application of transverse forces is not the only action that can introduce fixed-end forces. For instance, consider the temperature gradient shown in Fig. 13.6a. As was shown in Eq. 11.16, this gradient introduces a constant curvature of $\alpha\ \Delta T/h$, as depicted in Fig. 13.6b, where α is the coefficient of thermal expansion, h is the member depth, and ΔT is the difference in temperature between the bottom and top beam fibers.

Returning to Eq. 13.5 and replacing the load term with the temperature-induced curvature, we obtain

$$\frac{d^2y}{dx^2} = \frac{M_{AB}}{EI} + \frac{V_{AB}x}{EI} + \frac{\alpha\ \Delta T}{h} \tag{13.31}$$

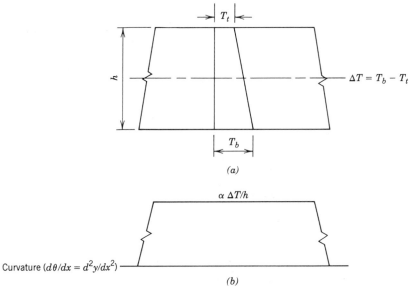

Fig. 13.6 *Temperature gradient on beam. (a) Specified gradient. (b) Induced curvature.*

Proceeding through the detailed derivation as before, we obtain

$$\{F\} = [k]\{\delta\} + \{F\}^f \tag{13.32}$$

where $\{F\}$, $[k]$, and $\{\delta\}$ are given in an expanded form in Eq. 13.14. However, in this case, $\{F\}^f$ takes the form

$$\{F\}^f = \begin{Bmatrix} -\dfrac{EI\alpha\ \Delta T}{h} \\ 0 \\ \dfrac{EI\alpha\ \Delta T}{h} \\ 0 \end{Bmatrix} \tag{13.33}$$

This reveals that there are no fixed-end shears associated with the temperature gradient. According to the notation introduced in Eq. 13.19, the fixed-end moments are given by

$$FEM_{nf} = -EI\alpha\ \Delta T/h \tag{13.34}$$
$$FEM_{fn} = EI\alpha\ \Delta T/h$$

Axial deformations associated with temperature variations induce length changes if the element is not restrained axially. However, if the element is restrained, axial forces are induced. In beam-type problems this is readily included; however, for frame-type problems the basic force–deformation relation of Eq. 13.14 should be broadened to include axial effects. These effects are touched upon in Sections 13.7.1 and are treated more fully in Chapter 17.

13.7.1 Example problem   Determine the end moments and the maximum bending stress that results from the indicated temperature gradient on the beam structure shown.

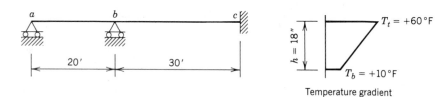

$$E = 30 \times 10^3 \text{ ksi}; \quad I = 1,500 \text{ in.}^4; \quad EI = 45 \times 10^6 \text{ k-in.}^2$$
$$\alpha = 0.000,006,5/^\circ\text{F}; \quad \Delta T = T_b - T_t = 10 - 60 = -50^\circ\text{F}$$

Compatibility and Boundary Conditions:

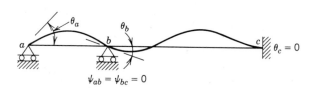

Moment Equations:

$$M_{nf} = 2EK_{nf}(2\theta_n + 6\theta_f - 3\psi_{nf}) + FEM_{nf} \qquad (13.19)$$

$$K_{ab} = I/20 = K; \quad K_{bc} = I/30 = 0.67K$$

$$FEM_{ab} = FEM_{bc} = -\frac{EI\alpha\ \Delta T}{h} \qquad (13.33)$$

$$= -\frac{(45 \times 10^6)(6.5 \times 10^{-6})(-50)}{18}$$

$$= +812.5 \text{ in.-kips} = +66.7 \text{ ft-kips}$$

$$FEM_{ba} = FEM_{cb} = -66.7 \text{ ft-kips}$$

$$M_{ab} = 2EK(2\theta_a + \theta_b) + 66.7$$

$$M_{ba} = 2EK(2\theta_b + \theta_a) - 66.7$$

$$M_{bc} = 2E(0.67K)(2\theta_b) + 66.7$$

$$M_{cb} = 2E(0.67K)(\theta_b) - 66.7$$

Equilibrium Equations:

$$M_{ab} = 0$$

$$M_{ba} + M_{bc} = 0$$

Substitution of the moment equations into the equilibrium equations yields

$$4EK\theta_a + 2EK\theta_b = -66.7$$

$$2EK\theta_a + 6.67EK\theta_b = 0$$

Solution for Displacements: Solution of the above equations yields

$$EK\theta_a = -19.61; \quad EK\theta_b = +5.89 \qquad \text{(ft-kips)}$$

Final Moments: Substitution of the $EK\theta$ values into the moment equations gives the final moments

$$M_{ab} = 4(-19.61) + 2(5.89) + 66.7 = 0$$

$$M_{ba} = 4(5.89) + 2(-19.61) - 66.7 = -82.4'^{-k}$$

$$M_{bc} = 2.67(5.89) + 66.7 = +82.4'^{-k}$$

$$M_{cb} = 1.33(5.89) - 66.7 = -58.9'^{-k}$$

Maximum Bending Stress: Assume neutral axis of the member to be at mid-depth ($c = h/2 = 9$ inches).

$$f = \frac{Mc}{I} = \frac{82.4 \times 12 \times 9}{1{,}500} = 5.93 \text{ ksi}$$

Moment Diagram and Deflected Structure:

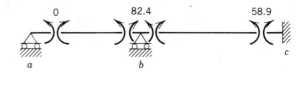

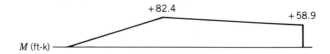

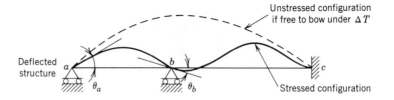

Notes:

1. The average temperature variation over the depth of the cross section is $+35°F$. Since this structure is not restrained against axial deformations, the beam will elongate by an amount given by

$$\Delta L = \alpha L T_{avg} = (0.000,006,5)(50 \times 12)(35) = 0.137 \text{ inch}$$

2. If the structure were restrained against horizontal movement at point a, then an axial compression force would be induced of magnitude

$$F = EA\alpha T_{avg} = (30 \times 10^3)A(0.000,006,5)(35) = 6.83A \text{ kips}$$

where A is the cross-sectional area in square inches.

3. It is evident that this axial force can be significant. This force, coupled with the beam deflections, will affect the bending moments along the beam. These are considered to be higher-order effects.

13.8

Matrix Formulation of Equilibrium Method for Beam Problems

The slope deflection method is an equilibrium method that is limited to the consideration of flexural deformations. This is a special case of the more general formulations of the equilibrium method where axial, shear, and torsional deformations are included. The generalized formulations are most conveniently expressed in terms of matrices, and, thus, the slope deflection method will be developed in terms of matrices. It is important to note that the same basic steps must be included as before; however, the matrix formulation permits them to be accomplished in a more general manner.

To illustrate the matrix formulation of the slope deflection method, the example beam problem treated in Section 13.4 will be used. The beam and loading are shown again in Fig. 13.7a. Again, it is useful to consider a qualitative sketch of the deflected structure as shown in Fig. 13.7b. This reflects the constraints of the boundary conditions and shows that the beam is free to rotate at each support.

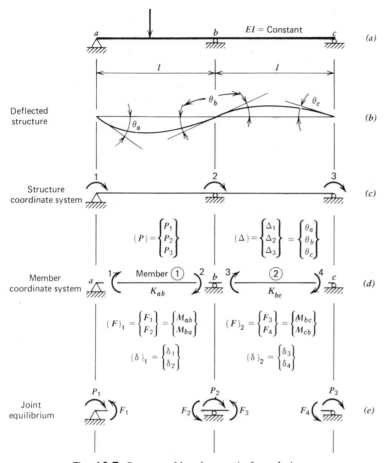

Fig. 13.7 *Beam problem by matrix formulation.*

Two separate coordinate systems are introduced as shown in Figs. 13.7c and 13.7d. The first one is a *structure coordinate system*. This involves the three rotational degrees of freedom for the structure as a whole. Corresponding to each of these degrees of freedom, there is a force (moment) and displacement (rotation). In vector form, these forces $\{P\}$ and displacements $\{\Delta\}$ in the structure coordinate system are

$$\{P\} = \begin{Bmatrix} P_1 \\ P_2 \\ P_3 \end{Bmatrix}; \qquad \{\Delta\} = \begin{Bmatrix} \Delta_1 \\ \Delta_2 \\ \Delta_3 \end{Bmatrix} = \begin{Bmatrix} \theta_a \\ \theta_b \\ \theta_c \end{Bmatrix} \qquad (13.35)$$

The second set of coordinates is a *member coordinate system*. This system enumerates the displacement degrees of freedom for each member. For the ith member, there are a set of forces (moments) $\{F\}_i$ and a set of displacements (rotations) $\{\delta\}_i$. For member 1,

$$\{F\}_1 = \begin{Bmatrix} F_1 \\ F_2 \end{Bmatrix} = \begin{Bmatrix} M_{ab} \\ M_{ba} \end{Bmatrix}; \qquad \{\delta\}_1 = \begin{Bmatrix} \delta_1 \\ \delta_2 \end{Bmatrix} \qquad (13.36)$$

and for member 2,

$$\{F\}_2 = \begin{Bmatrix} F_3 \\ F_4 \end{Bmatrix} = \begin{Bmatrix} M_{bc} \\ M_{cb} \end{Bmatrix}; \qquad \{\delta\}_2 = \begin{Bmatrix} \delta_3 \\ \delta_4 \end{Bmatrix} \tag{13.37}$$

Since there are no ψ terms present, the slope deflection equations (Eqs. 13.16 and 13.17) for the ith member can be written in matrix form as

$$\{F\}_i = [k]_i\{\delta\}_i + \{F\}_i^f \tag{13.38}$$

and $[k]_i$ is the stiffness matrix relating member-end moments to member-end rotations for the ith member and $\{F\}_i^f$ is the vector of fixed-end moments for the ith member. For member 1 of our example problem, Eq. 13.38 becomes

$$\{F\}_1 = \begin{Bmatrix} F_1 \\ F_2 \end{Bmatrix} = \begin{bmatrix} 4EK_{ab} & 2EK_{ab} \\ 2EK_{ab} & 4EK_{ab} \end{bmatrix} \begin{Bmatrix} \delta_1 \\ \delta_2 \end{Bmatrix} + \begin{Bmatrix} FEM_1 \\ FEM_2 \end{Bmatrix} \tag{13.39}$$

and for the second member

$$\{F\}_2 = \begin{Bmatrix} F_3 \\ F_4 \end{Bmatrix} = \begin{bmatrix} 4EK_{bc} & 2EK_{bc} \\ 2EK_{bc} & 4EK_{bc} \end{bmatrix} \begin{Bmatrix} \delta_3 \\ \delta_4 \end{Bmatrix} + \begin{Bmatrix} FEM_3 \\ FEM_4 \end{Bmatrix} \tag{13.40}$$

The member rotations must be related to the structure rotations to ensure that the members are compatibly connected. This is done by use of a compatibility or connectivity matrix $[\beta]_i$ for each member. For the ith member,

$$\{\delta\}_i = [\beta]_i\{\Delta\} \tag{13.41}$$

where $[\beta]_i$ relates the rotations of the ith member $\{\delta\}_i$ to the overall structure rotations $\{\Delta\}$. For member 1 of our example, compatibility requires that

$$\delta_1 = \Delta_1 \tag{13.42}$$
$$\delta_2 = \Delta_2$$

In the form of Eq. 13.41, these compatibility conditions become

$$\begin{Bmatrix} \delta_1 \\ \delta_2 \end{Bmatrix} = \begin{bmatrix} 1 & 0 & 0 \\ 0 & 1 & 0 \end{bmatrix} \begin{Bmatrix} \Delta_1 \\ \Delta_2 \\ \Delta_3 \end{Bmatrix} \tag{13.43}$$

from which $[\beta]_1$ can be extracted as

$$[\beta]_1 = \begin{bmatrix} 1 & 0 & 0 \\ 0 & 1 & 0 \end{bmatrix} \tag{13.44}$$

For member 2, the compatibility conditions are

$$\begin{Bmatrix} \delta_3 \\ \delta_4 \end{Bmatrix} = \begin{bmatrix} 0 & 1 & 0 \\ 0 & 0 & 1 \end{bmatrix} \begin{Bmatrix} \Delta_1 \\ \Delta_2 \\ \Delta_3 \end{Bmatrix} \tag{13.45}$$

$$[\beta]_2 = \begin{bmatrix} 0 & 1 & 0 \\ 0 & 0 & 1 \end{bmatrix} \tag{13.46}$$

The member-end moments can be expressed in terms of the structure rotations by substituting Eq. 13.41 into Eq. 13.38 to obtain

$$\{F\}_i = [k]_i[\beta]_i\{\Delta\} + \{F\}_i^f \tag{13.47}$$

For the example problem, application of this expression to member 1 leads to

$$\{F\}_1 = \begin{Bmatrix} F_1 \\ F_2 \end{Bmatrix} = \begin{bmatrix} 4EK_{ab} & 2EK_{ab} \\ 2EK_{ab} & 4EK_{ab} \end{bmatrix} \begin{bmatrix} 1 & 0 & 0 \\ 0 & 1 & 0 \end{bmatrix} \begin{Bmatrix} \Delta_1 \\ \Delta_2 \\ \Delta_3 \end{Bmatrix} + \begin{Bmatrix} FEM_1 \\ FEM_2 \end{Bmatrix}$$

$$= \begin{bmatrix} 4EK_{ab} & 2EK_{ab} \\ 2EK_{ab} & 4EK_{ab} \end{bmatrix} \begin{Bmatrix} \Delta_1 \\ \Delta_2 \end{Bmatrix} + \begin{Bmatrix} FEM_1 \\ FEM_2 \end{Bmatrix} \tag{13.48}$$

and for member 2,

$$\{F\}_2 = \begin{Bmatrix} F_3 \\ F_4 \end{Bmatrix} = \begin{bmatrix} 4EK_{bc} & 2EK_{bc} \\ 2EK_{bc} & 4EK_{bc} \end{bmatrix} \begin{bmatrix} 0 & 1 & 0 \\ 0 & 0 & 1 \end{bmatrix} \begin{Bmatrix} \Delta_1 \\ \Delta_2 \\ \Delta_3 \end{Bmatrix} + \begin{Bmatrix} FEM_3 \\ FEM_4 \end{Bmatrix}$$

$$= \begin{bmatrix} 4EK_{bc} & 2EK_{bc} \\ 2EK_{bc} & 4EK_{bc} \end{bmatrix} \begin{Bmatrix} \Delta_2 \\ \Delta_3 \end{Bmatrix} + \begin{Bmatrix} FEM_3 \\ FEM_4 \end{Bmatrix} \tag{13.49}$$

Equations 13.48 and 13.49 are the slope deflection equations in terms of the rotations of the structure coordinate system.

Equality of energy requires that the work done as measured in the structure coordinates be equal to the work done as measured in the member coordinates. For the entire structure

$$\tfrac{1}{2}\{P\}^T\{\Delta\} = \sum_{i=1}^{m} \tfrac{1}{2}\{F\}_i^T\{\delta\}_i \tag{13.50}$$

where m represents the number of members. For this case, $m = 2$, and substitution of Eq. 13.41 into Eq. 13.50 with some simplification yields

$$\{P\}^T\{\Delta\} = \sum_{i=1}^{2} \{F\}_i^T [\beta]_i \{\Delta\} \tag{13.51}$$

Transposition of each side and rearrangement of terms leads to

$$\{\Delta\}^T \left(\{P\} - \sum_{i=1}^{2} [\beta]_i^T \{F\}_i \right) = 0 \tag{13.52}$$

Since $\{\Delta\}^T$ can be independently controlled, the term in parentheses must be zero. This leads to the relation

$$\{P\} = \sum_{i=1}^{2} [\beta]_i^T \{F\}_i \tag{13.53}$$

This equation expresses the equilibrium condition between the member moments $\{F\}_i$ and the structure moments $\{P\}$. For our example problem,

$$\begin{Bmatrix} P_1 \\ P_2 \\ P_3 \end{Bmatrix} = \begin{bmatrix} 1 & 0 \\ 0 & 1 \\ 0 & 0 \end{bmatrix} \begin{Bmatrix} F_1 \\ F_2 \end{Bmatrix} + \begin{bmatrix} 0 & 0 \\ 1 & 0 \\ 0 & 1 \end{bmatrix} \begin{Bmatrix} F_3 \\ F_4 \end{Bmatrix}$$

or

$$\begin{aligned} P_1 &= F_1 \\ P_2 &= F_2 + F_3 \\ P_3 &= F_4 \end{aligned} \tag{13.54}$$

These are the joint equilibrium equations, and they are consistent with the free-body diagrams shown in Fig. 13.7e.

The equilibrium equations expressed by Eq. 13.53 can be modified by substituting Eq. 13.47 for the member moments $\{F\}_i$. This leads to

$$\{P\} = \sum_{i=1}^{2} [\beta]_i^T [k]_i [\beta]_i \{\Delta\} + \sum_{i=1}^{2} [\beta]_i^T \{F\}_i^f \qquad (13.55)$$

For the example problem being considered, since there are no externally applied moments at the joints, $\{P\} = \{0\}$. Also, since there is no loading on member 2, $\{F\}_2^f = \{0\}$, and Eq. 13.55 becomes

$$\begin{bmatrix} 4EK_{ab} & 2EK_{ab} & 0 \\ 2EK_{ab} & 4EK_{ab} & 0 \\ 0 & 0 & 0 \end{bmatrix} \begin{Bmatrix} \Delta_1 \\ \Delta_2 \\ \Delta_3 \end{Bmatrix} + \begin{bmatrix} 0 & 0 & 0 \\ 0 & 4EK_{bc} & 2EK_{bc} \\ 0 & 2EK_{bc} & 4EK_{bc} \end{bmatrix} \begin{Bmatrix} \Delta_1 \\ \Delta_2 \\ \Delta_3 \end{Bmatrix} + \begin{Bmatrix} FEM_1 \\ FEM_2 \\ 0 \end{Bmatrix} = \{0\}$$

$$(13.56)$$

or, through the combination of terms,

$$\begin{bmatrix} 4EK_{ab} & 2EK_{ab} & 0 \\ 2EK_{ab} & (4EK_{ab} + 4EK_{bc}) & 2EK_{bc} \\ 0 & 2EK_{bc} & 4EK_{bc} \end{bmatrix} \begin{Bmatrix} \Delta_1 \\ \Delta_2 \\ \Delta_3 \end{Bmatrix} = \begin{Bmatrix} -FEM_1 \\ -FEM_2 \\ 0 \end{Bmatrix} \qquad (13.57)$$

Precisely the same equation was developed earlier by the conventional slope deflection approach and was denoted as Eq. 13.24. The only differences are in the coordinate systems. Equations 13.57 gives the joint rotations in terms of the structure coordinates of Fig. 13.7c and the fixed-end moments in terms of the member coordinates of Fig. 13.7d instead of in the slope deflection convention of Fig. 13.3.

Equation 13.57 is then solved for the structure rotations $\{\Delta\}$, and the member rotations $\{\delta\}_i$ are determined from Eq. 13.41. Finally, the individual member-end moments $\{F\}_i$ are determined from either Eq. 13.38 or Eq. 13.47.

13.9

Matrix Formulation of Equilibrium Method for Frame Problems

The matrix formulation for frame problems follows the same general procedure as for beam problems. However, in this case there may be translational displacements in addition to joint rotations. To illustrate the matrix approach, the frame with side sway that was discussed in Section 13.5 will be considered. The frame is shown in Fig. 13.8a, and Fig. 13.8b illustrates the general shape of the deflected structure. This figure shows the constraints of the boundary conditions and also shows that joints b and c are free to rotate and that the frame is free to sway.

The structure coordinate system and the member coordinate system are shown in Figs. 13.8c and 13.8d. The vectors of structure forces $\{P\}$ and structure displacements $\{\Delta\}$ are

$$\{P\} = \begin{Bmatrix} P_1 \\ P_2 \\ P_3 \end{Bmatrix}; \qquad \{\Delta\} = \begin{Bmatrix} \Delta_1 \\ \Delta_2 \\ \Delta_3 \end{Bmatrix} = \begin{Bmatrix} \theta_b \\ \theta_c \\ \Delta \end{Bmatrix} \qquad (13.58)$$

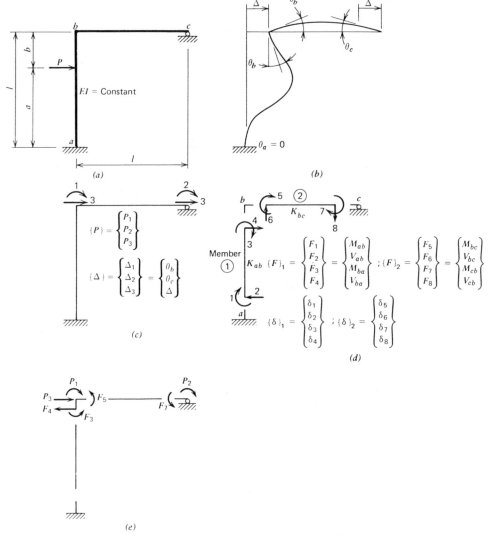

Fig. 13.8 *Frame problem by matrix formulation. (a) Frame. (b) Deflected structure.
(c) Structure coordinate system. (d) Member coordinate system. (e) Joint equilibrium.*

The vectors of member forces $\{F\}_i$ and member displacements $\{\delta\}_i$ for member 1
are

$$\{F\}_1 = \begin{Bmatrix} F_1 \\ F_2 \\ F_3 \\ F_4 \end{Bmatrix} = \begin{Bmatrix} M_{ab} \\ V_{ab} \\ M_{ba} \\ V_{ba} \end{Bmatrix} ; \quad \{\delta\}_1 = \begin{Bmatrix} \delta_1 \\ \delta_2 \\ \delta_3 \\ \delta_4 \end{Bmatrix} \qquad (13.59)$$

For member 2, these vectors are

$$\{F\}_2 = \begin{Bmatrix} F_5 \\ F_6 \\ F_7 \\ F_8 \end{Bmatrix} = \begin{Bmatrix} M_{bc} \\ V_{bc} \\ M_{cb} \\ V_{cb} \end{Bmatrix} ; \quad \{\delta\}_2 = \begin{Bmatrix} \delta_5 \\ \delta_6 \\ \delta_7 \\ \delta_8 \end{Bmatrix} \qquad (13.60)$$

For each member, the member forces are given by

$$\{F\}_i = [k]_i\{\delta\}_i + \{F\}_i^f \tag{13.61}$$

where the stiffness matrices for this equation are given in Eq. 13.14. For member 1,

$$\{F\}_1 = 2EK_{ab}\begin{bmatrix} 2 & -3/l & 1 & -3/l \\ -3/l & 6/l^2 & -3/l & 6/l^2 \\ 1 & -3/l & 2 & -3/l \\ -3/l & 6/l^2 & -3/l & 6/l^2 \end{bmatrix}\begin{Bmatrix} \delta_1 \\ \delta_2 \\ \delta_3 \\ \delta_4 \end{Bmatrix} + \{F\}_1^f \tag{13.62}$$

and for member 2,

$$\{F\}_2 = 2EK_{bc}\begin{bmatrix} 2 & -3/l & 1 & -3/l \\ -3/l & 6/l^2 & -3/l & 6/l^2 \\ 1 & -3/l & 2 & -3/l \\ -3/l & 6/l^2 & -3/l & 6/l^2 \end{bmatrix}\begin{Bmatrix} \delta_5 \\ \delta_6 \\ \delta_7 \\ \delta_8 \end{Bmatrix} + \{F\}_2^f \tag{13.63}$$

The compatible connection of members is ensured by selecting a connectivity matrix $[\beta]_i$ for each member such that

$$\{\delta\}_i = [\beta]_i\{\Delta\} \tag{13.64}$$

For member 1, $\delta_3 = \Delta_1$ and $\delta_4 = \Delta_3$, and thus

$$[\beta]_1 = \begin{bmatrix} 0 & 0 & 0 \\ 0 & 0 & 0 \\ 1 & 0 & 0 \\ 0 & 0 & 1 \end{bmatrix} \tag{13.65}$$

and for member 2, $\delta_5 = \Delta_1$ and $\delta_7 = \Delta_2$, which yields

$$[\beta]_2 = \begin{bmatrix} 1 & 0 & 0 \\ 0 & 0 & 0 \\ 0 & 1 & 0 \\ 0 & 0 & 0 \end{bmatrix} \tag{13.66}$$

Substitution of Eq. 13.64 into Eq. 13.61 yields an expression giving the member forces in terms of the structure displacements,

$$\{F\}_i = [k]_i[\beta]_i\{\Delta\} + \{F\}_i^f \tag{13.67}$$

For the example frame problem, application of this equation to member 1 yields

$$\{F\}_1 = 2EK_{ab}\begin{bmatrix} 1 & 0 & -3/l \\ -3/l & 0 & 6/l^2 \\ 2 & 0 & -3/l \\ -3/l & 0 & +6/l^2 \end{bmatrix}\begin{Bmatrix} \Delta_1 \\ \Delta_2 \\ \Delta_3 \end{Bmatrix} + \{F\}_1^f \tag{13.68}$$

and for member 2,

$$\{F\}_2 = 2EK_{bc}\begin{bmatrix} 2 & 1 & 0 \\ -3/l & -3/l & 0 \\ 1 & 2 & 0 \\ -3/l & -3/l & 0 \end{bmatrix}\begin{Bmatrix} \Delta_1 \\ \Delta_2 \\ \Delta_3 \end{Bmatrix} + \{F\}_2^f \tag{13.69}$$

At each joint, the same energy considerations used in Section 13.8 produce the equilibrium condition

$$\{P\} = \sum_{i=1}^{2} [\beta]_i^T \{F\}_i \tag{13.70}$$

For our example, Eq. 13.70 becomes

$$\begin{Bmatrix} P_1 \\ P_2 \\ P_3 \end{Bmatrix} = \begin{bmatrix} 0 & 0 & 1 & 0 \\ 0 & 0 & 0 & 0 \\ 0 & 0 & 0 & 1 \end{bmatrix} \begin{Bmatrix} F_1 \\ F_2 \\ F_3 \\ F_4 \end{Bmatrix} + \begin{bmatrix} 1 & 0 & 0 & 0 \\ 0 & 0 & 1 & 0 \\ 0 & 0 & 0 & 0 \end{bmatrix} \begin{Bmatrix} F_5 \\ F_6 \\ F_7 \\ F_8 \end{Bmatrix} \tag{13.71}$$

This equation reduces to

$$\begin{aligned} P_1 &= F_3 + F_5 \\ P_2 &= F_7 \\ P_3 &= F_4 \end{aligned} \tag{13.72}$$

These expressions correspond to the joint equilibrium conditions shown in Fig. 13.8e.

Substitution of Eq. 13.67 into Eq. 13.70 yields the following equilibrium equation:

$$\{P\} = \sum_{i=1}^{2} [\beta]_i^T [k]_i [\beta]_i \{\Delta\} + \sum_{i=1}^{2} [\beta]_i^T \{F\}_i^f \tag{13.73}$$

For our example frame problem, there are no external structure forces, and thus $\{P\} = \{0\}$. Also, since member 2 is not loaded, $\{F\}_2^f = \{0\}$, and thus Eq. 13.73 becomes

$$2EK_{ab} \begin{bmatrix} 2 & 0 & -3/l \\ 0 & 0 & 0 \\ -3/l & 0 & 6/l^2 \end{bmatrix} \begin{Bmatrix} \Delta_1 \\ \Delta_2 \\ \Delta_3 \end{Bmatrix} + 2EK_{bc} \begin{bmatrix} 2 & 1 & 0 \\ 1 & 2 & 0 \\ 0 & 0 & 0 \end{bmatrix} \begin{Bmatrix} \Delta_1 \\ \Delta_2 \\ \Delta_3 \end{Bmatrix} + \begin{Bmatrix} F_3^f \\ 0 \\ F_4^f \end{Bmatrix} = \{0\}$$

$$\tag{13.74}$$

or

$$\begin{bmatrix} (4EK_{ab} + 4EK_{bc}) & 2EK_{bc} & \dfrac{-6EK_{ab}}{l} \\[2mm] 2EK_{bc} & 4EK_{bc} & 0 \\[2mm] \dfrac{-6EK_{ab}}{l} & 0 & \dfrac{12EK_{ab}}{l^2} \end{bmatrix} \begin{Bmatrix} \Delta_1 \\ \Delta_2 \\ \Delta_3 \end{Bmatrix} = \begin{Bmatrix} -F_3^f \\ 0 \\ -F_4^f \end{Bmatrix} \tag{13.75}$$

This same set of equilibrium equations was developed by the standard slope deflection method and was given as Eq. 13.30. The only differences are that Eq. 13.75 gives the displacements in terms of the structure coordinates of Fig. 13.8c and the fixed-end forces in terms of the member coordinates of Fig. 13.8d, whereas Eq. 13.30 uses the slope deflection convention of Fig. 13.5.

The fixed-end forces on the right-hand side of Eq. 13.75, F_3^f and F_4^f, are shown in Fig. 13.8d to be the fixed-end moment and fixed-end shear at the b end of member 1. Referring to Eq. 13.14, we can evaluate the vector on the right-hand side of Eq. 13.75 to be

$$\begin{Bmatrix} -F_3^f \\ 0 \\ -F_4^f \end{Bmatrix} = \begin{Bmatrix} -Pa^2b/l^2 \\ 0 \\ Pa^2(l + 2b)/l^3 \end{Bmatrix} \tag{13.76}$$

This agrees with the right-hand side of Eq. 13.30 of the conventional slope deflection method.

The solution of Eq. 13.75 yields the structure displacements $\{\Delta\}$, and the individual member displacements $\{\delta\}_i$ result from applying Eq. 13.64. Finally, the member forces can be determined from Eq. 13.61 or Eq. 13.67.

13.10

Application of the Matrix Formulation

In the examples discussed in Sections 13.8 and 13.9, overall equations of equilibrium for the structure evolve in the form

$$\{P\} = \sum_{i=1}^{m} [\beta]_i^T [k]_i [\beta]_i \{\Delta\} + \sum_{i=1}^{m} [\beta]_i^T \{F\}_i^f \tag{13.77}$$

where m is the number of individual members in the structure. This equation can be written as

$$\{P\} = [K]\{\Delta\} + \sum_{i=1}^{m} [\beta]_i^T \{F\}_i^f \tag{13.78}$$

where

$$[K] = \sum_{i=1}^{m} [\beta]_i^T [k]_i [\beta]_i \tag{13.79}$$

The matrix $[K]$ is the structure stiffness matrix that relates the structure forces $\{P\}$ to the structure displacements $\{\Delta\}$. Equation 13.79 has the same form as Eq. 12.21, and it shows that the overall structure stiffness matrix results from the synthesis of the m individual stiffness matrices $[k]_i$.

The application of the procedures outlined in the preceding sections to numerical problems is highly systematic. In each case, a qualitative sketch of the deflected structure is used to show the pattern of deformation and the restraints of the boundary conditions. Based on the deformation pattern, a structure coordinate system and member coordinate systems are selected. Compatibility considerations are used to relate the structure displacements to the member displacements through the connectivity matrices $[\beta]_i$. Using the $[\beta]_i$ matrices, along with the individual member stiffness matrices $[k]_i$, we can generate the overall structure stiffness matrix $[K]$ from Eq. 13.79. The solution of Eq. 13.78 yields the structure displacements $\{\Delta\}$. Once $\{\Delta\}$ is known, the member displacements and the member forces can readily be determined.

The two example problems that follow were previously solved by the conventional slope deflection method in Section 13.6. It is instructive to compare the two approaches.

Determine the end moments for the structure of Example 13.6.1.

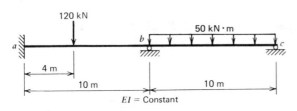

Compatibility and Boundary Conditions:

Coordinate Systems:

Structure:

$$\{P\} = \begin{Bmatrix} P_1 \\ P_2 \end{Bmatrix} = \begin{Bmatrix} 0 \\ 0 \end{Bmatrix} ; \quad \{\Delta\} = \begin{Bmatrix} \Delta_1 \\ \Delta_2 \end{Bmatrix} = \begin{Bmatrix} \theta_b \\ \theta_c \end{Bmatrix}$$

Members:

$$\{F\}_1 = \begin{Bmatrix} F_1 \\ F_2 \end{Bmatrix} = \begin{Bmatrix} M_{ab} \\ M_{ba} \end{Bmatrix} \qquad \{F\}_2 = \begin{Bmatrix} F_3 \\ F_4 \end{Bmatrix} = \begin{Bmatrix} M_{bc} \\ M_{cb} \end{Bmatrix}$$

$$\{\delta\}_1 = \begin{Bmatrix} \delta_1 \\ \delta_2 \end{Bmatrix} \qquad \{\delta\}_2 = \begin{Bmatrix} \delta_3 \\ \delta_4 \end{Bmatrix}$$

Member Forces: In this case, member-end forces are moments.

$$\{F\}_i = [k]_i \{\delta\}_i + \{F\}_i^f \tag{13.38}$$

Member 1 ($i = 1$):

$$K_{ab} = \frac{I}{10 \text{ m}} = K; \qquad [k]_1 = \begin{bmatrix} 4EK & 2EK \\ 2EK & 4EK \end{bmatrix} \tag{13.16, 17}$$

$$\{F\}_1^f = \begin{Bmatrix} -172.8 \\ +115.2 \end{Bmatrix} \text{ in kN} \cdot \text{m, see calculations in Example 13.6.1}$$

$$\therefore \begin{Bmatrix} M_{ab} \\ M_{ba} \end{Bmatrix} = \begin{bmatrix} 4EK & 2EK \\ 2EK & 4EK \end{bmatrix} \begin{Bmatrix} \delta_1 \\ \delta_2 \end{Bmatrix} + \begin{Bmatrix} -172.8 \\ +115.2 \end{Bmatrix} \quad (\text{kN} \cdot \text{m})$$

Member 2 ($i = 2$):

$$K_{bc} = \frac{I}{10 \text{ m}} = K; \qquad [k]_2 = \begin{bmatrix} 4EK & 2EK \\ 2EK & 4EK \end{bmatrix} \tag{13.16, 17}$$

$$\{F\}_2^f = \begin{Bmatrix} -416.7 \\ +416.7 \end{Bmatrix} \text{ in kN} \cdot \text{m, see calculations in Example 13.6.1}$$

$$\therefore \begin{Bmatrix} M_{bc} \\ M_{cb} \end{Bmatrix} = \begin{bmatrix} 4EK & 2EK \\ 2EK & 4EK \end{bmatrix} \begin{Bmatrix} \delta_3 \\ \delta_4 \end{Bmatrix} + \begin{Bmatrix} -416.7 \\ +416.7 \end{Bmatrix} \quad (\text{kN} \cdot \text{m})$$

Structure Connectivity:

$$\{\delta\}_i = [\beta]_i\{\Delta\} \tag{13.41}$$

Member 1 (i = 1):

$$\delta_2 = \Delta_1 = \theta_b$$

$$\begin{Bmatrix} \delta_1 \\ \delta_2 \end{Bmatrix} = \begin{bmatrix} 0 & 0 \\ 1 & 0 \end{bmatrix} \begin{Bmatrix} \theta_b \\ \theta_c \end{Bmatrix} \quad \therefore [\beta]_1 = \begin{bmatrix} 0 & 0 \\ 1 & 0 \end{bmatrix}$$

Member 2 (i = 2):

$$\delta_3 = \Delta_1 = \theta_b; \qquad \delta_4 = \Delta_2 = \theta_c$$

$$\begin{Bmatrix} \delta_3 \\ \delta_4 \end{Bmatrix} = \begin{bmatrix} 1 & 0 \\ 0 & 1 \end{bmatrix} \begin{Bmatrix} \theta_b \\ \theta_c \end{Bmatrix} \quad \therefore [\beta]_2 = \begin{bmatrix} 1 & 0 \\ 0 & 1 \end{bmatrix}$$

Structure Stiffness:

$$[K] = \sum_{i=1}^{2} [\beta]_i^T [k]_i [\beta]_i \tag{13.79}$$

$$[K] = \begin{bmatrix} 0 & 1 \\ 0 & 0 \end{bmatrix} \begin{bmatrix} 4EK & 2EK \\ 2EK & 4EK \end{bmatrix} \begin{bmatrix} 0 & 0 \\ 1 & 0 \end{bmatrix} + \begin{bmatrix} 1 & 0 \\ 0 & 1 \end{bmatrix} \begin{bmatrix} 4EK & 2EK \\ 2EK & 4EK \end{bmatrix} \begin{bmatrix} 1 & 0 \\ 0& 1 \end{bmatrix}$$

$$[K] = \begin{bmatrix} 8EK & 2EK \\ 2EK & 4EK \end{bmatrix}$$

Equations of Equilibrium:

$$\{P\} = [K]\{\Delta\} + \sum_{i=1}^{2} [\beta]_i^T \{F\}_i^f \tag{13.78}$$

$$\begin{Bmatrix} 0 \\ 0 \end{Bmatrix} = \begin{bmatrix} 8EK & 2EK \\ 2EK & 4EK \end{bmatrix} \begin{Bmatrix} \theta_b \\ \theta_c \end{Bmatrix} + \begin{bmatrix} 0 & 1 \\ 0 & 0 \end{bmatrix} \begin{Bmatrix} -172.8 \\ 115.2 \end{Bmatrix} + \begin{bmatrix} 1 & 0 \\ 0 & 1 \end{bmatrix} \begin{Bmatrix} -416.7 \\ +416.7 \end{Bmatrix}$$

$$\begin{bmatrix} 8EK & 2EK \\ 2EK & 4EK \end{bmatrix} \begin{Bmatrix} \theta_b \\ \theta_c \end{Bmatrix} = \begin{Bmatrix} 301.5 \\ -416.7 \end{Bmatrix}$$

These are the same equations that were developed in Example 13.6.1

Solution for Displacements:

$$\{\Delta\} = \begin{Bmatrix} \Delta_1 \\ \Delta_2 \end{Bmatrix} = \begin{Bmatrix} \theta_b \\ \theta_c \end{Bmatrix} = \frac{1}{EK} \begin{Bmatrix} 72.8 \\ -140.6 \end{Bmatrix}$$

$$\{\delta\}_1 = [\beta]_1\{\Delta\}; \quad \begin{Bmatrix} \delta_1 \\ \delta_2 \end{Bmatrix} = \begin{bmatrix} 0 & 0 \\ 1 & 0 \end{bmatrix} \begin{Bmatrix} \theta_b \\ \theta_c \end{Bmatrix} = \begin{Bmatrix} 0 \\ \theta_b \end{Bmatrix} = \frac{1}{EK} \begin{Bmatrix} 0 \\ 72.8 \end{Bmatrix}$$

$$\{\delta\}_2 = [\beta]_2\{\Delta\}; \quad \begin{Bmatrix} \delta_3 \\ \delta_4 \end{Bmatrix} = \begin{bmatrix} 1 & 0 \\ 0 & 1 \end{bmatrix} \begin{Bmatrix} \theta_b \\ \theta_c \end{Bmatrix} = \begin{Bmatrix} \theta_b \\ \theta_c \end{Bmatrix} = \frac{1}{EK} \begin{Bmatrix} 72.8 \\ -140.6 \end{Bmatrix}$$

Final Member Forces:

$$\{F\}_i = [k]_i\{\delta\}_i + \{F\}_i^f \tag{13.38}$$

Member 1:

$$\begin{Bmatrix} M_{ab} \\ M_{ba} \end{Bmatrix} = \begin{bmatrix} 4EK & 2EK \\ 2EK & 4EK \end{bmatrix} \cdot \frac{1}{EK} \begin{Bmatrix} 0 \\ 72.8 \end{Bmatrix} + \begin{Bmatrix} -172.8 \\ +115.2 \end{Bmatrix} = \begin{Bmatrix} -27.2 \text{ kN} \cdot \text{m} \\ +406.5 \text{ kN} \cdot \text{m} \end{Bmatrix}$$

Member 2:

$$\begin{Bmatrix} M_{bc} \\ M_{cb} \end{Bmatrix} = \begin{bmatrix} 4EK & 2EK \\ 2EK & 4EK \end{bmatrix} \cdot \frac{1}{EK} \begin{Bmatrix} 72.8 \\ -140.6 \end{Bmatrix} + \begin{Bmatrix} -416.7 \\ +416.7 \end{Bmatrix} = \begin{Bmatrix} -406.5 \text{ kN} \cdot \text{m} \\ 0 \end{Bmatrix}$$

Determine the end shears and end moments for the structure of Example 13.6.3.   **13.10.2 Example problem**

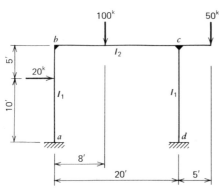

$I_1 = 500 \text{ in.}^4$
$I_2 = 2000 \text{ in.}^4$

Compatibility and Boundary Conditions:

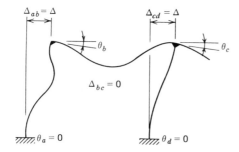

Coordinate Systems:

Structure:

$$\{P\} = \begin{Bmatrix} P_1 \\ P_2 \\ P_3 \end{Bmatrix} = \begin{Bmatrix} 0 \\ 250 \\ 0 \end{Bmatrix}$$

$$\{\Delta\} = \begin{Bmatrix} \Delta_1 \\ \Delta_2 \\ \Delta_3 \end{Bmatrix} = \begin{Bmatrix} \theta_b \\ \theta_c \\ \Delta \end{Bmatrix}$$

Members:

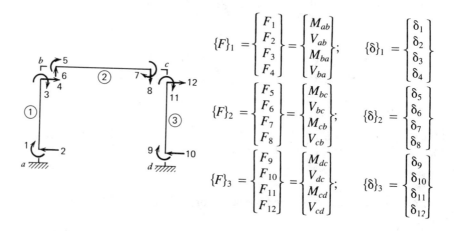

$$\{F\}_1 = \begin{Bmatrix} F_1 \\ F_2 \\ F_3 \\ F_4 \end{Bmatrix} = \begin{Bmatrix} M_{ab} \\ V_{ab} \\ M_{ba} \\ V_{ba} \end{Bmatrix}; \qquad \{\delta\}_1 = \begin{Bmatrix} \delta_1 \\ \delta_2 \\ \delta_3 \\ \delta_4 \end{Bmatrix}$$

$$\{F\}_2 = \begin{Bmatrix} F_5 \\ F_6 \\ F_7 \\ F_8 \end{Bmatrix} = \begin{Bmatrix} M_{bc} \\ V_{bc} \\ M_{cb} \\ V_{cb} \end{Bmatrix}; \qquad \{\delta\}_2 = \begin{Bmatrix} \delta_5 \\ \delta_6 \\ \delta_7 \\ \delta_8 \end{Bmatrix}$$

$$\{F\}_3 = \begin{Bmatrix} F_9 \\ F_{10} \\ F_{11} \\ F_{12} \end{Bmatrix} = \begin{Bmatrix} M_{dc} \\ V_{dc} \\ M_{cd} \\ V_{cd} \end{Bmatrix}; \qquad \{\delta\}_3 = \begin{Bmatrix} \delta_9 \\ \delta_{10} \\ \delta_{11} \\ \delta_{12} \end{Bmatrix}$$

Member Forces:

$$\{F\}_i = [k]_i\{\delta\}_i + \{F\}_i^f \tag{13.61}$$

Member 1 ($i = 1$):

$$K_{ab} = \frac{I_1}{15} = K; \qquad [k]_1 = 2EK \begin{bmatrix} 2 & -3/l & 1 & -3/l \\ -3/l & 6/l^2 & -3/l & 6/l^2 \\ 1 & -3/l & 2 & -3/l \\ -3/l & 6/l^2 & -3/l & 6/l^2 \end{bmatrix}_{l=15'} = 2EK[\bar{k}]_{l=15'}$$

$$\{F\}_1^f = \begin{Bmatrix} -22.2'^{-k} \\ +5.19^k \\ +44.4'^{-k} \\ -14.81^k \end{Bmatrix} \quad \begin{array}{l} \text{from formulae in Table 13.1} \\ \text{(also, see Example 13.6.3)} \end{array}$$

$$\begin{Bmatrix} M_{ab} \\ V_{ab} \\ M_{ba} \\ V_{ba} \end{Bmatrix} = 2EK[\bar{k}]_{l=15'} \cdot \begin{Bmatrix} \delta_1 \\ \delta_2 \\ \delta_3 \\ \delta_4 \end{Bmatrix} + \begin{Bmatrix} -22.2 \\ +5.19 \\ +44.4 \\ -14.81 \end{Bmatrix} \tag{13.14}$$

Member 2 ($i = 2$):

$$K_{bc} = \frac{4I_1}{20} = 3K; \qquad [k]_2 = 2E \cdot 3K[\bar{k}]_{l=20'}$$

$$\{F\}_2^f = \begin{Bmatrix} -288'^{-k} \\ +64.8^k \\ +192'^{-k} \\ -35.2^k \end{Bmatrix} \quad \begin{array}{l} \text{from formulae in Table 13.1} \\ \text{(also, see Example 13.6.3)} \end{array}$$

$$\begin{Bmatrix} M_{bc} \\ V_{bc} \\ M_{cb} \\ V_{cb} \end{Bmatrix} = 6EK[\bar{k}]_{l=20'} \cdot \begin{Bmatrix} \delta_5 \\ \delta_6 \\ \delta_7 \\ \delta_8 \end{Bmatrix} + \begin{Bmatrix} -288 \\ +64.8 \\ +192 \\ -35.2 \end{Bmatrix} \tag{13.14}$$

Member 3 ($i = 3$):

$$K_{cd} = \frac{I_1}{15} = K; \qquad [k]_3 = 2EK[\bar{k}]_{l=15'}$$

$$\{F\}_3^f = \{0\}$$

$$\begin{Bmatrix} M_{dc} \\ V_{dc} \\ M_{cd} \\ V_{cd} \end{Bmatrix} = 2EK \, [\bar{k}]_{l=15'} \cdot \begin{Bmatrix} \delta_9 \\ \delta_{10} \\ \delta_{11} \\ \delta_{12} \end{Bmatrix} + \begin{Bmatrix} 0 \\ 0 \\ 0 \\ 0 \end{Bmatrix} \qquad (13.14)$$

Structure Connectivity:

$$\{\delta\}_i = [\beta]_i\{\Delta\} \qquad (13.64)$$

Member 1 ($i= 1$):

$$\delta_3 = \Delta_1 = \theta_b; \qquad \delta_4 = \Delta_3 = \Delta$$

$$\begin{Bmatrix} \delta_1 \\ \delta_2 \\ \delta_3 \\ \delta_4 \end{Bmatrix} = \begin{bmatrix} 0 & 0 & 0 \\ 0 & 0 & 0 \\ 1 & 0 & 0 \\ 0 & 0 & 1 \end{bmatrix} \begin{Bmatrix} \theta_b \\ \theta_c \\ \Delta \end{Bmatrix} \qquad \therefore \ [\beta]_1 = \begin{bmatrix} 0 & 0 & 0 \\ 0 & 0 & 0 \\ 1 & 0 & 0 \\ 0 & 0 & 1 \end{bmatrix}$$

Member 2 ($i = 2$):

$$\delta_5 = \Delta_1 = \theta_b; \qquad \delta_7 = \delta_2 = \theta_c$$

$$\begin{Bmatrix} \delta_5 \\ \delta_6 \\ \delta_7 \\ \delta_8 \end{Bmatrix} = \begin{bmatrix} 1 & 0 & 0 \\ 0 & 0 & 0 \\ 0 & 1 & 0 \\ 0 & 0 & 0 \end{bmatrix} \begin{Bmatrix} \theta_b \\ \theta_c \\ \Delta \end{Bmatrix} \qquad \therefore \ [\beta]_2 = \begin{bmatrix} 1 & 0 & 0 \\ 0 & 0 & 0 \\ 0 & 1 & 0 \\ 0 & 0 & 0 \end{bmatrix}$$

Member 3($i = 3$):

$$\delta_{11} = \Delta_2 = \theta_c; \qquad \delta_{12} = \Delta_3 = \Delta$$

$$\begin{Bmatrix} \delta_9 \\ \delta_{10} \\ \delta_{11} \\ \delta_{12} \end{Bmatrix} = \begin{bmatrix} 0 & 0 & 0 \\ 0 & 0 & 0 \\ 0 & 1 & 0 \\ 0 & 0 & 1 \end{bmatrix} \begin{Bmatrix} \theta_b \\ \theta_c \\ \Delta \end{Bmatrix} \qquad \therefore \ [\beta]_3 = \begin{bmatrix} 0 & 0 & 0 \\ 0 & 0 & 0 \\ 0 & 1 & 0 \\ 0 & 0 & 1 \end{bmatrix}$$

Structure Stiffness:

$$[K] = \sum_{i=1}^{3} [\beta]_i^T[k]_i[\beta]_i \qquad (13.79)$$

$$[K] = 2EK \begin{bmatrix} 2 & 0 & -3/l \\ 0 & 0 & 0 \\ -3/l & 0 & 6/l^2 \end{bmatrix}_{l=15'} + 6EK \begin{bmatrix} 2 & 1 & 0 \\ 1 & 2 & 0 \\ 0 & 0 & 0 \end{bmatrix}$$

$$+ 2EK \begin{bmatrix} 0 & 0 & 0 \\ 0 & 2 & -3/l \\ 0 & -3/l & 6/l^2 \end{bmatrix}_{l=15'}$$

$$[K] = EK \begin{bmatrix} 16 & 6 & -6/l \\ 6 & 16 & -6/l \\ -6/l & -6/l & 24/l^2 \end{bmatrix}_{l=15'}$$

Equation of Equilibrium:

$$\{P\} = [K]\{\Delta\} = \sum_{i=1}^{3}[\beta]_i^T\{F\}_i^f \qquad (13.78)$$

$$\begin{Bmatrix} 0 \\ 250 \\ 0 \end{Bmatrix} = EK \begin{bmatrix} 16 & 6 & -6/l \\ 6 & 16 & -6/l \\ -6/l & -6/l & 24/l^2 \end{bmatrix} \begin{Bmatrix} \theta_b \\ \theta_c \\ \Delta \end{Bmatrix} + \begin{Bmatrix} -243.6 \\ +192.0 \\ -14.81 \end{Bmatrix}$$

or

$$EK \begin{bmatrix} 16 & 6 & -6 \\ 6 & 16 & -6 \\ -6 & -6 & 24 \end{bmatrix} \begin{Bmatrix} \theta_b \\ \theta_c \\ \Delta/l \end{Bmatrix} = \begin{Bmatrix} 243.6 \\ 58.0 \\ 222.2 \end{Bmatrix}$$

where $l = 15'$

Solution for Displacements:

$$\begin{Bmatrix} \theta_b \\ \theta_c \\ \Delta \end{Bmatrix} = \frac{1}{EK} \begin{Bmatrix} 20.14 \\ 1.58 \\ 14.69l \end{Bmatrix}$$

where $l = 15'$

$$\{\delta\}_1 = [\beta]_1\{\Delta\}; \qquad \begin{Bmatrix} \delta_1 \\ \delta_2 \\ \delta_3 \\ \delta_4 \end{Bmatrix} = \begin{bmatrix} 0 & 0 & 0 \\ 0 & 0 & 0 \\ 1 & 0 & 0 \\ 0 & 0 & 1 \end{bmatrix} \begin{Bmatrix} \theta_b \\ \theta_c \\ \Delta \end{Bmatrix} = \begin{Bmatrix} 0 \\ 0 \\ \theta_b \\ \Delta \end{Bmatrix} = \frac{1}{EK} \begin{Bmatrix} 0 \\ 0 \\ 20.14 \\ 14.69l \end{Bmatrix}$$

$$\{\delta\}_2 = [\beta]_2\{\Delta\}; \qquad \begin{Bmatrix} \delta_5 \\ \delta_6 \\ \delta_7 \\ \delta_8 \end{Bmatrix} = \begin{bmatrix} 1 & 0 & 0 \\ 0 & 0 & 0 \\ 0 & 1 & 0 \\ 0 & 0 & 0 \end{bmatrix} \begin{Bmatrix} \theta_b \\ \theta_c \\ \Delta \end{Bmatrix} = \begin{Bmatrix} \theta_b \\ 0 \\ \theta_c \\ 0 \end{Bmatrix} = \frac{1}{EK} \begin{Bmatrix} 20.14 \\ 0 \\ 1.58 \\ 0 \end{Bmatrix}$$

$$\{\delta\}_3 = [\beta]_3\{\Delta\}; \qquad \begin{Bmatrix} \delta_9 \\ \delta_{10} \\ \delta_{11} \\ \delta_{12} \end{Bmatrix} = \begin{bmatrix} 0 & 0 & 0 \\ 0 & 0 & 0 \\ 0 & 1 & 0 \\ 0 & 0 & 1 \end{bmatrix} \begin{Bmatrix} \theta_b \\ \theta_c \\ \Delta \end{Bmatrix} = \begin{Bmatrix} 0 \\ 0 \\ \theta_c \\ \Delta \end{Bmatrix} = \frac{1}{EK} \begin{Bmatrix} 0 \\ 0 \\ 1.58 \\ 14.69l \end{Bmatrix}$$

Final Member Forces:

Member 1:

$$\begin{Bmatrix} M_{ab} \\ V_{ab} \\ M_{ba} \\ V_{ba} \end{Bmatrix} = 2EK \begin{bmatrix} 2 & -3/l & 1 & -3/l \\ -3/l & 6/l^2 & -3/l & 6/l^2 \\ 1 & -3/l & 2 & -3/l \\ -3/l & 6/l^2 & -3/l & 6/l^2 \end{bmatrix}_{l=15'}$$

$$\cdot \frac{1}{EK} \begin{Bmatrix} 0 \\ 0 \\ 20.14 \\ 14.69l \end{Bmatrix} + \begin{Bmatrix} -22.2 \\ +5.19 \\ +44.4 \\ -14.81 \end{Bmatrix} = \begin{Bmatrix} -70.06'^{-k} \\ +8.88^k \\ +36.82'^{-k} \\ -11.12^k \end{Bmatrix}$$

Member 2:

$$
\begin{Bmatrix} M_{bc} \\ V_{bc} \\ M_{cb} \\ V_{cb} \end{Bmatrix} = 6EK \begin{bmatrix} 2 & -3/l & 1 & -3/l \\ -3/l & 6/l^2 & -3/l & 6/l^2 \\ 1 & -3/l & 2 & -3/l \\ -3/l & 6/l^2 & -3/l & 6/l^2 \end{bmatrix}_{l=20'}
$$

$$
\cdot \frac{1}{EK} \begin{bmatrix} 20.14 \\ 0 \\ 1.58 \\ 0 \end{bmatrix} + \begin{Bmatrix} -288 \\ +64.8 \\ +192 \\ -35.2 \end{Bmatrix} = \begin{Bmatrix} -36.84'^{-k} \\ +45.3^k \\ +331.80'^{-k} \\ -54.7^k \end{Bmatrix}
$$

Member 3:

$$
\begin{Bmatrix} M_{dc} \\ V_{dc} \\ M_{cd} \\ V_{cd} \end{Bmatrix} = 2EK \begin{bmatrix} 2 & -3/l & 1 & -3/l \\ -3/l & 6/l^2 & -3/l & 6/l^2 \\ 1 & -3/l & 2 & -3/l \\ -3/l & 6/l^2 & -3/l & 6/l^2 \end{bmatrix}_{l=15'}
$$

$$
\cdot \frac{1}{EK} \begin{bmatrix} 0 \\ 0 \\ 1.58 \\ 14.69l \end{bmatrix} + \begin{Bmatrix} 0 \\ 0 \\ 0 \\ 0 \end{Bmatrix} = \begin{Bmatrix} -84.98'^{-k} \\ +11.12^k \\ -81.82'^{-k} \\ -11.12^k \end{Bmatrix}
$$

Shear and Moment Diagrams: The member-end forces as given above represent a typical form of output from modern computer programs. From these data, the shear and moment diagrams are readily constructed. For instance, consider member *bc*. Recall that the signs of the individual forces are in accordance with the convention established in Section 13.2.

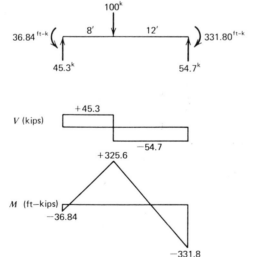

In the numerical problems, the structure displacements {Δ} were determined by solving Eq. 13.78. This solution can be shown in the form

$$
\{\Delta\} = [K]^{-1}\{P\} - [K]^{-1} \sum_{i=1}^{m} [\beta]_i^T \{F\}_i^f \tag{13.80}
$$

In most cases, the inverse of $[K]$ is not actually determined, but instead some numerical technique is employed to solve for $\{\Delta\}$. However, for any method that is used, the effect is to determine a solution that can be represented by Eq. 13.80.

Substitution of Eq. 13.80 into Eq. 13.41 or 13.64 yields the member displacements $\{\delta\}_i$. If these member displacements are then substituted into Eq. 13.38 or 13.61, the member forces are the ith member result. The final expression takes the form

$$\{F\}_i = [k]_i[\beta]_i[K]^{-1}\{P\} + \left([I] - [k]_i[\beta]_i[K]^{-1}\sum_{i=1}^{m}[\beta]_i^T\right)\{F\}_i^f \quad (13.81)$$

where $[I]$ is the identity matrix. Although this is not a practical equation for use, it does show that the final member forces are dependent on the structural characteristics ($[k]_i$, $[K]$, and $[\beta]_i$) and on the loading ($\{P\}$ and $\{F\}_i^f$).

13.11

Determination of Reactions and Effects of Support Settlements

The procedure outlined in Section 12.7 for treating support settlements and for determining the reaction components is valid here also. In the present case, a slight modification is necessary to account for the effects of loads that are applied along the member lengths.

Equation 13.78 can be written in the form

$$[K]\{\Delta\} = \{P\} - \sum_{i=1}^{m}[\beta]_i^T\{F\}_i^f = \{P\} - \{P\}^e \quad (13.82)$$

where $\{P\}^e$ represents the equivalent set of structure forces that result from the member fixed-end forces. Here, the structure coordinate system must have components that include the desired reaction components and the designated support settlements. This equation is now partitioned according to the procedure described in Eqs. 12.23 and 12.24. Thus, we have

$$\begin{bmatrix} [K]_{11} & [K]_{12} \\ [K]_{21} & [K]_{22} \end{bmatrix} \begin{Bmatrix} \{\Delta\}_1 \\ \{\Delta\}_2 \end{Bmatrix} = \begin{Bmatrix} \{P\}_1 - \{P\}_1^e \\ \{P\}_2 - \{P\}_2^e \end{Bmatrix} \quad (13.83)$$

In the above equation, $\{\Delta\}_1$ represents the unknown structure displacements whose elements correspond to the applied $\{P\}_1$ structure forces, and $\{P\}_2$ includes the unknown reaction components that correspond to the specified displacements, $\{\Delta\}_2$. The vector $\{P\}^e$ is partitioned into $\{P\}_1^e$ and $\{P\}_2^e$ in the same fashion that $\{P\}$ is partitioned into $\{P\}_1$ and $\{P\}_2$.

Expanding Eq. 13.83 into two separate matrix equations and solving the first one for $\{\Delta\}_1$, we obtain

$$\{\Delta\}_1 = [K]_{11}^{-1}(\{P\}_1 - \{P\}_1^e - [K]_{12}\{\Delta\}_2) \quad (13.84)$$

Substituting Eq. 13.84 into the second equation from Eq. 13.83 gives

$$\{P\}_2 = [K]_{21}[K]_{11}^{-1}(\{P\}_1 - \{P\}_1^e - [K]_{12}\{\Delta\}_2) + [K]_{22}\{\Delta\}_2 + \{P\}_2^e \quad (13.85)$$

Equation 13.84 gives the unknown structure displacements $\{\Delta\}_1$ that result from the loading on the structure $\{P\}_1$ and $\{P\}_1^e$ and the imposed displacements $\{\Delta\}_2$. The displacements $\{\Delta\}_1$ are then used to obtain the member displacements and member forces in the usual fashion. Equation 13.85 gives the reaction components $\{P\}_2$, and it is clear that these depend on the loading on the structure, as given by $\{P\}_1$, $\{P\}_1^e$, and $\{P\}_2^e$ and on the imposed displacements $\{\Delta\}_2$.

Determine the structure displacements for the structure shown below if, in addition to the loading given, support b settles 0.03 m downward. Verify that the results correspond to those that come from a superposition of the results of Examples 13.6.1 and 13.6.2. **13.11.1 Example problem**

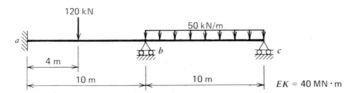

Compatibility and Boundary Conditions:

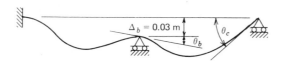

$$\Delta_b = 0.03 \text{ m} \qquad \theta_b \qquad \theta_c$$

Coordinate Systems:

Structure:

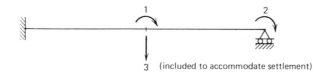

3 (included to accommodate settlement)

$$\{P\} = \left\{ \begin{matrix} \{P\}_1 \\ \{P\}_2 \end{matrix} \right\} = \left\{ \begin{matrix} P_1 \\ P_2 \\ P_3 \end{matrix} \right\} = \left\{ \begin{matrix} 0 \\ 0 \\ P_3 \end{matrix} \right\} \qquad \{\Delta\} = \left\{ \begin{matrix} \{\Delta\}_1 \\ \{\Delta\}_2 \end{matrix} \right\} = \left\{ \begin{matrix} \Delta_1 \\ \Delta_2 \\ \Delta_3 \end{matrix} \right\} = \left\{ \begin{matrix} \theta_b \\ \theta_c \\ \Delta_b \end{matrix} \right\} = \left\{ \begin{matrix} \theta_b \\ \theta_c \\ 0.03 \end{matrix} \right\}$$

Members:

$$\{F\}_1 = \left\{ \begin{matrix} F_1 \\ F_2 \\ F_3 \\ F_4 \end{matrix} \right\} = \left\{ \begin{matrix} M_{ab} \\ V_{ab} \\ M_{ba} \\ V_{ba} \end{matrix} \right\}; \quad \{\delta\}_1 = \left\{ \begin{matrix} \delta_1 \\ \delta_2 \\ \delta_3 \\ \delta_4 \end{matrix} \right\} \qquad \{F\}_2 = \left\{ \begin{matrix} F_5 \\ F_6 \\ F_7 \\ F_8 \end{matrix} \right\} = \left\{ \begin{matrix} M_{bc} \\ V_{bc} \\ M_{cb} \\ V_{cb} \end{matrix} \right\}; \quad \{\delta\}_2 = \left\{ \begin{matrix} \delta_5 \\ \delta_6 \\ \delta_7 \\ \delta_8 \end{matrix} \right\}$$

Member Forces:

$$\{F\}_i = [k]_i \{\delta\}_i + \{F\}_i^f \qquad (13.38)$$

Member 1 ($i = 1$):

$$K_{ab} = \frac{I}{10 \text{ m}} = K$$

$$[k]_1 = 2EK \begin{Bmatrix} 2 & -3/l & 1 & -3/l \\ -3/l & 6/l^2 & -3/l & 6/l^2 \\ 1 & -3/l & 2 & -3/l \\ -3/l & 6/l^2 & -3/l & 6/l^2 \end{Bmatrix}_{l=10m} = 2EK[\bar{k}]_{l=10m} \qquad (13.14)$$

$$\{F\}_1^f = \begin{Bmatrix} -172.8 \\ +77.76 \\ +115.2 \\ -42.24 \end{Bmatrix} \quad \text{from formulae in Table 13.1}$$

$$\begin{Bmatrix} M_{ab} \\ V_{ab} \\ M_{ba} \\ V_{ba} \end{Bmatrix} = 2EK[\bar{k}]_{l=10m} \begin{Bmatrix} \delta_1 \\ \delta_2 \\ \delta_3 \\ \delta_4 \end{Bmatrix} + \begin{Bmatrix} -172.8 \\ +77.76 \\ +115.2 \\ -42.24 \end{Bmatrix}$$

Member 2 (i = 2):

$$K_{bc} = \frac{I}{10 \text{ m}} = K; \qquad [k]_2 = 2EK[\bar{k}]_{l=10m}$$

$$\{F\}_2^f = \begin{Bmatrix} -416.7 \\ +250.0 \\ +416.7 \\ -250.0 \end{Bmatrix} \quad \text{from formulae in Table 13.1}$$

$$\begin{Bmatrix} M_{bc} \\ V_{bc} \\ M_{cb} \\ V_{cb} \end{Bmatrix} = 2EK[\bar{k}]_{l=10m} \begin{Bmatrix} \delta_5 \\ \delta_6 \\ \delta_7 \\ \delta_8 \end{Bmatrix} + \begin{Bmatrix} -416.7 \\ +250.0 \\ +416.7 \\ -250.0 \end{Bmatrix}$$

Structure Connectivity:

$$\{\delta\}_i = [\beta]_i\{\Delta\} \qquad (13.41)$$

Member 1 (i = 1):

$$\delta_3 = \Delta_1 = \theta_b; \qquad \delta_4 = \Delta_3 = \Delta_b$$

$$\begin{Bmatrix} \delta_1 \\ \delta_2 \\ \delta_3 \\ \delta_4 \end{Bmatrix} = \begin{bmatrix} 0 & 0 & 0 \\ 0 & 0 & 0 \\ 1 & 0 & 0 \\ 0 & 0 & 1 \end{bmatrix} \begin{Bmatrix} \theta_b \\ \theta_c \\ \Delta_b \end{Bmatrix}$$

Member 2 (i = 2):

$$\delta_5 = \Delta_1 = \theta_b; \qquad \delta_6 = -\Delta_3 = -\Delta_b; \qquad \delta_7 = \Delta_2 = \theta_c$$

$$\begin{Bmatrix} \delta_5 \\ \delta_6 \\ \delta_7 \\ \delta_8 \end{Bmatrix} = \begin{bmatrix} 1 & 0 & 0 \\ 0 & 0 & -1 \\ 0 & 1 & 0 \\ 0 & 0 & 0 \end{bmatrix} \begin{Bmatrix} \theta_b \\ \theta_c \\ \Delta_b \end{Bmatrix}$$

Structure Stiffness:

$$[K] = \sum_{i=1}^{2} [\beta]_i^T[k]_i[\beta]_i \qquad (13.79)$$

$$[K] = 2EK \begin{bmatrix} 2 & 0 & -3/l \\ 0 & 0 & 0 \\ -3/l & 0 & 6/l^2 \end{bmatrix} + 2EK \begin{bmatrix} 2 & 1 & 3/l \\ 1 & 2 & 3/l \\ 3/l & 3/l & 6/l^2 \end{bmatrix}$$

$$= 2EK \left[\begin{array}{cc|c} 4 & 1 & 0 \\ 1 & 2 & 3/l \\ \hline 0 & 3/l & 12/l^2 \end{array} \right]_{l=10m} = \left[\begin{array}{c|c} [K]_{11} & [K]_{12} \\ \hline [K]_{21} & [K]_{22} \end{array} \right]$$

Determination of {P}ᵉ:

$$\{P\}^e = \sum_{i=1}^{2} [\beta]_i^T \{F\}_i^f \tag{13.82}$$

$$\{P\}^e = \begin{bmatrix} 0 & 0 & 1 & 0 \\ 0 & 0 & 0 & 0 \\ 0 & 0 & 0 & 1 \end{bmatrix} \begin{Bmatrix} -172.8 \\ + 77.76 \\ +115.2 \\ - 42.24 \end{Bmatrix} + \begin{bmatrix} 1 & 0 & 0 & 0 \\ 0 & 0 & 1 & 0 \\ 0 & -1 & 0 & 0 \end{bmatrix} \begin{Bmatrix} -416.7 \\ +250.0 \\ +416.7 \\ -250.0 \end{Bmatrix}$$

$$= \begin{Bmatrix} -301.5 \\ +416.7 \\ \hline -292.2 \end{Bmatrix} = \begin{Bmatrix} \{P\}_1^e \\ \hline \{P\}_2^e \end{Bmatrix}$$

Determination of Structure Displacements:

$$\{\Delta\}_1 = [K]_{11}^{-1} (\{P\}_1 - \{P\}_1^e - [K]_{12}\{\Delta\}_2) \tag{13.84}$$

$$\{\Delta\}_1 = \frac{1}{14EK} \begin{bmatrix} 2 & -1 \\ -1 & 4 \end{bmatrix} \left(\begin{Bmatrix} 0 \\ 0 \end{Bmatrix} - \begin{Bmatrix} -301.5 \\ 416.7 \end{Bmatrix} - 2EK \begin{Bmatrix} 0 \\ 3/l \end{Bmatrix}_{l=10m} (0.03) \right) \tag{0.03}$$

$$= \frac{1}{EK} \begin{Bmatrix} 72.8 \\ -140.6 \end{Bmatrix} + \begin{Bmatrix} 0.001\ 29 \\ -0.005\ 14 \end{Bmatrix}$$

or,

$$EK\{\Delta\}_1 = \begin{Bmatrix} 72.8 \\ -140.6 \end{Bmatrix} + EK \begin{Bmatrix} 0.001\ 29 \\ -0.005\ 14 \end{Bmatrix}$$

for $EK = 40$ MN $\cdot$ m $= 40\ 000$ kN $\cdot$ m, we obtain

$$EK\{\Delta\}_1 = \begin{Bmatrix} 72.8 \\ -140.6 \end{Bmatrix} + \begin{Bmatrix} 51.6 \\ -205.6 \end{Bmatrix} = \begin{Bmatrix} 124.4 \\ -346.2 \end{Bmatrix} \text{(kN} \cdot \text{m)}$$

Agrees with Example 13.6.1 ↗ Agrees with Example 13.6.2

$$\{\Delta\} = \begin{Bmatrix} \{\Delta\}_1 \\ \hline \{\Delta\}_2 \end{Bmatrix} = \begin{Bmatrix} \Delta_1 \\ \Delta_2 \\ \hline \Delta_3 \end{Bmatrix} = \begin{Bmatrix} \theta_b \\ \theta_c \\ \hline \Delta_b \end{Bmatrix} = \begin{Bmatrix} 0.003\ 11 \text{ rad.} \\ -0.008\ 66 \text{ rad.} \\ 0.03 \text{ m} \end{Bmatrix}$$

The member displacements can be determined from Eq. 13.41, and the member forces can then be found from Eq. 13.38.

13.12

Temperature Variation Problems

The flexural effects caused by temperature variations were introduced in Section 13.7. It was shown that a linear temperature gradient acts to introduce a set of fixed-end moments without the inducement of any accompanying fixed-end shears. These fixed-end forces are given by Eq. 13.33, in which the order of terms is the same as that given by Eq. 13.14.

For beam-type problems in which axial effects are not included, the solution proceeds according to the steps outlined in Section 13.10 with the fixed-end forces for the ith member, $\{F\}_i^f$, given by Eq. 13.33. The procedure is straightforward, and no example problems are included.

In frame-type problems, where axial effects are important, the member force–displacement relations must include axial components and the vector $\{F\}_i^f$ will include fixed-end axial forces. This topic is not treated here because the slope deflection method customarily includes only flexural effects; however, it is treated in detail in Chapter 17.

13.13

Problem Variations

In this section, a few variations in the problems already discussed will be treated. As a first example, the two-story frame shown in Fig. 13.9a is considered. For this structure, there are six unknown displacements, as shown in Fig. 13.9b.

The slope deflection method requires that the slope deflection equation be written for each member end in the structure. These equations give the end moments in terms of θ_b, θ_c, θ_d, θ_e, $\psi_{ab} = \psi_{fe} = \Delta_1/h_1$, and $\psi_{bc} = \psi_{ed} = (\Delta_2 - \Delta_1)/h_2$. The solution requires six equilibrium equations—four of these involve moment equilibrium at the joints where rotation occurs and two are related to shear equilibrium for each story that is free to sway. The first sway equation requires that the base shears in columns ab and fe equal 30^k, and the second sway equation requires that the base shears in columns bc and de are equal to 10^k.

In the matrix formulation, no special measures have to be taken to handle the two-story frame. The method illustrated in Problem 13.10.2 is applicable, and it would yield the total displacements with respect to the original position. These displacements would then lead to the final member-end forces.

As a second example, consider the bent with sloping legs shown in Fig. 13.10a. The deflected structure shown in Fig. 13.10b shows that there are three unknown displacement quantities.

A solution by the slope deflection method requires that a ψ term be evaluated for each member. As the frame sways laterally through the displacement Δ, each member experiences relative end displacements. Since axial distortions are ignored in the slope deflection method, the top of each column must displace normal to its original direction through a horizontal displacement of Δ, as shown in Fig.

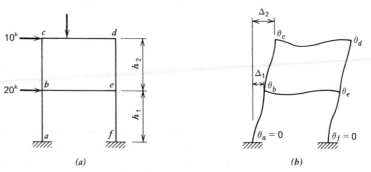

(a)   (b)

Fig. 13.9 *Two-story frame.*

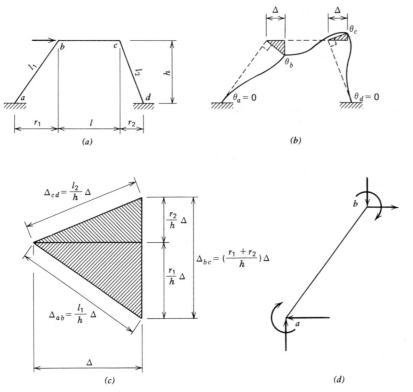

Fig. 13.10 *Sloping leg bent.*

13.10*b*. Thus, the shaded triangles at points *b* and *c* are similar to the triangles formed by the sloping columns and their projections. These shaded triangles are shown in an enlarged view in Fig. 13.10*c*, from which the following ψ terms are determined:

$$\psi_{ab} = \frac{\Delta_{ab}}{l_1} = \frac{(l_1/h) \cdot \Delta}{l_1} = \frac{\Delta}{h}$$

$$\psi_{cd} = \frac{\Delta_{cd}}{l^2} = \frac{(l_2/h) \cdot \Delta}{l^2} = \frac{\Delta}{h} \qquad (13.86)$$

$$\psi_{bc} = \frac{\Delta_{bc}}{l} = \frac{[(r_1 + r_2)/h]\Delta}{l} = \frac{r_1 + r_2}{lh} \Delta \quad \text{(negative)}$$

It is thus seen that all three ψ terms are a function of the sway displacement Δ.

Three equations of equilibrium are required for the solution of the three displacements. Two of these equations stem from moment equilibrium at joints *b* and *c*, and the third equation requires that the horizontal reactions at points *a* and *d* satisfy the requirements of statics for the entire structure.

This latter equation results from taking a free-body diagram of each column and expressing the horizontal base reaction in terms of the end moments on the column and the vertical load carried by the column. Figure 13.10*d* shows such a free-body diagram for column *ab*. The vertical load on the column is determined from the end shear on member *bc*.

The matrix method can be applied to the bent with sloping legs. However, the connectivity matrices for the columns [β]$_i$ must reflect the fact that the member-end displacements for the columns are not all codirectional with the structure coordinate displacements of Figs. 13.10b. Problems of this type are discussed in greater detail in Part III.

13.14

Multiple Loading Cases

The equilibrium analysis of multiple loading cases for pin-connected frameworks was discussed in Section 12.11, and the governing equation for such an analysis was given as Eq. 12.39. For beam and frame structures, the situation is complicated somewhat by the admittance of loads along the length of the members.

The solution for the structural displacements for r separate loading conditions can be given by an extension of Eq. 13.80 in the form

$$[\{\Delta\}_1\{\Delta\}_2 \cdots \{\Delta\}_j \cdots \{\Delta\}_r] = [K]^{-1}[\{P\}_1\{P\}_2 \cdots \{P\}_j \cdots \{P\}_r]$$

$$- [K]^{-1} \sum_{i=1}^{m} [\beta]_i^T [\{F\}_{i1}^f \{F\}_{i2}^f \cdots \{F\}_{ij}^f \cdots \{F\}_{ir}^f] \quad (13.87)$$

In this equation, the jth column of the rectangular array of displacements on the left-hand side of the equation gives the structure displacements $\{\Delta\}_j$ associated with the joint loads $\{P\}_j$ and the fixed-end forces $\{F\}_{ij}^f$. Once these displacements are determined, all other response quantities can be determined for each of the r separate loading cases.

13.15

Additional Reading

Ghali, A., and Neville, A. M., *Structural Analysis,* Chapters 3 and 11, Intext Educational Publishers, Scranton, Pa., 1972.

Hsieh, Yuan-Yu, *Elementary Theory of Structures,* Chapter 12, Prentice–Hall, Englewood Cliffs, N.J., 1970.

Kinney, J. S., *Indeterminate Structural Analysis,* Chapter 11, Addison–Wesley, Reading, Mass., 1959.

Parcel, J. I., and Moorman, R. B. B., *Analysis of Statically Indeterminate Structures,* Chapters 5 and 8, Wiley, New York, 1955.

Wang, Chu-Kia, *Matrix Methods of Structural Analysis,* 2nd Ed., Chapters 4 through 7, International Textbook, Scranton, Pa., 1970.

13.16

Suggested Problems

1 through 32. Analyze the given statically indeterminate structures by the slope deflection method and construct the shear and moment diagrams for all members. Take E as a constant for all problems.

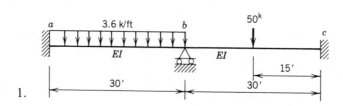

1.

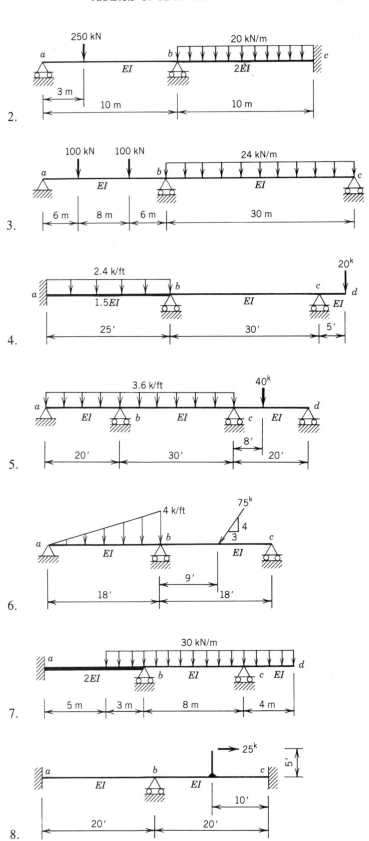

2.

3.

4.

5.

6.

7.

8.

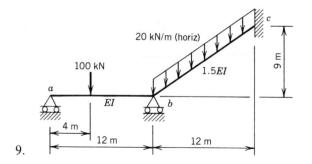

9.

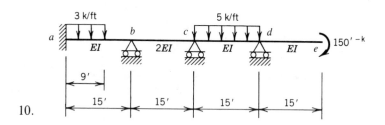

10.

11. The structure of Problem 4 with a settlement of 1.5 inches at support b if $E = 30 \times 10^3$ ksi and $I = 1,000$ in.^{4} Disregard the loads of Problem 4.

12. The structure and loading of Problem 6 with settlements at support a and support c of 0.75 and 0.35 inch, respectively, if $E = 30 \times 10^3$ ksi and $I = 1,200$ in.^{4}

13. The structure of Problem 9 if the support at point c settles 15 mm downward and rotates 0.000 7 radian counterclockwise. Table $EI = 500\ 000$ kN · m^2.

14.

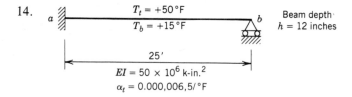

15. The structure of Problem 1 with the following temperature gradients:

   a. $T_t = -40°F$, $T_b = -10°F$ for span ab only.

   b. $T_t = -40°F$, $T_b = -10°F$ for both spans.
   For both cases, $E = 30 \times 10^6$ psi, $I = 1,500$ in.^{4}, $\alpha_t = 0.000,006,5/°F$ and $h = 15$ inches.

16. The structure of Problem 9 with temperature changes for span bc of $T_t = +30°C$ and $T_b = +5°C$. Take $I = 2\ 500 \times 10^{-6}$ m^4, $E = 200$ GPa, $\alpha_t = 0.000\ 012/°C$, and $h = 0.25$ m.

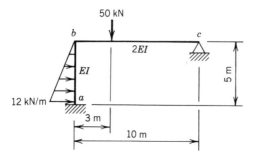

17.

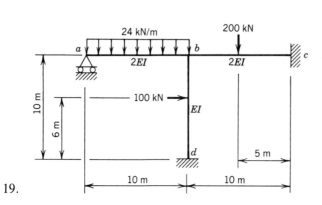

18.

19.

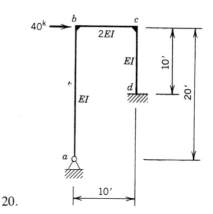

20.

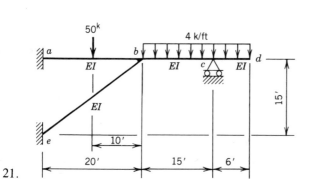

21.

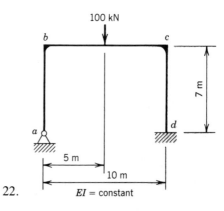

22.

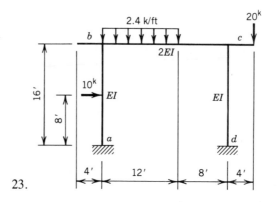

23.

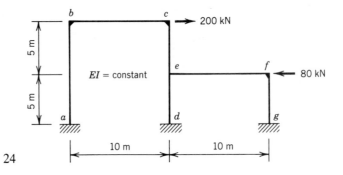

24

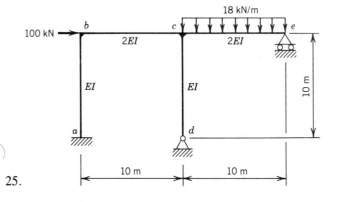

25.

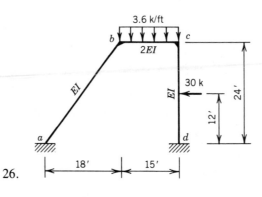

26.

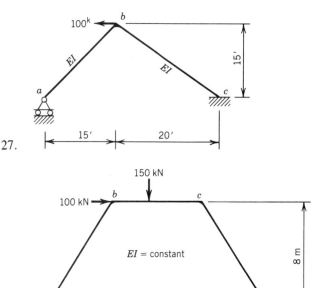

27.

28.

29. The structure of Problem 19 with a clockwise rotation of 0.001 radian at support c and a settlement of 0.03 m at support d if $E = 200 \times 10^9$ Pa and $I = 1\,000 \times 10^{-6}$ m⁴. Disregard the loads of Problem 19.

30. The structure of Problem 20 if, in addition to the loads, point a shifts 0.5 inch downward and 0.35 inch to the right. Take $EI = 35 \times 10^6$ k-in².

31. The structure of Problem 22 with the following temperature changes for members ab and bc:

$$T = +30\,°C$$
$$T = 25\,°C$$
$$T = +10\,°C$$

$$E = 200 \times 10^9 \text{ Pa}$$
$$I = 2\,000 \times 10^{-6} \text{ m}^4$$
$$\alpha_t = 0.000\,012/°C$$
Member depths $= 0.20$ m

32. The structure of Problem 18 with the following temperature changes for span ab:

$$T_t = -50°F, \qquad T_b = -15°F$$

Disregard the loads given in Problem 18, and take $EI = 50 \times 10^6$ k-in², $\alpha_t = 0.000,006,5/°F$, and $h = 18$ inches.

33 through 38. Use the matrix formulation of the slope deflection method to analyze the designated structures.

33. The structure and loading of Problem 1.

34. The structure and loading of Problem 4.

35. The structure and loading of Problem 11.

36. The structure and loading of Problem 19.

37. The structure and loading of Problem 23.

38. The structure and loading of Problem 32.

39 and 40. Generate the structure stiffness matrix [K] for the designated structures according to the matrix formulation.

39. The structure of Problem 24.

40. The structure of Problem 28.

41. Develop a computer program for the solution of statically indeterminate beams and frames by the method of Sections 13.8 and 13.9. Use the program to solve Problems 1, 3, 18, 22, and 25.

CHAPTER 14

*Sears Tower,
Chicago, Illinois
(courtesy of
Skidmore, Owings &
Merrill, architects/
engineers, photo by
Timothy Hursley).*

ANALYSIS OF BEAM AND FRAME STRUCTURES BY EQUILIBRIUM METHODS— ITERATIVE SOLUTIONS

14.1

Solution Techniques for Equilibrium Methods

In Chapter 13, equilibrium methods were considered for the analysis of beam and frame problems. For both the slope deflection method and the equivalent matrix formulation, a set of equilibrium equations resulted which were expressed in terms of deformation quantities. In solving these equations, the analyst is seeking the set of deformations that simultaneously satisfies all of the equilibrium conditions.

The methods for solving the equilibrium equations for the desired deformations were not discussed in Chapter 13; it was merely implied that a solution existed. In essence, the solution process requires inverting the stiffness matrix, or solving the associated simultaneous equations. For small problems, where there are only a few unknowns, simple methods of elimination or Cramer's rule can be applied. For large problems, numerous techniques are available for reaching a solution. Some of these methods are direct in that they produce the correct results after applying the appropriate algorithm. Other methods are iterative; that is, they are based on a step-by-step procedure. The initial step gives an approximation to the solution, and each subsequent step acts to refine the solution. The analyst can stop whenever the desired degree of accuracy is achieved.

In the above discussion, it is implied that the complete set of equilibrium equations is generated, and the solution centers on the problem of matrix inversion. However, there are iterative methods that lead to a solution without actually developing the complete set of equilibrium equations. It is this type of iterative solution that is considered in this chapter.

14.2

Iterative Solutions

As stated in the previous section, an iterative solution results from the application of a step-by-step procedure in which each step acts to refine the results from the previous step. In many cases, it can be shown that the iterative process converges to the correct solution if enough steps are taken. However, the analyst has the option of terminating the procedure when the desired accuracy has been achieved.

To illustrate an iterative solution, consider the continuous beam problem of Section 13.6.1. This beam is shown again in Fig. 14.1 along with its deflected structure, and the governing equations of equilibrium are

$$8EK\theta_b + 2EK\theta_c = 301.5 \tag{14.1}$$
$$2EK\theta_b + 4EK\theta_c = -416.7 \tag{14.2}$$

The solution requires a set of joint rotations θ_b and θ_c that will simultaneously satisfy the equilibrium equations.

The iterative technique that will be employed requires that each of the equilibrium equations be used to solve for one deformation quantity based on assumed values for the other deformation quantities. For convenience, Eqs. 14.1 and 14.2 are rewritten as follows:

$$EK\theta_b = 37.7 - 0.25(EK\theta_c) \tag{14.3}$$
$$EK\theta_c = -104.2 - 0.5(EK\theta_b) \tag{14.4}$$

As an initial step, it is assumed that $EK\theta_c = 0$ and Eq. 14.3 is used to solve for $EK\theta_b = 37.7$. Using $EK\theta_b = 37.7$, we then use Eq. 14.4 to solve for $EK\theta_c = -123.1$. This new value for $EK\theta_c$ is now used in Eq. 14.3 to obtain a new value for $EK\theta_b$. The procedure continues, as summarized in Table 14.1, until the same

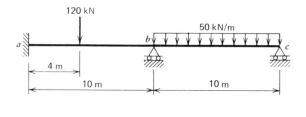

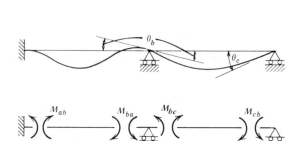

Fig. 14.1 *Example problem for iterative solution.*

$EK\theta$ values satisfy both equilibrium equations. The final results of $EK\theta_b = 72.8$ and $EK\theta_c = -140.6$ agree with those determined in Section 13.6.1.

The same procedure can be used when there are more than two unknowns. For instance, if there are n unknown deformations, each of the n equations should be solved for one of the unknowns in terms of the remaining $(n - 1)$ deformation quantities. Then each equation is used, in turn, to determine the value of one deformation quantity from the latest values of the other $(n - 1)$ deformations. The process continues until a consistent set of deformation quantities is obtained that satisfies all of the governing equilibrium equations.

The procedure outlined here is the so-called *Gauss–Seidel iteration method*. There are, of course, other techniques that can be employed for an iterative solution, and there are procedures that will hasten convergence. It will be seen in subsequent sections that the technique illustrates here has a convenient physical interpretation that attaches a special value to it.

TABLE 14.1 Example Iterative Solution

		Assumed Value		Solve for	
Step	Eq. No.	$EK\theta_b$	$EK\theta_c$	$EK\theta_b$	$EK\theta_c$
1	14.3	—	0	37.7	—
2	14.4	37.7	—	—	− 123.1
3	14.3	—	− 123.1	68.5	—
4	14.4	68.5	—	—	− 138.5
5	14.3	—	− 138.5	72.3	—
6	14.4	72.3	—	—	− 140.4
7	14.3	—	− 140.4	72.8	—
8	14.4	72.8	—	—	− 140.6
9	14.3	—	− 140.6	72.8	—

14.3

Physical Interpretation of Iterative Solutions

Each step of the iterative process developed in Section 14.2 has a physical significance. To illustrate this point, we will recall some of the equations used in the example problem of Section 13.6.1. The moment equations are

$$M_{ab} = 2(EK\theta_b) - 172.8$$
$$M_{ba} = 4(EK\theta_b) + 115.2$$
$$M_{bc} = 4(EK\theta_b) + 2(EK\theta_c) - 416.7 \tag{14.5}$$
$$M_{cb} = 4(EK\theta_c) + 2(EK\theta_b) + 416.7$$

where these are member-end moments, as shown in Fig. 14.1, and the equilibrium equations are

$$M_{ba} + M_{bc} = 0 \tag{14.6}$$
$$M_{cb} = 0 \tag{14.7}$$

Table 14.2 gives the values for $EK\theta_b$ and $EK\theta_c$ for each step of the iteration process as determined in Table 14.1. Also shown in Table 14.2 are the moments corresponding to each step as determined by Eqs. 14.5 with the appropriate $EK\theta$ values. An examination of the joint rotations and moments in Table 14.2 reveals the following:

- Initially, neither Eq. 14.6 nor Eq. 14.7 is satisfied. That is, $M_{ba} \neq -M_{bc}$ and $M_{cb} \neq 0$.

- From the initial step to Step 1, joint b is allowed to rotate while joint c remains clamped. In the process, Eq. 14.6 ($M_{ba} = -M_{bc}$) is satisfied[1] while Eq. 14.7 remains unsatisfied.

- From Step 1 to Step 2, joint b is clamped in a rotated position while joint c is permitted to rotate. As a result, Eq. 14.7 ($M_{bc} = 0$) is satisfied[1], and Eq. 14.6 is not satisfied.

- From Step 2 to Step 3, joint c is clamped in its rotated position while joint b is again permitted to rotate. This action satisfies[1] Eq. 14.6 ($M_{ba} = -M_{bc}$), but Eq. 14.7 is, again, not satisfied.

TABLE 14.2 Moments for Iterative Solution

Step	$EK\theta_b$	$EK\theta_c$	M_{ab}	M_{ba}	M_{bc}	M_{cb}
Initial	0	0	−172.8	+115.2	−416.7	+416.7
1	37.7	0	−97.4	+266.0	−265.9	+492.1
2	37.7	−123.1	−97.4	+266.0	−512.1	−0.3
3	68.5	−123.1	−35.8	+389.2	−388.9	+61.3
4	68.5	−138.5	−35.8	+389.2	−419.7	−0.3
5	72.3	−138.5	−28.2	+404.4	−404.5	+7.3
6	72.3	−140.4	−28.2	+404.4	−408.3	−0.3
7	72.8	−140.4	−27.2	+406.4	−406.3	+0.7
8,9	72.8	−140.6	−27.2	+406.4	−406.7	−0.1

[1]Values of appropriate moments are enclosed within broken lines of Table 14.2. Satisfaction is within a few tenths of a kN · m. Exact satisfaction is not acheived because of round-off error.

This pattern continues: At each step, one of the equilibrium conditions is satisfied, and one is not. In going to the next step, the joint where equilibrium is not satisfied is allowed to rotate; this produces equilibrium at this joint, but disturbs equilibrium at the other. The process continues until both joints are in equilibrium, at which time the correct moments are obtained.

This process of alternatively clamping and releasing joints until equilibrium is achieved at each joint is the physical process that is followed according to the iterative solution. This suggests a physical analog to the iterative procedure. Such a technique is called the *moment distribution method,* and it is developed in the next section.

14.4

Moment Distribution Method for Beam Problems

When the iterative-type solution outlined in the previous sections is carried out with emphasis on the physical process of alternatively clamping and releasing joints until equilibrium is achieved, it is known as the *moment distribution method.* This method was developed by Professor Hardy Cross, who taught it to his students at the University of Illinois in the late 1920s. The method was formally presented by Cross in published form in 1930. In the 1930s, the moment distribution method emerged as the dominant method for frame analysis, and it enjoyed this position through the 1950s. With the advent of the digital computer, direct solutions became feasible. Moment distribution remains an important method for hand solutions for small problems.

To illustrate the essential features of the moment distribution method, consider the two-span continuous beam shown in Fig. 14.2a. This beam is clamped at points a and c, but is free to rotate at point b. However, it is initially assumed

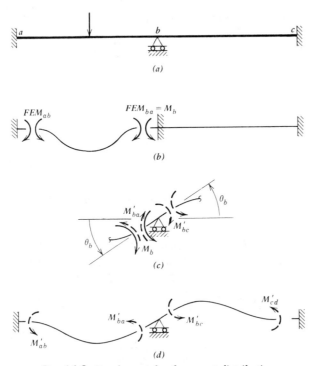

Fig. 14.2 *Fundamentals of moment distribution.*

that joint b is also clamped, and thus the loading on span ab causes fixed-end moments at ends a and b as shown in Fig. 14.2b. If joint b is released, it will rotate under the unbalanced moment M_b. As rotation occurs, end moments develop at the b end of members ab and bc. Rotation continues as shown in Fig. 14.2c until the summation of the end moments is equal and opposite to the unbalanced moment M_b. That is,

$$M'_{ba} + M'_{bc} = -M_b \tag{14.8}$$

where M'_{ba} and M'_{bc} are the end moments that develop as a result of the rotation θ_b at joint b.

The fundamental slope deflection equation, Eq. 13.19, can be used to express the end moments that develop from member-end rotations in the form

$$M'_{nf} = 2EK_{nf}(2\theta_n + \theta_f) \tag{14.9}$$

Since $\theta_a = \theta_c = 0$, the moments that develop at the b end of members ab and bc as joint b rotates through the angle θ_b are given by

$$\begin{aligned} M'_{ba} &= 2EK_{ba}\,(2\theta_b) = 4EK_{ba}\theta_b \\ M'_{bc} &= 2EK_{bc}\,(2\theta_b) = 4EK_{bc}\theta_b \end{aligned} \tag{14.10}$$

Substitution of Eq. 14.10 into Eq. 14.8 yields

$$4E\theta_b(K_{ba} + K_{bc}) = -M_b \tag{14.11}$$

from which

$$\theta_b = \frac{-M_b}{4E(K_{ba} + K_{bc})}$$

Substituting this expression for θ_b into Equations 14.10, we obtain

$$M'_{ba} = -M_b\left(\frac{K_{ba}}{K_{ba} + K_{bc}}\right) \tag{14.12}$$

and

$$M'_{bc} = -M_b\left(\frac{K_{bc}}{K_{ba} + K_{bc}}\right) \tag{14.13}$$

or, in general,

$$M'_{bi} = -M_b\left(\frac{K_{bi}}{\sum\limits_{j} K_{bj}}\right) \tag{14.14}$$

where i is the far end of a specific member framing into joint b, and j sums over all members connected to joint b. Equation 14.14 can be rewritten as

$$M'_{bi} = -M_b \cdot D_{bi} \tag{14.15}$$

where

$$D_{bi} = \frac{K_{bi}}{\sum\limits_{j} K_{bj}} \tag{14.16}$$

The quantity D_{bi} relates the stiffness of member bi to the summation of the stiffnesses of all members framing into joint b. This quantity is called the *distribution factor* for member bi, and it indicates the portion of the unbalanced moment M_b that is distributed to member bi as joint b is allowed to rotate. The negative sign on the right-hand side of Eq. 14.15 simply indicates that the moment M'_{bi} acts in opposition to the unbalanced moment M_b.

As joint b rotates and moments develop at the b end of all members framing into joint b, moments also develop at points a and c as shown in Fig. 14.2*d*. Using Eq. 14.9 and recalling that $\theta_a = \theta_c = 0$, we get

$$M'_{ab} = 2EK_{ab}(\theta_b) \qquad (14.17)$$
$$M'_{cb} = 2EK_{cb}(\theta_b)$$

Since $K_{bi} = K_{ib}$, a comparison of Eqs. 14.10 and 14.17 shows that

$$M'_{ab} = \tfrac{1}{2}M'_{ba} \qquad (14.18)$$
$$M'_{cb} = \tfrac{1}{2}M'_{bc}$$

or, in general,

$$M'_{ib} = \tfrac{1}{2}M'_{bi} \qquad (14.19)$$

Equation 14.19 can be expressed as

$$M'_{ib} = C_{bi} \cdot M'_{bi} \qquad (14.20)$$

where C_{bi} is called the *carry-over factor*. This expresses the factor that must be applied to M'_{bi} to determine the moment M'_{ib} that is "carried over" to the i end of member bi as joint b rotates. In this case,

$$C_{bi} = \tfrac{1}{2} \qquad (14.21)$$

For nonprismatic members and other special cases, the carry-over factor may have other values.

14.5 Numerical Moment Distribution Problems for Beam Structures

The analysis of a statically indeterminate beam by the moment distribution method follows a very systematic procedure. The initial step is to determine the stiffness, $K = I/l$, for each member throughout the structure. These stiffnesses are then used to determine the distribution factors in accordance with Eq. 14.16 at each joint that will be released during the moment distribution process. It is clear from Eq. 14.16 that only relative stiffness quantities are needed to determine the distribution factors. These relative stiffnesses simply show the relative magnitudes of the member stiffnesses for the members that frame into a joint. The fixed-end moments are then determined. These moments correspond to the situation in which all support joints are fixed against rotation. If there are joints that should not be fixed, the moment distribution process of sequential release, balance, and carry-over is followed until each released joint is in equilibrium.

The moment distribution accounting process registers member-end moments in accordance with the slope deflection sign convention. There are, of course, companion moments that act on the joints, but the method focuses on the member-end moments.

It should be noted that moment distribution is not an approximate method of analysis. If enough cycles are employed, the procedure will converge to the exact solution. In most cases, however, the analyst will stop when the carry-over moments are small when compared with the original fixed-end moments. In all cases, the procedure should not be terminated until all joints are balanced.

14.5.1 Example problem Determine the end moments for the structure of Example 13.6.1. This problem was also considered earlier in this chapter when iteration methods were discussed. The structure and loading are shown.

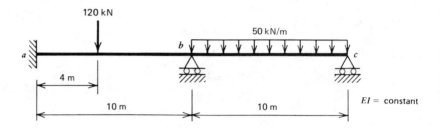

Stiffnesses and Relative Stiffnesses: Stiffness $K = I/l$; relative stiffness is circled.

$$K_{ab} = K_{ba} = \left(\frac{I}{l}\right)_{ab} = \left(\frac{I}{10}\right) = \textcircled{K}$$

$$K_{bc} = K_{cb} = \left(\frac{I}{l}\right)_{bc} = \left(\frac{I}{10}\right) = \textcircled{K}$$

Distribution Factors:

$$D_{bi} = \frac{K_{bi}}{\sum\limits_{j} K_{bj}} \tag{14.16}$$

At joint b:

$$D_{ba} = \frac{K_{ba}}{K_{ba} + K_{bc}} = \frac{K}{K + K} = 0.500$$

$$D_{bc} = \frac{K_{bc}}{K_{ba} + K_{bc}} = \frac{K}{K + K} = 0.500$$

At joint c:

$$D_{cb} = \frac{K_{cb}}{K_{cb}} = \frac{K}{K} = 1.000$$

Fixed-End Moments: (see Section 13.6.1)

$$FEM_{ab} = -172.8 \text{ kN} \cdot \text{m}$$
$$FEM_{ba} = +115.2 \text{ kN} \cdot \text{m}$$
$$FEM_{bc} = -416.7 \text{ kN} \cdot \text{m}$$
$$FEM_{cb} = +416.7 \text{ kN} \cdot \text{m}$$

Moment Distribution Process: The information computed above is displayed in a tabular arrangement as follows:

Rel. *K*			ⓚ				ⓚ	
D	Not Released			0.500	0.500		1.000	
FEM (kN · m)	− 172.8			+ 115.2	− 416.7		+ 416.7	

At this stage, all joints are fixed, and thus the support conditions at points *b* and *c* are not satisfied. If joint *b* is released, there is an unbalanced moment of (115.2 − 416.7) = − 301.5. Balancing moments develop in members *ba* and *bc* according to Eq. 14.15, and these are shown in our tabulation in the following manner:

———		0.500	0.500		1.000	
− 172.8		+ 115.2	− 416.7		+ 416.7	
		+ 150.7	+ 150.8			

Since the unbalanced moment is negative, the balancing moments are positive. A line is drawn beneath the balancing moments to indicate that the joint is balanced and that the sum of all the moments above the line at joint *b* is zero. As the balancing moments develop at the *b* end of the members framing into joint *b*, carry-over moments are introduced at the member ends at point *a* and *c* in accordance with Eq. 14.20. These carry-over moments are shown in the tabulation at the terminal of the arrows:

———		0.500	0.500		1.000	
− 172.8		+ 115.2	− 416.7		+ 416.7	
+ 75.4 ←		+ 150.7	+ 150.8 →		+ 75.4	

The equilibrium condition at point *b* is now satisfied, but joint *c* remains erroneously fixed. Joint *b* is now fixed in its rotated position, and joint *c* is released. The accumulated unbalanced moment at joint *c* is (+416.7 + 75.4) = +492.1; it is distributed according to Eq. 14.16, and a carry-over moment is induced at the *b* end of member *bc* as shown below.

———		0.500	0.500		1.000	
− 172.8		+ 115.2	− 416.7		+ 416.7	
+ 75.4 ———		+ 150.7	+ 150.8 →		+ 75.4	
			+ 246.1 ←		+ 492.1	

A line is drawn under the balancing moment to indicate that joint *c* is balanced. However, joint *b* is again unbalanced—the new unbalance of − 246.1 will be distributed to members *ba* and *bc* when joint *b* is released. In this case, for the distribution factors at joint *b* to be correct, joint *c* must be reclamped before joint *b* is released.

The process continues until joints b and c are both in a released state and there is no unbalance at either joint. At this stage, the governing support conditions and equilibrium conditions are satisfied, and the final member-end moments are determined by summing the individual contributions.

The complete moment distribution is shown below:

Rel. K D	——	Ⓚ 0.500	0.500	Ⓚ 1.000
FEM (kN · m)	− 172.8	+ 115.2	− 416.7	+ 416.7
	+ 75.4 ←	+ 150.7	+ 150.8 →	+ 75.4
			− 246.1 ←	− 492.1
---	+ 61.5 ←	+ 123.0	+ 123.1 →	+ 61.5 ---
			− 30.8 ←	− 61.5
	+ 7.7 ←	+ 15.4	+ 15.4 →	+ 7.7
			− 3.8 ←	− 7.7
	+ 1.0 ←	+ 1.9	+ 1.9 →	+ 1.0
			− 0.5	− 1.0
	+ 0.1 ←	+ 0.2	+ 0.3	
Final moments (kN · m)	− 27.1	+ 406.4	− 406.4	0

These final moments agree with those determined earlier in the example problems of Sections 13.6.1 and 13.10.1. It is also of interest to compare the moment distribution procedure with the iterative solution given in Table 14.1. At any interim point in the moment distribution, the summation of the accumulated moments will correspond to the moments at some interim step in the iterative solution. For instance, if the moments are summed up to the dashed lines at the margins of the moment distribution, the moments are $M_{ab} = -35.9$, $M_{ba} = +388.9$, $M_{bc} = -388.9$, $M_{cb} = +61.5$. These agree very closely with those given at Step 3 in Table 14.2. The final moments given in Table 14.2 are essentially the same as those determined in the moment distribution.

In a typical solution, the shear and moment diagrams would follow the moment distribution as was done in Section 13.6.1.

14.5.2 Support settlements

The moment distribution process directly accounts only for the moments associated with member-end rotations. This is evident if Eq. 13.19 is written in the form

$$M_{nf} = 2EK_{nf}(2\theta_n + \theta_f) - \frac{6EK_{nf}\Delta_{nf}}{l} + FEM_{nf} \qquad (14.22)$$

where Δ_{nf} is the relative transverse member-end displacement for the member nf, which has the length l. In the problem just considered, Δ_{nf} was zero for all members. The FEM_{nf} terms were determined initially, and the subsequent moment distribution process yielded the additional moments, which resulted from the member-end rotations θ_n and θ_f for all members.

If settlement is present, then the members where Δ_{nf} is not zero should be given initial fixed-end moments that reflect the effects of both loading and settlement. That is, the total fixed-end moment, FEM^T_{nf}, is

$$FEM^T_{nf} = FEM^S_{nf} + FEM_{nf} \qquad (14.23)$$

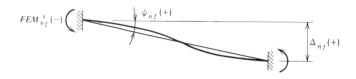

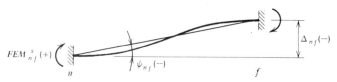

Fig. 14.3 *Fixed-end moments induced by settlement.*

where FEM^S_{nf} is the fixed-end moment induced by the relative settlement Δ_{nf} and is given by

$$FEM^S_{nf} = \frac{-6EK_{nf}\Delta_{nf}}{l} \tag{14.24}$$

and FEM_{nf} is the usual fixed-end moment that results from the loading on the member. The total fixed-end moments correspond to the condition in which the structure is fixed against rotation at all points but has the prescribed support settlements induced. The moment distribution process then accounts for the moments that result from the moment-end rotations.

It should be noted that Eqs. 14.22 and 14.24 indicate that negative end moments are induced by a positive Δ_{nf}, where a positive Δ_{nf} is associated with a clockwise chord rotation of $\psi_{nf} = \Delta_{nf}/l$. Conversely, a negative Δ_{nf}, which corresponds to a counterclockwise chord rotation ψ_{nf}, induces positive end moments. These relationships are illustrated in Fig. 14.3.

14.5.3 Example problem

Determine the end moments for the structure of Example 14.5.1 if, in addition to the loading, point b settles 0.03 m.

$$E = 200 \times 10^9 \text{ Pa} = 200 \text{ GN/m}^2; \quad I = 2\,000 \times 10^{-6} \text{ m}^4$$

Stiffnesses and Relative Stiffnesses: Same as for Example 14.5.1

Distribution Factors: Same as for Example 14.5.1

Fixed-End Moments:

From loading (same as for Example 14.5.1)

$$FEM_{ab} = -172.8 \text{ kN} \cdot \text{m}$$
$$FEM_{ba} = +115.2 \text{ kN} \cdot \text{m}$$
$$FEM_{bc} = -416.7 \text{ kN} \cdot \text{m}$$
$$FEM_{cb} = +416.7 \text{ kN} \cdot \text{m}$$

From settlement
Member ab:

$$K_{ab} = \left(\frac{I}{l}\right)_{ab} = \frac{2\ 000 \times 10^{-6}\ \text{m}^4}{10\ \text{m}} = 2\ 000 \times 10^{-7}\ \text{m}^3$$

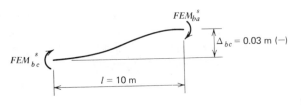

$l = 10\ \text{m}$

$$FEM_{ab}^S = FEM_{ba}^S = \frac{-6EK_{ab}\Delta_{ab}}{l}$$

$$= \frac{-6 \times 200 \times 10^9 \times 2\ 000 \times 10^{-7} \times 0.03}{10}$$

$$= -720\ 000\ \text{N} \cdot \text{m} = -720\ \text{kN} \cdot \text{m}$$

Member bc:

$$K_{bc} = \left(\frac{I}{l}\right)_{bc} = 2\ 000 \times 10^{-7}\ \text{m}^3$$

FEM_{ba}^s

$\Delta_{bc} = 0.03\ \text{m}\ (-)$

FEM_{bc}^s

$l = 10\ \text{m}$

$$FEM_{bc}^S = FEM_{cb}^S = \frac{-6EK_{bc}\Delta_{bc}}{l} = +720\ \text{kN} \cdot \text{m}$$

Moment Distribution Process:

		$\textcircled{K}$			$\textcircled{K}$	
Rel. K						
D	—		0.500	0.500		1.000
FEM	-172.8		$+115.2$	-416.7		$+\ 416.7$
FEM^S	-720.0		-720.0	$+720.0$		$+\ 720.0$
FEM^T(kN · m)	-892.8		-604.8	$+303.3$		$+1136.7$
				-568.4 ←		-1136.7
	$+217.5$ ←		$+435.0$	$+434.9$	→	$+\ 217.5$
				-108.8 ←		$-\ 217.5$
	$+\ 27.2$ ←		$+\ 54.4$	$+\ 54.4$	→	$+\ \ \ 27.2$
				$-\ 13.6$ ←		$-\ \ \ 27.2$
	$+\ \ \ 3.4$ ←		$+\ \ \ 6.8$	$+\ \ \ 6.8$	→	$+\ \ \ \ 3.4$
				$-\ \ \ 1.7$ ←		$-\ \ \ \ 3.4$
			$+\ \ \ 0.9$	$+\ \ \ 0.8$		
Final						
moments	-644.7		-108.7	$+108.7$		0
(kN · m)						

These final moments compare favorably with those obtained from superimposing the results of Example 13.6.1 and 13.6.2.

The example problems presented thus far have followed the distribution procedure **14.5.4** based on the sequence of releasing one joint at a time. Actually, it is possible to **Alternative** release several joints simultaneously, balance each, and then make multiple carry- **distribution** overs. This procedure does not represent a definite sequence of physical opera- **sequence** tions, but it does possess the proper mathematical convergence.

The technique is illustrated by repeating the distribution process for the problem presented in Section 14.5.1. Here, the moment distribution takes the following form:

		0.500	0.500		1.000
FEM(kN · n)	− 172.8	+ 115.2	− 416.7		+ 416.7
	+ 75.4 ⟵	+ 150.7	+ 150.8		− 416.7
			− 208.4		+ 75.4
	+ 52.1 ⟵	+ 104.2	+ 104.2		− 75.4
			− 37.7		+ 52.1
	+ 9.5 ⟵	+ 18.9	+ 18.8		− 52.1
			− 26.1		+ 9.4
	+ 6.6 ⟵	+ 13.1	+ 13.0		− 9.4
			− 4.7		+ 6.5
	+ 1.2 ⟵	+ 2.4	+ 2.3		− 6.5
			− 3.3		+ 1.2
	+ 0.9 ⟵	+ 1.7	+ 1.6		− 1.2
			− 0.6		+ 0.8
Final		+ 0.3	+ 0.3		− 0.8
moments	− 27.1	+ 406.5	− 406.5		0
(kN · m)					

This sequence of distribution is sometimes referred to as the "cross-carry-over" procedure, for obvious reasons. In the example shown, the convergence of the cross-carry-over approach is slower than the normal procedure. However, there are cases where this approach will speed convergence, and this is the main reason for employing the method.

As was shown in Section 13.7, the flexural effects of temperature change are **14.5.5** manifested through the introduction of fixed-end moments. These moments are **Temperature** expressed in the form of Eq. 13.34 and are included with those from other effects, **effects** such as member loads and/or settlement, to determine the total fixed-end moments. The moment distribution procedure then advances according to the normal process to detemine the final member-end moments.

It should be noted that the axial effects of temperature change are not accounted for by this approach. For some structures, such as statically indeterminate frames, the axial distortions of some members induce flexure in other members, and the approach outlined above is not suitable.

14.6

Certain situations lend themselves to modifications in the moment distribution pro- **Modifications** cedure. One situation is which a modified approach is possible is where an exte- **in Moment** rior span is simply supported at its exterior support point. **Distribution**

Consider again the structure given in Fig. 14.2a, but this time let point c be **Procedure** simply supported as shown in Fig. 14.4a. If point b is initially clamped against rotation, fixed-end moments develop in span ab as shown in Fig. 14.4b. Once again, if joint b is released, it will rotate under the unbalanced moment M_b, and equilibrating moments M'_{ba} and M'_{bc} will be induced along with the corresponding carry-over moments as shown in Fig. 14.4c. However, in this case, $M'_{cb} = 0$, and thus Eq. 14.9 gives $M'_{cb} = 2EK_{bc}(2\theta_c + \theta_b) = 0$, from which $\theta_c = -\theta_b/2$. This value for θ_c, along with $\theta_a = 0$, when substituted into Eq. 14.9, gives

$$M'_{ba} = 2EK_{ba}(2\theta_b) = 4EK_{ba}\theta_b \tag{14.25}$$

and

$$M'_{bc} = 2EK_{bc}\left(2\theta_b - \frac{\theta_b}{2}\right) = 3EK_{bc}\theta_b \tag{14.26}$$

The moment M'_{bc} can be written in the form

$$M'_{bc} = 4E(\tfrac{3}{4}K_{bc})\theta_b = 4EK^m_{bc}\theta_b \tag{14.27}$$

where K^m_{bc} is the modified stiffness of member bc and is given by

$$K^m_{bc} = \tfrac{3}{4}K_{bc} \tag{14.28}$$

Since the unbalanced moment M_b is again equilibrated by M'_{ba} and M'_{bc}, Eq. 14.8 remains valid. Substitution of Eqs. 14.25 and 14.27 into Eq. 14.8 gives

$$4E\theta_b(K_{ba} + K^m_{bc}) = -M_b \tag{14.29}$$

Solving for θ_b and substituting the resulting expression into Eqs. 14.25 and 14.27, we obtain

$$M'_{ba} = -M_b\left(\frac{K_{ba}}{K_{ba} + K^m_{bc}}\right)$$

and $\tag{14.30}$

$$M'_{bc} = -M_b\left(\frac{K^m_{bc}}{K_{ba} + K^m_{bc}}\right)$$

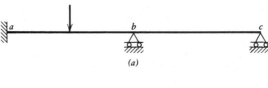

(a)

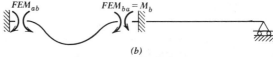

(b)

(c)

Fig. 14.4 *Modified stiffness.*

for the respective response functions in Fig. 15.1c. To construct the complete set of influence lines according to this procedure, a separate indeterminate analysis is essential for each position of the unit load. Thus, the computational effort required to generate influence lines for statically indeterminate structures is substantially greater than that required for statically determinate structures. Nevertheless, influence lines are extremely useful, especially in bridge design, and modern computers provide the computational power needed to carry out the repetitious analyses required to produce influence lines by this fundamental approach. However, other methods are available that are computationally less intensive and conceptually more elegant. The alternative approaches that are used are developed in the sections that follow.

Another point that is worthy of note concerns the shape of the influence lines. In Chapter 7, it was seen that the influence lines for statically determinate structures are composed entirely of straight-line segments. However, the influence lines for statically indeterminate structures have curved segments, as shown in Fig. 15.1c. The reason for this difference will become clear as subsequent sections are studied.

15.2 Müller-Breslau Principle

The *Müller-Breslau principle* was developed in Section 7.4 and it was applied there to generate influence lines for statically determinate structures. The principle evolves from a direct application of Betti's law, which stems from a unique application of virtual work concepts. Recall the formal statement of the principle: *The influence line for any response function is given by the deflection curve that results when the restraint corresponding to that response function is removed and a unit displacement is introduced in its place.*

In Chapter 7, since statically determinate structures were exclusively treated, to remove a restraint was to render the structure unstable. Thus, the unit displacement could be induced without any resistance; the unstable structure merely responded with the rigid-body motion of a linkage-type mechanism. For this reason, the influence lines for statically determinate structures are composed of straight-line segments. However, when a restraint of a statically indeterminate structure is removed, the degree of indeterminacy is reduced by one, but the remaining structure is a stable structure. In this case, the inducement of the unit displacement is resisted. As the structure responds, the deflected shape is curved; thus, the influence lines have curved segments.

In Chapter 7, the Müller-Breslau principle was used primarily as a qualitative tool for verifying the shapes of influence lines. The qualitative value remains as one of the major features of the principle for statically indeterminate structures. This is illustrated by the three influence lines in Fig. 15.1c. If the restraints associated with R_A, V_D, and M_D are removed, in turn, and if a unit displacement is introduced in their place, the shapes of the respective influence lines are verified. However, in addition to the qualitative property, the Müller-Breslau principle can be used in a quantitative fashion. This can be done by actually building a model of the structure and mechanically generating an influence line by introducing the appropriate unit displacement and measuring the ordinates of the deflected structure. Or, analytic procedures may be used to calculate the ordinates of the deflected structure that result when the unit displacement is introduced. Here, any of the methods that have been studied for determining the deflected configuration of a structure is appropriate; however, certain methods have inherent advantages for certain classes of structures, as will be illustrated in the following section.

As was explained in Section 7.4, it is imperative that the induced deflected shape contain no displacement discontinuities other than the unit displacement corresponding to the response function for which the influence line is being constructed.

15.3

Influence Lines for Continuous Beams by the Müller-Breslau Principle

The application of the Müller-Breslau principle to a continuous beam reduces to the determination of the ordinates to the deflected structure. Since influence line ordinates are generally required at regular intervals along the structure, it is essential that a deflection analysis method be selected to meet these needs. One convenient approach is to couple the conjugate beam method with the tabular procedure suggested by Newmark. An example of this approach was given in Section 6.7.

Before proceeding with a numerical problem, the method will be described conceptually. Consider the structure shown in Fig. 15.2a, and assume that the influence line for R_A is desired. According to the provisions of the Müller-Breslau principle, the restraint at point A is removed and a unit displacement is introduced in its place. This unit displacement is induced through the application of an upward force at point A, as shown in Fig. 15.2b. This loading gives rise to the moment diagram of Fig. 15.2c, which is converted to the M/EI loading on the conjugate beam of Fig. 15.2d. The moment diagram for this conjugate beam and loading is given in Fig. 15.2e, which is actually the deflection diagram for the given beam and loading of Fig. 15.2b. Since there is no way of knowing in advance the magnitude of the load that is required in Fig. 15.2b to produce the desired unit displacement, some arbitrary value is assumed. This is normally accomplished by assuming some arbitrary value for M_B in Fig. 15.2c. The resulting deflection diagram, which has the deflection Δ_A at point A, is normalized by dividing all ordinates by Δ_A to produce the desired influence line for R_A.

Since this continuous beam is statically indeterminate to the first degree, once the influence line for R_A has been established, it can be used to obtain the influence lines for other response functions. For instance, consider the influence line for R_B. The structure is redrawn in Fig. 15.3a, with a unit load acting at point x. Taking moments about point C, we have

$$R_A(l_1 + l_2) + R_B(l_2) - 1(l_1 + l_2 - x) = 0 \tag{15.1}$$

from which

$$R_B = \frac{(l_1 + l_2 - x)}{l_2} - R_A \frac{(l_1 + l_2)}{l_2} \tag{15.2}$$

Thus, the influence line for R_B is formed by a superposition of a straight line and the influence line for R_A times a constant, as is illustrated in Fig. 15.3b.

Similarly, the influence lines for V_D and M_D can be determined from the influence line for R_A. Specifically, for a free-body diagram left of point D, we have

$$V_D = R_A - 1 \tag{15.3}$$

for $x < d$, or

$$V_D = R_A \tag{15.4}$$

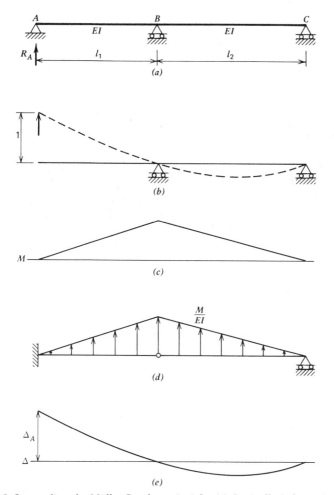

Fig. 15.2 *Influence lines by Müller-Breslau principle.* (a) *Statically indeterminate beam.*
(b) *Displacement induced in direction of* R_A. (c) *Moment diagram for induced
displacement.* (d) *Conjugate beam and loading.* (e) *Moment diagram for conjugate beam.*

where $x \gtrless d$. For the same free-body diagram, M_D is given by

$$M_D = d\left[R_A - \left\{1 - \frac{x}{d}\right\}\right] \tag{15.5}$$

where the quantity $\{1 - x/d\}$ is taken as zero if it is negative. It follows that the
influence lines for V_D and M_D are formed by superposition relationships that involve the influence line for R_A. These combinations are shown in Figs. 15.3c and
15.3d.

Equations 15.2 through 15.5 can be used to determine the ordinates for the
desired influence lines through a simple extension of Newmark's numerical table.
This procedure is illustrated in the first of the example problems that follow.

Of course, the influence lines for R_B, V_D, and M_D could be determined
through a direct application of the Müller-Breslau principle. The second of the
example problems that follow illustrates such an application.

Structures with additional spans can also be treated. In such cases, the structure that remains after a support is removed is statically indeterminate. Thus, a

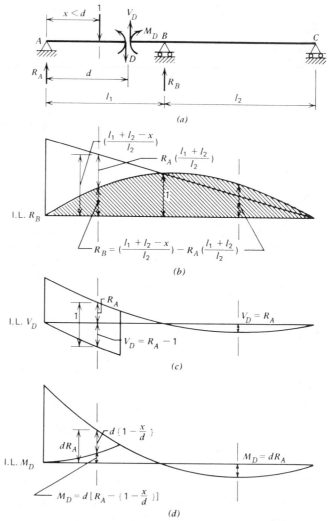

Fig. 15.3 *Formulation of influence lines. (a) Beam with unit loading. (b) Influence line for R_B. (c) Influence line for V_D. (d) Influence line for M_D.*

complete indeterminate analysis is necessary to determine the loading on the conjugate beam. The conjugate beam, however, is always determinate, and thus the deflected structure (influence line) can readily be determined. For such cases, the determination of one influence line is no longer sufficient for the determination of other influence lines. Instead, a number of influence lines must be established through the regular Müller-Breslau principle so that statics can be used to obtain the others that are required.

The reader is also reminded that the quantity EI need not be constant. Variations in EI must be reflected in the M/EI diagram; however, the basic procedure remains unaltered.

15.3.1 Example problem   Use the Müller-Breslau principle to determine the influence line for the reaction at point A for the two-span beam shown below. Use the resulting influence line to determine the influence lines for the reaction at point B and the shear and moment at point D. It is required that the ordinates for each influence line be determined at 10-foot intervals.

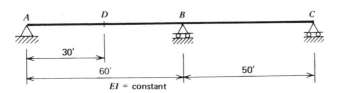

EI = constant

Influence Line for Reaction at Point A: To determine the influence line for the reaction at point A, the vertical restraint is removed at that point, and a unit displacement is induced through the introduction of an upward concentrated load. Thus, the structure whose deflected shape must be determined is the following:

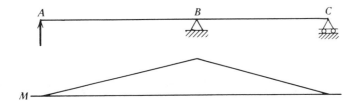

The shape of the moment diagram is established by the loading. The magnitude of the moment at point B, $M_B = +60$ ft-k, is arbitrarily selected.

													Units
M	0	+10	+20	+30	+40	+50	+60	+48	+36	+24	+12	0	ft-k

Conjugate Beam and Loading:

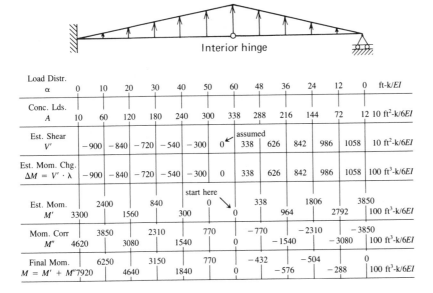

Interior hinge

Load Distr. α	0	10	20	30	40	50	60	48	36	24	12	0	ft-k/EI
Conc. Lds. A	10	60	120	180	240	300	338	288	216	144	72	12	10 ft²-k/6EI
Est. Shear V'		−900	−840	−720	−540	−300	0 assumed	338	626	842	986	1058	10 ft²-k/6EI
Est. Mom. Chg. $\Delta M = V' \cdot \lambda$		−900	−840	−720	−540	−300	0	338	626	842	986	1058	100 ft³-k/6EI
Est. Mom. M'	3300	2400	1560	840	300	0 start here	0	338	964	1806	2792	3850	100 ft³-k/6EI
Mom. Corr M''	4620	3850	3080	2310	1540	770	0	−770	−1540	−2310	−3080	−3850	100 ft³-k/6EI
Final Mom. $M = M' + M''$	7920	6250	4640	3150	1840	770	0	−432	−576	−504	−288	0	100 ft³-k/6EI

The final moments for the conjugate beam are the deflections of the real beam. These deflections are the ordinates of the influence line for the reaction at point A when they are normalized so that the ordinate at point A is unity. Thus, dividing by 7,920 (100 ft^3-k/6EI), we have

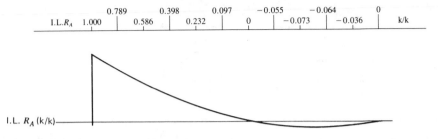

I.L. R_A (k/k)

Influence Line for Reaction at Point B:

$$R_B = \left(\frac{100 - x}{50}\right) - R_A\left(\frac{110}{50}\right) \tag{15.2}$$

where x is measured from point A.

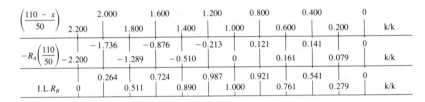

I.L. R_B (k/k)

Influence Line for Shear at Point D:

$$V_D = R_A - 1 \qquad (x < 30 \text{ ft}) \tag{15.3}$$
$$V_D = R_A \qquad (x \geq 30 \text{ ft}) \tag{15.4}$$

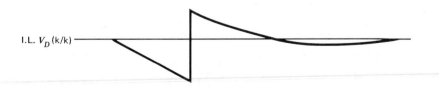

I.L. V_D (k/k)

Influence Line for Moment at Point D:

$$M_D = 30\left[R_A - \left\{1 - \frac{x}{30}\right\}\right] \tag{15.5}$$

zero if negative

$30R_A$	30.00	23.67 17.58	11.94 6.96	2.91 0	-1.65 -2.19	-1.92 -1.08	0	ft-k/k
$-30\left\{1 - \dfrac{x}{30}\right\}$	-30.00	-20.00 -10.00	0.00					ft-k/k
I.L.M_D	0	3.67 7.58	11.94 6.96	2.91 0	-1.65 -2.19	-1.92 -1.08	0	ft-k/k

I.L. M_D (ft-k/k)

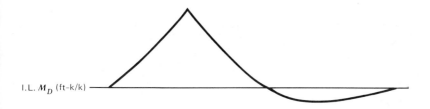

Use a direct application of the Müller-Breslau principle to determine the influence **15.3.2 Example** line for the moment at point D for the given structure. Calculate the ordinates at **problem** 10-foot intervals, and compare the results with those determined in Example 15.3.1.

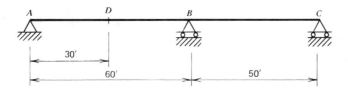

Influence Line for Moment at Point D: To determine the influence line for the moment at point D, the rotation restraint is removed at that point, and a unit rotational displacement is induced (without any relative transverse displacement) through the introduction of an internal moment and companion shear. Thus, the structure whose deflected shape must be determined is the following:

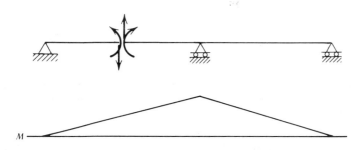

The shape of the moment diagram is established by the loading. The magnitude of the moment at point B, $M_B = +60$ ft-k, is arbitrarily selected. This is the same moment diagram that was used in Example 15.3.1.

M	0	10	20	30	40	50	60	48	36	24	12	0	ft-k

Conjugate Beam and Loading:

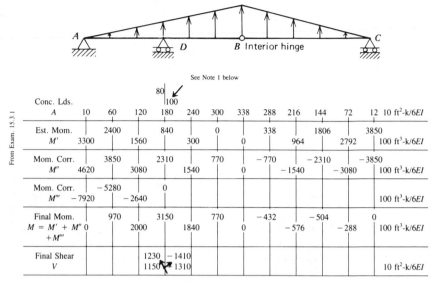

See Note 1 below

From Exam. 15.3.1

Conc. Lds.	A	10	60	120	180 / 80,100	240	300	338	288	216	144	72	12	10 ft²-k/6EI
Est. Mom.		2400		840		0		338		1806		3850		10 ft³-k/6EI
M'	3300		1560		300		0		964		2792			100 ft³-k/6EI
Mom. Corr.		3850		2310		770		-770		-2310		-3850		100 ft³-k/6EI
M"	4620		3080		1540		0		-1540		-3080			100 ft³-k/6EI
Mom. Corr.	-5280		0											
M'''	-7920		-2640											100 ft³-k/6EI
Final Mom.		970		3150		770		-432		-504		0		
M = M' + M"	0		2000		1840		0		-576		-288			100 ft³-k/6EI
+ M'''														
Final Shear				1230 / -1410										
V				1150 ⤢ 1310										10 ft²-k/6EI

ΔV = -2,640(10 ft²-k/6EI) See Note 2 below.

Notes:

1. At point D, total concentrated load of $+180$ is composed of $+80$ from panel to left and $+100$ from panel to right.

2. Final average panel shears are $+1,150$ and $-1,310$. The concentrated loads of $+80$ and $+100$ are applied to obtain shears of 1,230 just left of point D and $-1,410$ just right of point D. Thus, there is an abrupt change in shear of $-2,640$ at point D.

The final moments for the conjugate beam are the deflections of the real beam. These deflections are the ordinates of the influence line for the moment at D when they are normalized so that the angle change (shear change on conjugate beam) at point D is negative one. Thus, dividing by 2,640 (10ft²-k/6EI), we have

I.L.M_D	0	3.67	7.58	11.93	6.97	2.92	0	-1.64	-2.18	-1.91	-1.09	0	ft-k/k

These ordinates agree well with those determined in Example 15.3.1.

15.4

Influence Lines for Continuous Trusses

Influence lines for statically indeterminate trusses can be determined in several ways. None of these methods is developed here in detail; instead, a brief description of the various approaches is given. In all instances, it is convenient to identify the redundant reaction components and member forces and to determine the influence lines for these response functions. The influence lines for the remaining reaction components and member forces can then be determined from statics.

One approach is based on the fundamental definition of an influence line. That is, a unit load is sequentially placed at each panel point along the load way to which the live load is applied, and the redundants are determined for each separate unit load case. The determination of the redundants proceeds according to the methods outlined in Chapter 10. This approach is somewhat tedious, especially for large trusses, but it can be greatly simplified through an orderly arrangement of the calculations.

Another approach is based on the Müller-Breslau principle. Here, the restraint associated with the response quantity for which the influence line is being constructed is removed, and a unit displacement is induced in its place. The resulting displacement configuration of the load way gives the influence line. The procedures given in Chapter 5 are used to determine the required displacement quantities, and if the released structure is indeterminate, the member forces must be established by the methods given in Chapter 10. It can be shown that, although conceptually different, this method involves precisely the same calculations as does the fundamental approach that was previously described.

A modification of the Müller-Breslau principle, coupled with the conjugate beam method, provides yet another approach that is particularly useful for externally redundant trusses. It is recalled that the loading on a conjugate beam is composed of angle changes. There are techniques available, such as the one given by Johnson, Bryan, and Turneaure (see Section 15.8), that enable the analyst to determine the angle changes which occur at the panel points along the load way of the truss. Thus, a reaction component of the original truss is released, and a unit displacement is induced in its place. The panel point angle changes are computed and are applied to the corresponding conjugate truss, and the resulting deflection pattern of the load way gives the desired reaction influence line.

For certain kinds of continuous trusses, which are fairly uniform in terms of member arrangement, it is possible to establish an equivalent moment of inertia for the truss and to use the methods that have already been described in Section 15.3 for continuous beams. The moment of inertia may vary along the span as the individual chord members change, but this will be reflected in the M/EI values of the conjugate beam loading.

15.5 Matrix Formulation for the Generation of Influence Lines

In Section 7.5, a simple matrix method was given for the construction of influence lines for statically determinate structures. The fundamental equation has the form

$$\{F_R\} = [b]_{F_R P}\{P\} \qquad (15.6)$$

where $\{F_R\}$ represents specific response functions, $\{P\}$ is the load vector, and $[b]_{F_R P}$ is the influence matrix that relates the loads to the response functions. Extracting the equation for F_{Ri} from the above, we have

$$F_{Ri} = [b_{i1} \quad b_{i2} \quad \cdots \quad b_{ij} \quad \cdots \quad b_{ir}] \begin{Bmatrix} P_1 \\ P_2 \\ \cdot \\ \cdot \\ \cdot \\ P_j \\ \cdot \\ \cdot \\ \cdot \\ P_r \end{Bmatrix} \qquad (15.7)$$

In this form, it is clear that the ith row of $[b]_{F_R P}$ contains the influence line ordinates for F_{Ri} as the unit load occupies points corresponding to the load positions 1 through r. This fundamental approach remains valid for statically indeterminate structures; however, in such cases, statics is no longer sufficient for the determination of the elements of the influence matrix.

The present discussion will be restricted to continuous beam structures, and the elements of $\{F_R\}$ will be limited to member-end moments. These are not considered to be serious constraints. Most structures that are subjected to moving-load arrangements are beam-type structures, and once the influence lines for the member-end moments are determined, all other influence lines of interest can be found.

Consider the n-span continuous beam shown in Fig. 15.4. There are $2n$ member-end moments, of which M_i is typical. Collectively, these moments form the $2n$ elements of the moment vector $\{M\}$. The influence line ordinates are desired at the r load points indicated on the sketch; thus, there are r possible load points, and the loads corresponding to these points are assembled in the load vector $\{P\}$. The objective is to formulate the expression

$$\{M\} = [b]_{MP}\{P\} \tag{15.8}$$

where the ith row of $[b]_{MP}$ consists of the influence line for the end moment M_i corresponding to the unit load's placement at the load points 1 through r.

The ith member-end moment that is caused by P_j is given by the relationship

$$M_{ij} = [m_{i1} \quad m_{i2} \quad \cdots \quad m_{ik} \quad \cdots \quad m_{i,2n}] \begin{Bmatrix} M^f_{1j} \\ M^f_{2j} \\ \cdot \\ \cdot \\ M^f_{kj} \\ \cdot \\ \cdot \\ M^f_{2n,j} \end{Bmatrix} \tag{15.9}$$

where m_{ik} is the value of the ith member-end moment that is induced through the introduction of a unit fixed-end moment corresponding to the kth member-end

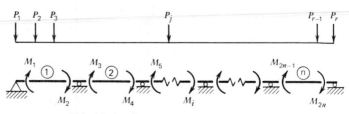

Fig. 15.4 *Moment and load identification.*

moment, and M^f_{kj} is the actual kth fixed-end moment that is caused by the load P_j. For the full array of end moments, Eq. 15.9 becomes

$$
\begin{Bmatrix} M_{1j} \\ M_{2j} \\ \cdot \\ \cdot \\ \cdot \\ M_{ij} \\ \cdot \\ \cdot \\ \cdot \\ M_{2n,j} \end{Bmatrix}
=
\begin{bmatrix}
m_{11} & m_{12} & \cdots & m_{1k} & \cdots & m_{1,2n} \\
m_{21} & m_{22} & \cdots & m_{2k} & \cdots & m_{2,2n} \\
\cdot & \cdot & & & & \cdot \\
\cdot & \cdot & & & & \cdot \\
\cdot & \cdot & & & & \cdot \\
m_{i1} & m_{i2} & \cdots & m_{ik} & \cdots & m_{i,2n} \\
\cdot & \cdot & & & & \cdot \\
\cdot & \cdot & & & & \cdot \\
\cdot & \cdot & & & & \cdot \\
m_{2n,1} & m_{2n,2} & \cdots & m_{2n,k} & \cdots & m_{2n,2n}
\end{bmatrix}
\begin{Bmatrix} M^f_{1j} \\ M^f_{2i} \\ \cdot \\ \cdot \\ \cdot \\ M^f_{kj} \\ \cdot \\ \cdot \\ \cdot \\ M^f_{2n,j} \end{Bmatrix}
\qquad (15.10)
$$

In an abbreviated form, Eq. 15.10 becomes

$$\{M_j\} = [m]\{M^f_j\} \qquad (15.11)$$

where $\{M_j\}$ gives the final member-end moments that result from the load P_j, which introduces the fixed-end moments $\{M^f_j\}$. The square matrix $[m]$ gives the member-end moments that result from unit fixed-end moments. For instance, the kth column gives the $2n$ member-end moments that result from a simple moment distribution that starts with a unit value of the fixed-end moment at the kth member end.

The fixed-end moment vector $\{M^f_j\}$ can be given in the form

$$
\{M^f_j\} =
\begin{Bmatrix} m^f_{1j} \\ m^f_{2j} \\ \cdot \\ \cdot \\ \cdot \\ m^f_{kj} \\ \cdot \\ \cdot \\ \cdot \\ m^f_{2n,} \end{Bmatrix}
P_j = \{m^f_j\}P_j
\qquad (15.12)
$$

where m^f_{kj} is the kth fixed-end moment that results from a unit value of P_j, and the vector $\{m^f_j\}$ represents the complete set of fixed-end moments that are caused by a unit value of P_j.

Substitution of Eq. 15.12 into Eq. 15.11 gives

$$\{M_j\} = [m]\{m^f_j\}P_j \qquad (15.13)$$

This equation expresses the final member-end moments that result from the load P_j in terms of that load.

For all loads on the structure, Eq. 15.12 can be expanded to

$$\begin{Bmatrix} M_1^f \\ M_2^f \\ \cdot \\ \cdot \\ M_k^f \\ \cdot \\ M_{2n}^f \end{Bmatrix} = \begin{bmatrix} m_{11}^f & m_{12}^f & \cdots & m_{1j}^f & \cdots & m_{1r}^f \\ m_{21}^f & m_{22}^f & \cdots & m_{2j}^f & \cdots & m_{2r}^f \\ \cdot & \cdot & & \cdot & & \cdot \\ \cdot & \cdot & & \cdot & & \cdot \\ m_{k1}^f & m_{k2}^f & \cdots & m_{kj}^f & \cdots & m_{kr}^f \\ \cdot & \cdot & & \cdot & & \cdot \\ m_{2n,1}^f & m_{2n,2}^f & \cdots & m_{2n,j}^f & \cdots & m_{2n,r}^f \end{bmatrix} \begin{Bmatrix} P_1 \\ P_2 \\ \cdot \\ \cdot \\ P_j \\ \cdot \\ P_r \end{Bmatrix} \qquad (15.14)$$

Or, in a simplified form, this relationship becomes

$$\{M^f\} = [m^f]\{P\} \qquad (15.15)$$

where $\{M^f\}$ contains the total fixed-end moments that result from all loads on the structure, and the jth column of $[m^f]$ is $\{m_j^f\}$.

Therefore, the expanded form of Eq. 15.13 becomes

$$\{M\} = [m][m^f]\{P\} \qquad (15.16)$$

where $\{M\}$ includes the member-end moments that result from all the loads on the structure.

Comparing Eqs. 15.8 and 15.16, we see that

$$[b]_{MP} = [m][m^f] \qquad (15.17)$$

This is the desired influence matrix, which includes the influence line ordinates for the $2n$ member-end moment influence lines.

Once the end-moment influence lines have been determined, other influence line ordinates are readily determined from statics. This technique is illustrated in the example problem that follows.

15.5.1 Example problem Determine the influence line ordinates for the member-end moments for the structure shown. Once these are determined, obtain the ordinates for the influence lines for the reaction at A and the moment at D. All ordinates should be determined at 10-foot intervals.

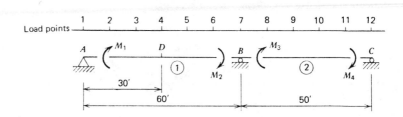

Formulation of* [m] *Matrix: The moment distribution procedure is used.

Unit Fixed-end Moment Corresponding to M_1:

D'	1.000		0.455	0.545		1.000
FEM	1.000		0	0		0
	-1.000		-0.500			
			$+0.228$	$+2.272$		
Final mom.	$0 = m_{11}$		$m_{21} = -0.272$	$+0.272 = m_{31}$		$m_{41} = 0$

Unit Fixed-end Moment Corresponding to M_2:

D'	1.000		0.455	0.545		1.000
FEM	0		1.000	0		0
			-0.455	-0.545		
Final mom.	$0 = m_{12}$		$m_{22} = +0.545$	$-0.545 = m_{32}$		$m_{42} = 0$

Unit Fixed-end Moment Corresponding to M_3:

D'	1.000		0.455	0.545		1.000
FEM	0		0	1.000		0
			-0.455	-0.545		
Final mom.	$0 = m_{13}$		$m_{23} = -0.455$	$+0.455 = m_{33}$		$m_{43} = 0$

Unit fixed-end moment corresponding to M_4:

D'	1.000		0.455	0.545		1.000
FEM	0		0	0		1.000
				-0.500		-1.000
			$+0.228$	$+0.272$		
Final mom.	$0 = m_{14}$		$m_{24} = +0.228$	$-0.228 = m_{34}$		$m_{44} = 0$

$$[m] = \begin{bmatrix} 0 & 0 & 0 & 0 \\ -0.272 & +0.545 & -0.455 & +0.228 \\ +0.272 & -0.545 & +0.455 & -0.228 \\ 0 & 0 & 0 & 0 \end{bmatrix}$$

Determination of* [mᶠ] *Matrix: From Table 13.1

$$P_j = 1$$

$$m^f_{2i-1,j} = -\frac{a_j b_j^2}{l_i^2} = -b_j\left(\frac{a_j}{l_i}\right)\left(\frac{b_j}{l_i}\right)$$

$$m^f_{2i,j} = +\frac{a_j^2 b_j}{l_i^2} = a_j\left(\frac{a_j}{l_i}\right)\left(\frac{b_j}{l_i}\right)$$

Ld. Pt. j	Span i	a_j	b_j	l_i	$\dfrac{a_j}{l_i}$	$\dfrac{b_j}{l_i}$	$m^f_{2i-1,j}$	$m^f_{2i,j}$	m^f_{1j}	m^f_{2j}	m^f_{3j}	m^f_{4j}
1	1	0	60	60	0	1.000	0	0	0	0	0	0
2	1	10	50	60	0.167	0.833	−6.96	1.39	−6.96	1.39	0	0
3	1	20	40	60	0.333	0.667	−8.88	4.44	−8.88	4.44	0	0
4	1	30	30	60	0.500	0.500	−7.50	7.50	−7.50	7.50	0	0
5	1	40	20	60	0.667	0.333	−4.44	8.88	−4.44	8.88	0	0
6	1	50	10	60	0.833	0.167	−1.39	6.96	−1.39	6.96	0	0
7	1	60	0	60	1.000	0	0	0	0	0	0	0
8	2	10	40	50	0.200	0.800	−6.40	1.60	0	0	−6.40	1.60
9	2	20	30	50	0.400	0.600	−7.20	4.80	0	0	−7.20	4.80
10	2	30	20	50	0.600	0.400	−4.80	7.20	0	0	−4.80	7.20
11	2	40	10	50	0.800	0.200	−1.60	6.40	0	0	−1.60	6.40
12	2	50	0	50	1.000	0	0	0	0	0	0	0

$$[m^f] = \begin{bmatrix} 0 & -6.96 & -8.88 & -7.50 & -4.44 & -1.39 & 0 & 0 & 0 & 0 & 0 & 0 \\ 0 & 1.39 & 4.44 & 7.50 & 8.88 & 6.96 & 0 & 0 & 0 & 0 & 0 & 0 \\ 0 & 0 & 0 & 0 & 0 & 0 & 0 & -6.40 & -7.20 & -4.80 & -1.60 & 0 \\ 0 & 0 & 0 & 0 & 0 & 0 & 0 & 1.60 & 4.80 & 7.20 & 6.40 & 0 \end{bmatrix}$$

Generation of Influence Matrix:

$$[b]_{MP} = [m][m^f] \tag{15.17}$$

$$[b]_{MP} = \begin{bmatrix} 0 & 0 & 0 & 0 & 0 & 0 & 0 & 0 & 0 & 0 & 0 & 0 \\ 0 & 2.65 & 4.84 & 6.13 & 6.04 & 4.17 & 0 & 3.28 & 4.37 & 3.82 & 2.19 & 0 \\ 0 & -2.65 & -4.84 & -6.13 & -6.04 & -4.17 & 0 & -3.28 & -4.37 & -3.82 & -2.19 & 0 \\ 0 & 0 & 0 & 0 & 0 & 0 & 0 & 0 & 0 & 0 & 0 & 0 \end{bmatrix}$$

Plotted Influence Lines for Member-End Moments: Influence line for M_i is the ith row of $[b]_{MP}$

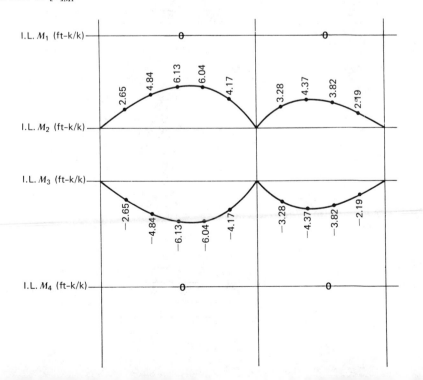

Influence Line for $\mathbf{R_A}$*:*

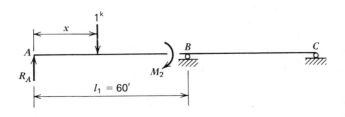

$$\sum M_B = 0; \qquad R_A = \frac{-M_2}{l_1} + \underbrace{\left\{1 - \frac{x}{l_1}\right\}}_{\text{zero if negative}}$$

$-M_2/l_1$	0	-0.044 -0.081	-0.102 -0.101	-0.070	-0.055 0	-0.064 -0.073	-0.037	0	k/k
$\left\{1 - \dfrac{x}{l_1}\right\}$	1.0	0.833 0.667	0.500 0.333	0.167	0				k/k
I.L.R_A	1.0	0.789 0.586	0.398 0.232	0.097	-0.055 0	-0.064 -0.073	-0.037	0	k/k

These ordinates agree with those determined in Example 15.3.1.

Influence Line for $\mathbf{M_D}$*:* Having determined the influence line for R_A, we would now proceed to obtain the influence line for M_D in the manner illustrated in Example 15.3.1.

15.6
Influence Line Equations for Continuous Beams

The Müller-Breslau principle, which is presented in Section 15.3, and the matrix formulation, which is described in Section 15.5, are both numerical procedures that yield influence line ordinates at a discrete number of points along the structure. It is apparent that it would be convenient to be able to write equations for the individual segments of the influence lines. If this were possible, then the ordinates could be accurately determined at any point along the span, the maximum and minimum ordinates could easily be determined along with their locations, and the area under any segment of the influence line could readily be established. Such equations for the influence lines of statically determinate structures are easily determined since these lines are composed of straight-line segments. For continuous statically indeterminate beams, it is difficult to establish the influence line equations. Many authors show that it can be done for simple cases but avoid the complex situations. This subject will not be treated here because it is beyond the scope of this book. Instead, your attention is called to the work of Mozingo, who goes well beyond the simple cases in his published compilation of influence line expressions for continuous beams. His work, which is cited in Section 15.8, is quite comprehensive and extremely useful for those cases which it covers.

15.7

Use of Influence Lines

Influence lines for statically indeterminate structures are used in precisely the same fashion as they are for statically determinate ones. For this reason, the student is urged to review the material given in Section 7.6. There, both the qualitative and quantitative aspects of influence line usage are developed in detail.

A specific qualitative application of influence lines would follow the procedure used in the example problems of Section 7.6. The major difference would involve determination of the areas under the influence lines. For statically determinate structures, the areas are easily obtained because the influence lines are composed of straight-line segments. However, for statically indeterminate structures, the influence lines are curved. Numerous approximate methods of integration could be used to obtain the area under the individual influence line segments. However, in many cases, the curved influence line can be replaced by a series of straight-line segments between the points where the ordinates are given. Trapezoidal formulae are then used to determine the desired areas.

For continuous beams, the qualitative use of influence lines is extremely helpful. Here, the Müller-Breslau principle is used to generate the qualitative influence lines, which are then used to establish the load patterns that will maximize any given response function.

Consider the continuous beam shown in Fig. 15.5a, which is subjected to the indicated uniform dead load *(DL)* and live load *(LL)*. The dead load is applied over the entire structure, whereas the live load can occupy any portion of the structure. For design purposes, it is necessary to determine the maximum positive and negative moments that can occur at the end points as well as at the centerline of any span. Since the center of the span is primarily subjected to a positive moment, whereas the ends of the span are generally subjected to negative moments, the influence lines will be constructed for these signed moments. Thus, for span *AB*, the centerline moment influence line will be constructed for positive moment $M_{\mathbb{C}}^{+}$, while the member-end influence lines will be constructed for negative moments M_{AB}^{-} and M_{BA}^{-}. These moments are defined as shown in Fig. 15.5a.

The qualitative influence line for $M_{\mathbb{C}}^{+}$ is shown in Fig. 15.5b. Thus, for maximum $M_{\mathbb{C}}^{+}$, the loading arrangement of Fig. 15.5c should be applied. The moment diagram for this case is superimposed on the sketch of the beam, with ① identifying the portion of the moment diagram in span *AB*. The centerline moments in the spans to the left and right of span *AB* can be either positive or negative, depending on the relative magnitudes of the dead and live loads. For the case shown, it is assumed that these spans are subjected to negative moment throughout, which corresponds to a high live to dead load ratio. The loading condition for minimum $M_{\mathbb{C}}^{+}$ (or maximum $M_{\mathbb{C}}^{-}$ in the case shown) is given in Fig. 15.5d, and the corresponding moment diagram is shown, with ② indicating the segment of the moment diagram for span *AB*.

The qualitative influence line for M_{AB}^{-} is given in Fig. 15.5e, and the load distribution that will maximize M_{AB}^{-} is shown in Fig. 15.5f along with the moment diagram. The symbol ③ marks that portion of the moment diagram for span *AB*. Similarly, the loading that will minimize M_{AB}^{-} (or maximize M_{AB}^{+} in the case shown) is displayed in Fig. 15.5g. The resulting moment diagram is also given, with ④ indicating the moment diagram in span *AB*.

The four individual moment diagram segments are superimposed in Fig. 15.6. Also, the curves ③' and ④' are shown, which gives the moment diagrams for maximum M_{BA}^{-} and maximum M_{BA}^{+}, respectively. These curves are deduced from

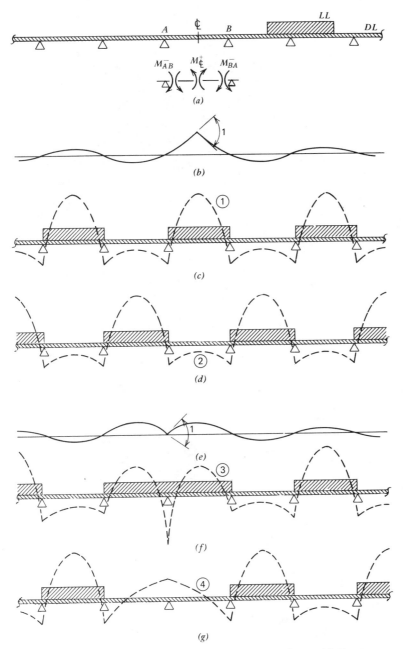

Fig. 15.5 *Load patterns for maximum moments in continuous beam.* (a) *Continuous beam and definition of* M_{AB}^-, M_{BA}^-, *and* $M_{\mathbb{C}}^+$. (b) *Influence line for* $M_{\mathbb{C}}^+$. (c) *Loading pattern and moment diagram for maximum* $M_{\mathbb{C}}^+$. (d) *Loading pattern and moment diagram for minimum* $M_{\mathbb{C}}^+$ *or maximum* $M_{\mathbb{C}}^-$ *(shown).* (e) *Influence line for* M_{AB}^-. (f) *Loading pattern and moment diagram for maximum* M_{AB}^-. (g) *Loading pattern and moment diagram for minimum* M_{AB}^- *or maximum* M_{AB}^+ *(shown.)*

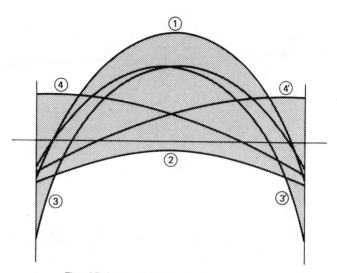

Fig. 15.6 *Envelope of maximum moments.*

symmetry and curves ③ and ④. The total shaded region of Fig. 15.6 gives the envelope of maximum effects for which member *AB* must be designed. Of course, a quantitative influence line analysis is necessary in order to quantify the ordinates of the envelope; however, the qualitative use of the influence line is necessary to establish the various loading patterns.

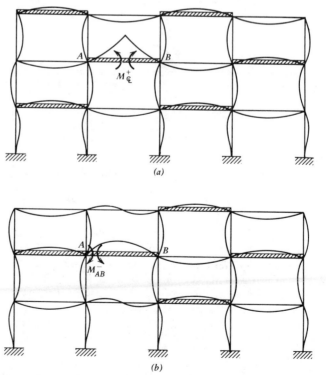

Fig. 15.7 *Frame loading patterns.* (a) *Influence line for* $M_{\mathbb{C}}^{+}$ *and load arrangement to maximize* $M_{\mathbb{C}}^{+}$. (b) *Influence line for* M_{AB}^{-} *and load arrangement to maximize* M_{AB}^{-}.

The Müller-Breslau principle can likewise be used to great advantage in deciding how frames should be loaded to maximize certain response functions. For instance, consider the three-story, four-bay frame shown in Fig. 15.7. Figure 15.7a shows the qualitative influence line for $M_{\mathfrak{L}}^{+}$ of span AB and the bays that must be subjected to live load in order to maximize this moment. Similarly, Fig. 15.7b gives the qualitative influence line for M_{AB}^{-} and the load pattern that will maximize this negative moment. Only those spans that have positive ordinates throughout are loaded. The resulting load patterns for maximizing $M_{\mathfrak{L}}^{+}$ and M_{AB}^{-} are referred to as "checkerboard" loadings because of the alternate nature of the load pattern. For any given member, an envelope of maximum effects could be developed by using the procedure that was discussed for the continuous beam.

15.8
Additional Reading

Cross, H., and Morgan, N. D., *Continuous Frames of Reinforced Concrete*, Chapters 6 and 8, Wiley, New York, 1932.

Ghali, A., and Neville, A. M., *Structural Analysis*, Chapter 13, Intext Educational Publishers, Scranton, Pa., 1972.

Hseigh, Yuan Yu, *Elementary Theory of Structures*, Chapter 11, Prentice–Hall, Englewood Cliffs, N.J., 1970.

Johnson, J. B., Bryan, C. W., and Turneaure, F. E., *The Theory and Practice of Modern Framed Structures*, Part II, 10th Ed., Chapter VII, Wiley, New York, 1929.

Mozingo, R. R., *Influence Line Expressions for Continuous Beams*, Engineering Research Bulletin B-89, Pennsylvania State University, University Park, Pa., 1964.

Newmark, N. M., "Numerical Procedure for Computing Deflections, Moments, and Buckling Loads," *Trans. ASCE*, **108**, p. 1161, 1943.

15.9
Suggested Problems

1. Use the Müller-Breslau principle to determine the influence line for the reaction at point A for the continuous beam shown below. Use the resulting influence line to determine the influence lines for the reaction at point B and the shear and moment at point D. All influence line ordinates should be determined at 10-foot intervals.

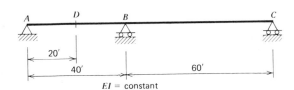

EI = constant

2. By a direct application of the Müller-Breslau principle, determine the influence lines for the reaction at point B, the shear at point D, and the moment at point D for the structure of Problem 1.

3 through 6. Use the Müller-Breslau principle to determine the indicated influence lines for each of the structures given.

3. Influence line for reaction at point B (ordinates at 5-m intervals).

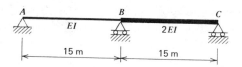

4. Influence line for moment at point E (ordinates at 2.5-m intervals).

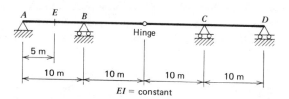

5. Influence line for reaction at point C (ordinates at 10-foot intervals).

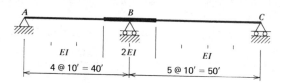

6. Influence line for the reaction at point B (ordinates at 10-foot intervals).

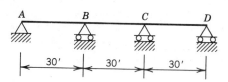

7. Use the matrix formulation of Section 15.5 to determine the influence line ordinates at 5-m intervals for the member-end moments for the structure of Problem 3.

8. Use the matrix formulation to determine the influence line ordinates at 10-foot intervals for the member-end moments for the structure of Problem 6.

PART III

GENERALIZED
MATRIX METHODS

I79 Bridge, North of Pittsburgh, Pa. (courtesy of Pennsylvania Department of Transportation).

FORCE–DEFORMATION RELATIONS FOR ELASTIC MEMBERS

16.1

In Chapters 2, 3, 4, and 8 focus was given to the external and internal forces that were required to equilibrate the whole or part of a structure. Subsequently, in Chapters 5, 6, and 8, consideration was given to how structures deform from the forces that they carry.

Importance of Member Interactive Relationships

For structures as a whole, there is an important distinction between statically determinate and indeterminate systems. For statically determinate cases, the forces can be determined by considering static equilibrium alone, and the deformations are then determined as a function of these forces. Each member deforms according to its own force–deformation relationships, and the overall deformation of the structure will result from the accumulation of the individual member effects. However, for statically indeterminate situations, the determination of forces and deformations must be coupled since they are mutually dependent through the combined requirements of static equilibrium and compatibility. In the latter case, as was evident in Chapters 9 through 14, the analysis requires systematic synthesis of the member force–deformations.

Therefore, in either statically determinate on indeterminate structures, the member force–deformation relationships must be known. These relationships have previously been established in a very limited and somewhat conceptualized fashion. In Section 5.3 the axial force–deformation relationships were established. Here a differential element was considered, and it served as a building block for the study of deflections for pin-connected frameworks in Chapter 5. Similarly, flexural force–deformation relationships were derived in Section 6.2. Again, these were based on a differential element, and they provided the basis for the study of deflections for beam- and frame-type structures in Chapter 6.

Presently a more generalized approach to structural analysis is to be developed, and the fundamental building blocks are to be the individual members. Therefore, it is necessary to establish the complete interactive relationships between member-end forces and the associated member-end displacements.

As has been stipulated in earlier sections of the text, we will limit our treatment to members that are not stressed beyond the elastic range.

16.2

Structure and Member Coordinate Systems

A portion of a typical three-dimensional frame structure is shown in Fig. 16.1a along with the right-hand XYZ structure coordinate system. At each joint, the loading condition is specified by six force components, and the response is described by six displacement components. These force and displacement components are shown in their positive directions at joint i of Fig. 16.1a. The first three force and displacement components correspond to the direct forces and translational displacements, with the positive directions being along the positive coordinate axes. The remaining force and displacement components represent the moment forces and rotational displacements. Here, the right-hand rule is employed to establish the positive directions. In matrix form, the joint forces and displacements can be expressed, respectively, as

$$\{P\}_i = \begin{Bmatrix} P_1 \\ P_2 \\ P_3 \\ P_4 \\ P_5 \\ P_6 \end{Bmatrix}_i \qquad \{\Delta\}_i = \begin{Bmatrix} \Delta_1 \\ \Delta_2 \\ \Delta_3 \\ \Delta_4 \\ \Delta_5 \\ \Delta_6 \end{Bmatrix}_i \qquad (16.1)$$

where the i subscript on a vector indicates that each element of the array is associated with joint i.

Figure 16.1b shows a typical member ij along with the positive member coordinate system at each end of the member. Each of these xyz sets of axes is a right-hand coordinate system. The figure also shows the right-hand XYZ structure coordinate system, which becomes important when the individual member characteristics are combined to establish the overall structure characteristics.

At the i end of the member, the local xyz axes are shown along with the positive member-end forces and displacements. The first three forces and displacements correspond to direct forces and translational displacements along the three coordinate axes, whereas the remaining three forces and displacements are moments and rotational displacements about the three coordinate axes. The positive

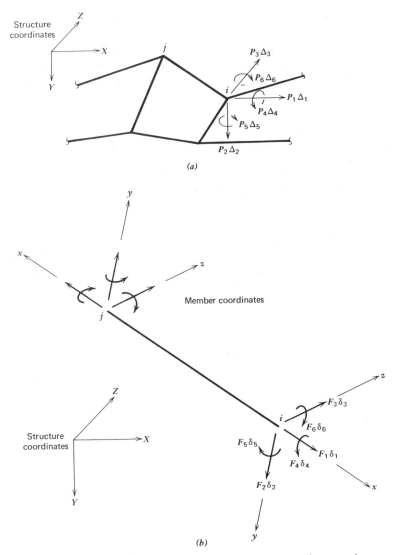

Fig. 16.1 *Structure and member coordinate systems. (a) Structure forces and displacements. (b) Member forces and displacements.*

directions for the moments and rotations are based on the right-hand rule. These forces and displacements can be written in matrix form as

$$\{F\}_{ij} = \left\{ \begin{matrix} F_1 \\ F_2 \\ F_3 \\ F_4 \\ F_5 \\ F_6 \end{matrix} \right\}_{ij} \; ; \quad \{\delta\}_{ij} = \left\{ \begin{matrix} \delta_1 \\ \delta_2 \\ \delta_3 \\ \delta_4 \\ \delta_5 \\ \delta_6 \end{matrix} \right\}_{ij} \tag{16.2}$$

where the double subscript indicates that these arrays refer to the i end of member ij.

A similar set of forces and displacements is present at the j end of the member. These are defined in precisely the same manner as those at the i end, and the vector representation would be the same as those given by Eqs. 16.2 with the ij subscript replaced by ji.

In subsequent sections of this chapter, the individual elements may carry a single member subscript, not the double subscript—that is, the axial component at the i end of member ij is represented by F_{1i} and not $(F_1)_{ij}$. This is done only when it is clear from the context that this force is at the i end of member ij.

16.3

Member Stiffnesses

The concept of stiffness was introduced in Section 5.3 and 6.2 as it is related to separate axial and flexural actions. This concept will now be broadened to express the interactive relationships between member-end forces and deformations for the member as a whole.

Consider the member shown in Fig. 16.2a, which represents a much simplified case when compared to the general one given in Fig. 16.1. Here, only axial forces and bending moments about the z axis are considered along with the corresponding member-end deformations. In addition, the forces and displacements are ordered and numbered differently for this simplified case than for the case

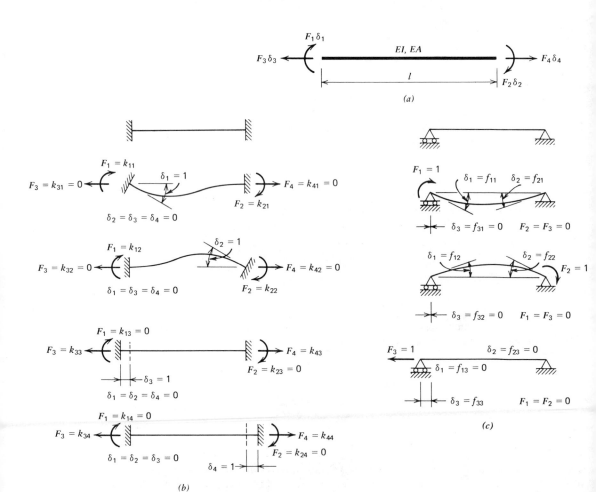

Fig. 16.2 Member stiffnesses and flexibilities. (a) Member actions and deformations. (b) Member stiffnesses. (c) Member flexibilities.

given in Fig. 16.1. This represents a temporary departure from the general coordinate system described in Section 16.2 that is necessary for a clear understanding of the concepts that are being introduced in this section.

In order to get member stiffnesses, the member is initially clamped at both ends. From this condition, the member-end displacements can be introduced one at a time and the corresponding member-end forces can be determined. Thus, for the member shown in Fig. 16.2a, the four forces are related to the four displacements by the expressions

$$
\begin{aligned}
F_1 &= k_{11}\delta_1 + k_{12}\delta_2 + k_{13}\delta_3 + k_{14}\delta_4 \\
F_2 &= k_{21}\delta_1 + k_{22}\delta_2 + k_{23}\delta_3 + k_{24}\delta_4 \\
F_3 &= k_{31}\delta_1 + k_{32}\delta_2 + k_{33}\delta_3 + k_{34}\delta_4 \\
F_4 &= k_{41}\delta_1 + k_{42}\delta_2 + k_{43}\delta_3 + k_{44}\delta_4
\end{aligned}
\tag{16.3}
$$

where k_{ij} is the ith member-end force associated with a unit value of the jth member-end displacement. These quantities are referred to as *stiffness coefficients*.

In matrix form, Eqs. 16.3 can be written as

$$
\begin{Bmatrix} F_1 \\ F_2 \\ F_3 \\ F_4 \end{Bmatrix} =
\begin{bmatrix}
k_{11} & k_{12} & k_{13} & k_{14} \\
k_{21} & k_{22} & k_{23} & k_{24} \\
k_{31} & k_{32} & k_{33} & k_{34} \\
k_{41} & k_{42} & k_{43} & k_{44}
\end{bmatrix}
\begin{Bmatrix} \delta_1 \\ \delta_2 \\ \delta_3 \\ \delta_4 \end{Bmatrix}
\tag{16.4}
$$

or

$$
\{F\} = [k]\{\delta\}
\tag{16.5}
$$

where $\{F\}$ is the vector of member-end forces, $\{\delta\}$ is the vector of member-end displacements, and $[k]$ is the *member stiffness matrix*.

The individual stiffness coefficients can be determined in a very orderly fashion. For instance, the jth column of the stiffness matrix gives the four member-end forces that are associated with a unit value of the jth member-end displacement while all other member-end displacements are zero. Thus, each of the separate conditions in Fig. 16.2b gives one column of the stiffness matrix.

The details concerning the determination of the stiffness coefficients are covered in Section 16.7. Upon determination of these coefficients, Eq. 16.4 can be written in the form

$$
\begin{Bmatrix} F_1 \\ F_2 \\ F_3 \\ F_4 \end{Bmatrix} =
\begin{bmatrix}
\dfrac{4EI}{l} & \dfrac{2EI}{l} & 0 & 0 \\[2mm]
\dfrac{2EI}{l} & \dfrac{4EI}{l} & 0 & 0 \\[2mm]
0 & 0 & \dfrac{EA}{l} & \dfrac{EA}{l} \\[2mm]
0 & 0 & \dfrac{EA}{l} & \dfrac{EA}{l}
\end{bmatrix}
\begin{Bmatrix} \delta_1 \\ \delta_2 \\ \delta_3 \\ \delta_4 \end{Bmatrix}
\tag{16.6}
$$

From this equation, it is clear that the end moments F_1 and F_2 are solely dependent on the end rotations δ_1 and δ_2. The coupling stiffness coefficients are in terms of the quantity EI/l, which reflects the flexural stiffness of the member. Similarly, the axial end forces F_3 and F_4 are dependent only on the axial displacements δ_3 and δ_4, and the coupling stiffness coefficients are in terms of EA/l, which reflects the axial stiffness of the member. These general observations, as well as the specific makeup of the stiffness matrix, confirm the earlier findings of Sections 5.3 and 6.2.

16.4

Member Flexibilities

Reference is again made to the member of Fig. 16.2a. For the determination of the member flexibilities, the member must be supported in a fashion that makes it stable and statically determinate. In this case, the member is pin-connected at its right end and supported by a roller at its left end. The resulting force–deformation relations involve only the first three forces and deformations shown in Fig. 16.2a since the support condition eliminates δ_4. Thus, the force–deformation relations take the form

$$
\begin{aligned}
\delta_1 &= f_{11}F_1 + f_{12}F_2 + f_{13}F_3 \\
\delta_2 &= f_{21}F_1 + f_{22}F_2 + f_{23}F_3 \\
\delta_3 &= f_{31}F_1 + f_{32}F_2 + f_{33}F_3
\end{aligned}
\tag{16.7}
$$

where f_{ij} is the ith member-end displacement that results from a unit value of the jth member-end force. These quantities are called the *flexibility coefficients*.

Equation 16.7 can be written in matrix form as

$$
\begin{Bmatrix} \delta_1 \\ \delta_2 \\ \delta_3 \end{Bmatrix} = \begin{bmatrix} f_{11} & f_{12} & f_{13} \\ f_{21} & f_{22} & f_{23} \\ f_{31} & f_{32} & f_{33} \end{bmatrix} \begin{Bmatrix} F_1 \\ F_2 \\ F_3 \end{Bmatrix}
\tag{16.8}
$$

or

$$
\{\delta\}_r = [f]\{F\}_r
\tag{16.9}
$$

where $\{\delta\}_r$ and $\{F\}_r$ are reduced versions of the displacement and force matrices that were defined earlier, and $[f]$ is the *member flexibility matrix*.

The individual flexibility coefficients can be systematically determined. For example, the jth column of the flexibility matrix gives the three member-end displacements that are induced by a unit valve of the jth member-end force while all other member-end forces are zero. Thus, each of the separate conditions illustrated in Fig. 16.2c corresponds to one column of the flexibility matrix.

The detailed determination of the flexibility coefficients is given in Section 16.6. Using those methods, Eq. 16.8 could be determined as

$$
\begin{Bmatrix} \delta_1 \\[1em] \delta_2 \\[1em] \delta_3 \end{Bmatrix} = \begin{bmatrix} \dfrac{l}{3EI} & -\dfrac{l}{6EI} & 0 \\[1em] -\dfrac{l}{6EI} & \dfrac{l}{3EI} & 0 \\[1em] 0 & 0 & \dfrac{l}{EA} \end{bmatrix} \begin{Bmatrix} F_1 \\[1em] F_2 \\[1em] F_3 \end{Bmatrix}
\tag{16.10}
$$

As was the case with the stiffness approach, it is again clear that the end rotations depend on the end moments only and the axial displacement depends on the axial force only. The coupling flexibility coefficients are in terms of l/EI for flexural effects and l/EA for axial effects, which is consistent with the findings of Sections 5.3 and 6.2 for the differential element.

16.5

Relationship
between
Stiffness and
Flexibility
Matrices

Figure 16.3a shows the simplified member of Fig. 16.2a along with its member-end forces and displacements. Equations 16.4 and 16.5 indicate that for any set of displacements $\{\delta\}$, there is a unique set of forces $\{F\}$. However, if one were to attempt to solve for the displacements by forming the expression $\{\delta\} = [k]^{-1}\{F\}$, it would be found that $[k]^{-1}$ does not exist. That is, the member stiffness matrix is singular and does not possess an inverse. This makes sense, since there is no unique set of displacements that corresponds to a given set of forces. For instance, a given set of forces would not be altered by a set of displacements which corresponds to a rigid-body motion.

To preclude rigid-body motion, the member must be supported in a manner that will render it stable. One way to support the structure is shown in Fig. 16.3b, in which the member is restrained as a statically determinate beam-type structure. In this case, $\delta_4 = 0$ while all other δ quantities are allowed to exist. Equation 16.4 can be partitioned in the form

$$\begin{Bmatrix} F_1 \\ F_2 \\ F_3 \\ \hline F_4 \end{Bmatrix} = \left[\begin{array}{ccc:c} k_{11} & k_{12} & k_{13} & k_{14} \\ k_{21} & k_{22} & k_{23} & k_{24} \\ k_{31} & k_{32} & k_{33} & k_{34} \\ \hdashline k_{41} & k_{42} & k_{43} & k_{44} \end{array}\right] \begin{Bmatrix} \delta_1 \\ \delta_2 \\ \delta_3 \\ \hline \delta_4 \end{Bmatrix} \qquad (16.11)$$

where the partitioning separates the restrained point force and displacement from the others. Since $\delta_4 = 0$, we can write Eq. 16.11 as

$$\begin{Bmatrix} F_1 \\ F_2 \\ F_3 \end{Bmatrix} = \begin{bmatrix} k_{11} & k_{12} & k_{13} \\ k_{21} & k_{22} & k_{23} \\ k_{31} & k_{32} & k_{33} \end{bmatrix} \begin{Bmatrix} \delta_1 \\ \delta_2 \\ \delta_3 \end{Bmatrix} \qquad (16.12)$$

and

$$F_4 = k_{41}\delta_1 + k_{42}\delta_2 + k_{43}\delta_3 \qquad (16.13)$$

Equation 16.12 may be rewritten as

$$\{F\}_r = [k]_r\{\delta\}_r \qquad (16.14)$$

where $[k]_r$ is the reduced stiffness matrix which gives the relationship between the reduced member-end force and displacement vectors $\{F\}_r$ and $\{\delta\}_r$, respectively.

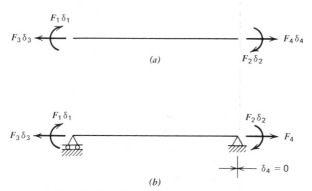

Fig. 16.3 *Typical member.* (a) *Member-end forces and displacements.* (b) *Member-end forces and displacements for restrained member.*

Since rigid-body displacements are not possible for the structure of Fig. 16.3b, it is possible to form the relationship

$$\{\delta\}_r = [k]_r^{-1}\{F\}_r \tag{16.15}$$

Comparing Eqs. 16.9 and 16.15, we see that

$$[f] = [k]_r^{-1} \tag{16.16}$$

If both sides of Eq. 16.16 are premultiplied by $[k]_r$, then

$$[k]_r[f] = [I] \tag{16.17}$$

It must be noted, however, that $[k]_r$ is not the total member stiffness matrix, but rather is the reduced stiffness matrix that corresponds to the structure stiffness matrix for the member after it is transformed into a stable structure. Equation 16.17 can be verified by substituting $[k]_r$ and $[f]$ from Eqs. 16.6 and 16.10, respectively. That is,

$$[k]_r[f] = \begin{bmatrix} \dfrac{4EI}{l} & \dfrac{2EI}{l} & 0 \\ \dfrac{2EI}{l} & \dfrac{4EI}{l} & 0 \\ 0 & 0 & \dfrac{EA}{l} \end{bmatrix} \begin{bmatrix} \dfrac{l}{3EI} & -\dfrac{l}{6EI} & 0 \\ -\dfrac{l}{6EI} & \dfrac{l}{3EI} & 0 \\ 0 & 0 & \dfrac{l}{EA} \end{bmatrix} = \begin{bmatrix} 1 & 0 & 0 \\ 0 & 1 & 0 \\ 0 & 0 & 1 \end{bmatrix} \tag{16.18}$$

For the case cited here, the structure was transformed to a statically determinate stable structure by restraining δ_4. In general, the member can be transformed in other ways. In any case, $[k]_r$ is formed by deleting the rows and columns of the displacement components that are restrained.

16.6

Member Flexibility Matrix

We now return to the general member coordinate system introduced in Section 16.2 with the full six-component member-end force and deformation vectors at the i and j ends as shown in Fig. 16.1b.

As noted in the previous section, the member flexibility matrix can be determined only when the member is restrained against rigid-body motion. This can be done by fixing the member at the j end and developing the relationships between the displacement and forces at the i end of the member.

For our present development, consider a slightly simplified version of the member shown in Fig. 16.1b. Here, we take a member in the xy plane and consider only the forces and displacements shown in Fig. 16.4a.

The total strain energy for this member can be expressed as

$$U = \int_{x=0}^{x=l} \left(\frac{F^2}{2EA} + \frac{M^2}{2EI_z} + \frac{\lambda V^2}{2GA} + \frac{T^2}{2GJ} \right) dx \tag{16.19}$$

The first two terms within the parentheses correspond to the strain energy associated with axial and flexural deformations as previously expressed in Eqs. 5.44 and 6.61, respectively. Here, the quantities E, A, and I_z represent the modulus of elasticity, cross-sectional area, and moment of inertia about the z axis, respectively. The last two terms within the parentheses give the strain energies of shearing and torsional deformations, where λ is a constant that depends on the shape of the cross section, G is the shear modulus, and J is the torsional constant. The

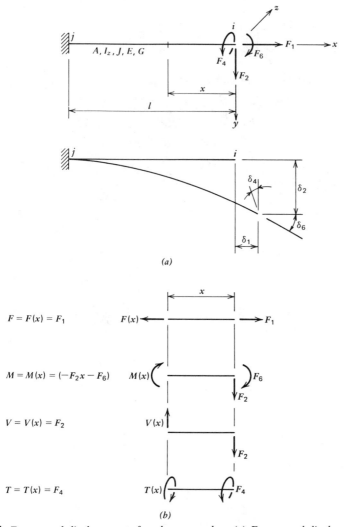

Fig. 16.4 *Forces and displacements for planar member.* (a) *Forces and displacements.* (b) *Internal forces at* x.

strain energy term corresponding to torsion ignores the effects of warping resistance, which can be significant in some cases. This is consistent with the treatment in Section 8.16.

The quantities F, M, V, and T in Eq. 16.19 represent the axial force, bending moment, shear force, and torsional moment, respectively, and each must be expressed in terms of the variable x as shown in Fig. 16.4b.

Castigliano's second theorem is now used to determine the desired displacement–force relationships. For example, using Eqs. 5.40 and 16.19, we obtain

$$\delta_1 = \frac{\partial U}{\partial F_1} = \frac{\partial}{\partial F_1} \left[\int_0^l \left(\frac{F^2}{2EA} + \cdots \right) dx \right] \tag{16.20}$$

Substitution into Eq. 16.20 of the expression for F given in Fig. 16.4b gives

$$\delta_1 = \frac{\partial}{\partial F_1} \left[\int_0^l \left(\frac{F_1^2}{2EA} \right) dx \right] = \frac{F_1 l}{EA} \tag{16.21}$$

Similarly,

$$\delta_2 = \frac{\partial}{\partial F_2}\left[\int_0^1 \left(\cdots + \frac{M^2}{2EI_z} + \frac{\lambda V^2}{2GA} + \cdots\right) dx\right] \tag{16.22}$$

Using the expressions for M and V from Fig. 16.4b, we can express Eq. 16.22 as

$$\delta_2 = \frac{\partial}{\partial F_2}\left[\int_0^l \left(\frac{[-F_2 x - F_6]^2}{2EI_z} + \frac{\lambda F_2^2}{2GA}\right) dx\right] \tag{16.23}$$

$$= \frac{F_2 l^3}{3EI_z} + \frac{F_6 l^2}{2EI_z} + \frac{\lambda F_2 l}{GA} \tag{16.24}$$

In like fashion

$$\delta_4 = \frac{F_4 l}{GJ} \tag{16.25}$$

and

$$\delta_6 = \frac{F_2 l^2}{2EI_z} + \frac{F_6 l}{EI_z} \tag{16.26}$$

Equations 16.21, 16.24, 16.25, and 16.26 can be capsulized in a single matrix equation in the form

$$\begin{Bmatrix} \delta_1 \\ \delta_2 \\ \delta_4 \\ \delta_6 \end{Bmatrix}_{ij} = \begin{bmatrix} \dfrac{l}{EA} & 0 & 0 & 0 \\ 0 & \dfrac{l^3}{3EI_z} + \dfrac{\lambda l}{GA} & 0 & \dfrac{l^2}{2EI_z} \\ 0 & 0 & \dfrac{l}{GJ} & 0 \\ 0 & \dfrac{l^2}{2EI_z} & 0 & \dfrac{l}{EI_z} \end{bmatrix}_{ij} \begin{Bmatrix} F_1 \\ F_2 \\ F_4 \\ F_6 \end{Bmatrix}_{ij} \tag{16.27}$$

For most members, the quantity $\lambda l/GA$ is very much smaller than $l^3/3EI_z$ and is thus ignored. Throughout this book, shearing deformations will be neglected by setting $\lambda l/GA$ equal to zero.

For the complete member shown in Fig. 16.1b, which includes bending about the y axis and shearing forces in the xz plane, Eq. 16.27 can be expanded to the form

$$\begin{Bmatrix} \delta_1 \\ \delta_2 \\ \delta_3 \\ \delta_4 \\ \delta_5 \\ \delta_6 \end{Bmatrix}_{ij} = \begin{bmatrix} \dfrac{l}{EA} & 0 & 0 & 0 & 0 & 0 \\ 0 & \dfrac{l^3}{3EI_z} & 0 & 0 & 0 & \dfrac{l^2}{2EI_z} \\ 0 & 0 & \dfrac{l^3}{3EI_y} & 0 & \dfrac{-l^2}{2EI_y} & 0 \\ 0 & 0 & 0 & \dfrac{l}{GJ} & 0 & 0 \\ 0 & 0 & \dfrac{-l^2}{2EI_y} & 0 & \dfrac{l^3}{EI_y} & 0 \\ 0 & \dfrac{l^2}{2EI_z} & 0 & 0 & 0 & \dfrac{l}{EI_z} \end{bmatrix} \begin{Bmatrix} F_1 \\ F_2 \\ F_3 \\ F_4 \\ F_5 \\ F_6 \end{Bmatrix}_{ij} \tag{16.28}$$

This equation can be written in the abbreviated form

$$\{\delta\}_{ij} = [f]_{ii}^{j}\{F\}_{ij} \qquad (16.29)$$

where $[f]_{ii}^{j}$ is the flexibility matrix relating the displacements at the i end to the forces at the i end for a member that connects points i and j. This is referred to as the *direct flexibility matrix* for member ij.

16.7

Member Stiffness Matrix

As was done in the previous section for the derivation of the member flexibility matrix, a simplified case is initially considered in developing the member stiffness matrix. Figure 16.5a shows a member in the xy plane along with the forces and

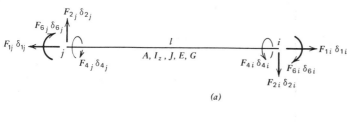

(a)

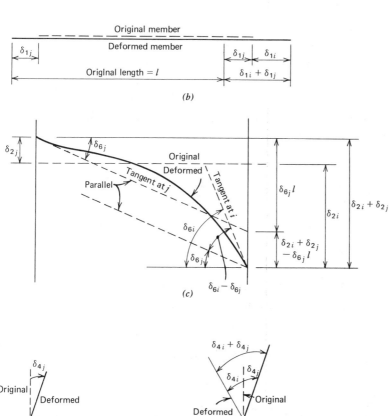

Fig. 16.5 *Forces and displacements for planar member.* (a) *Forces and displacements.* (b) *Axial action.* (c) *Flexural action.* (d) *Torsional action.*

displacements that are to be included. This is the same member that was shown in Fig. 16.4a; however, in the present case, the member is not restrained at point j.

From Eq. 16.21

$$F_1 = \frac{EA}{l}\delta_1 \tag{16.30}$$

where F_1 is the axial force at the i end of the member and δ_1 is the axial displacement at the i end of the member relative to the j end as shown in Fig. 16.4a. In the present case, Figs. 16.5a and 16.5b show that $F_1 = F_{1i}$ and $\delta_1 = (\delta_{1i} + \delta_{1j})$, and thus Eq. 16.30 becomes

$$F_{1i} = \frac{EA}{l}\delta_{1i} + \frac{EA}{l}\delta_{1j} \tag{16.31}$$

From equilibrium considerations,

$$F_{1j} = F_{1i} = \frac{EA}{l}\delta_{1i} + \frac{EA}{l}\delta_{1j} \tag{16.32}$$

Solving Eqs. 16.24 and 16.26 simultaneously for F_2 and F_6 and recalling that $\gamma l/GA$ can be ignored, we obtain

$$F_2 = \frac{12EI_z}{l^3}\delta_2 - \frac{6EI_z}{l^2}\delta_6$$
$$F_6 = -\frac{6EI_z}{l^2}\delta_2 + \frac{4EI_z}{l}\delta_6 \tag{16.33}$$

where F_2 and F_6 are the shearing force and the moment, respectively, at the i end of the member and δ_2 and δ_6 are the corresponding displacements at the i end of the member relative to the tangent at the j end as shown in Fig. 16.4a. Figures 16.5a and 16.5c show that in the present case $F_2 = F_{2i}$, $F_6 = F_{6i}$, $\delta_2 = (\delta_{2i} + \delta_{2j} - \delta_{6j}l)$, and $\delta_6 = (\delta_{6i} - \delta_{6j})$, and thus Eq. 16.33 becomes

$$F_{2i} = \frac{12EI_z}{l^3}\delta_{2i} + \frac{12EI_z}{l^3}\delta_{2j} - \frac{6EI_z}{l^2}\delta_{6i} - \frac{6EI_z}{l^2}\delta_{6j} \tag{16.34}$$

$$F_{6i} = -\frac{6EI_z}{l^2}\delta_{2i} - \frac{6EI_z}{l^2}\delta_{2j} + \frac{4EI_z}{l}\delta_{6i} + \frac{2EI_z}{l}\delta_{6j} \tag{16.35}$$

Equilibrium then requires that

$$F_{2j} = F_{2i} = \frac{12EI_z}{l^3}\delta_{2i} + \frac{12EI_z}{l^3}\delta_{2j} - \frac{6EI_z}{l^2}\delta_{6i} - \frac{6EI_z}{l^2}\delta_{6j} \qquad (16.36)$$

and

$$
\begin{aligned}
F_{6j} &= -(F_{6i} + F_{2i}l) \\
&= -\frac{6EI_z}{l^2}\delta_{2i} - \frac{6EI_z}{l^2}\delta_{2j} + \frac{2EI_z}{l}\delta_{6i} + \frac{4EI_z}{l}\delta_{6j} \qquad (16.37)
\end{aligned}
$$

From Eq. 16.25,

$$F_4 = \frac{GJ}{l}\delta_4 \qquad (16.38)$$

in which F_4 is the torsional force at the i end of the member and δ_4 is the rotational displacement at the i end of the member relative to the j end as illustrated in Fig. 16.4d. In the present situation, Figs. 16.5a and 16.5d show that $F_4 = F_{4i}$ and $\delta_4 = \delta_{4i} + \delta_{4j}$. Accordingly, Eq. 16.38 takes the form

$$F_{4i} = \frac{GJ}{l}\delta_{4i} + \frac{GJ}{l}\delta_{4j} \qquad (16.39)$$

Equilibrium requires that

$$F_{4j} = F_{4i} = \frac{GJ}{l}\delta_{4i} + \frac{GJ}{l}\delta_{4j} \qquad (16.40)$$

Equations 16.31, 16.32, 16.34 through 16.37, 16.39, and 16.40 can be collected in a single matrix equation as

$$
\left\{
\begin{array}{c}
\left\{\begin{array}{c} F_1 \\ F_2 \\ F_4 \\ F_6 \end{array}\right\}_{ij} \\
\left\{\begin{array}{c} F_1 \\ F_2 \\ F_4 \\ F_6 \end{array}\right\}_{ji}
\end{array}
\right\}
=
\left[
\begin{array}{cccc:cccc}
\dfrac{EA}{l} & 0 & 0 & 0 & \dfrac{EA}{l} & 0 & 0 & 0 \\[6pt]
0 & \dfrac{12EI_z}{l^3} & 0 & -\dfrac{6EI_z}{l^2} & 0 & \dfrac{12EI_z}{l^3} & 0 & -\dfrac{6EI_z}{l^2} \\[6pt]
0 & 0 & \dfrac{GJ}{l} & 0 & 0 & 0 & \dfrac{GJ}{l} & 0 \\[6pt]
0 & -\dfrac{6EI_z}{l^2} & 0 & \dfrac{4EI_z}{l} & 0 & -\dfrac{6EI_z}{l^2} & 0 & \dfrac{2EI_z}{l} \\[6pt]
\hdashline
\dfrac{EA}{l} & 0 & 0 & 0 & \dfrac{EA}{l} & 0 & 0 & 0 \\[6pt]
0 & \dfrac{12EI_z}{l^3} & 0 & -\dfrac{6EI_z}{l^2} & 0 & \dfrac{12EI_z}{l^3} & 0 & -\dfrac{6EI_z}{l^2} \\[6pt]
0 & 0 & \dfrac{GJ}{l} & 0 & 0 & 0 & \dfrac{GJ}{l} & 0 \\[6pt]
0 & -\dfrac{6EI_z}{l^2} & 0 & \dfrac{2EI_z}{l} & 0 & -\dfrac{6EI_z}{l^2} & 0 & \dfrac{4EI_z}{l}
\end{array}
\right]
\left\{
\begin{array}{c}
\left\{\begin{array}{c} \delta_1 \\ \delta_2 \\ \delta_4 \\ \delta_6 \end{array}\right\}_{ij} \\
\left\{\begin{array}{c} \delta_1 \\ \delta_2 \\ \delta_4 \\ \delta_6 \end{array}\right\}_{ji}
\end{array}
\right\}
\qquad (16.41)
$$

For the complete three-dimensional member shown in Fig. 16.1b, Eq. 16.41 takes on the following expanded form:

$$
\left\{
\begin{array}{c}
\left\{\begin{array}{c} F_1 \\ F_2 \\ F_3 \\ F_4 \\ F_5 \\ F_6 \end{array}\right\}_{ij} \\[6pt]
\left\{\begin{array}{c} F_1 \\ F_2 \\ F_3 \\ F_4 \\ F_5 \\ F_6 \end{array}\right\}_{ji}
\end{array}
\right\}
=
\left[
\begin{array}{cccccc|cccccc}
\frac{EA}{l} & & & & & & \frac{EA}{l} & & & & & \\[4pt]
& \frac{12EI_z}{l^3} & & & & -\frac{6EI_z}{l^2} & & \frac{12EI_z}{l^3} & & & & -\frac{6EI_z}{l^2} \\[4pt]
& & \frac{12EI_y}{l^3} & & \frac{6EI_y}{l^2} & & & & -\frac{12EI_y}{l^3} & & -\frac{6EI_y}{l^2} & \\[4pt]
& & & \frac{GJ}{l} & & & & & & \frac{GJ}{l} & & \\[4pt]
& & \frac{6EI_y}{l^2} & & \frac{4EI_y}{l} & & & & -\frac{6EI_y}{l^2} & & -\frac{2EI_y}{l} & \\[4pt]
& -\frac{6EI_z}{l^2} & & & & \frac{4EI_z}{l} & & -\frac{6EI_z}{l^2} & & & & \frac{2EI_z}{l} \\[4pt]
\hline
\frac{EA}{l} & & & & & & \frac{EA}{l} & & & & & \\[4pt]
& \frac{12EI_z}{l^3} & & & & -\frac{6EI_z}{l^2} & & \frac{12EI_z}{l^3} & & & & -\frac{6EI_z}{l^2} \\[4pt]
& & -\frac{12EI_y}{l^3} & & -\frac{6EI_y}{l^2} & & & & \frac{12EI_y}{l^3} & & \frac{6EI_y}{l^2} & \\[4pt]
& & & \frac{GJ}{l} & & & & & & \frac{GJ}{l} & & \\[4pt]
& & -\frac{6EI_y}{l^2} & & -\frac{2EI_y}{l} & & & & \frac{6EI_y}{l^2} & & \frac{4EI_y}{l} & \\[4pt]
& -\frac{6EI_z}{l^2} & & & & \frac{2EI_z}{l} & & -\frac{6EI_z}{l^2} & & & & \frac{4EI_z}{l}
\end{array}
\right]
\left\{
\begin{array}{c}
\left\{\begin{array}{c} \delta_1 \\ \delta_2 \\ \delta_3 \\ \delta_4 \\ \delta_5 \\ \delta_6 \end{array}\right\}_{ij} \\[6pt]
\left\{\begin{array}{c} \delta_1 \\ \delta_2 \\ \delta_3 \\ \delta_4 \\ \delta_5 \\ \delta_6 \end{array}\right\}_{ji}
\end{array}
\right\}
\qquad (16.42)
$$

In an abbreviated form, Eq. 16.42 becomes

$$
\left\{
\begin{array}{c}
\{F\}_{ij} \\ \hline \{F\}_{ji}
\end{array}
\right\}
=
\left[
\begin{array}{c|c}
[k]^j_{ii} & [k]_{ij} \\ \hline
[k]_{ji} & [k]^i_{jj}
\end{array}
\right]
\left\{
\begin{array}{c}
\{\delta\}_{ij} \\ \hline \{\delta\}_{ji}
\end{array}
\right\}
\qquad (16.43)
$$

The matrix $[k]^j_{ii}$ is a *direct stiffness matrix* for member ij. It relates the forces at the i end to the displacements at the i end for the member that connects points i and j. The matrix $[k]^i_{jj}$ is the direct stiffness matrix for member ij relating forces and displacements at the j end, and it is clear from Eq. 16.42 that $[k]^j_{ii} = [k]^i_{jj}$. The matrix $[k]_{ij}$ is a *cross stiffness matrix* which relates the forces at the i end to the displacements at the j end. Similarly, $[k]_{ji}$ is a cross stiffness matrix relating the forces at the j end to the displacements at the i end.

Equation 16.43 can be further simplified as

$$
\{F\} = [k]\{\delta\}
\qquad (16.44)
$$

which is the form given earlier as Eq. 16.5. As was explained in Section 6.5, Eq. 16.44 cannot be solved for the displacements because $[k]^{-1}$ does not exist. Physically, this means that there is no unique set of displacements associated with a given set of forces. For instance, the forces shown in Fig. 16.5a would not be changed by a rigid-body motion. However, if a member is restrained to preclude rigid-body movement, then there is a unique set of displacements associated with

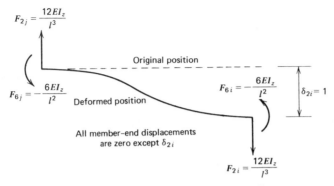

Fig. 16.6 *Physical meaning of stiffness coefficients.*

a prescribed set of forces. For instance, if the member shown in Fig. 16.5a is fixed at point j, then $\{\delta\}_{ji} = \{0\}$, and

$$\{F\}_{ij} = [k]^j_{ii}\{\delta\}_{ij} \tag{16.45}$$

Premultiplication by $[k]^{j-1}_{ii}$ gives

$$\{\delta\}_{ij} = [k]^{j-1}_{ii}\{F\}_{ij} \tag{16.46}$$

Comparing Eqs. 16.29 and 16.46, we see that

$$[k]^{j-1}_{ii} = [f]^j_{ii} \tag{16.47}$$

which can be verified by multiplication of the appropriate matrices from Eqs. 16.28 and 16.42. According to the terminology of Section 16.5, $[k]^j_{ii}$ is the reduced stiffness matrix, $[k]_r$.

One final note regarding the member stiffness matrix concerns the physical meaning of the individual stiffness coefficients. Each column of the stiffness matrix of Eq. 16.42 gives the end forces associated with a unit value of a single end displacement. For instance, the second column gives the array of member-end forces that are caused by a unit value of δ_{2i}, while other displacements are zero. This condition is shown in Fig. 16.6, and the results are verifiable by a number of methods.

16.8

Some Observations on the Stiffness Matrix

Each of the flexibility coefficients that make up the flexibility matrix is a specified displacement quantity. That is, f_{rs} is the rth member-end displacement that results from the application of a unit value of the sth member-end force. Based on Maxwell's law, which is given by Eq. 5.54,

$$f_{sr} = f_{rs} \tag{16.48}$$

Therefore, the flexibility matrix of Eq. 16.28, $[f]^j_{ii}$, is a symmetrical matrix. Taking the inverse of both sides of Eq. 16.47 yields

$$[k]^j_{ii} = [f]^{j-1}_{ii} \tag{16.49}$$

and since the inverse of a symmetrical matrix is itself a symmetrical matrix, we conclude that $[k]^j_{ii}$ is symmetrical and that

$$k_{sr} = k_{rs} \tag{16.50}$$

Of course, this holds true over the range of the elements for $[k]^j_{ii}$, the so-called reduced stiffness matrix, which includes the kinematic degrees of freedom of the restrained member. However, the member could be restrained by inhibiting different displacement components and, therefore, eventually all components could be included as part of the reduced stiffness matrix. Therefore, Eq. 16.50 applies for all combinations of r and s, and this proves that the entire stiffness matrix is symmetrical.

Examination of Eq. 16.43, therefore, clearly shows that the direct stiffness matrices, $[k]^j_{ii}$ and $[k]^i_{jj}$, are symmetrical matrices and that the cross stiffness matrices have a transpose relationship. That is,

$$[k]_{ij} = [k]^T_{ji}$$

and   (16.51)

$$[k]_{ji} = [k]^T_{ij}$$

16.9

Flexibility–Stiffness Transformations

Sections 16.6 and 16.7 traced the systematic development of the flexibility and stiffness matrices for member ij. The procedure used is as follows: The member was rendered stable by imposing displacement constraints at end j; for the resulting statically determinate structure, the displacements at end i were determined as a function of the forces at end i, thus establishing the flexibility matrix for the restrained structure; displacement compatibilities and equilibrium conditions were applied to the entire member to form the complete set of member-end force–deformation relationships from which the member stiffness matrix was extracted.

This procedure will now be accomplished through a formal matrix approach. The member introduced in Fig. 16.4a is used as a basis for the discussion. Figure 16.7a shows the member with the complete set of member coordinates, and Fig. 16.7b gives the restrained member, which is clamped at end j.

The member-end forces associated with the kinematic degrees of freedom for

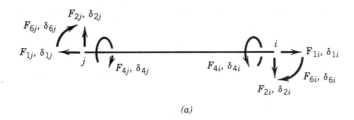

(a)

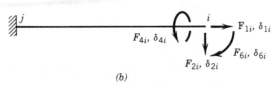

(b)

Fig. 16.7 *Member-end forces and displacements. (a) All member-end components. (b) Components corresponding to kinematic degrees of freedom.*

the structure of Fig. 16.7b, $\{F\}_I$, are separated from the member-end forces that correspond to the support reactions $\{F\}_{II}$. That is,

$$\{F\}_I = \begin{Bmatrix} F_{1i} \\ F_{2i} \\ F_{4i} \\ F_{6i} \end{Bmatrix}; \qquad \{F\}_{II} = \begin{Bmatrix} F_{1j} \\ F_{2j} \\ F_{4j} \\ F_{6j} \end{Bmatrix} \tag{16.52}$$

The member-end displacement components are similarly separated into the vectors $\{\delta\}_I$ and $\{\delta\}_{II}$. According to this separation, Eq. 16.44 can be written in a partitioned form as

$$\begin{Bmatrix} \{F\}_I \\ \{F\}_{II} \end{Bmatrix} = \begin{bmatrix} [k]_{I,I} & [k]_{I,II} \\ [k]_{II,I} & [k]_{II,II} \end{bmatrix} \begin{Bmatrix} \{\delta\}_I \\ \{\delta\}_{II} \end{Bmatrix} \tag{16.53}$$

where the subscripts on the $[k]$ submatrices designate, respectively, the $\{F\}$ vector that is related to the $\{\delta\}$ vector.

Since $\{\delta\}_{II} = \{0\}$, Eq. 16.53 can be expanded to

$$\{F\}_I = [k]_{I,I}\{\delta\}_I \tag{16.54}$$

and

$$\{F\}_{II} = [k]_{II,I}\{\delta\}_I \tag{16.55}$$

where $[k]_{I,I}$ is the reduced stiffness matrix, which was designated as $[k]_r$ in Section 16.5.

Since the restrained member of Fig. 16.7b is secure against rigid-body motions, $[k]_{I,I}$ is the inverse of the flexibility matrix $[f]$. That is,

$$[k]_{I,I} = [f]^{-1} \tag{16.56}$$

As was stated earlier, the structure of Fig. 16.7b is a statically determinate structure. Therefore, the support forces (reactions) can be determined as a function of the applied forces through the application of the equations of equilibrium. That is, we may write

$$\{F\}_{II} = [b]\{F\}_I \tag{16.57}$$

where $[b]$ is the equilibrium matrix. Substitution of Eq. 16.56 into Eq. 16.54 and of the resulting expression into Eq. 16.57 gives

$$\{F\}_{II} = [b][f]^{-1}\{\delta\}_I \tag{16.58}$$

Comparing Eqs. 16.55 and 16.58, we see that

$$[k]_{II,I} = [b][f]^{-1} \tag{16.59}$$

Since the total stiffness matrix is symmetrical, as was verified in Section 16.8, we have

$$[k]_{I,II} = [k]_{II,I}^T = ([b][f]^{-1})^T = [f]^{-1}[b]^T \tag{16.60}$$

In the above matrix manipulation, it is noted that $([f]^{-1})^T = [f]^{-1}$ because of the symmetrical nature of $[f]$.

In forming $[k]_{II,II}$, we return to Eq. 16.57 and note that $[b]$ is the equilibrium matrix that relates the $\{F\}_{II}$ forces to the $\{F\}_I$ forces. Thus, based on the interpretation of the subscripts on the $[k]$ submatrices, it is clear that

$$[k]_{II,II} = [b][k]_{I,II} \tag{16.61}$$

or, substituting Eq. 16.60 into Eq. 16.61, we have

$$[k]_{\text{II,II}} = [b][f]^{-1}[b]^T \qquad (16.62)$$

Collection of the individual submatrices from Eqs. 16.56, 16.59, 16.60, and 16.62 according to the requirements of Eq. 16.53 leads to

$$[k] = \left[\begin{array}{c|c} [f]^{-1} & [f]^{-1}[b]^T \\ \hline [b][f]^{-1} & [b][f]^{-1}[b]^T \end{array} \right] \qquad (16.63)$$

This equation expresses the general transformation from the flexibility matrix for the restrained structure to the total stiffness matrix that includes all degrees of freedom, including those corresponding to rigid-body motion. The number of components in $\{F\}_{\text{II}}$ is limited by the requirements that the restrained structure be statically determinate; however, there is no restriction on the number of components in $\{F\}_{\text{I}}$.

In the presentation given here, the focus was on an individual member. However, this member is treated as a structure in itself, and therefore the procedure is easily extended to more complex structures in which the total structure stiffness matrix is determined from the flexibility matrix for the stable system.

16.9.1 Example problem   Use the flexibility matrix for the restrained structure of Fig. 16.4a (Eq. 16.27) to develop the total stiffness matrix for the member shown below.

Flexibility Matrix: The member is fixed at point j, and the flexibility matrix relates $\{\delta\}_{\text{I}}$ to $\{F\}_{\text{I}}$ as follows:

$$\{\delta\}_{\text{I}} = [f]\{F\}_{\text{I}}$$

$$\left\{ \begin{array}{c} \delta_1 \\ \delta_2 \\ \delta_4 \\ \delta_6 \end{array} \right\}_i = \left[\begin{array}{cccc} \dfrac{l}{EA} & 0 & 0 & 0 \\ 0 & \dfrac{l^3}{3EI_z} & 0 & \dfrac{l^2}{2EI_z} \\ 0 & 0 & \dfrac{l}{GJ} & 0 \\ 0 & \dfrac{l^2}{2EI_z} & 0 & \dfrac{l}{EI_z} \end{array} \right]_{ij} \left\{ \begin{array}{c} F_1 \\ F_2 \\ F_4 \\ F_6 \end{array} \right\}_i \qquad (16.27)$$

Note:
$\{F\}_{\text{I}}$ and $\{\delta\}_{\text{I}}$ are defined by Eq. 16.52. The quantity $\lambda l / GA$ is set equal to zero.

Equilibrium Matrix: The equilibrium matrix, $[b]$, relates the support forces, $\{F\}_{\text{II}}$, to the free-end forces, $\{F\}_{\text{I}}$.

$$\{F\}_{\text{II}} = [b]\{F\}_{\text{I}} \qquad (16.57)$$

$$
\begin{aligned}
F_{1j} &= F_{1i} \\
F_{2j} &= F_{2i} \\
F_{4j} &= F_{4i} \\
F_{6j} &= -F_{2i}l - F_{6i}
\end{aligned}
\qquad
\begin{Bmatrix} F_1 \\ F_2 \\ F_4 \\ F_6 \end{Bmatrix}_j
=
\begin{bmatrix}
1 & 0 & 0 & 0 \\
0 & 1 & 0 & 0 \\
0 & 0 & 1 & 0 \\
0 & -l & 0 & -1
\end{bmatrix}_{ij}
\begin{Bmatrix} F_1 \\ F_2 \\ F_4 \\ F_6 \end{Bmatrix}_i
$$

Note:

$\{F\}_\mathrm{I}$ and $\{F\}_\mathrm{II}$ are defined by Eq. 16.52.

Stiffness Matrices:

$$
[k]_{\mathrm{I,I}} = [f]^{-1} =
\begin{bmatrix}
\dfrac{EA}{l} & 0 & 0 & 0 \\[2mm]
0 & \dfrac{12EI_z}{l^3} & 0 & \dfrac{-6EI_z}{l^2} \\[2mm]
0 & 0 & \dfrac{GJ}{l} & 0 \\[2mm]
0 & \dfrac{-6EI_z}{l^2} & 0 & \dfrac{4EI_z}{l}
\end{bmatrix}
\tag{16.56}
$$

$$
[k]_{\mathrm{II,I}} = [b][f]^{-1} =
\begin{bmatrix}
\dfrac{EA}{l} & 0 & 0 & 0 \\[2mm]
0 & \dfrac{12EI_z}{l^3} & 0 & \dfrac{-6EI_z}{l^2} \\[2mm]
0 & 0 & \dfrac{GJ}{l} & 0 \\[2mm]
0 & \dfrac{-6EI_z}{l^2} & 0 & \dfrac{2EI_z}{l}
\end{bmatrix}
\tag{16.59}
$$

$$
[k]_{\mathrm{I,II}} = [f]^{-1}[b]^T = [k]_{\mathrm{II,I}}^T
\tag{16.60}
$$

$$
[k]_{\mathrm{II,II}} = [b][f]^{-1}[b]^T =
\begin{bmatrix}
\dfrac{EA}{l} & 0 & 0 & 0 \\[2mm]
0 & \dfrac{12EI_z}{l^3} & 0 & \dfrac{-6EI_z}{l^2} \\[2mm]
0 & 0 & \dfrac{GJ}{l} & 0 \\[2mm]
0 & -\dfrac{6EI_z}{l^2} & 0 & \dfrac{4EI_z}{l}
\end{bmatrix}
\tag{16.62}
$$

The total stiffness matrix is assembled according to Eq. 16.63, and it is seen to agree with Eq. 16.41.

16.10

Laursen, H. I., *Matrix Analysis of Structures*, Chapter 2, McGraw–Hill, New York, 1966. **Additional**

Livesley, R. K., *Matrix Methods of Structural Analysis*, Chapter 5, Pergamon Press, Mac- **Reading**
   millan Co., New York, 1964.

McGuire, W., and Gallagher, R. H., *Matrix Structural Analysis*, Chapters 2 and 4, Wiley,
   New York, 1979.

16.11

**Suggested
Problems**

1. Use the virtual work method to determine the complete flexibility matrix for the structure shown with the prescribed boundary conditions. Use the coordinate system indicated, and compare your results with Eq. 16.10.

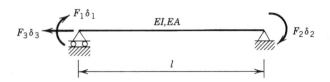

2. Use Castigliano's second theorem to determine the complete stiffness matrix for the structure shown using the given coordinate system. Compare your results with Eq. 16.6. *Hint:* Generate force–displacement equations and solve for the forces (stiffness coefficients) for the appropriate unit displacements.

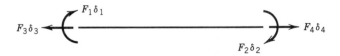

3 and 4. Consider the member shown below with the indicated member-end actions and deformations.

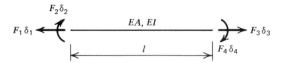

3. Use Castigliano's second theorem to determine the flexibility coefficients f_{11}, f_{12}, f_{21}, and f_{22} for the given member subject to the support conditions shown below. Assemble these flexibility coefficients into the flexibility matrix $[f]$.

4. Rewrite the stiffness matrix $[k]$, which is given as Eq. 16.6, in accordance with the member-end actions and deformations noted for the given member. Extract the appropriate reduced stiffness matrix $[k]_r$ for the restrained member of Problem 3, and show that $[k]_r[f] = [I]$, where $[f]$ is the flexibility matrix of Problem 3.

5 through 7. Consider the member shown below with the indicated member-end actions and deformations.

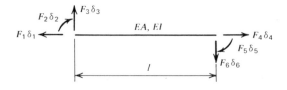

5. Use the virtual work method to determine the flexibility coefficients f_{11}, f_{12}, f_{13}, f_{21}, f_{22}, f_{23}, f_{31}, f_{32}, and f_{33} for the given member subject to the support conditions shown below. Assemble these coefficients into the flexibility matrix $[f]$. Consider only the bending deformations associated with the shear forces F_3 and F_6.

6. Use Castigliano's second theorem to determine the full 6×6 stiffness matrix $[k]$ for the given member. Again, consider only flexural deformations associated with the shear forces.

7. Extract the reduced stiffness matrix $[k]_r$ from the results of Problem 6 for the restrained member of Problem 5. Show that $[k]_r[f] = [I]$, where $[f]$ is the flexibility matrix of Problem 5.

8. Use the flexibility matrix for the restrained structure of Problem 1 (Eq. 16.10) and the transformation procedures of Section 16.9 to establish the total stiffness matrix for the structure of Problem 2 (Eq. 16.6).

9. Use the flexibility matrix for the restrained structure of Problem 5 and the transformation procedures of Section 16.9 to generate the full 6×6 stiffness matrix called for in Problem 6.

CHAPTER 17

*John Hancock
Center, Chicago,
Illinois (courtesy of
Skidmore, Owing &
Merrill, architects/
engineers, photo by
Ezra Stoller Esto).*

STIFFNESS METHOD: GENERAL FORM OF EQUILIBRIUM METHOD

17.1

Fundamental Concepts of Stiffness Method

Throughout this textbook, the importance of equilibrium has been stressed. It plays a central role in the analysis of any structure, whether statically determinate or indeterminate. In those methods that have been specifically identified as equilibrium methods, the solution results from simultaneously satisfying a number of equations that represent equilibrium throughout the structure. The solution of these equilibrium equations yields displacement quantities; thus, these techniques are sometimes identified as displacement methods.

When equilibrium methods are formulated in a general matrix format, the resulting equations contain the structure stiffness matrix, which is derived from a synthesis of the individual member stiffness matrices. For this reason, the matrix method, which results from a generalization of the equilibrium approach, is referred to as the *stiffness method*.

Consider the framed structure given in Fig. 17.1a in which loads are admissible
only at the points that are identified as nodes or joints on the structure. Figure
17.1b shows the structure forces and displacements that are operative at each of
the four node points. The relationship between these structure forces and displacements is given by

$$\{P\} = [K]\{\Delta\} \tag{17.1}$$

where $\{P\}$ and $\{\Delta\}$ contain all of the structure forces and displacements, with the
individual components as defined in Section 16.2. The matrix $[K]$ is the *total
structure stiffness matrix*. In an expanded form, Eq. 17.1 can be written as

$$
\begin{Bmatrix}
(P_1)_1 \\
(P_2)_1 \\
(P_6)_1 \\
\cdot \\
\cdot \\
\cdot \\
\cdot \\
\cdot \\
\cdot \\
(P_1)_4 \\
(P_2)_4 \\
(P_6)_4
\end{Bmatrix}
= [K]
\begin{Bmatrix}
(\Delta_1)_1 \\
(\Delta_2)_1 \\
(\Delta_6)_1 \\
\cdot \\
\cdot \\
\cdot \\
\cdot \\
\cdot \\
\cdot \\
(\Delta_1)_4 \\
(\Delta_2)_4 \\
(\Delta_6)_4
\end{Bmatrix}
\tag{17.2}
$$

This matrix equation represents 12 simultaneous equations relating the 12 structure
forces to the 12 structure displacements. Figure 17.1a shows that there are some
structure displacements that are constrained by the boundary conditions. Separat-
ing these displacements into the vector $\{\Delta\}_{II}$ and collecting the remaining free
displacements into the vector $\{\Delta\}_I$, we have

$$\{\Delta\}_I^T = \{(\Delta_1)_1 \ (\Delta_2)_1 \ (\Delta_6)_1 \ (\Delta_1)_3 \ (\Delta_2)_3 \ (\Delta_6)_3 \ (\Delta_2)_4\} \tag{17.3}$$
$$\{\Delta\}_{II}^T = \{(\Delta_1)_2 \ (\Delta_2)_2 \ (\Delta_6)_2 \ (\Delta_1)_4 \ (\Delta_6)_4\}$$

The load vector can similarly be separated into the applied loads $\{P\}_I$, which correspond, element for element, with the free displacements, and the reaction forces
$\{P\}_{II}$, which are associated with the constrained boundary displacements. These
vectors take the form

$$\{P\}_I^T = \{(P_1)_2 \ (P_2)_2 \ (P_6)_2 \ (P_1)_4 \ (P_6)_4\} \tag{17.4}$$
$$\{P\}_{II}^T = \{(P_1)_1 \ (P_2)_1 \ (P_6)_1 \ (P_1)_3 \ (P_2)_3 \ (P_6)_3 \ (P_2)_4\}$$

Rearranging the rows and columns of Eq. 17.2 in accordance with the ordering of
Eqs. 17.3 and 17.4, we have

$$
\begin{Bmatrix}
\{P\}_I \\
\hline
\{P\}_{II}
\end{Bmatrix}
=
\begin{bmatrix}
[K]_{I,I} & [K]_{I,II} \\
\hline
[K]_{II,I} & [K]_{II,II}
\end{bmatrix}
\begin{Bmatrix}
\{\Delta\}_I \\
\hline
\{\Delta\}_{II}
\end{Bmatrix}
\tag{17.5}
$$

It should be noted that Eq. 17.5 is not restricted to the illustrative problem of
Fig. 17.1—it would apply for any structure in which $\{P\}_I$ and $\{\Delta\}_I$ represent the
applied forces and free displacements, respectively, and $\{P\}_{II}$ and $\{\Delta\}_{II}$ give the
reaction forces and the constrained displacements, respectively.

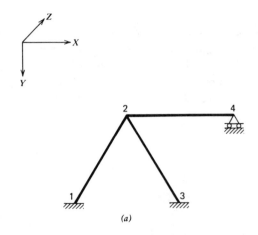

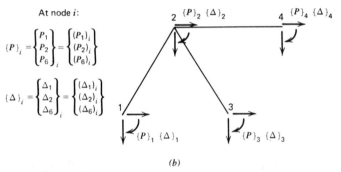

Fig. 17.1 *Typical planar frame structure.* (a) *Given structure and node points.* (b) *Structure loads and displacements.*

Expansion of Eq. 17.5 gives

$$\{P\}_{\mathrm{I}} = [K]_{\mathrm{I,I}}\{\Delta\}_{\mathrm{I}} + [K]_{\mathrm{I,II}}\{\Delta\}_{\mathrm{II}} \qquad (17.6)$$

$$\{P\}_{\mathrm{II}} = [K]_{\mathrm{II,I}}\{\Delta\}_{\mathrm{I}} + [K]_{\mathrm{II,II}}\{\Delta\}_{\mathrm{II}} \qquad (17.7)$$

However, from the boundary conditions

$$\{\Delta\}_{\mathrm{II}} = \{0\} \qquad (17.8)$$

and thus Eq. 17.6 reduces to

$$\{P\}_{\mathrm{I}} = [K]_{\mathrm{I,I}}\{\Delta\}_{\mathrm{I}} \qquad (17.9)$$

This equation gives the relationship between the applied forces and the free displacements, and $[K]_{\mathrm{I,I}}$ is the *reduced stiffness matrix*.

Equation 17.1 shows that a given set of structure displacements $\{\Delta\}$ produces a unique set of structure forces $\{P\}$. However, any attempt to solve this equation for $\{\Delta\}$ will fail because $[K]$ does not possess an inverse. That is, no unique set of displacements is associated with a given set of forces. This fact is illustrated by noting that a rigid-body motion of the entire structure will alter the displacements without changing the forces. This argument is similar to the one used in Sections 16.5 and 16.7, in which the member stiffness matrix was found to be singular. However, Eq. 17.9 gives the force–displacement relations for a structure that is kinematically stable and is restrained against rigid-body motion; thus, $[K]_{\mathrm{I,I}}$ possesses an inverse, and

$$\{\Delta\}_{\mathrm{I}} = [K]_{\mathrm{I,I}}^{-1}\{P\}_{\mathrm{I}} \qquad (17.10)$$

If the reaction forces are desired, Eqs. 17.8 and 17.10 are substituted into Eq. 17.7 to obtain

$$\{P\}_{\mathrm{II}} = [K]_{\mathrm{II,I}}\{\Delta\}_{\mathrm{I}} = [K]_{\mathrm{II,I}}[K]_{\mathrm{I,I}}^{-1}\{P\}_{\mathrm{I}} \qquad (17.11)$$

Before the analysis can develop in the direction presented in this section, it is necessary to generate the total structure stiffness matrix $[K]$ from which the reduced structure stiffness matrix $[K]_{\mathrm{I,I}}$ is extracted. The following section of this chapter presents the technique for generating the structure stiffness matrix from the individual member stiffness matrices.

Actually, it is not necessary that the boundary conditions prescribe zero displacements as is indicated by Eq. 17.8. Instead, a set of boundary conditions might be prescribed that includes some support settlements. In this case,

$$\{\Delta\}_{\mathrm{II}} = \{\Delta\}_{S} \qquad (17.12)$$

For this case, Eq. 17.10 is replaced by

$$\{\Delta\}_{\mathrm{I}} = [K]_{\mathrm{I,I}}^{-1}(\{P\}_{\mathrm{I}} - [K]_{\mathrm{I,II}}\{\Delta\}_{S}) \qquad (17.13)$$

The reaction forces are then determined from

$$\{P\}_{\mathrm{II}} = [K]_{\mathrm{II,I}}[K]_{\mathrm{I,I}}^{-1}(\{P\}_{\mathrm{I}} - [K]_{\mathrm{I,II}}\{\Delta\}_{S}) + [K]_{\mathrm{II,II}}\{\Delta\}_{S} \qquad (17.14)$$

17.3
Generation of Structure Stiffness Matrix

The total array of equilibrium equations for the entire structure is given by Eq. 17.1. This equation can be written in the form

$$\begin{Bmatrix} \{P\}_1 \\ \vdots \\ \vdots \\ \{P\}_i \\ \vdots \\ \vdots \\ \{P\}_n \end{Bmatrix} = \begin{bmatrix} & & & & \\ \cdots & [K]_{im} & \cdots & [K]_{ii} & \cdots & [K]_{ir} & \cdots \\ & & & & \end{bmatrix} \begin{Bmatrix} \vdots \\ \{\Delta\}_m \\ \vdots \\ \{\Delta\}_i \\ \vdots \\ \{\Delta\}_r \\ \vdots \end{Bmatrix} \qquad (17.15)$$

In this form, $\{P\}_i$ and $\{\Delta\}_i$ are the structure force and displacement vectors at node i, respectively, for a structure containing n nodes. The ith node of the structure is shown in Fig. 17.2a with members $mi \cdots ji \cdots ri$ framing into joint i. The matrix $[K]_{im}$ is a structure stiffness submatrix that relates the forces at node i to the displacements at node m. Figure 17.2b shows that the elements of this matrix will depend on the stiffness characteristics of the deformed member im. The sub-

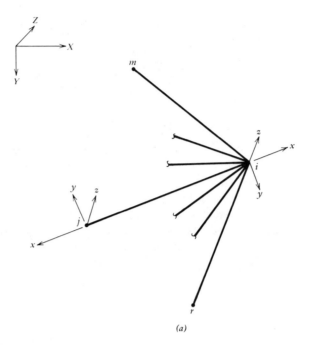

(a)

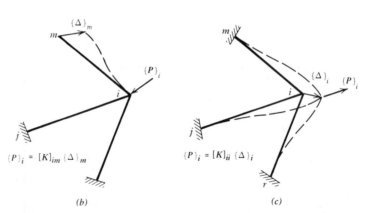

(b)

(c)

Fig. 17.2 *Stiffness relations for the ith node.* (a) *Members framing into node* i.
(b) $\{P\}_i = [K]_{im}\{\Delta\}_m$. (c) $\{P\}_i = [K]_{ii}\{\Delta\}_i$.

matrix $[K]_{ii}$ gives the forces at node i that are associated with the displacements at the same node. Figure 17.2c shows that these displacements induce deformations into all of the members that frame into node i, and thus the elements of this submatrix will contain a superposition of the stiffness characteristics of all the members. This superposition can be expressed in the form

$$[K]_{ii} = [K]_{ii}^m + \cdots + [K]_{ii}^j + \cdots + [K]_{ii}^r$$

$$= \sum_{j=m}^{r} [K]_{ii}^j$$

(17.16)

where $[K]_{ii}^j$ relates the structure forces induced at node i, through the deformations of member ij, to the displacements of node i. The structure force vector at any given node can be expressed in terms of the structure displacements throughout the structure. For instance, extraction of the expression for $\{P\}_i$ from Eq. 17.15 and substitution of Eq. 17.16 gives

$$\{P\}_i = \sum_{j=m}^r ([K]_{ii}^j\{\Delta\}_i + [K]_{ij}\{\Delta\}_j) \tag{17.17}$$

From Eq. 16.43, the member-end forces $\{F\}_{ij}$ can be expressed in terms of the member-end displacements $\{\delta\}_{ij}$ and $\{\delta\}_{ji}$ in the form

$$\{F\}_{ij} = [k]_{ii}^j\{\delta\}_{ij} + [k]_{ij}\{\delta\}_{ji} \tag{17.18}$$

where $[k]_{ii}^j$ and $[k]_{ij}$ are member stiffness submatrices as defined in Section 16.7. If the individual members are to fit compatibly into the structure, there must be a relationship between the member-end displacements and the structure displacements. For instance, at node i

$$\{\delta\}_{ij} = [\beta]_{ij}\{\Delta\}_i \tag{17.19}$$

and at node j

$$\{\delta\}_{ji} = [\beta]_{ji}\{\Delta\}_j \tag{17.20}$$

where $[\beta]_{ij}$ and $[\beta]_{ji}$ are the *compatibility* or *connectivity matrices* at ends i and j, respectively. Substitution of Eqs. 17.19 and 17.20 into Eq. 17.18 gives

$$\{F\}_{ij} = [k]_{ii}^j[\beta]_{ij}\{\Delta\}_i + [k]_{ij}[\beta]_{ji}\{\Delta\}_j \tag{17.21}$$

From energy considerations, the work done at joint i must be the same whether expressed in structure quantities or member quantities. Thus, we can write

$$\tfrac{1}{2}\{P\}_i^T\{\Delta\}_i = \sum_{j=m}^r \tfrac{1}{2}\{F\}_{ij}^T\{\delta\}_{ij} \tag{17.22}$$

Substitution of Eq. 17.19 and cancellation of $\tfrac{1}{2}$ gives

$$\{P\}_i^T\{\Delta\}_i = \sum_{j=m}^r \{F\}_{ij}^T[\beta]_{ij}\{\Delta\}_i$$

Transposition of both sides and subsequent rearrangement give

$$\{\Delta\}_i^T\left(\{P\}_i - \sum_{j=m}^r [\beta]_{ij}^T\{F\}_{ij}\right) = 0 \tag{17.23}$$

Since $\{\Delta\}_i^T$ can be arbitrarily prescribed, the term in parentheses must vanish. This leads to

$$\{P\}_i = \sum_{j=m}^r [\beta]_{ij}^T\{F\}_{ij} \tag{17.24}$$

which states the condition of equilibrium at the ith node point. Substituting Eq. 17.21 into Eq. 17.24, we obtain

$$\{P\}_i = \sum_{j=m}^r ([\beta]_{ij}^T[k]_{ii}^j[\beta]_{ij}\{\Delta\}_i + [\beta]_{ij}^T[k]_{ij}[\beta]_{ji}\{\Delta\}_j) \tag{17.25}$$

A comparison of Eqs. 17.17 and 17.25 reveals that

$$[K]_{ii}^{j} = [\beta]_{ij}^{T}[k]_{ii}^{j}[\beta]_{ij} \tag{17.26}$$

$$[K]_{ij} = [\beta]_{ij}^{T}[k]_{ij}[\beta]_{ji} \tag{17.27}$$

Application of Eqs. 17.26 and 17.27 for all i and j along with the application of Eq. 17.16 leads to the submatrices that make up the total structure stiffness matrix of Eq. 17.15.

17.4

Compatibility Matrices

Equations 17.19 and 17.20 express relationships between the member-end displacements at ends i and j of member ij and the structure displacements of nodes i and j, respectively. In each case, the $[\beta]$ matrix provides the connectivity between the member end and the structure node to which it is attached and ensures that the structure remains compatibly intact as it deforms. In the terminology of matrix algebra, $[\beta]$ is a transformation matrix, and its precise nature depends on the member type and the member orientation with regard to the structure coordinate system.

Consider the planar truss-type member shown in Fig. 17.3a. For this case, Eq. 17.19 becomes

$$(\delta_1)_{ij} = [\beta]_{ij} \begin{Bmatrix} \Delta_1 \\ \Delta_2 \end{Bmatrix}_i \tag{17.28}$$

This equation relates the axial member-end displacement to the two independent structure displacements at node i, and it can be further expanded to

$$(\delta_1)_{ij} = [l_{ix} \quad m_{ix}] \begin{Bmatrix} \Delta_1 \\ \Delta_2 \end{Bmatrix}_i \tag{17.29}$$

In this form, l and m are the direction cosines relating the member axis to the structure axes. Specifically, l_{ix} is the cosine of the angle between the X axis and the x axis, and m_{ix} is the cosine of the angle between the Y axis and the x axis—both angles being measured clockwise at the i end of the member. Reference to Fig. 17.3a shows that

$$l_{ix} = \cos \theta \tag{17.30}$$

$$m_{ix} = \cos \phi = \sin \theta$$

and thus Eq. 17.29 becomes

$$(\delta_1)_{ij} = [\cos \theta \quad \sin \theta] \begin{Bmatrix} \Delta_1 \\ \Delta_2 \end{Bmatrix}_i \tag{17.31}$$

For a full three-dimensional truss-type member, Eq. 17.29 takes the form

$$(\delta_1)_{ij} = [l_{ix} \quad m_{ix} \quad n_{ix}] \begin{Bmatrix} \Delta_1 \\ \Delta_2 \\ \Delta_6 \end{Bmatrix}_i \tag{17.32}$$

where n_{ix} is the cosine of the angle between the Z axis and the x axis, measured clockwise at the i end of the member. The direction cosines in Eqs. 17.29 and 17.32 are defined in general terms as

$$l_{ix} = \frac{X_i - X_j}{L_{ij}}; \quad m_{ix} = \frac{Y_i - Y_j}{L_{ij}}; \quad n_{ix} = \frac{Z_i - Z_j}{L_{ij}} \tag{17.33}$$

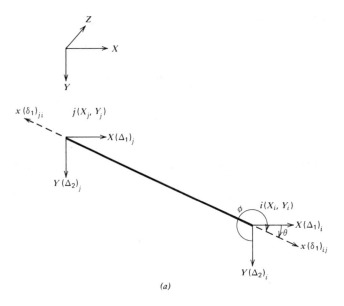

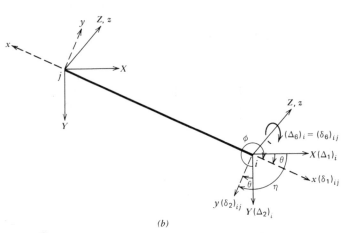

Fig. 17.3 *Relations between structure and member axes.* (a) *Planar truss-type member.*
(b) *Planar beam-type member.*

in which (X_i, Y_i, Z_i) and (X_j, Y_j, Z_j) are the structure coordinates of nodes i and j,
respectively, and

$$L_{ij} = \sqrt{(X_i - X_j)^2 + (Y_i - Y_j)^2 + (Z_i - Z_j)^2} \qquad (17.34)$$

At the j end of the truss member, the structure coordinate system remains the
same, while the member x axis is reversed. Thus, $[\beta]_{ji} = -[\beta]_{ij}$.

For a planar beam-type member, the compatibility matrix takes on a more
complicated format. Reference to Fig. 17.3b shows that in this case, Eq. 17.19
becomes

$$\left\{ \begin{array}{c} \delta_1 \\ \delta_2 \\ \delta_6 \end{array} \right\}_{ij} = [\beta]_{ij} \left\{ \begin{array}{c} \Delta_1 \\ \Delta_2 \\ \Delta_6 \end{array} \right\}_i \qquad (17.35)$$

In terms of direction cosines, this relationship can be expressed as

$$\begin{Bmatrix} \delta_1 \\ \delta_2 \\ \delta_6 \end{Bmatrix}_{ij} = \begin{bmatrix} l_{ix} & m_{ix} & n_{ix} \\ l_{iy} & m_{iy} & n_{iy} \\ l_{iz} & m_{iz} & n_{iz} \end{bmatrix} \begin{Bmatrix} \Delta_1 \\ \Delta_2 \\ \Delta_6 \end{Bmatrix}_i \tag{17.36}$$

where l_{iq}, m_{iq}, and n_{iq} are the direction cosines for the angles between the structure X, Y, and Z axes, respectively, and the member q axis. For the planar member shown in Fig. 17.3b, we have

$$l_{ix} = \frac{X_i - X_j}{L_{ij}} = \cos\theta; \qquad m_{ix} = \frac{Y_i - Y_j}{L_{ij}} = \cos\phi = \sin\theta; \qquad n_{ix} = \frac{Z_i - Z_j}{L_{ij}} = 0$$

$$l_{iy} = \cos\eta = -\sin\theta; \qquad m_{iy} = \cos\theta; \qquad n_{iy} = 0 \tag{17.37}$$

$$l_{iz} = 0; \qquad m_{iz} = 0; \qquad n_{iz} = 1$$

Using these direction cosines, we can write Eq. 17.36 as

$$\begin{Bmatrix} \delta_1 \\ \delta_2 \\ \delta_6 \end{Bmatrix}_{ij} = \begin{bmatrix} \cos\theta & \sin\theta & 0 \\ -\sin\theta & \cos\theta & 0 \\ 0 & 0 & 1 \end{bmatrix} \begin{Bmatrix} \Delta_1 \\ \Delta_2 \\ \Delta_6 \end{Bmatrix}_i \tag{17.38}$$

At the j end of a beam-type member, the structure coordinate system remains the same, as does the member z axis. However, the member x and y axes are reversed, and thus the direction cosines related to these axes have their signs reversed.

For a full three-dimensional arrangement of a beam-type member, Eq. 17.19 becomes

$$\begin{Bmatrix} \delta_1 \\ \delta_2 \\ \delta_3 \\ \delta_4 \\ \delta_5 \\ \delta_6 \end{Bmatrix}_{ij} = \begin{bmatrix} l_{ix} & m_{ix} & n_{ix} & & & \\ l_{iy} & m_{iy} & n_{iy} & & 0 & \\ l_{iz} & m_{iz} & n_{iz} & & & \\ & & & l_{ix} & m_{ix} & n_{ix} \\ & 0 & & l_{iy} & m_{iy} & n_{iy} \\ & & & l_{iz} & m_{iz} & n_{iz} \end{bmatrix} \begin{Bmatrix} \Delta_1 \\ \Delta_2 \\ \Delta_3 \\ \Delta_4 \\ \Delta_5 \\ \Delta_6 \end{Bmatrix}_i \tag{17.39}$$

The detailed procedures for determining the direction cosines for the three-dimensional situation are expanded in textbooks on matrix structural analysis.

17.5

Application of Stiffness Method

The stiffness method is applied in accordance with a very orderly procedure. The step-by-step approach follows:

1. Number the nodal points of the structure. These nodes must include all load points, support points, and points where two or more members join together. This process implicitly introduces the structure forces and displacements in the form of Eq. 16.1 and the member-end forces and displacements are defined by Eq. 16.2.

2. Form the individual compatibility matrices $[\beta]_{ij}$, which relate the member-end displacements of member ij to the structure displacements at node i for all combinations of i and j. The precise format of these matrices will depend on the arrays of member-end displacements and nodal displacements that must be connected, as explained in Section 17.4.

3. Establish the individual member stiffness matrices $[k]_{ii}^j$ and $[k]_{ij}$ in accordance with the designations of Eqs. 16.43 and 16.42 for all combinations of i and j

throughout the structure. The makeup of these stiffness matrices will depend on the member-end forces and displacements that are assumed operable for the individual members.

4. Generate the structure stiffness submatrices $[K]^j_{ii}$ and $[K]_{ij}$ in accordance with Eqs. 17.26 and 17.27. These submatrices are combined to form the total structure stiffness matrix, as outlined by Eqs. 17.16 and 17.15.

5. Extract the reduced stiffness matrix $[K]_{I,I}$ from the total stiffness matrix by deleting the rows and columns corresponding to the displacement components that are controlled through the boundary conditions.

6. Determine the free displacements $\{\Delta\}_I$ from Eq. 17.10 or 17.13. If desired, the member-end displacements may be determined from Eq. 17.19.

7. Calculate the final member-end forces from Eq. 17.21 or Eq. 17.18 and the reaction forces from Eq. 17.11 or 17.14.

The example problems that follow in this section illustrate the step-by-step procedure for analyzing a structure by the stiffness method.

It should be noted that the stiffness method makes no distinction between statically determinate and indeterminate structures. Thus, the method eliminates the need for the analyst to select redundant forces, and this precludes all of the problems that attend this decision-making process.

Determine the member-end forces and the displacements at the loaded point of the structure shown. Assume EA is the same for each member. The joint coordinates are given in parentheses. **17.5.1 Example problem**

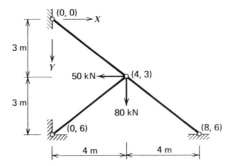

Selection of Node Points:

Structure forces and displacements:

$$\{P\}_1 = \begin{Bmatrix} P_1 \\ P_2 \end{Bmatrix}_1; \qquad \{\Delta\}_1 = \begin{Bmatrix} \Delta_1 \\ \Delta_2 \end{Bmatrix}_1$$

$$\{P\}_2 = \begin{Bmatrix} P_1 \\ P_2 \end{Bmatrix}_2 = \begin{Bmatrix} -50 \\ 80 \end{Bmatrix}; \qquad \{\Delta\}_2 = \begin{Bmatrix} \Delta_1 \\ \Delta_2 \end{Bmatrix}_2$$

$$\{P\}_4 = \begin{Bmatrix} P_1 \\ P_2 \end{Bmatrix}_4; \qquad \{\Delta\}_4 = \begin{Bmatrix} \Delta_1 \\ \Delta_2 \end{Bmatrix}_4$$

$$\{P\}_3 = \begin{Bmatrix} P_1 \\ P_2 \end{Bmatrix}_3; \qquad \{\Delta\}_3 = \begin{Bmatrix} \Delta_1 \\ \Delta_2 \end{Bmatrix}_3$$

Member forces and displacements:

$$(F_1)_{ij}; \quad (\delta_1)_{ij} \longleftarrow \overset{i}{\underline{\hspace{5cm}}} \overset{j}{\longrightarrow} (F_1)_{ji}; \quad (\delta_1)_{ji}$$

Compatibility Matrices:

$$(\delta_1)_{ij} = [\beta]_{ij} \begin{Bmatrix} \Delta_1 \\ \Delta_2 \end{Bmatrix}_i = [l_{ix} \quad m_{ix}] \begin{Bmatrix} \Delta_1 \\ \Delta_2 \end{Bmatrix}_i \tag{17.29}$$

where

$$l_{ix} = \frac{X_i - X_j}{L_{ij}}, \qquad m_{ix} = \frac{Y_i - Y_j}{L_{ij}} \tag{17.33}$$

Member 1–2: $l_{1x} = \dfrac{0 - 4}{5} = -0.8;\quad m_{1x} = \dfrac{0 - 3}{5} = -0.6$

$[\beta]_{12} = [l_{1x} \quad m_{1x}] = [-0.8 \quad -0.6]$

$[\beta]_{21} = [0.8 \quad 0.6]$

Member 2–3: $l_{2x} = \dfrac{4 - 0}{5} = 0.8;\quad m_{2x} = \dfrac{3 - 6}{5} = -0.6$

$[\beta]_{23} = [0.8 \quad -0.6]$

$[\beta]_{32} = [-0.8 \quad 0.6]$

Member 2–4: $l_{2x} = \dfrac{4 - 8}{5} = -0.8;\quad m_{2x} = \dfrac{3 - 6}{5} = -0.6$

$[\beta]_{24} = [-0.8 \quad -0.6]$

$[\beta]_{42} = [0.8 \quad 0.6]$

Member Stiffnesses:

$$\{F\}_{ij} = [k]_{ii}^{j}\{\delta\}_{ij} + [k]_{ij}\{\delta\}_{ji} \tag{16.43}$$

In this case

$$[k]_{ii}^{j} = [k]_{ij} = \left(\frac{EA}{l}\right)_{ij} \tag{16.42}$$

Member 1–2: $[k]_{11}^{2} = [k]_{12} = \left(\dfrac{EA}{l}\right)_{12} = 0.2EA$

$[k]_{22}^{1} = [k]_{21} = \left(\dfrac{EA}{l}\right)_{21} = 0.2EA$

Member 2–3: $[k]_{22}^{3} = [k]_{23} = [k]_{33}^{2} = [k]_{32} = 0.2EA$

Member 2–4: $[k]_{22}^{4} = [k]_{24} = [k]_{44}^{2} = [k]_{42} = 0.2EA$

Structure Stiffness Matrix:

Total stiffness equations:

$$\begin{Bmatrix} \{P\}_1 \\ \{P\}_2 \\ \{P\}_3 \\ \{P\}_4 \end{Bmatrix} = \begin{bmatrix} [K]_{11} & [K]_{12} & & \\ [K]_{21} & [K]_{22} & [K]_{23} & [K]_{24} \\ & [K]_{32} & [K]_{33} & \\ & [K]_{42} & & [K]_{44} \end{bmatrix} \begin{Bmatrix} \{\Delta\}_1 \\ \{\Delta\}_2 \\ \{\Delta\}_3 \\ \{\Delta\}_4 \end{Bmatrix} \tag{17.15}$$

Reduced stiffness equations:

$$\{P\}_1 = [K]_{1,1}\{\Delta\}_1 \rightarrow \{P\}_2 = [K]_{22}\{\Delta\}_2 \tag{17.9}$$

since

$$\{\Delta\}_1 = \{\Delta\}_3 = \{\Delta\}_4 = \{0\}$$

Required submatrices:

$$[K]_{ii}^{j} = [\beta]_{ij}^{T}[k]_{ii}^{j}[\beta]_{ij} \tag{17.26}$$

$$[K]_{ii} = \sum_{j} [K]_{ii}^{j} \tag{17.16}$$

$$[K]_{22}^{1} = [\beta]_{21}^{T}[k]_{22}^{1}[\beta]_{21} = \begin{bmatrix} 0.8 \\ 0.6 \end{bmatrix}(0.2EA)[0.8 \quad 0.6] = EA\begin{bmatrix} 0.128 & 0.096 \\ 0.096 & 0.072 \end{bmatrix}$$

$$[K]_{22}^{3} = [\beta]_{23}^{T}[k]_{22}^{3}[\beta]_{23} = \begin{bmatrix} 0.8 \\ -0.6 \end{bmatrix}(0.2EA)[0.8 \quad -0.6] = EA\begin{bmatrix} 0.128 & -0.096 \\ -0.096 & 0.072 \end{bmatrix}$$

$$[K]_{22}^{4} = [\beta]_{24}^{T}[k]_{22}^{4}[\beta]_{24} = \begin{bmatrix} -0.8 \\ -0.6 \end{bmatrix}(0.2EA)[-0.8 \quad -0.6] = EA\begin{bmatrix} 0.128 & 0.096 \\ 0.096 & 0.072 \end{bmatrix}$$

$$[K]_{22} = [K]_{22}^{1} + [K]_{22}^{3} + [K]_{22}^{4} = EA\begin{bmatrix} 0.384 & 0.096 \\ 0.096 & 0.216 \end{bmatrix} = \frac{EA}{1000}\begin{bmatrix} 384 & 96 \\ 96 & 216 \end{bmatrix}$$

Determination of Displacements:

$$\{\Delta\}_1 = [K]_{1,1}^{-1}\{P\}_1 \rightarrow \{\Delta\}_2 = [K]_{22}^{-1}\{P\}_2 \tag{17.10}$$

$$\{\Delta\}_2 = \frac{1}{73.728EA}\begin{bmatrix} 216 & -96 \\ -96 & 384 \end{bmatrix}\begin{Bmatrix} -50 \\ 80 \end{Bmatrix} = \frac{1}{EA}\begin{Bmatrix} -250.65 \\ 481.77 \end{Bmatrix}$$

Member Forces:

$$\{F\}_{ij} = [k]_{ii}^{j}[\beta]_{ij}\{\Delta\}_i + [k]_{ij}[\beta]_{ji}\{\Delta\}_j \tag{17.21}$$

Member 1–2: $\{F\}_{12} = [k]_{12}[\beta]_{21}\{\Delta\}_2 = (0.2EA)[0.8 \quad 0.6]\dfrac{1}{EA}\begin{Bmatrix} -250.65 \\ 481.77 \end{Bmatrix}$

$$= +17.71^{k}$$

Member 3–2: $\{F\}_{32} = [k]_{32}[\beta]_{23}\{\Delta\}_2 = (0.2EA)[0.8 \quad -0.6]\dfrac{1}{EA}\begin{Bmatrix} -250.65 \\ 481.77 \end{Bmatrix}$

$$= -97.92^{k}$$

Member 4–2: $\{F\}_{42} = [k]_{42}[\beta]_{24}\{\Delta\}_2 = (0.2EA)[-0.8 \quad -0.6]\dfrac{1}{EA}\begin{Bmatrix} -250.65 \\ 481.77 \end{Bmatrix}$

$$= -17.71^{k}$$

17.5.2 Example problem   Determine the member-end forces for the structure shown, and construct the shear and moment diagrams. The joint coordinates are given in parentheses.

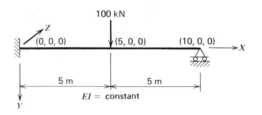

Selection of Node Points:

Structure forces and displacements:

$$\{P\}_i = \begin{Bmatrix} P_2 \\ P_6 \end{Bmatrix}_i = \begin{Bmatrix} (P_2)_i \\ (P_6)_i \end{Bmatrix} \quad \{P\}_1 \quad \{P\}_2 \quad \{P\}_3$$

$$\{\Delta\}_i = \begin{Bmatrix} \Delta_2 \\ \Delta_6 \end{Bmatrix}_i = \begin{Bmatrix} (\Delta_2)_i \\ (\Delta_6)_i \end{Bmatrix} \quad \{\Delta\}_1 \quad \{\Delta\}_2 \quad \{\Delta\}_3$$

Member forces and displacements:

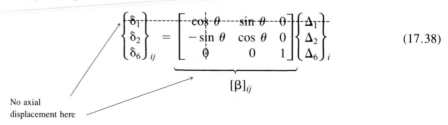

$$\{F\}_{ij} = \begin{Bmatrix} F_2 \\ F_6 \end{Bmatrix}_{ij}; \quad \{F\}_{ji} = \begin{Bmatrix} F_2 \\ F_6 \end{Bmatrix}_{ji}; \quad \{\delta\}_{ij} = \begin{Bmatrix} \delta_2 \\ \delta_6 \end{Bmatrix}_{ij}; \quad \{\delta\}_{ji} = \begin{Bmatrix} \delta_2 \\ \delta_6 \end{Bmatrix}_{ji}$$

Member 1–2: $\{F\}_{12} \quad \{F\}_{21} \quad \{\delta\}_{12} \quad \{\delta\}_{21}$

Member 2–3: $\{F\}_{23} \quad \{F\}_{32} \quad \{\delta\}_{23} \quad \{\delta\}_{32}$

Compatibility Matrices:

$$\begin{Bmatrix} \delta_1 \\ \delta_2 \\ \delta_6 \end{Bmatrix}_{ij} = \begin{bmatrix} \cos\theta & \sin\theta & 0 \\ -\sin\theta & \cos\theta & 0 \\ 0 & 0 & 1 \end{bmatrix} \begin{Bmatrix} \Delta_1 \\ \Delta_2 \\ \Delta_6 \end{Bmatrix}_i \qquad (17.38)$$

$$\underbrace{\hspace{4cm}}_{[\beta]_{ij}}$$

No axial
displacement here

Member 1–2:

$$\cos \theta = 1_{1x} = \frac{X_1 - X_2}{L_{12}} = \frac{0 - 5}{5} = -1; \qquad \sin \theta = 0 \qquad (17.37)$$

$$[\beta]_{12} = \begin{bmatrix} -1 & 0 & 0 \\ 0 & -1 & 0 \\ 0 & 0 & 1 \end{bmatrix}$$

$$[\beta]_{21} = \begin{bmatrix} 1 & 0 & 0 \\ 0 & 1 & 0 \\ 0 & 0 & 1 \end{bmatrix}$$

Member 2–3:

$$\cos \theta = l_{2x} = \frac{X_2 - X_3}{L_{23}} = \frac{5 - 10}{5} = -1; \qquad \sin \theta = 0 \qquad (17.37)$$

$$[\beta]_{23} = \begin{bmatrix} -1 & 0 & 0 \\ 0 & -1 & 0 \\ 0 & 0 & 1 \end{bmatrix}$$

$$[\beta]_{32} = \begin{bmatrix} 1 & 0 & 0 \\ 0 & 1 & 0 \\ 0 & 0 & 1 \end{bmatrix}$$

Member Stiffnesses: (from Eqs. 16.42 and 16.43)

$$\text{Member 1–2: } [k]_{11}^2 = [k]_{22}^1 = \begin{bmatrix} \dfrac{12EI_z}{l^3} & -\dfrac{6EI_z}{l^2} \\ -\dfrac{6EI_z}{l^2} & \dfrac{4EI_z}{l} \end{bmatrix}_{12} = EI \begin{bmatrix} 0.096 & -0.24 \\ -0.24 & 0.80 \end{bmatrix}$$

$$[k]_{12} = [k]_{21} = \begin{bmatrix} \dfrac{12EI_z}{l^3} & -\dfrac{6EI_z}{l^2} \\ -\dfrac{6EI_z}{l^2} & \dfrac{2EI_z}{l} \end{bmatrix}_{12} = EI \begin{bmatrix} 0.096 & -0.24 \\ -0.24 & 0.40 \end{bmatrix}$$

$$\text{Member 2–3: } [k]_{22}^3 = [k]_{33}^2 = EI \begin{bmatrix} 0.096 & -0.24 \\ -0.24 & 0.80 \end{bmatrix}$$

$$[k]_{23} = [k]_{32} = EI \begin{bmatrix} 0.096 & -0.24 \\ -0.24 & 0.40 \end{bmatrix}$$

Structure Stiffness Matrix:

Total stiffness equations:

$$\begin{Bmatrix} \begin{Bmatrix} P_2 \\ P_6 \end{Bmatrix}_1 \\ \begin{Bmatrix} P_2 \\ P_6 \end{Bmatrix}_2 \\ \begin{Bmatrix} P_2 \\ P_6 \end{Bmatrix}_3 \end{Bmatrix} = \begin{bmatrix} [K]_{11} & [K]_{12} & [0] \\ [K]_{21} & [K]_{22} & [K]_{23} \\ [0] & [K]_{32} & [K]_{33} \end{bmatrix} \begin{Bmatrix} \begin{Bmatrix} \Delta_2 \\ \Delta_6 \end{Bmatrix}_1 \\ \begin{Bmatrix} \Delta_2 \\ \Delta_6 \end{Bmatrix}_2 \\ \begin{Bmatrix} \Delta_2 \\ \Delta_6 \end{Bmatrix}_3 \end{Bmatrix} \qquad (17.15)$$

Reduced stiffness equations:

$$\{P\}_I = [K]_{I,I}\{\Delta\}_I \rightarrow \left\{\begin{Bmatrix} P_2 \\ P_6 \end{Bmatrix}_2 \\ \{P_6\}_3 \end{Bmatrix}\right\} = [K]_{I,I} \left\{\begin{Bmatrix} \Delta_2 \\ \Delta_6 \end{Bmatrix}_2 \\ \{\Delta_6\}_3 \end{Bmatrix}\right\} \tag{17.9}$$

since

$$(\Delta_2)_1 = (\Delta_6)_1 = (\Delta_2)_3 = 0$$

Required submatrices:

$$[K]^j_{ii} = [\beta]^T_{ij}[k]^j_{ii}[\beta]_{ij} \tag{17.26}$$

$$[K]_{ii} = \sum_j [K]^j_{ii} \tag{17.16}$$

$$[K]^1_{22} = [\beta]^T_{21}[k]^1_{22}[\beta]_{21} = \begin{bmatrix} 1 & 0 \\ 0 & 1 \end{bmatrix} EI \begin{bmatrix} 0.096 & -0.24 \\ -0.24 & 0.80 \end{bmatrix} \begin{bmatrix} 1 & 0 \\ 0 & 1 \end{bmatrix}$$

$$= EI \begin{bmatrix} 0.096 & -0.24 \\ -0.24 & 0.80 \end{bmatrix}$$

$$[K]^3_{22} = [\beta]^T_{23}[k]^3_{22}[\beta]_{23} = \begin{bmatrix} -1 & 0 \\ 0 & 1 \end{bmatrix} EI \begin{bmatrix} 0.096 & -0.24 \\ -0.24 & 0.80 \end{bmatrix} \begin{bmatrix} -1 & 0 \\ 0 & 1 \end{bmatrix}$$

$$= EI \begin{bmatrix} 0.096 & 0.24 \\ 0.24 & 0.80 \end{bmatrix}$$

$$[K]_{22} = [K]^1_{22} + [K]^3_{22} = EI \begin{bmatrix} 0.192 & 0 \\ 0 & 1.60 \end{bmatrix}$$

$$[K]_{33} = [K]^2_{33} = [\beta]^T_{32}[k]^2_{33}[\beta]_{32} = \begin{bmatrix} 1 & 0 \\ 0 & 1 \end{bmatrix} EI \begin{bmatrix} 0.096 & -0.24 \\ -0.24 & 0.80 \end{bmatrix} \begin{bmatrix} 1 & 0 \\ 0 & 1 \end{bmatrix}$$

$$= EI \begin{bmatrix} 0.096 & -0.24 \\ -0.24 & 0.80 \end{bmatrix}$$

$$[K]_{ij} = [\beta]^T_{ij}[k]_{ij}[\beta]_{ji} \tag{17.27}$$

$$[K]_{23} = [\beta]^T_{23}[k]_{23}[\beta]_{32} = \begin{bmatrix} -1 & 0 \\ 0 & 1 \end{bmatrix} EI \begin{bmatrix} 0.096 & -0.24 \\ -0.24 & 0.40 \end{bmatrix} \begin{bmatrix} 1 & 0 \\ 0 & 1 \end{bmatrix}$$

$$= EI \begin{bmatrix} -0.096 & 0.24 \\ -0.24 & 0.40 \end{bmatrix}$$

$$[K]_{32} = [\beta]^T_{32}[k]_{32}[\beta]_{23} = [K]^T_{23} = EI \begin{bmatrix} -0.096 & -0.24 \\ 0.24 & 0.40 \end{bmatrix}$$

Thus, the reduced stiffness matrix becomes

$$[K]_{I,I} = EI \begin{bmatrix} 0.192 & 0 & 0.24 \\ 0 & 1.60 & 0.40 \\ 0.24 & 0.40 & 0.80 \end{bmatrix}$$

Determination of Displacements:

$$\{\Delta\}_I = [K]_{I,I}^{-1}\{P\}_I \tag{17.10}$$

$$\left\{\begin{Bmatrix}\Delta_2\\\Delta_6\end{Bmatrix}_2\\\{\Delta_6\}_3\right\} = \frac{1}{EI}\begin{bmatrix}0.192 & 0 & 0.24\\0 & 1.60 & 0.40\\0.24 & 0.40 & 0.80\end{bmatrix}^{-1}\left\{\begin{Bmatrix}P_2\\P_6\end{Bmatrix}_2\\\{P_6\}_3\right\}$$

$$= \frac{1}{EI}\begin{bmatrix}9.1146 & 0.7813 & -3.125\\0.7813 & 0.7813 & -0.6250\\-3.125 & -0.6250 & 2.500\end{bmatrix}\begin{Bmatrix}100\\0\\0\end{Bmatrix}$$

$$= \frac{1}{EI}\begin{Bmatrix}911.46\\78.13\\-312.50\end{Bmatrix}$$

Member Forces:

$$\{F\}_{ij} = [k]_{ii}^{j}[\beta]_{ij}\{\Delta\}_i + [k]_{ij}[\beta]_{ji}\{\Delta\}_j \tag{17.21}$$

Member 1–2: $\{\Delta\}_1 = \{0\}$; $\{\Delta\}_2 = \dfrac{1}{EI}\begin{Bmatrix}911.46\\78.13\end{Bmatrix}$

$$\{F\}_{12} = [k]_{12}[\beta]_{21}\{\Delta\}_2$$

$$= EI\begin{bmatrix}0.096 & -0.24\\-0.24 & 0.40\end{bmatrix}\begin{bmatrix}1 & 0\\0 & 1\end{bmatrix}\frac{1}{EI}\begin{Bmatrix}911.46\\78.13\end{Bmatrix}$$

$$= \begin{Bmatrix}68.75\ \text{kN}\\-187.50\ \text{kN}\cdot\text{m}\end{Bmatrix}$$

$$\{F\}_{21} = [k]_{22}^{1}[\beta]_{21}\{\Delta\}_2$$

$$= EI\begin{bmatrix}0.096 & -0.24\\-0.24 & 0.80\end{bmatrix}\begin{bmatrix}1 & 0\\0 & 1\end{bmatrix}\frac{1}{EI}\begin{Bmatrix}911.46\\78.13\end{Bmatrix}$$

$$= \begin{Bmatrix}68.75\ \text{kN}\\-156.25\ \text{kN}\cdot\text{m}\end{Bmatrix}$$

Member 2–3: $\{\Delta\}_2 = \dfrac{1}{EI}\begin{Bmatrix}911.46\\78.13\end{Bmatrix}$; $\{\Delta\}_3 = \dfrac{1}{EI}\begin{Bmatrix}0\\-312.50\end{Bmatrix}$

$$\{F\}_{23} = [k]_{22}^{3}[\beta]_{23}\{\Delta\}_2 + [k]_{23}[\beta]_{32}\{\Delta\}_3$$

$$= EI\begin{bmatrix}0.096 & -0.24\\-0.24 & 0.80\end{bmatrix}\begin{bmatrix}-1 & 0\\0 & 1\end{bmatrix}\frac{1}{EI}\begin{Bmatrix}911.46\\78.13\end{Bmatrix}$$

$$+ EI\begin{bmatrix}0.096 & -0.24\\-0.24 & 0.40\end{bmatrix}\begin{bmatrix}1 & 0\\0 & 1\end{bmatrix}\frac{1}{EI}\begin{Bmatrix}0\\-312.50\end{Bmatrix}$$

$$= \begin{Bmatrix}-106.25\\281.25\end{Bmatrix} + \begin{Bmatrix}75.00\\-125.00\end{Bmatrix}$$

$$= \begin{Bmatrix}-31.25\ \text{kN}\\156.25\ \text{kN}\cdot\text{m}\end{Bmatrix}$$

$$\{F\}_{32} = [k]_{33}^2[\beta]_{32}\{\Delta\}_3 + [k]_{32}[\beta]_{23}\{\Delta\}_2$$

$$= EI\begin{bmatrix} 0.096 & -0.24 \\ -0.24 & 0.80 \end{bmatrix}\begin{bmatrix} 1 & 0 \\ 0 & 1 \end{bmatrix}\frac{1}{EI}\begin{Bmatrix} 0 \\ -312.50 \end{Bmatrix}$$

$$+ EI\begin{bmatrix} 0.096 & -0.24 \\ -0.24 & 0.40 \end{bmatrix}\begin{bmatrix} -1 & 0 \\ 0 & 1 \end{bmatrix}\frac{1}{EI}\begin{Bmatrix} 911.46 \\ 78.13 \end{Bmatrix}$$

$$= \begin{Bmatrix} 75.00 \\ -250.00 \end{Bmatrix} + \begin{Bmatrix} -106.25 \\ +250.00 \end{Bmatrix}$$

$$= \begin{Bmatrix} -31.25 \text{ kN} \\ 0 \quad \text{kN} \cdot \text{m} \end{Bmatrix}$$

Shear and Moment Diagrams:

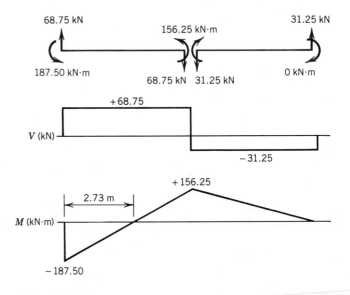

Note:

We start with the forces at the left end of each member and use the static relations for shear and moment to obtain the forces at the right end of the member. These should be consistent with the computed forces at the right end.

The sign convention used for the shear and moment diagrams is that which was introduced in Chapter 4.

Determine the member-end forces for the given frame structure. The joint coordinates are noted in parentheses.

17.5.3 Example problem

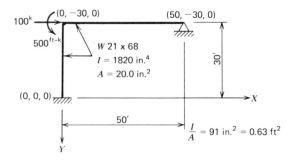

Selection of Node Points:

Structure forces and displacements:

$$\{P\}_i = \begin{Bmatrix} P_1 \\ P_2 \\ P_6 \end{Bmatrix}_i = \begin{Bmatrix} (P_1)_i \\ (P_2)_i \\ (P_6)_i \end{Bmatrix}$$

$$\{\Delta\}_i = \begin{Bmatrix} \Delta_1 \\ \Delta_2 \\ \Delta_6 \end{Bmatrix}_i = \begin{Bmatrix} (\Delta_1)_i \\ (\Delta_2)_i \\ (\Delta_6)_i \end{Bmatrix}$$

$$\{P\}_2 = \begin{Bmatrix} 100 \\ 0 \\ -500 \end{Bmatrix}$$

$$\{P\}_3 \quad \{\Delta\}_3 = \begin{Bmatrix} 0 \\ 0 \\ (\Delta_6)_3 \end{Bmatrix}$$

$$\{P\}_1 \quad \{\Delta\}_1 = \begin{Bmatrix} 0 \\ 0 \\ 0 \end{Bmatrix}$$

Member forces and displacements:

$$\{F\}_{ij} = \begin{Bmatrix} F_1 \\ F_2 \\ F_6 \end{Bmatrix}_{ij} ; \quad \{F\}_{ji} = \begin{Bmatrix} F_1 \\ F_2 \\ F_6 \end{Bmatrix}_{ji} ; \quad \{\delta\}_{ij} = \begin{Bmatrix} \delta_1 \\ \delta_2 \\ \delta_6 \end{Bmatrix}_{ij} ; \quad \{\delta\}_{ji} = \begin{Bmatrix} \delta_1 \\ \delta_2 \\ \delta_6 \end{Bmatrix}_{ji}$$

Member 1–2: $\{F\}_{12}$ $\{F\}_{21}$, $\{\delta\}_{12}$ $\{\delta\}_{21}$

Member 2–3: $\{F\}_{23}$ $\{F\}_{32}$ $\{\delta\}_{23}$ $\{\delta\}_{32}$

Compatibility Matrices:

$$\begin{Bmatrix} \delta_1 \\ \delta_2 \\ \delta_6 \end{Bmatrix}_{ij} = \underbrace{\begin{bmatrix} \cos\theta & \sin\theta & 0 \\ -\sin\theta & \cos\theta & 0 \\ 0 & 0 & 1 \end{bmatrix}}_{[\beta]_{ij}} \begin{Bmatrix} \Delta_1 \\ \Delta_2 \\ \Delta_6 \end{Bmatrix}_i \qquad (17.38)$$

Member 1–2:

$$\cos \theta = l_{1x} = \frac{X_1 - X_2}{L_{12}} = \frac{0 - 0}{30} = 0$$

$$\sin \theta = m_{1x} = \frac{Y_1 - Y_2}{L_{12}} = \frac{0 - (-30)}{30} = 1 \tag{17.37}$$

$$[\beta]_{12} = \begin{bmatrix} 0 & 1 & 0 \\ -1 & 0 & 0 \\ 0 & 0 & 1 \end{bmatrix}; \qquad [\beta]_{21} = \begin{bmatrix} 0 & -1 & 0 \\ 1 & 0 & 0 \\ 0 & 0 & 1 \end{bmatrix}$$

Member 2–3:

$$\cos \theta = l_{2x} = \frac{X_2 - X_3}{L_{23}} = \frac{0 - 50}{50} = -1$$

$$\sin \theta = m_{2x} = \frac{Y_2 - Y_3}{L_{23}} = \frac{-30 - (-30)}{50} = 0 \tag{17.37}$$

$$[\beta]_{23} = \begin{bmatrix} -1 & 0 & 0 \\ 0 & -1 & 0 \\ 0 & 0 & 1 \end{bmatrix}; \qquad [\beta]_{32} = \begin{bmatrix} 1 & 0 & 0 \\ 0 & 1 & 0 \\ 0 & 0 & 1 \end{bmatrix}$$

Member Stiffnesses: (Extract from Eqs. 16.42 and 16.43)

$$[k]_{ii}^{j} = \begin{bmatrix} \dfrac{EA}{l} & 0 & 0 \\[2mm] 0 & \dfrac{12EI_z}{l^3} & \dfrac{-6EI_z}{l^2} \\[2mm] 0 & \dfrac{-6EI_z}{l_2} & \dfrac{4EI_z}{l} \end{bmatrix}_{ij}$$

$$\tag{16.42, 16.43}$$

$$[k]_{ij} = \begin{bmatrix} \dfrac{EA}{l} & 0 & 0 \\[2mm] 0 & \dfrac{12EI_z}{l^3} & \dfrac{-6EI_z}{l^2} \\[2mm] 0 & \dfrac{-6EI_z}{l^2} & \dfrac{2EI_z}{l} \end{bmatrix}_{ij}$$

Member 1–2: $l = 30'$

$$[k]_{11}^{2} = [k]_{22}^{1} = EI \begin{bmatrix} 0.052,9 & 0 & 0 \\ 0 & 0.000,444 & -0.006,67 \\ 0 & -0.006,67 & 0.133 \end{bmatrix}$$

$$[k]_{12} = [k]_{21} = EI \begin{bmatrix} 0.052,9 & 0 & 0 \\ 0 & 0.000,444 & -0.006,67 \\ 0 & -0.006,67 & 0.066,7 \end{bmatrix}$$

Member 2–3: $l = 50'$

$$[k]_{22}^{3} = [k]_{33}^{2} = EI \begin{bmatrix} 0.031,7 & 0 & 0 \\ 0 & 0.000,096,0 & -0.002,40 \\ 0 & -0.002,40 & 0.080,0 \end{bmatrix}$$

$$[k]_{23} = [k]_{32} = EI \begin{bmatrix} 0.031,7 & 0 & 0 \\ 0 & 0.000,096,0 & -0.002,40 \\ 0 & -0.002,40 & 0.040,0 \end{bmatrix}$$

Structure Stiffness Matrix:

Total stiffness equations:

$$
\left\{
\begin{array}{c}
\left\{\begin{array}{c} P_1 \\ P_2 \\ P_6 \end{array}\right\}_1 \\
\left\{\begin{array}{c} P_1 \\ P_2 \\ P_6 \end{array}\right\}_2 \\
\left\{\begin{array}{c} P_1 \\ P_2 \\ P_6 \end{array}\right\}_3
\end{array}
\right\}
=
\left[
\begin{array}{c|c|c}
[K]_{11} & [K]_{12} & [0] \\
\hline
[K]_{21} & [K]_{22} & [K]_{23} \\
\hline
[0] & [K]_{32} & [K]_{33}
\end{array}
\right]
\left\{
\begin{array}{c}
\left\{\begin{array}{c} \Delta_1 \\ \Delta_2 \\ \Delta_6 \end{array}\right\}_1 \\
\left\{\begin{array}{c} \Delta_1 \\ \Delta_2 \\ \Delta_6 \end{array}\right\}_2 \\
\left\{\begin{array}{c} \Delta_1 \\ \Delta_2 \\ \Delta_6 \end{array}\right\}_3
\end{array}
\right\}
\qquad (17.15)
$$

Reduced stiffness equations:

$$
\{P\}_I = [K]_{I,I}\{\Delta\}_I \Rightarrow
\left\{
\begin{array}{c}
\left\{\begin{array}{c} P_1 \\ P_2 \\ P_6 \end{array}\right\}_2 \\
\{P_6\}_3
\end{array}
\right\}
= [K]_{I,I}
\left\{
\begin{array}{c}
\left\{\begin{array}{c} \Delta_1 \\ \Delta_2 \\ \Delta_6 \end{array}\right\}_2 \\
\{\Delta_6\}_3
\end{array}
\right\}
\qquad (17.9)
$$

since $\{\Delta\}_1 = [0]$ and $(\Delta_1)_3 = (\Delta_2)_3 = 0$

Required submatrics:

$$
[K]_{ii}^j = [\beta]_{ij}^T [k]_{ii}^j [\beta]_{ij} \qquad (17.26)
$$

$$
[K]_{ii} = \sum_j [K]_{ii}^j \qquad (17.16)
$$

$[K]_{22}^1 = [\beta]_{21}^T [k]_{22}^1 [\beta]_{21}$

$$
=
\begin{bmatrix} 0 & 1 & 0 \\ -1 & 0 & 0 \\ 0 & 0 & 1 \end{bmatrix}
\cdot EI
\begin{bmatrix} 0.052,9 & 0 & 0 \\ 0 & 0.000,444 & -0.006,67 \\ 0 & -0.006,67 & 0.133 \end{bmatrix}
\begin{bmatrix} 0 & -1 & 0 \\ 1 & 0 & 0 \\ 0 & 0 & 1 \end{bmatrix}
$$

$$
= EI
\begin{bmatrix} 0.000,444 & 0 & -0.006,67 \\ 0 & 0.052,9 & 0 \\ -0.006,67 & 0 & 0.133 \end{bmatrix}
$$

$[K]_{22}^3 = [\beta]_{23}^T [k]_{22}^3 [\beta]_{23}$

$$
=
\begin{bmatrix} -1 & 0 & 0 \\ 0 & -1 & 0 \\ 0 & 0 & 1 \end{bmatrix}
\cdot EI
\begin{bmatrix} 0.031,7 & 0 & 0 \\ 0 & 0.000,096,0 & -0.002,40 \\ 0 & -0.002,40 & 0.080,0 \end{bmatrix}
\begin{bmatrix} -1 & 0 & 0 \\ 0 & -1 & 0 \\ 0 & 0 & 1 \end{bmatrix}
$$

$$
= EI
\begin{bmatrix} 0.031,7 & 0 & 0 \\ 0 & 0.000,096,0 & 0.002,40 \\ 0 & 0.002,40 & 0.080,0 \end{bmatrix}
$$

$[K]_{22} = [K]_{22}^1 + [K]_{22}^3$

$$
= EI
\begin{bmatrix} 0.032,14 & 0 & -0.006,67 \\ 0 & 0.053,0 & 0.002,40 \\ -0.006,67 & 0.002,40 & 0.213 \end{bmatrix}
$$

$$[K]_{33} = [K]_{33}^2 = [\beta]_{32}^T[k]_{33}^2[\beta]_{32}$$

$$= \begin{bmatrix} 1 & 0 & 0 \\ 0 & 1 & 0 \\ 0 & 0 & 1 \end{bmatrix} \cdot EI \begin{bmatrix} 0.031,7 & 0 & 0 \\ 0 & 0.000,096,0 & -0.002,40 \\ 0 & -0.002,40 & 0.080,0 \end{bmatrix} \begin{bmatrix} 1 & 0 & 0 \\ 0 & 1 & 0 \\ 0 & 0 & 1 \end{bmatrix}$$

$$= EI \begin{bmatrix} 0.031,7 & 0 & 0 \\ 0 & 0.000,096,0 & -0.002,40 \\ 0 & -0.002,40 & 0.080,0 \end{bmatrix}$$

$$[K]_{ij} = [\beta]_{ij}^T[k]_{ij}[\beta]_{ji} \tag{17.27}$$

$$[K]_{23} = [\beta]_{23}^T[k]_{23}[\beta]_{32}$$

$$= \begin{bmatrix} -1 & 0 & 0 \\ 0 & -1 & 0 \\ 0 & 0 & 1 \end{bmatrix} \cdot EI \begin{bmatrix} 0.031,7 & 0 & 0 \\ 0 & 0.000,096,0 & -0.002,40 \\ 0 & -0.002,40 & 0.040,0 \end{bmatrix} \begin{bmatrix} 1 & 0 & 0 \\ 0 & 1 & 0 \\ 0 & 0 & 1 \end{bmatrix}$$

$$= EI \begin{bmatrix} -0.031,7 & 0 & 0 \\ 0 & -0.000,096,0 & 0.002,40 \\ 0 & -0.002,40 & 0.040,0 \end{bmatrix}$$

$$[K]_{32} = [\beta]_{32}^T[k]_{32}[\beta]_{23} = [K]_{23}^T$$

$$= EI \begin{bmatrix} -0.031,7 & 0 & 0 \\ 0 & -000,096,0 & -0.002,40 \\ 0 & 0.002,40 & 0.040,0 \end{bmatrix}$$

Thus, the reduced stiffness matrix becomes

$$[K]_{I,I} = EI \begin{bmatrix} 0.032,14 & 0 & -0.006,67 & 0 \\ 0 & 0.053,0 & 0.002,40 & 0.002,40 \\ -0.006,67 & 0.002,40 & 0.213 & 0.040,0 \\ 0 & 0.002,40 & 0.040,0 & 0.080,0 \end{bmatrix}$$

Determination of Displacements:

$$\{\Delta\}_I = [K]_{I,I}^{-1}\{P\} \tag{17.10}$$

$$\left\{ \begin{Bmatrix} \Delta_1 \\ \Delta_2 \\ \Delta_6 \end{Bmatrix}_2 \\ \{\Delta_6\}_3 \right\} = [K]_{I,I}^{-1} \begin{Bmatrix} 100 \\ 0 \\ -500 \\ 0 \end{Bmatrix} = \frac{1}{EI} \begin{Bmatrix} 2,592 \\ 56.71 \\ -2,501 \\ 1,249 \end{Bmatrix}$$

Member Forces:

$$\{F\}_{ij} = [k]_{ii}^j[\beta]_{ij}\{\Delta\}_i + [k]_{ij}[\beta]_{ji}\{\Delta\}_j \tag{17.21}$$

Member 1–2: $\{\Delta\}_1 = \{0\}$ $\{\Delta\}_2 = \dfrac{1}{EI} \begin{Bmatrix} 2,592 \\ 56.71 \\ -2,501 \end{Bmatrix}$

$$\{F\}_{12} = [k]_{12}[\beta]_{21}\{\Delta\}_2$$

$$= EI \begin{bmatrix} 0.052,9 & 0 & 0 \\ 0 & 0.000,444 & -0.006,67 \\ 0 & -0.006,67 & 0.066,7 \end{bmatrix} \begin{bmatrix} 0 & -1 & 0 \\ 1 & 0 & 1 \\ 0 & 0 & 1 \end{bmatrix} \frac{1}{EI} \begin{Bmatrix} 2,592 \\ 56.71 \\ -2,501 \end{Bmatrix}$$

$$= \begin{Bmatrix} -3.00^k \\ 17.83^k \\ -184.00'^{-k} \end{Bmatrix}$$

$$\{F\}_{21} = [k]_{22}^{1}[\beta]_{21}\{\Delta\}_2$$

$$= EI \begin{bmatrix} 0.052,9 & 0 & 0 \\ 0 & 0.000,444 & -0.006,67 \\ 0 & -0.006,67 & 0.133 \end{bmatrix} \begin{bmatrix} 0 & -1 & 0 \\ 1 & 0 & 0 \\ 0 & 0 & 1 \end{bmatrix} \frac{1}{EI} \begin{Bmatrix} 2,592 \\ 56.71 \\ -2,501 \end{Bmatrix}$$

$$= \begin{Bmatrix} -3.00^k \\ 17.83^k \\ -349.92'^{-k} \end{Bmatrix}$$

$$\text{Member 2–3: } \{\Delta\}_2 = \frac{1}{EI} \begin{Bmatrix} 2,592 \\ 56.71 \\ -2,501 \end{Bmatrix} \qquad \{\Delta\}_3 = \frac{1}{EI} \begin{Bmatrix} 0 \\ 0 \\ 1,249 \end{Bmatrix}$$

$$\{F\}_{23} = [k]_{22}^{3}[\beta]_{23}\{\Delta\}_2 + [k]_{23}[\beta]_{32}\{\Delta\}_3$$

$$= EI \begin{bmatrix} 0.031,7 & 0 & 0 \\ 0 & 0.000,096,0 & -0.002,40 \\ 0 & -0.002,40 & 0.080,0 \end{bmatrix} \begin{bmatrix} -1 & 0 & 0 \\ 0 & -1 & 0 \\ 0 & 0 & 1 \end{bmatrix} \frac{1}{EI} \begin{Bmatrix} 2,592 \\ 56.71 \\ -2,501 \end{Bmatrix}$$

$$+ EI \begin{bmatrix} 0.031,7 & 0 & 0 \\ 0 & 0.000,096,0 & -0.002,40 \\ 0 & -0.002,40 & 0.040,0 \end{bmatrix} \begin{bmatrix} 1 & 0 & 0 \\ 0 & 1 & 0 \\ 0 & 0 & 1 \end{bmatrix} \frac{1}{EI} \begin{Bmatrix} 0 \\ 0 \\ 1,249 \end{Bmatrix}$$

$$= \begin{Bmatrix} -82.17 + 0 \\ 6.00 - 3.00 \\ -199.95 + 49.96 \end{Bmatrix} = \begin{Bmatrix} -82.17^k \\ 3.00^k \\ -149.99'^{-k} \end{Bmatrix}$$

$$\{F\}_{32} = [k]_{33}^{2}[\beta]_{32}\{\Delta\}_3 + [k]_{32}[\beta]_{23}\{\Delta\}_2$$

$$= EI \begin{bmatrix} 0.031,7 & 0 & 0 \\ 0 & 0.000,096,0 & -0.002,40 \\ 0 & -0.002,40 & 0.080,0 \end{bmatrix} \begin{bmatrix} 1 & 0 & 0 \\ 0 & 1 & 0 \\ 0 & 0 & 1 \end{bmatrix} \frac{1}{EI} \begin{Bmatrix} 0 \\ 0 \\ 1,249 \end{Bmatrix}$$

$$+ EI \begin{bmatrix} 0.031,7 & 0 & 0 \\ 0 & 0.000,096,0 & -0.002,40 \\ 0 & -0.002,40 & 0.040,0 \end{bmatrix} \begin{bmatrix} -1 & 0 & 0 \\ 0 & -1 & 0 \\ 0 & 0 & 1 \end{bmatrix} \frac{1}{EI} \begin{Bmatrix} 2,592 \\ 56.21 \\ -2,501 \end{Bmatrix}$$

$$= \begin{Bmatrix} 0 - 82.17 \\ -3.00 + 6.00 \\ 99.92 - 99.91 \end{Bmatrix} = \begin{Bmatrix} -82.17^k \\ +3.00^k \\ 0.01'^{-k} \end{Bmatrix}$$

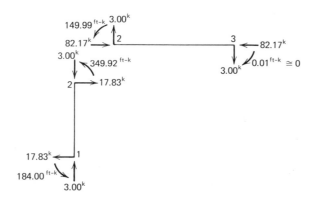

Notes:

- The shear and moment diagrams could be constructed in the normal fashion.
- If the normal frame analysis was used in which axial deformations are ignored, such as the slope deflection method, the member-end forces would be

$$\{F\}_{12} = \begin{Bmatrix} -3.10^k \\ 17.25^k \\ -172.5'^{-k} \end{Bmatrix}; \quad \{F\}_{23} = \begin{Bmatrix} -82.75^k \\ -3.10^k \\ -155.0'^{-k} \end{Bmatrix}$$

17.6

Loads Applied between Node Points— Equivalent Structure Forces

The stiffness method as it has been presented thus far admits the application of loads only at the structure nodes. This constitutes a serious constraint because it forces the analyst to assign a node to each point where a concentrated load is applied, and it precludes the application of distributed loads along the members. This limitation will now be relaxed.

Consider span ij of the planar beam in Fig. 17.4a with a transverse load acting along the member. Assume that the member is temporarily fixed at its ends as shown in Fig. 17.4b. The fixed-end member forces corresponding to this condition are determined in accordance with the formulae given in Table 13.1 and are arranged in the fixed-end member force vectors $\{F\}^f_{ij}$ and $\{F\}^f_{ji}$. The cases given in Table 13.1 are somewhat restrictive, and more complete treatments are available in standard matrix analysis textbooks. The cases given, however, are sufficient for our present considerations.

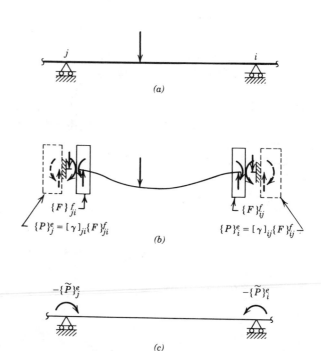

Fig. 17.4 *Loads applied along member lengths. (a) Span* ij *and loading. (b) Member fixed-end forces and equivalent structure forces. (c) Application of negative equivalent structure forces.*

Equilibrium requires that the fixed-end forces at the i end of member ij be accompanied by a set of equivalent structure forces at node i, $\{P\}_i^e$. These are shown in Fig. 17.4b and are expressed in the form

$$\{P\}_i^e = [\gamma]_{ij}\{F\}_{ij}^f \qquad (17.40)$$

where $[\gamma]_{ij}$ evolves from considerations of static equilibrium. Comparing Eq. 17.40 with Eq. 17.24, we can see that

$$[\gamma]_{ij} = [\beta]_{ij}^T \qquad (17.41)$$

where $[\beta]_{ij}$ is the compatibility matrix introduced in Section 17.4.

However, the given structure may not be capable of sustaining *all* of the required equivalent structure forces. For instance, in Fig. 17.4b, the vertical components of the equivalent forces at ends i and j can be sustained by the supports; however, the moment components cannot be sustained. Thus, those equivalent force components that cannot be sustained must be removed through the application of the negative of these components. These forces are denoted as $-\{\bar{P}\}_i^e$ and are shown in Fig. 17.4c. The final solution is given by the superposition of Figs. 17.4b and 17.4c.

If there are applied loads at note i that correspond to the components of $\{\bar{P}\}_i^e$, they must be included in the analysis. Thus, the stiffness analysis must consider the net load $\{P\}_i$ at each node as given by

$$\{P\}_i = \{P\}_i^a - \{\bar{P}\}_i^e \qquad (17.42)$$

where $\{P\}_i^a$ are the applied loads.

In an actual structure, more than one loaded span may join at node i, and thus

$$\{P\}_i^e = \sum_j [\beta]_{ij}^T\{F\}_{ij}^f \qquad (17.43)$$

where j ranges over all of the members framing into joint i.

The detailed application of the procedure is given in the example problem that follows. The general approach is the same as that outlined in Section 17.5 with the exceptions that the nodal numbering need not include all load points and the final member-end forces are given by

$$\{F\}_{ij} = [k]_{ii}^j[\beta]_{ij}\{\Delta\}_i + [k]_{ij}[\beta]_{ji}\{\Delta\}_j + \{F\}_{ij}^f \qquad (17.44)$$

Determine the member-end forces for the structure shown below.

17.6.1 Example problem

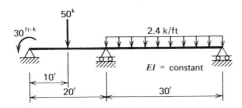

Selection of Node Points:

Structure forces and displacements:

$$\{P\}_i = \begin{Bmatrix} P_2 \\ P_6 \end{Bmatrix}_i \qquad \{P\}_1 \qquad \{P\}_2 \qquad \{P\}_3$$

$$\{\Delta\}_i = \begin{Bmatrix} \Delta_2 \\ \Delta_6 \end{Bmatrix}_i \qquad \{\Delta\}_1 \qquad \{\Delta\}_2 \qquad \{\Delta\}_3$$

Member forces and displacements:

$(F_2)_{ji}$

$(F_6)_{ji}$ j i $(F_6)_{ij}$

$(F_2)_{ij}$

Member 1–2: $\{F\}_{12} = \begin{Bmatrix} F_2 \\ F_6 \end{Bmatrix}_{12}; \quad \{\delta\}_{12} = \begin{Bmatrix} \delta_2 \\ \delta_6 \end{Bmatrix}_{12}$

Member 2–3: $\{F\}_{23} = \begin{Bmatrix} F_2 \\ F_6 \end{Bmatrix}_{23}; \quad \{\delta\}_{23} = \begin{Bmatrix} \delta_2 \\ \delta_6 \end{Bmatrix}_{23}$

Compatibility Matrices:

$$\begin{Bmatrix} \delta_2 \\ \delta_6 \end{Bmatrix}_{ij} = \begin{bmatrix} \cos\theta & 0 \\ 0 & 1 \end{bmatrix} \begin{Bmatrix} \Delta_2 \\ \Delta_6 \end{Bmatrix}_i \qquad (17.38)$$

Member 1–2: $\cos\theta = l_{1x} = \dfrac{X_1 - X_2}{L_{12}} = \dfrac{0 - 20}{20} = -1$

$$[\beta]_{12} = \begin{bmatrix} -1 & 0 \\ 0 & 1 \end{bmatrix} \qquad [\beta]_{21} = \begin{bmatrix} 1 & 0 \\ 0 & 1 \end{bmatrix}$$

Member 2–3: $\cos\theta = l_{2x} = \dfrac{X_2 - X_3}{L_{23}} = \dfrac{20 - 50}{30} = -1$

$$[\beta]_{23} = \begin{bmatrix} -1 & 0 \\ 0 & 1 \end{bmatrix} \qquad [\beta]_{32} = \begin{bmatrix} 1 & 0 \\ 0 & 1 \end{bmatrix}$$

Fixed-End Forces and Equivalent Structure Forces:

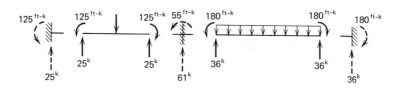

$$\{F\}^f_{12} = \left\{ \begin{array}{c} +25 \\ -125 \end{array} \right\} \qquad \{F\}^f_{21} = \left\{ \begin{array}{c} -25 \\ +125 \end{array} \right\}$$

$$\{F\}^f_{23} = \left\{ \begin{array}{c} +36 \\ -180 \end{array} \right\} \qquad \{F\}^f_{32} = \left\{ \begin{array}{c} -36 \\ +180 \end{array} \right\}$$

$$\{P\}^e_i = \sum_j [\beta]^T_{ij}\{F\}^f_{ij} \qquad\qquad (17.43)$$

$$\{P\}^e_1 = [\beta]^T_{12}\{F\}^f_{12} = \begin{bmatrix} -1 & 0 \\ 0 & 1 \end{bmatrix} \left\{ \begin{array}{c} +25 \\ -125 \end{array} \right\} = \left\{ \begin{array}{c} -25 \\ -125 \end{array} \right\}$$

$$\{P\}^e_2 = [\beta]^T_{21}\{F\}^f_{21} + [\beta]_{23}\{F\}^f_{23}$$

$$= \begin{bmatrix} 1 & 0 \\ 0 & 1 \end{bmatrix} \left\{ \begin{array}{c} -25 \\ +125 \end{array} \right\} + \begin{bmatrix} -1 & 0 \\ 0 & 1 \end{bmatrix} \left\{ \begin{array}{c} 36 \\ -180 \end{array} \right\} = \left\{ \begin{array}{c} -61 \\ -55 \end{array} \right\}$$

$$\{P\}^e_3 = [\beta]^T_{32}\{F\}^f_{32} = \begin{bmatrix} 1 & 0 \\ 0 & 1 \end{bmatrix} \left\{ \begin{array}{c} -36 \\ +180 \end{array} \right\} = \left\{ \begin{array}{c} -36 \\ +180 \end{array} \right\}$$

Components of $\{P\}^e_i$ sustained by supports:

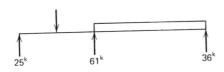

Loads for stiffness analysis:

$$\{P\}_i = \{P\}^a_i - \{\tilde{P}\}^e_i \qquad\qquad (17.42)$$

$125 - 30 = 95^{\text{ft-k}} \qquad 55^{\text{ft-k}} \qquad\qquad 180^{\text{ft-k}}$

Member Stiffnesses: (extract from Eqs. 16.42 and 16.43)

Member 1–2:

$$[k]^2_{11} = [k]^1_{22} = \begin{bmatrix} \dfrac{12EI_z}{l^3} & -\dfrac{6EI_z}{l^2} \\[2mm] -\dfrac{6EI_z}{l^2} & \dfrac{4EI_z}{l} \end{bmatrix}_{l=20'} = EI \begin{bmatrix} 0.001{,}5 & -0.015 \\ -0.015 & 0.200 \end{bmatrix}$$

$$[k]_{12} = [k]_{21} = \begin{bmatrix} \dfrac{12EI_z}{l^3} & -\dfrac{6EI_z}{l^2} \\[2mm] -\dfrac{6EI_z}{l^2} & \dfrac{2EI_z}{l} \end{bmatrix}_{l=20'} = EI \begin{bmatrix} 0.001{,}5 & -0.015 \\ -0.015 & 0.100 \end{bmatrix}$$

Member 2–3:

$$[k]^3_{22} = [k]^2_{33} = \begin{bmatrix} \dfrac{12EI_z}{l^3} & -\dfrac{6EI_z}{l^2} \\[2mm] -\dfrac{6EI_z}{l^2} & \dfrac{4EI_z}{l} \end{bmatrix}_{l=30'} = EI \begin{bmatrix} 0.000{,}444 & -0.006{,}67 \\ -0.006{,}67 & 0.133 \end{bmatrix}$$

$$[k]_{23} = [k]_{32} = \begin{bmatrix} \dfrac{12EI_z}{l^3} & \dfrac{-6EI_z}{l^2} \\ \dfrac{-6EI_z}{l^2} & \dfrac{2EI_z}{l} \end{bmatrix}_{l=30'} = EI \begin{bmatrix} 0.000,444 & -0.006,67 \\ -0.006,67 & 0.066,7 \end{bmatrix}$$

Structure Stiffness Matrix:

Total stiffness equations:

$$\begin{Bmatrix} \begin{Bmatrix} P_2 \\ P_6 \end{Bmatrix}_1 \\ \begin{Bmatrix} P_2 \\ P_6 \end{Bmatrix}_2 \\ \begin{Bmatrix} P_2 \\ P_6 \end{Bmatrix}_3 \end{Bmatrix} = \begin{bmatrix} [K]_{11} & [K]_{12} & [0] \\ [K]_{21} & [K]_{22} & [K]_{23} \\ [0] & [K]_{32} & [K]_{33} \end{bmatrix} \begin{Bmatrix} \begin{Bmatrix} \Delta_2 \\ \Delta_6 \end{Bmatrix}_1 \\ \begin{Bmatrix} \Delta_2 \\ \Delta_6 \end{Bmatrix}_2 \\ \begin{Bmatrix} \Delta_2 \\ \Delta_6 \end{Bmatrix}_3 \end{Bmatrix} \qquad (17.15)$$

Reduced stiffness equations:

$$[P]_I = [K]_{I,I}\{\Delta\}_I \Rightarrow \begin{Bmatrix} (P_6)_1 \\ (P_6)_2 \\ (P_6)_3 \end{Bmatrix} = [K]_{I,I} \begin{Bmatrix} (\Delta_6)_1 \\ (\Delta_6)_2 \\ (\Delta_6)_3 \end{Bmatrix} \qquad (17.9)$$

since $(\Delta_2)_1 = (\Delta_2)_2 = (\Delta_2)_3 = 0$

Required submatrices:

$$[K]_{ii}^j = [\beta]_{ij}^T[k]_{ii}^j[\beta]_{ij} \qquad (17.26)$$
$$[K]_{ii} = \sum_j [K]_{ii}^j \qquad (17.16)$$

$$[K]_{11}^2 = [\beta]_{12}^T[k]_{11}^2[\beta]_{12}$$

$$= \begin{bmatrix} -1 & 0 \\ 0 & 1 \end{bmatrix} \cdot EI \begin{bmatrix} 0.001,5 & -0.015 \\ -0.015 & 0.200 \end{bmatrix} \begin{bmatrix} -1 & 0 \\ 0 & 1 \end{bmatrix}$$

$$= EI \begin{bmatrix} 0.001,5 & 0.015 \\ 0.015 & 0.200 \end{bmatrix}$$

$$[K]_{22}^1 = [\beta]_{21}^T[k]_{22}^1[\beta]_{21}$$

$$= \begin{bmatrix} 1 & 0 \\ 0 & 1 \end{bmatrix} \cdot EI \begin{bmatrix} 0.001,5 & -0.015 \\ -0.015 & 0.200 \end{bmatrix} \begin{bmatrix} 1 & 0 \\ 0 & 1 \end{bmatrix}$$

$$= EI \begin{bmatrix} 0.001,5 & -0.015 \\ -0.015 & 0.200 \end{bmatrix}$$

$$[K]_{22}^3 = [\beta]_{23}^T[k]_{22}^3[\beta]_{23}$$

$$= \begin{bmatrix} -1 & 0 \\ 0 & 1 \end{bmatrix} \cdot EI \begin{bmatrix} 0.000,444 & -0.006,67 \\ -0.006,67 & 0.133 \end{bmatrix} \begin{bmatrix} -1 & 0 \\ 0 & 1 \end{bmatrix}$$

$$= EI \begin{bmatrix} 0.000,444 & 0.006,67 \\ 0.006,67 & 0.133 \end{bmatrix}$$

$$[K]_{33}^2 = [\beta]_{32}^T[k]_{33}^2[\beta]_{32}$$

$$= \begin{bmatrix} 1 & 0 \\ 0 & 1 \end{bmatrix} \cdot EI \begin{bmatrix} 0.000,444 & -0.006,67 \\ -0.006,67 & 0.133 \end{bmatrix} \begin{bmatrix} 1 & 0 \\ 0 & 1 \end{bmatrix}$$

$$= EI \begin{bmatrix} 0.000,444 & -0.006,67 \\ -0.006,67 & 0.133 \end{bmatrix}$$

$$[K]_{11} = [K]_{11}^2 = EI \begin{bmatrix} 0.001,5 & 0.015 \\ 0.015 & 0.200 \end{bmatrix}$$

$$[K]_{22} = [K]_{22}^1 + [K]_{22}^3 = EI \begin{bmatrix} 0.001,94 & -0.008,33 \\ -0.008,33 & 0.333 \end{bmatrix}$$

$$[K]_{33} = [K]_{33}^2 = EI \begin{bmatrix} 0.000,444 & -0.006,67 \\ -0.006,67 & 0.133 \end{bmatrix}$$

$$[K]_{ij} = [\beta]_{ij}^T[k]_{ij}[\beta]_{ji} \qquad (17.27)$$

$$[K]_{12} = [\beta]_{12}^T[k]_{12}[\beta]_{21}$$

$$= \begin{bmatrix} -1 & 0 \\ 0 & 1 \end{bmatrix} \cdot EI \begin{bmatrix} 0.001,5 & -0.015 \\ -0.015 & 0.100 \end{bmatrix} \begin{bmatrix} 1 & 0 \\ 0 & 1 \end{bmatrix}$$

$$= EI \begin{bmatrix} -0.001,5 & 0.015 \\ -0.015 & 0.100 \end{bmatrix}$$

$$[K]_{21} = [\beta]_{21}^T[k]_{21}[\beta]_{12} = [K]_{12}^T = EI \begin{bmatrix} -0.001,5 & -0.015 \\ 0.015 & 0.100 \end{bmatrix}$$

$$[K]_{23} = [\beta]_{23}^T[k]_{23}[\beta]_{32}$$

$$= \begin{bmatrix} -1 & 0 \\ 0 & 1 \end{bmatrix} \cdot EI \begin{bmatrix} 0.000,444 & -0.006,67 \\ -0.006,67 & 0.066,7 \end{bmatrix} \begin{bmatrix} 1 & 0 \\ 0 & 1 \end{bmatrix}$$

$$= EI \begin{bmatrix} -0.000,444 & 0.006,67 \\ -0.006,67 & 0.066,7 \end{bmatrix}$$

$$[K]_{32} = [\beta]_{32}^T[k]_{32}[\beta]_{23} = [K]_{23}^T = EI \begin{bmatrix} -0.000,444 & -0.006,67 \\ 0.006,67 & 0.066,7 \end{bmatrix}$$

Thus, the reduced stiffness matrix becomes

$$[K]_{I,I} = EI \begin{bmatrix} 0.200 & 0.100 & 0 \\ 0.100 & 0.333 & 0.066,7 \\ 0 & 0.066,7 & 0.133 \end{bmatrix}$$

Determination of Displacements:

$$\{\Delta\}_I = [K]_{I,I}^{-1}\{P\}_I \qquad (17.10)$$

$$\begin{Bmatrix} (\Delta_6)_1 \\ (\Delta_6)_2 \\ (\Delta_6)_3 \end{Bmatrix} = \frac{1}{EI} \begin{bmatrix} 0.200 & 0.100 & 0 \\ 0.100 & 0.333 & 0.066,7 \\ 0 & 0.066,7 & 0.133 \end{bmatrix}^{-1} \begin{Bmatrix} (P_6)_1 \\ (P_6)_2 \\ (P_6)_3 \end{Bmatrix}$$

$$= \frac{1}{EI} \begin{bmatrix} 6.002 & -2.004 & 1.005 \\ -2.004 & 4.007 & -2.009 \\ 1.005 & -2.009 & 8.527 \end{bmatrix} \begin{Bmatrix} 95 \\ 55 \\ -180 \end{Bmatrix}$$

$$= \frac{1}{EI} \begin{Bmatrix} 279.07 \\ 391.63 \\ -1,549.88 \end{Bmatrix}$$

Member Forces:

$$\{F\}_{ij} = [k]^j_{ii}[\beta]_{ij}\{\Delta\}_i + [k]_{ij}[\beta]_{ji}\{\Delta\}_j + \{F\}^f_{ij} \tag{17.44}$$

$$\text{Member 1–2: } \{\Delta\}_1 = \frac{1}{EI}\begin{Bmatrix} 0 \\ 279.07 \end{Bmatrix} \quad \{\Delta\}_2 = \frac{1}{EI}\begin{Bmatrix} 0 \\ 391.63 \end{Bmatrix}$$

$$\{F\}_{12} = [k]^2_{11}[\beta]_{12}\{\Delta\}_1 + [k]_{12}[\beta]_{21}\{\Delta\}_2 + \{F\}^f_{12}$$

$$= EI\begin{bmatrix} 0.001,5 & -0.015 \\ -0.015 & 0.200 \end{bmatrix}\begin{bmatrix} -1 & 0 \\ 0 & 1 \end{bmatrix}\frac{1}{EI}\begin{Bmatrix} 0 \\ 279.07 \end{Bmatrix}$$

$$+ EI\begin{bmatrix} 0.001,5 & -0.015 \\ -0.015 & 0.100 \end{bmatrix}\begin{bmatrix} 1 & 0 \\ 0 & 1 \end{bmatrix}\frac{1}{EI}\begin{Bmatrix} 0 \\ 391.63 \end{Bmatrix} + \begin{Bmatrix} +25 \\ -125 \end{Bmatrix}$$

$$= \begin{Bmatrix} -4.19 \\ 55.81 \end{Bmatrix} + \begin{Bmatrix} -5.87 \\ 39.16 \end{Bmatrix} + \begin{Bmatrix} 25 \\ -125 \end{Bmatrix} = \begin{Bmatrix} 14.94^k \\ -30.03'^{-k} \end{Bmatrix}$$

$$\{F\}_{21} = [k]^1_{22}[\beta]_{21}\{\Delta\}_2 + [k]_{21}[\beta]_{12}\{\Delta\}_1 + \{F\}^f_{21}$$

$$= \begin{Bmatrix} -5.87 \\ 78.33 \end{Bmatrix} + \begin{Bmatrix} -4.19 \\ 27.91 \end{Bmatrix} + \begin{Bmatrix} -25 \\ 125 \end{Bmatrix} = \begin{Bmatrix} -35.06^k \\ 231.24'^{-k} \end{Bmatrix}$$

$$\text{Member 2–3: } \{\Delta\}_2 = \frac{1}{EI}\begin{Bmatrix} 0 \\ 391.63 \end{Bmatrix} \quad \{\Delta\}_3 = \frac{1}{EI}\begin{Bmatrix} 0 \\ -1{,}549.88 \end{Bmatrix}$$

$$\{F\}_{23} = [k]^3_{22}[\beta]_{23}\{\Delta\}_2 + [k]_{23}[\beta]_{32}\{\Delta\}_3 + \{F\}^f_{23}$$

$$= \begin{Bmatrix} -2.61 \\ 52.09 \end{Bmatrix} + \begin{Bmatrix} 10.34 \\ -103.38 \end{Bmatrix} + \begin{Bmatrix} 36 \\ -180 \end{Bmatrix} = \begin{Bmatrix} 43.73^k \\ -231.29'^{-k} \end{Bmatrix}$$

$$\{F\}_{32} = [k]^2_{33}[\beta]_{32}\{\Delta\}_3 + [k]_{32}[\beta]_{23}\{\Delta\}_2 + \{F\}^f_{32}$$

$$= \begin{Bmatrix} 10.34 \\ -206.13 \end{Bmatrix} + \begin{Bmatrix} -2.61 \\ 26.12 \end{Bmatrix} + \begin{Bmatrix} -36 \\ 180 \end{Bmatrix} = \begin{Bmatrix} -28.27^k \\ 0'^{-k} \end{Bmatrix}$$

14.94^k   35.06^k   43.73^k   28.27^k

30.03 ^{ft-k}   231.24 ^{ft-k}   231.29 ^{ft-k}   0 ^{ft-k}

The shear and moment diagrams are now easily constructed, as shown in Section 17.5.2.

17.7
Self-Straining Problems

Self-straining occurs when a structure is subjected to internal strains and a resulting state of stress in the absence of externally applied loads. An example of self-straining is the support settlement that was described in Section 17.2. Other self-straining situations result from temperature variations or member lack of fit, in which a member of erroneous length or alignment is forced to fit during the fabrication process.

For all of these situations, there is a distinct difference in the nature of the response for statically determinate as opposed to statically indeterminate structures. In the statically determinate case, the induced movements are uninhibited, and the structure merely assumes a distorted configuration without the inducement of stresses. For example, consider the simply supported beam shown in Fig. 17.5a. Figures 17.5b, c, and d show, respectively, the responses of the structure

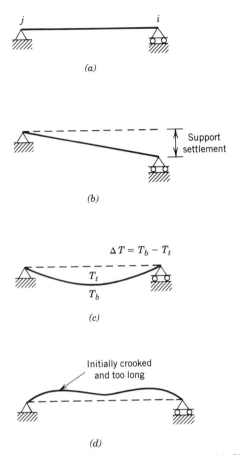

$\Delta T = T_b - T_t$

Fig. 17.5 *Induced movements for statically determinate beam.* (a) *Simply supported beam.* (b) *Support settlement.* (c) *Temperature gradient.* (d) *Lack-of-fit.*

to support settlement, temperature gradient, and fabrication error. In each case, the structure is deformed, but there are no integral stresses and no reactions are induced. For a statically indeterminate case, however, the response is quite different. The propped cantilever beam of Fig. 17.6a serves as an example; Figs. 17.6b, c, and d depict the responses to support settlement, temperature gradient, and fabrication error, respectively. Here, internal stresses result, and in each case there is an associated set of self-equilibrating external reactions. The structure itself is serving to inhibit the deformation; the structure is "straining against itself," or it is a *self-straining problem*.

A comparison of Figs. 17.5 and 17.6 shows why the reactions and the associated internal stresses are induced for the statically indeterminate system. For each of the determinate cases given in Fig. 17.5, there are no forces induced through the induced movement. However, for each of the indeterminate cases, an end moment, along with a set of equilibrating vertical reactions, has to be introduced in order to impose the zero-rotation boundary condition at point *j* for the structure of Fig. 17.6.

The formal analysis approach for the self-straining problem resembles that which is used when loads are applied between node points. We will limit our present consideration to planar structures without torsion, and Fig. 17.7a shows a

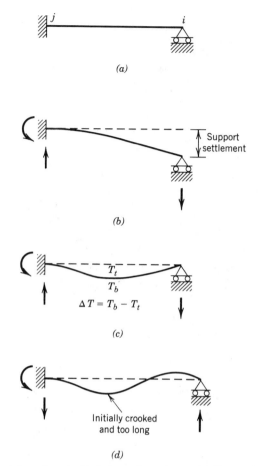

Fig. 17.6 *Self straining of statically indeterminate beam.* (a) *Propped cantilever beam.* (b) *Support settlement.* (c) *Temperature gradient.* (d) *Lack-of-fit.*

typical member ij. It is initially assumed that the span is fixed at its ends as shown in Fig. 17.7b. Of course, this condition does not satisfy the prescribed boundary conditions. The member is now severed at end i and the member-end displacements caused by the self-straining action are allowed to accrue at end i. These are shown in Fig. 17.7c as the vector $\{\delta_s\}_{ij}$. If these displacements are restrained, that is, if the displacements $-\{\delta_s\}_{ij}$ are applied, then a set of fixed-end forces is imposed at joints i and j as shown in Fig. 17.7d. These fixed-end forces are given by

$$\{F_s\}^f_{ij} = -[k]^j_{ii}\{\delta_s\}_{ij} \qquad (17.45)$$

and

$$\{F_s\}^f_{ji} = -[k]_{ji}\{\delta_s\}_{ij} \qquad ((17.46)$$

where $\{F_s\}^f_{ij}$ and $\{F_s\}^f_{ji}$ are the fixed-end forces induced at ends i and j, respectively, through the removal of $\{\delta_s\}_{ij}$, and the $[k]$ matrices are as defined by Eqs. 16.42 and 16.43.

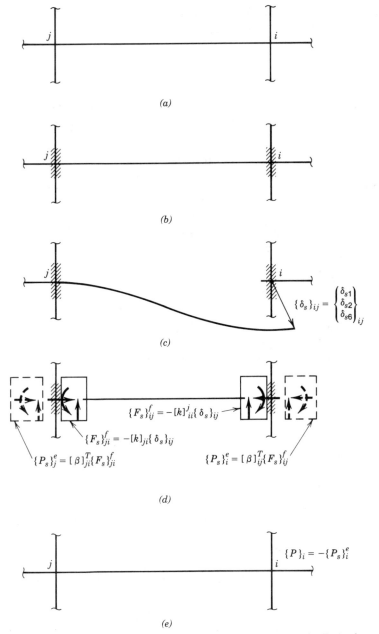

Fig. 17.7 *Loading for self-straining problem. (a) Planar member ij. (b) Fixity for member ij. (c) Accrued displacements at severed j end. (d) Member fixed-end forces and equivalent joint forces. (e) Joint loading for stiffness analysis.*

In order to sustain these fixed-end forces, there must be an equivalent set of structure forces as shown in Fig. 17.7d. According to the presentation of Section 17.6, these equivalent forces are

$$\{P_s\}_i^e = [\beta]_{ij}^T \{F_s\}_{ij}^f \tag{17.47}$$

where $\{P_s\}_i^e$ includes the equivalent sustaining forces at joint i, and $[\beta]_{ij}$ is the compatibility matrix. Of course, in the overall structure, more than one member may join at node i and, therefore,

$$\{P_s\}_i^e = \sum_j [\beta]_{ij}^T \{F_s\}_{ij}^f \qquad (17.48)$$

where j ranges over all members framing into joint i.

Since the given structure of Fig. 17.7a cannot provide these sustaining forces, they must be removed through a standard joint-loaded stiffness analysis, as noted in Fig. 17.7e, in which the loads at joint i are

$$\{P\}_i = -\{P_s\}_i^e \qquad (17.49)$$

If some components of $\{P_s\}_i^e$ can be sustained by the support at point i, then a reduced vector of structure forces corresponding to the kinematic degrees of freedom, $\{\bar{P}_s\}_i^e$, is applied as described in Section 17.6.

The general analysis procedure follows that given in Section 17.5, and the member-end forces are determined by Eq. 17.44 in which $\{F\}_{ij}^f$ is replaced by $\{F_s\}_{ij}^f$.

17.7.1
Fabrication error

For the fabrication error, or lack-of-fit, problem, the vector $\{\delta_s\}_{ij}$ includes the displacements at the i end relative to the j end that indicate the components of improper length or alignment. This vector of displacement introduces fixed-end forces at both ends of the faulty member in accordance with Eqs. 17.45 and 17.46.

17.7.2
Temperature variations

For temperature effects, $\{\delta_s\}_{ij}$ includes the displacements that accrue at the i end relative to the j end of Fig. 17.7c when the temperature variations occur. For the planar structure under consideration, these displacements are shown in Fig. 17.8c for the temperature gradient specified in Fig. 17.8a. Substitution of these displacements in Eqs. 17.45 and 17.46 along with appropriate $[k]$ matrices from Eq. 16.42 gives

$$\{F_s\}_{ij}^f = - \begin{bmatrix} \dfrac{EA}{l} & 0 & 0 \\ 0 & \dfrac{12EI_z}{l^3} & -\dfrac{6EI_z}{l^2} \\ 0 & -\dfrac{6EI_z}{l^2} & \dfrac{4EI_z}{l} \end{bmatrix} \begin{Bmatrix} T_{avg}\alpha l \\ -\dfrac{\alpha\Delta T l^2}{2h} \\ -\dfrac{\alpha\Delta T l}{h} \end{Bmatrix} = \begin{Bmatrix} -EA\alpha T_{avg} \\ 0 \\ \dfrac{EI_z\alpha\Delta T}{h} \end{Bmatrix} \qquad (17.50)$$

and

$$\{F_s\}_{ij}^f = - \begin{bmatrix} \dfrac{EA}{l} & 0 & 0 \\ 0 & \dfrac{12EI_z}{l^3} & -\dfrac{6EI_z}{l^2} \\ 0 & -\dfrac{6EI_z}{l^2} & \dfrac{2EI_z}{l} \end{bmatrix} \begin{Bmatrix} T_{avg}\alpha l \\ -\dfrac{\alpha\Delta T l^2}{2h} \\ -\dfrac{\alpha\Delta T l}{h} \end{Bmatrix} = \begin{Bmatrix} -EA\alpha T_{avg} \\ 0 \\ -\dfrac{EI_z\alpha\Delta T}{h} \end{Bmatrix} \qquad (17.51)$$

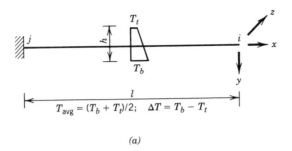

$$T_{avg} = (T_b + T_t)/2; \quad \Delta T = T_b - T_t$$

(a)

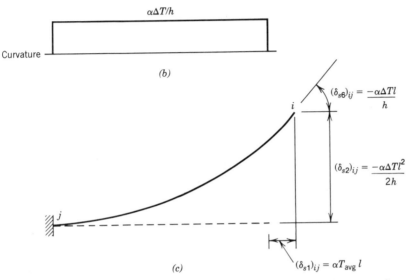

(c)

Fig. 17.8 *Temperature effects on planar member* ij. (a) *Imposed temperature gradient.*
(b) *Induced member curvatures.* (c) *Accrued displacements at end* j.

where $\Delta T = T_b - T_t$ and $T_{avg} = (T_b + T_t)/2$, in which T_b and T_t are the temperatures at the bottom and top flanges of the beam, respectively. In applying these equations to any member, T_b and T_t must be carefully interpreted with respect to the assigned i and j of the member ends in order to avoid a mistake in signs.

For the settlement problem, $\{\delta_s\}_{ij}$ includes the settlement displacement components **17.7.3** that are experienced by member ij at joint i. These displacements will cause fixed- **Settlement** end forces at both ends of all the members that frame into joint i; however, since these displacements are imposed rather than restrained, $-\{\delta_s\}_{ij}$ should be employed in Eqs. 17.45 and 17.46, which has the effect of changing the signs on the right-hand sides of these equations.

It should be recalled that the settlement problems can also be solved using the technique described in Section 17.2, where the globally induced settlement displacements were used.

17.7.4 Example problem Determine the member-end forces for the structure of Section 17.5.3 if the column is subjected to the temperature gradient shown.

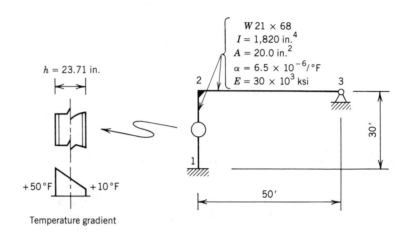

Temperature gradient

The solution proceeds as in the example of Section 17.5.3. The equations of equilibrium are as follows:

$$\{P\}_I = [K]_{I,I}\{\Delta\}_I \tag{17.9}$$

where

$$\{P\}_I = \left\{\begin{matrix} \left\{\begin{matrix} P_1 \\ P_2 \\ P_6 \end{matrix}\right\}_2 \\ \{P_6\}_3 \end{matrix}\right\}; \qquad \{\Delta\}_I = \left\{\begin{matrix} \left\{\begin{matrix} \Delta_1 \\ \Delta_2 \\ \Delta_6 \end{matrix}\right\}_2 \\ \{\Delta_6\}_3 \end{matrix}\right\}$$

$$[K]_{I,I} = \begin{bmatrix} 0.032{,}14 & 0 & -0.006{,}67 & 0 \\ 0 & 0.053{,}0 & 0.002{,}40 & 0.002{,}40 \\ -0.006{,}67 & 0.002{,}40 & 0.213 & 0.040{,}0 \\ 0 & 0.002{,}40 & 0.040{,}0 & 0.080{,}0 \end{bmatrix}$$

Self-Straining Fixed-End Forces and Equivalent Structure Forces:

Fixed-end forces:

$$\{F_s\}^f_{ij} = \left\{\begin{matrix} -EA\alpha T_{avg} \\ 0 \\ \dfrac{EI\alpha\Delta T}{h} \end{matrix}\right\}; \qquad \{F_s\}^f_{ji} = \left\{\begin{matrix} -EA\alpha T_{avg} \\ 0 \\ -\dfrac{EI\alpha\Delta T}{h} \end{matrix}\right\} \tag{17.50; 17.51}$$

Member 1–2: For member orientation with respect to the selection of i and j (see Fig. 17.8),

$$\Delta T = T_b - T_t = 50 - 10 = 40°F$$

$$T_{avg} = (50 + 10)/2 = 30°F$$

$$\{F_s\}_{12}^f = \begin{bmatrix} -30 \times 10^3 \times 20 \times 6.5 \times 10^{-6} \times 30 \\ 0 \\ \overline{30 \times 10^3 \times 1,820 \times 6.5 \times 10^{-6} \times 40} \\ 23.71 \end{bmatrix} = \begin{Bmatrix} -117.0^k \\ 0 \\ 598.73''^{-k} \end{Bmatrix}$$

$$= \begin{Bmatrix} -117.0^k \\ 0 \\ 49.89'^{-k} \end{Bmatrix}; \qquad \{F_s\}_{21}^f = \begin{Bmatrix} -117.0^k \\ 0 \\ -49.89'^{-k} \end{Bmatrix}$$

Member 2–3:

$$\{F_s\}_{23}^f = \{0\}; \qquad \{F_s\}_{32}^f = \{0\}$$

Equivalent structure forces:

$$\{P_s\}_i^e = \sum_j [\beta]_{ij}^T \{F_s\}_{ij}^f \qquad (17.48)$$

$$\{P_s\}_1^e = [\beta]_{12}^T \{F_s\}_{12}^f = \begin{bmatrix} 0 & -1 & 0 \\ 1 & 0 & 0 \\ 0 & 0 & 1 \end{bmatrix} \begin{Bmatrix} -117.0 \\ 0 \\ 49.89 \end{Bmatrix} = \begin{Bmatrix} 0 \\ -117.0^k \\ 49.89'^{-k} \end{Bmatrix}$$

$$\{P_s\}_2^e = [\beta]_{21}^T \{F_s\}_{21}^f = \begin{bmatrix} 0 & 1 & 0 \\ -1 & 0 & 0 \\ 0 & 0 & 1 \end{bmatrix} \begin{Bmatrix} -117.0 \\ 0 \\ -49.89 \end{Bmatrix} = \begin{Bmatrix} 0 \\ 117.0^k \\ -49.89'^{-k} \end{Bmatrix}$$

$$\{P_s\}_3^e = \{0\}$$

For the stiffness analysis,

$$\{P\}_i = -\{P_s\}_i^e \qquad (17.49)$$

where only the components corresponding to the kinematic degrees of freedom enter $\{P\}_I$ of Eq. 17.9.

Determination of Displacements:

$$\{\Delta\}_I = [K]_{I,I}^{-1} \{P\}_I \qquad (17.10)$$

$$\begin{Bmatrix} \begin{Bmatrix} \Delta_1 \\ \Delta_2 \\ \Delta_6 \end{Bmatrix}_2 \\ \{\Delta_6\}_3 \end{Bmatrix} = [K]_{I,I}^{-1} \begin{Bmatrix} 0 \\ -117.0 \\ 49.89 \\ 0 \end{Bmatrix} = \frac{1}{EI} \begin{Bmatrix} 56.91 \\ -2,216.77 \\ 274.25 \\ -70.62 \end{Bmatrix}$$

Member Forces:

$$\{F\}_{ij} = [k]_{ii}^j [\beta]_{ij} \{\Delta\}_i + [k]_{ij} [\beta]_{ji} \{\Delta\}_j + \{F_s\}_{ij}^f \qquad (17.44)$$

Member 1–2: $\{\Delta\}_1 = \{0\}$; $\{\Delta\}_2 = \dfrac{1}{EI} \begin{Bmatrix} 56.91 \\ -2,216.77 \\ 274.25 \end{Bmatrix}$

$$\{F\}_{12} = EI \begin{bmatrix} 0.052,9 & 0 & 0 \\ 0 & 0.000,444 & -0.006,67 \\ 0 & -0.006,67 & 0.066,7 \end{bmatrix} \begin{bmatrix} 0 & -1 & 0 \\ 1 & 0 & 0 \\ 0 & 0 & 1 \end{bmatrix} \cdot \frac{1}{EI} \left\{ \begin{matrix} 56.91 \\ -2,216.77 \\ 274.25 \end{matrix} \right\} + \left\{ \begin{matrix} -117.0 \\ 0 \\ 49.89 \end{matrix} \right\}$$

$$= \left\{ \begin{matrix} 117.27 - 117.00 \\ -1.80 + 0 \\ 17.91 + 49.89 \end{matrix} \right\} = \left\{ \begin{matrix} 0.27^k \\ -1.80^k \\ 67.80'^{-k} \end{matrix} \right\}$$

$$\{F\}_{21} = \left\{ \begin{matrix} 117.27 - -117.00 \\ -1.80 + 0 \\ 36.90 - -49.89 \end{matrix} \right\} = \left\{ \begin{matrix} 0.27^k \\ -1.80^k \\ -13.80'^{-k} \end{matrix} \right\}$$

Member 2–3: $\{\Delta\}_2 = \dfrac{1}{EI} \left\{ \begin{matrix} 56.91 \\ -2,216.77 \\ 274.25 \end{matrix} \right\}$; $\{\Delta\}_3 = \dfrac{1}{EI} \left\{ \begin{matrix} 0 \\ 0 \\ -70.62 \end{matrix} \right\}$

$$\{F\}_{23} = \left\{ \begin{matrix} -1.80^k \\ -0.27^k \\ 13.80'^{-k} \end{matrix} \right\}; \qquad \{F\}_{32} = \left\{ \begin{matrix} -1.80^k \\ -0.27^k \\ 0'^{-k} \end{matrix} \right\}$$

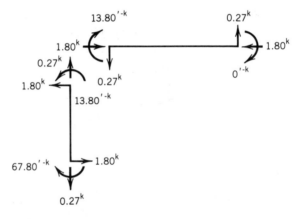

Deflected Structure:

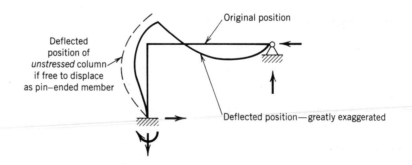

Original position

Deflected
position of
unstressed column
if free to displace
as pin–ended member

Deflected position—greatly exaggerated

Note:

The member forces and reactions result from forcing column from unstressed position to fit compatibly into frame.

Equations 17.26 and 17.27 give expressions for the structure stiffness submatrices that collectively make up the total structure stiffness matrix; however, these submatrices remain clearly identified with member ij. That is, they represent the contribution of member ij to the total structure stiffness matrix.

On the right-hand side of each equation, the $[k]$ matrix is sandwiched between a postmultiplication by a $[\beta]$ matrix and a premultiplication by a $[\beta]^T$ matrix. By tracing through the derivation, we find that the postmultiplier ensures compatibility at the end of the member corresponding to the second subscript on the stiffness matrix, whereas the premultiplier ensures equilibrium between the structure and member forces at the end of the member corresponding to the first subscript on the stiffness matrix. These operations on $[k]$ provide for the transformation of the member stiffnesses from the member coordinate system to the structure coordinate system. Therefore, the left-hand side of Eqs. 17.26 and 17.27 are member stiffness matrices that are expressed in the structure coordinate system.

In the so-called *direct stiffness method*, emphasis is placed on the member stiffness matrices in the structure coordinate system. In fact, the right-hand sides of Eqs. 17.26 and 17.27 are frequently expanded by substituting specific forms for $[k]$ and $[\beta]$ for different structure classifications. For example, for a planar truss member, $[k]^j_{ii}$ and $[k]_{ij}$ are extracted from Eq. 16.42, and $[\beta]_{ij}$ and $[\beta]_{ji}$ are taken from Section 17.4 to produce the following:

$$[K]^j_{ii} = [\beta]^T_{ij}[k]^j_{ii}[\beta]_{ij}$$

$$= \begin{bmatrix} \cos\theta \\ \sin\theta \end{bmatrix} \begin{bmatrix} EA \\ \overline{l} \end{bmatrix} [\cos\theta \ \sin\theta] = \frac{EA}{l} \begin{bmatrix} \cos^2\theta & \sin\theta\cos\theta \\ \sin\theta\cos\theta & \sin^2\theta \end{bmatrix} \quad (17.52)$$

$$[K]_{ij} = [\beta]^T_{ij}[k]_{ij}[\beta]_{ji}$$

$$= \begin{bmatrix} \cos\theta \\ \sin\theta \end{bmatrix} \begin{bmatrix} EA \\ \overline{l} \end{bmatrix} [-\cos\theta \ -\sin\theta] = \frac{EA}{l} \begin{bmatrix} -\cos^2\theta & -\sin\theta\cos\theta \\ -\sin\theta\cos\theta & -\sin^2\theta \end{bmatrix} \quad (17.53)$$

Similarly, for planar frame members, the appropriate $[k]$ matrices are taken from Eq. 16.42, and the $[\beta]$ matrices are taken from Section 17.4, from which the following equations result:

$$[K]^j_{ii} = [\beta]^T_{ij}[k]^j_{ii}[\beta]_{ij}$$

$$= \begin{bmatrix} \cos\theta & -\sin\theta & 0 \\ \sin\theta & \cos\theta & 0 \\ 0 & 0 & 1 \end{bmatrix} \begin{bmatrix} \dfrac{EA}{l} & 0 & 0 \\ 0 & \dfrac{12EI_z}{l^3} & \dfrac{-6EI_z}{l^2} \\ 0 & \dfrac{-6EI_z}{l^2} & \dfrac{4EI_z}{l} \end{bmatrix} \begin{bmatrix} \cos\theta & \sin\theta & 0 \\ -\sin\theta & \cos\theta & 0 \\ 0 & 0 & 1 \end{bmatrix}$$

$$= \begin{bmatrix} \dfrac{EA}{l}\cos^2\theta + \dfrac{12EI_z}{l^3}\sin^2\theta & \dfrac{EA}{l}\sin\theta\cos\theta - \dfrac{12EI_z}{l^3}\sin\theta\cos\theta & \dfrac{6EI_z}{l^2}\sin\theta \\ \dfrac{EA}{l}\sin\theta\cos\theta - \dfrac{12EI_z}{l^3}\sin\theta\cos\theta & \dfrac{EA}{l}\sin^2\theta + \dfrac{12EI_z}{l^3}\cos^2\theta & -\dfrac{6EI_z}{l^2}\cos\theta \\ \dfrac{6EI_z}{l^2}\sin\theta & -\dfrac{6EI_z}{l^2}\cos\theta & \dfrac{4EI_z}{l} \end{bmatrix}$$

$$(17.54)$$

$[K]_{ij} = [\beta]_{ij}^T[k]_{ij}[\beta]_{ji}$

$$
= \begin{bmatrix} \cos\theta & -\sin\theta & 0 \\ \sin\theta & \cos\theta & 0 \\ 0 & 0 & 1 \end{bmatrix} \begin{bmatrix} \dfrac{EA}{l} & 0 & 0 \\ 0 & \dfrac{12EI_z}{l^3} & \dfrac{-6EI_z}{l^2} \\ 0 & \dfrac{-6EI_z}{l^2} & \dfrac{2EI_z}{l} \end{bmatrix} \begin{bmatrix} -\cos\theta & -\sin\theta & 0 \\ \sin\theta & -\cos\theta & 0 \\ 0 & 0 & 1 \end{bmatrix}
$$

$$
= \begin{bmatrix} -\dfrac{EA}{l}\cos^2\theta - \dfrac{12EI_z}{l^3}\sin^2\theta & -\dfrac{EA}{l}\sin\theta\cos\theta + \dfrac{12EI_z}{l^3}\sin\theta\cos\theta & \dfrac{6EI_z}{l^2}\sin\theta \\ -\dfrac{EA}{l}\sin\theta\cos\theta + \dfrac{12EI_z}{l^3}\sin\theta\cos\theta & -\dfrac{EA}{l}\sin^2\theta - \dfrac{12EI_z}{l^3}\cos^2\theta & -\dfrac{6EI_z}{l^2}\cos\theta \\ -\dfrac{6EI_z}{l^2}\sin\theta & \dfrac{6EI_z}{l^2}\cos\theta & \dfrac{2EI_z}{l} \end{bmatrix}
$$

$$(17.55)$$

Similar equations can be developed for other structure types, such as space trusses, grids, and space frames.

Once the structure stiffness submatrices have been established, the total structure stiffness matrix is assembled according to Eqs. 17.15 and 17.16. In the direct stiffness method, these equations are seen as the mechanism for superimposing the contribution from each member to the whole of the structure stiffness matrix. Each contribution, as well as the total structure stiffness matrix, is expressed in the structure coordinate system.

For hand calculations, Eqs. 17.52 through 17.55 offer no real advantages over Eqs. 17.26 and 17.27. However, for computer applications, the expanded forms are best because the repetitive matrix multiplications are avoided.

17.9

Characteristics of Stiffness (Equilibrium) Equations

The arguments used in Section 16.8 to prove that a member stiffness matrix is symmetrical are applicable for a structure as a whole. Therefore, the total structure stiffness matrix given in Eq. 17.15 is a symmetrical matrix. The significance of this fact is mainly practical as it is related to computational considerations, and this will be treated in Section 17.10.

It is also clear from examining Eq. 17.15 that the stiffness matrix is not fully populated but is rather somewhat sparse or weakly populated. Specifically, consider the equations corresponding to $\{P\}_i$. These equations will possess only terms associated with the displacements $\{\Delta\}_i$ and the displacements at the distant ends of members framing into point i, which in this case include $\{\Delta\}_m \cdots \{\Delta\}_r$. Figure 17.9 demonstrates this arrangement based on a planar truss structure in which there are two kinematic degrees of freedom at each point. The range over which these terms are present in the stiffness matrix is referred to as the *bandwidth* as noted in Fig. 17.9. Because of the limited size of the matrix portrayed in Fig. 17.9, the relationship of the bandwidth to the total array of kinematic degrees of freedom is not evident; however, for large structures the sparseness of the stiffness matrix is more evident.

In solution techniques employed for computers, only the terms over the bandwidth are stored. That is, the stiffness coefficients from the first nonzero term through the last nonzero term are stored for each equation. Therefore, it is best to

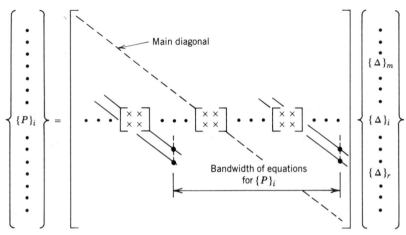

Fig. 17.9 *Composition of Equilibrium Equations.*

concentrate the nonzero terms as close to the main diagonal as possible. This can be done by numbering the kinematic degrees of freedom in such a manner as to minimize the bandwidth.

Bandwidth minimization coupled with recognition of the symmetric nature of the stiffness matrix is essential for efficient solution of the equilibrium equations of the stiffness method.

17.10 Generation of [K] in Computer Applications

As explained in the previous section, the structure stiffness matrix is symmetrical. Therefore, only the terms on the main diagonal and those that appear on one side of it must be computed and stored in the computer. All other terms can be deduced from symmetry considerations when they are needed.

A systematic approach to the generation of the structure stiffness matrix is based on taking full advantage of its symmetrical nature, and the detailed procedure is outlined below. In our present discussion, we limit our treatment to planar trusses and frames, but the approach is equally valid for any type of structure.

1. The n nodal points for the structure are numbered in the customary fashion ($1 \cdot \cdot \cdot r \cdot \cdot \cdot s \cdot \cdot \cdot n \cdot \cdot \cdot$) according to the structure coordinate system.

2. Within the allocated space in the computer, all elements of a $dn \times dn$ square matrix are initially set equal to zero, where d is the number of kinematic degrees of freedom at each joint.

3. For nodal point r, such that $r < s$, the structure stiffness submatrix $[K]_{rs}$ is determined and located in the total stiffness matrix as shown in Fig. 17.10. For a planar truss structure Eq. 17.53 is used, whereas for a planar frame Eq. 17.55 is applicable.

4. Through a rather simple adjustment to certain elements of $[K]_{rs}$, the submatrix $[K]_{rr}^s$ is determined. For a truss structure, a change of sign for all elements will convert $[K]_{rs}$ to $[K]_{rr}^s$, as can be seen by examining Eqs. 17.52 and 17.53. However, for a frame, once $[K]_{rs}$ is established, a change in sign for the first two *columns* and a doubling of the last element are needed to obtain $[K]_{rr}^s$, as is clear by examining Eqs. 17.54 and 17.55. This submatrix is then

accumulated on the main diagonal. According to a FORTRAN-type equation, $[K]_{rr}$ is determined by

$$[K]_{rr} = [K]_{rr} + [K]_{rr}^s \qquad (17.56)$$

and this matrix is positioned within the total matrix as shown in Fig. 17.10 with only the terms on and above the main diagonal being stored in the computer. The first time this equation is applied, the $[K]_{rr}$ on the right-hand side is composed of zeros, whereas in subsequent applications it would be composed of the accumulative contributions from all members previously considered which frame into joint r.

5. The submatrix $[K]_{sr}$ could be determined from $[K]_{rs}$ by recalling the symmetrical nature of the total stiffness matrix. That is,

$$[K]_{sr} = [K]_{rs}^T \qquad (17.57)$$

This would be positioned in the total stiffness matrix as shown in Fig. 17.10; however, because of symmetry, this submatrix is typically not determined or stored.

6. Again, through a rather simple adjustment in the elements of $[K]_{sr}$, the submatrix $[K]_{ss}^r$ could be determined. However, since $[K]_{sr}$ is not generally determined, it is necessary to establish $[K]_{ss}^r$ directly from $[K]_{rs}$. For a truss structure, a mere sign change in all elements will transform $[K]_{rs}$ to $[K]_{ss}^r$. For a frame structure, once $[K]_{rs}$ is known, a change of sign in the first two

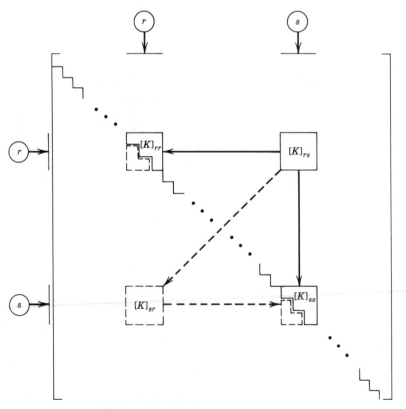

Fig. 17.10 *Generation of structure stiffness matrix.*

rows and a doubling of the last element will produce $[K]_{ss}^r$. These adjustments are clear from studying Eqs. 17.52 through 17.55 in the light of Eq. 17.57. This submatrix is then accumulated on the diagonal. Again, using a FORTRAN-type equation, we have

$$[K]_{ss} = [K]_{ss} + [K]_{ss}^r \qquad (17.58)$$

This array is located in the total array as noted in Fig. 17.10 with only the terms on or above the main diagonal being stored. The same arguments apply here as were advanced under Step 4 for the sequential applications of Eq. 17.56.

7. Steps 1 through 6 are applied for all combinations of $r < s$ over the full range of the n nodal points.

Figure 17.10 summarizes the sequence for a given r and s.

- $[K]_{rs}$ is first determined and stored.
- $[K]_{rr}^s$ is then determined from $[K]_{rs}$ and accumulated in $[K]_{rr}$; the terms on or above the main diagonal are stored.
- $[K]_{sr}$ could be determined from $[K]_{rs}$, and then $[K]_{ss}^r$ could be established from $[K]_{sr}$; however, since $[K]_{sr}$ is not stored, this path is not followed.
- $[K]_{ss}^r$ is determined from $[K]_{rs}$ and accumulated in $[K]_{ss}$; the terms on or above the main diagonal are stored.

In Fig. 17.10, the solid arrows indicate the actual paths that are followed; the dashed paths are not taken since the entire array is not stored. Also, the solid outlines embrace submatrices, or portions thereof, that are stored in the computer, whereas the dashed outlines indicate unstored quantities. The figure is based on a 3×3 submatrix, which corresponds to a planar frame or a grid. The size of this submatrix would vary depending on the structure type.

Generate the total stiffness matrix for the structure of Section 17.5.3 using the procedure outlined in Section 17.10. $r = 1$, $s = 2$: **17.10.1 Example problem**

$$[K]_{12} = EI \begin{bmatrix} -0.000{,}444 & 0 & 0.006{,}67 \\ 0 & -0.052{,}9 & 0 \\ -0.006{,}67 & 0 & 0.066{,}7 \end{bmatrix} \qquad (17.55)$$

Changing the signs of the first two *columns* and doubling the last entry of $[K]_{12}$, we obtain

$$[K]_{11}^2 = EI \begin{bmatrix} 0.000{,}444 & 0 & 0.006{,}67 \\ 0 & 0.052{,}9 & 0 \\ 0.006{,}67 & 0 & 0.133 \end{bmatrix}$$

Then

initially null

$$[K]_{11} = [K]_{11} + [K]_{11}^2 = EI \begin{bmatrix} 0.000{,}444 & 0 & 0.006{,}67 \\ 0 & 0.052{,}9 & 0 \\ 0.006{,}67 & 0 & 0.133 \end{bmatrix} \qquad (17.56)$$

Changing the signs of the first two *rows* and doubling the last entry of $[K]_{12}$, we have

$$[K]_{22}^1 = EI \begin{bmatrix} 0.000,444 & 0 & -0.006,67 \\ 0 & 0.052,9 & 0 \\ -0.006,67 & 0 & 0.133 \end{bmatrix}$$

initially null

$$[K]_{22} = [K]_{22} + [K]_{22}^1 = EI \begin{bmatrix} 0.000,444 & 0 & -0.006,67 \\ 0 & 0.052,9 & 0 \\ -0.006,67 & 0 & 0.133 \end{bmatrix} \quad (17.58)$$

$r = 1, s = 3$:

No member connects joints 1 and 3; therefore,

$$[K]_{13} = [K]_{11}^3 = [K]_{33}^1 = [0]$$

$r = 2, s = 3$:

$$[K]_{23} = EI \begin{bmatrix} -0.031,7 & 0 & 0 \\ 0 & -0.000,096,0 & 0.002,40 \\ 0 & -0.002,40 & 0.040,0 \end{bmatrix} \quad (17.55)$$

Changing signs of the first two *columns* and doubling the last entry, we have

$$[K]_{22}^3 = EI \begin{bmatrix} 0.031,7 & 0 & 0 \\ 0 & 0.000,096,0 & 0.002,40 \\ 0 & 0.002,40 & 0.080,0 \end{bmatrix}$$

$$[K]_{22} = [K]_{22} + [K]_{22}^3 = EI \begin{bmatrix} 0.032,14 & 0 & -0.006,67 \\ 0 & 0.053,0 & 0.002,40 \\ -0.006,67 & 0.002,40 & 0.213 \end{bmatrix} \quad (17.56)$$

Accumulated from previous r and s values

Changing the signs of the first two *rows* and doubling the last entry of $[K]_{23}$, we obtain

$$[K]_{33}^2 = EI \begin{bmatrix} 0.031,7 & 0 & 0 \\ 0 & 0.000,096,0 & -0.002,40 \\ 0 & -0.002,40 & 0.080,0 \end{bmatrix}$$

null at this point

$$[K]_{33} = [K]_{33} + [K]_{33}^2 = EI \begin{bmatrix} 0.031,7 & 0 & 0 \\ 0 & 0.000,096,0 & -0.002,40 \\ 0 & -0.002,40 & 0.080,0 \end{bmatrix} \quad (17.58)$$

Total Stiffness Matrix:

Assembling the submatrices, we have

$$[K] = \begin{bmatrix} 0.000,444 & 0 & 0.006,67 & -0.000,444 & 0 & 0.006,67 & 0 & 0 & 0 \\ & 0.052,9 & 0 & 0 & -0.052,9 & 0 & 0 & 0 & 0 \\ & & 0.133 & -0.006,67 & 0 & 0.066,7 & 0 & 0 & 0 \\ & & & 0.032,14 & 0 & -0.006,67 & -0.031,7 & 0 & 0 \\ & & & & 0.053,0 & 0.002,40 & 0 & -0.000,096,0 & 0.002,40 \\ & & & & & 0.213 & 0 & -0.002,40 & 0.040,0 \\ & & & & & & 0.031,7 & 0 & 0 \\ & & & & & & & 0.000,096,0 & -0.002,40 \\ & & & & & & & & 0.080,0 \end{bmatrix}$$

The selection of a member coordinate system is somewhat arbitrary. The one **Alternative** selected in this text, which was first introduced in Fig. 16.1b, is shown again in **Member** Fig. 17.11a. This system has the advantage that the member-end forces for a **Coordinate** planar structure (xy plane) are consistent with conventions established earlier in **Systems** the text. Specifically, tension forces are positive at both ends of the member; end shear forces are positive when they act down at the right end and up at the left end; and end moments are positive when they act clockwise on the member ends. This coordinate system also has the advantage that if the member is rotated end-for-end about the z axis, the member axes at each end retain the same orientation with respect to the global axes. One disadvantage is that there is a different [β] matrix at each end of the member, which has been observed in the example problems.

However, the adopted coordinate system is not the one that is most commonly used in computer applications. Instead, the one shown in Fig. 17.11b is employed. The major advantage of this system is that there is a common [β] matrix at each end of the member. When this system is used, care must be observed in interpreting the resulting member-end forces. For instance, a tension force would be given as a positive force at end i but a negative force at end j. This, of course, is not a serious disadvantage, but it does underscore the fact that computer results must be carefully examined and properly interpreted.

Both of the systems described in Fig. 17.11 are right-handed systems. Other variations are positive, but most systems employed are right-handed in nature.

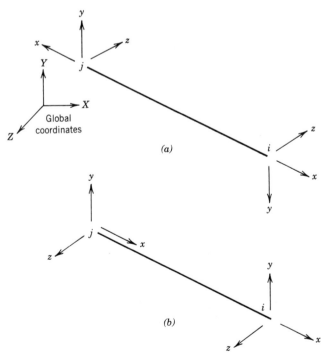

Fig. 17.11 *Member coordinate systems. (a) As used in this text. (b) Popular system for modern computer programs.*

17.12

Additional Reading

Gere, J. M., and Weaver, W., Jr., *Analysis of Framed Structures,* Chapter 4, Van Nostrand, Princeton, N.J., 1965.

Kardestuncer, H., *Elementary Matrix Analysis of Structures,* Chapters 5 through 10, McGraw–Hill, New York, 1974.

Laursen, H. I., *Matrix Analysis of Structures,* Chapter 3, McGraw–Hill, New York, 1966.

Livesley, R. K., *Matrix Methods of Structural Analysis.* Chapter 4, Pergamon Press, Macmillan Co., New York, 1964.

McGuire, W., and Gallagher, R. H., *Matrix Structural Analysis,* Chapters 3 through 5, Wiley, New York, 1979.

Rubinstein, M. F., *Computer Matrix Analysis of Structures,* Chapter 9, Prentice–Hall, Englewood Cliffs, N.J., 1966.

17.13

Suggested Problems

1 through 17. Use the generalized stiffness method to analyze each of the structures shown below for the loading conditions indicated. In each case, determine the structure displacements and the corresponding member forces for the given nodal numbering system.

1. The structure and loading of Problem 1 of Section 12.13.

2. The structure and loading of Problem 3 of Section 12.13.

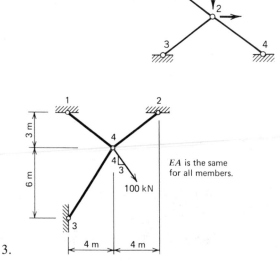

EA is the same for all members.

3.

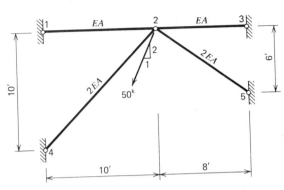

4.

5. The structure and loading of Problem 2 of Section 12.13.

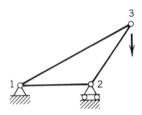

6. The structure and loading of Problem 1 of Section 10.14.

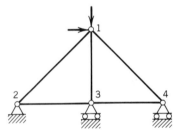

7. The structure and loading of Problem 2 of Section 10.14.

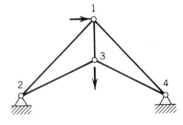

8. The structure and loading of Problem 4 of Section 10.14.

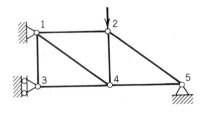

9. The structure and loading of Problem 20 of Section 10.14.

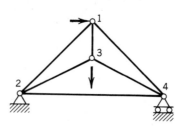

10. The structure and loading of Problem 21 of Section 10.14.

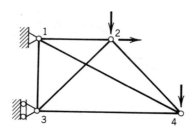

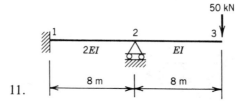

11.

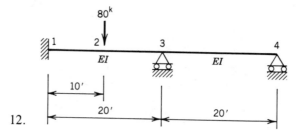

12.

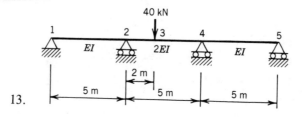

13.

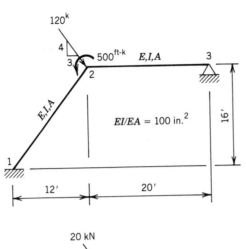

14.

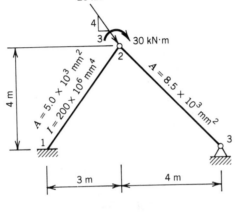

15.

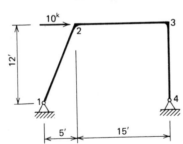

16.

Ignore axial flexibilities; EI is the same for each member.

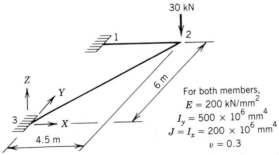

17.

Note: XY plane is horizontal and the local xy plane for each member lies within the global XY plane.

18. Determine the displacements of joint b if members ab and bc each have a torsional stiffness of GJ/L and flexural stiffnesses of EI/L for bending out of the plane of the structure. Assume $GJ = 0.25\ EI$ and $L = 6m$.

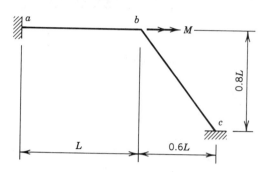

19. Consider the truss structure shown in which A and E are the same for all members.

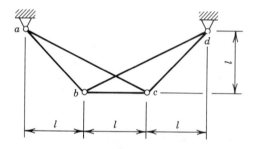

a. Assemble the total and reduced stiffness matrices.

b. For the application of equal vertical (downward) loads of magnitude P at points b and c, use symmetry considerations to reduce the order of the reduced stiffness matrix, and calculate the resulting displacements at points b and c.

c. For the application of equal horizontal (rightward) loads of magnitude H at points b and c, use conditions of antisymmetry to reduce the order of the reduced stiffness matrix, and calculate the resulting displacements at points b and c.

d. If members ac and bd are removed, demonstrate through a singularity test that the structure is unstable for general loading.

e. If the modified structure of part (d) is subjected to equal vertical (downward) loads of magnitude P at points b and c, show that a solution is possible and determine the resulting displacements at points b and c.

20. Consider the beam structure shown below.

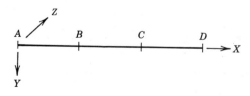

For each set of boundary conditions and prescribed load components, indicate the matrix format of the total and reduced equilibrium equations.

Use an "x" to identify each nonzero element of the stiffness matrix, express the individual displacement terms in symbolic form, and make the appropriate substitutions for the load terms.

Consider response in the *XY* plane only, including torsion about the *Z* axis. Comment on any decoupling of individual displacements which allows for the solution of a subset of equations.

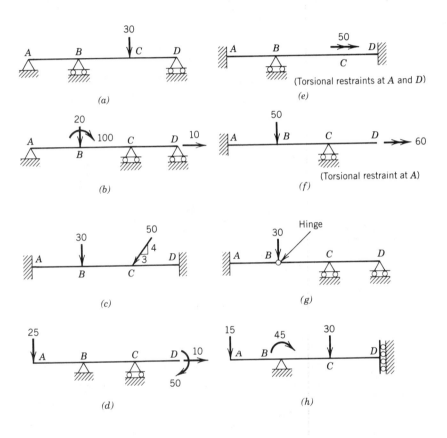

Note: Loads in kips, moments in ft-kips.

21. Consider the frame structure shown below.

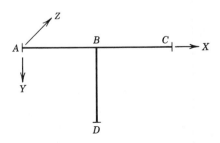

Apply the instructions that are given in Problem 20 to the structures and loading conditions shown below.

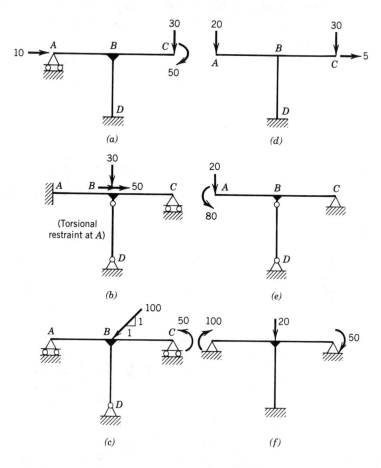

22 through 26. Use the generalized stiffness method to analyze each of the structures shown below for the loading conditions indicated. In each case, determine the structure displacements and the corresponding member forces for the given nodal numbering system.

22. The structure and loading of Problem 12 with the altered boundary condition at point 1 and the following nodal numbering system:

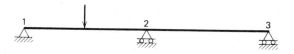

23. The structure and loading of Problem 13 with the following nodal numbering system:

24. The structure and loading of Problem 4 in Section 13.16 with the following nodal numbering system:

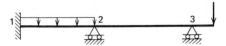

25. The structure and loading of Problem 19 in Section 13.16 with the following nodal numbering system:

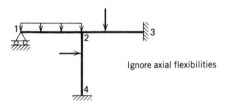

Ignore axial flexibilities

26. The structure and loading of Problem 21 in Section 13.16 with the following nodal numbering system:

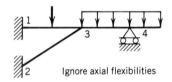

Ignore axial flexibilities

27 through 30. Determine the structure displacements, member forces, and reactions for each of the structures and conditions of support movements described below.

27. The structure of Problem 3 in Section 12.13 with the support movement given in Problem 17 of Section 12.13. Use the nodal numbering of Problem 2 in this section.

28. The structure of Problem 1 in Section 10.14 with the pattern of support movements given in Problem 12 of Section 10.14. Use the nodal numbering specified in Problem 6 of this section.

29. The structure of Problem 1 of Section 11.14 with the settlement conditions and loading described in Problem 11 of Section 11.14. Use the nodal numbering system shown below.

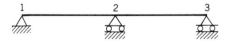

30. The structure of Problem 4 in Section 11.14 with the settlement pattern given in Problem 13 of Section 11.14. The nodal numbering shown below should be employed.

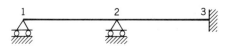

31 through 34. Determine the structure displacements and member forces for each of the structures and temperature variations given below.

31. The structure of Problem 3 of Section 12.13 using the temperature changes and supplemental data given in Problem 21 of Section 12.13. Apply the nodal numbering of Problem 2 of this section.

32. The structure of Problem 20 of Section 10.14 with the temperature variations and supplemental data presented in Problem 24 of Section 10.14. Employ the nodal numbering given in Problem 9 of this section.

33. The structure of Problem 1 of Section 11.14 with the temperature gradient and additional structural data given in Problem 20 of Section 11.14. The nodal numbering given in Problem 30 of this section should be used.

34. The structure of Problem 1 of Section 13.16 using the temperature gradients and supplemental structural data prescribed in Problem 15 of Section 13.16. Use the nodal numbering scheme assigned below.

35 through 38. Determine the structure displacements and member forces for each of the structures and lack-of-fit situations described below:

35. The structure of Problem 4 in Section 12.13, using the nodal numbering system shown below, for the lack-of-fit condition stated in Problem 23 of Section 12.13.

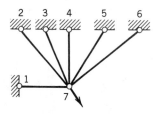

36. The structure of Problem 20 of Section 10.14 with the lack-of-fit situation given in Problem 32 of Section 10.14. Use the nodal numbering pattern described in Problem 9 of this section.

37. The structure of Problem 3 of Section 11.14 with the initial crookedness and supplemental structural information given in Problem 23 of Section 11.14. Employ the nodal numbering shown below.

38. The structure of Problem 4 of Section 11.14 with the initial crookedness and other structural data given in Problem 22 of Section 11.14. The nodal numbering shown below should be used.

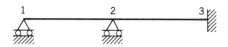

39. Develop a computer program for the synthesis of the total stiffness matrix for planar trusses and frames based on the procedures presented in Section 17.10.

40. Develop a computer program for the general stiffness analysis for joint-loaded planar trusses and frames. Use the synthesis program of Problem 39 as a subprogram.

41. Expand the program of Problem 40 to include the effects of member loads.

Shea Stadium, Flushing, N.Y. (courtesy New York Convention and Visitors Bureau).

FLEXIBILITY METHOD: GENERAL FORM OF COMPATIBILITY METHOD

18.1

Fundamental Concepts of Flexibility Method

It has already been seen that the concept of compatibility plays a vital role in the analysis of statically indeterminate structures. In fact, the satisfaction of the conditions of compatibility has been a requirement of each of the methods of indeterminate structural analysis that has been presented. In some methods, these conditions are satisfied in a subtle fashion; however, the so-called compatibility methods are clearly based on the simultaneous satisfaction of a number of equations that express the compatibility conditions throughout the structure. The solution of these compatibility equations results in the determination of the redundant forces. For this reason, these methods are sometimes referred to as force methods.

When compatibility methods are expressed in a general matrix formulation,

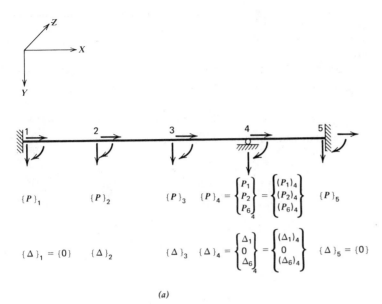

$$\{P\}_4 = \begin{Bmatrix} P_1 \\ P_2 \\ P_6 \end{Bmatrix}_4 = \begin{Bmatrix} (P_1)_4 \\ (P_2)_4 \\ (P_6)_4 \end{Bmatrix}$$

$$\{\Delta\}_4 = \begin{Bmatrix} \Delta_1 \\ 0 \\ \Delta_6 \end{Bmatrix}_4 = \begin{Bmatrix} (\Delta_1)_4 \\ 0 \\ (\Delta_6)_4 \end{Bmatrix}$$

$\{P\}_1$ $\{P\}_2$ $\{P\}_3$ $\{P\}_5$

$\{\Delta\}_1 = \{0\}$ $\{\Delta\}_2$ $\{\Delta\}_3$ $\{\Delta\}_5 = \{0\}$

(a)

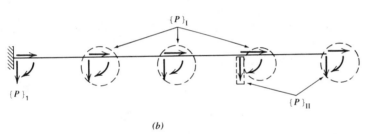

$\{P\}_{\mathrm{I}}$

$\{P\}_1$

$\{P\}_{\mathrm{II}}$

(b)

Fig. 18.1 *Typical statically indeterminate structure.* (a) *Structure loads and displacements.* (b) *Primary structure with applied forces and redundant forces.*

the governing equations involve the structure flexibility matrix, which in turn results from a synthesis of the individual member flexibility matrices. Thus, the matrix method, which evolves from a generalization of the compatibility formulation, is called the *flexibility method*.

18.2

Overview of the Flexibility Method

Consider the externally statically indeterminate beam structure shown in Fig. 18.1a. In this case, loads are applied only at points that are identified as nodes on the structure. The redundant support forces are represented by $\{P\}_{\mathrm{II}}$ so that

$$\{P\}_{\mathrm{II}}^T = \{(P_2)_4 (P_1)_5 (P_2)_5 (P_6)_5\} \tag{18.1}$$

When the constraints associated with these redundant forces are removed from the structure, the statically determinate primary structure shown in Fig. 18.1b remains, and the forces associated with the free displacements of the original structure are given by $\{P\}_{\mathrm{I}}$ in the form

$$\{P\}_{\mathrm{I}}^T = \{(P_1)_2 (P_2)_2 (P_6)_2 (P_1)_3 (P_2)_3 (P_6)_3 (P_1)_4 (P_6)_4\} \tag{18.2}$$

In Eqs. 18.1 and 18.2, the first subscript on each P term identifies the force component in accordance with Fig. 16.1, and the second subscript gives the node on the structure.

For the primary structure, the relationships between the displacements and the forces at the displaceable points (nodes 2 through 5) are given by

$$\{\Delta\} = [f]\{P\} \tag{18.3}$$

where $[f]$ is the structure flexibility matrix. This equation can be partitioned in the form

$$\left\{\begin{array}{c} \{\Delta\}_{\rm I} \\ \hline \{\Delta\}_{\rm II} \end{array}\right\} = \left[\begin{array}{c|c} [f]_{\rm I,I} & [f]_{\rm I,II} \\ \hline [f]_{\rm II,I} & [f]_{\rm II,II} \end{array}\right] \left\{\begin{array}{c} \{P\}_{\rm I} \\ \hline \{P\}_{\rm II} \end{array}\right\} \tag{18.4}$$

where $\{\Delta\}_{\rm I}$ and $\{\Delta\}_{\rm II}$ are defined so that

$$\{\Delta\}_{\rm I}^T = \{(\Delta_1)_2(\Delta_2)_2(\Delta_6)_2(\Delta_1)_3(\Delta_2)_3(\Delta_6)_3(\Delta_1)_4(\Delta_6)_4\} \tag{18.5}$$

$$\{\Delta\}_{\rm II}^T = \{(\Delta_2)_4(\Delta_1)_5(\Delta_2)_5(\Delta_6)_5\} \tag{18.6}$$

Once the redundant forces $\{P\}_{\rm II}$ are determined, they are used in conjunction with the applied loads $\{P\}_{\rm I}$ and the laws of statics to determine the reactions for the primary structure at node 1.

Equation 18.4 is not limited to the problem of Fig. 18.1, but is general, where $\{P\}_{\rm I}$ corresponds to the applied force components, $\{P\}_{\rm II}$ represents the redundant reaction components, and $\{\Delta\}_{\rm I}$ and $\{\Delta\}_{\rm II}$ give the corresponding displacement components, respectively.

Equation 18.4 can be expanded to give

$$\{\Delta\}_{\rm I} = [f]_{\rm I,I}\{P\}_{\rm I} + [f]_{\rm I,II}\{P\}_{\rm II} \tag{18.7}$$

$$\{\Delta\}_{\rm II} = [f]_{\rm II,I}\{P\}_{\rm I} + [f]_{\rm II,II}\{P\}_{\rm II} \tag{18.8}$$

Since the displacements at the redundant reactions are zero for the given structure, as shown in Fig. 18.1a, Eq. 18.8 becomes

$$\{0\} = [f]_{\rm II,I}\{P\}_{\rm I} + [f]_{\rm II,II}\{P\}_{\rm II} \tag{18.9}$$

This equation can now be solved for the redundant forces to give

$$\{P\}_{\rm II} = -[f]_{\rm II,II}^{-1}[f]_{\rm II,I}\{P\}_{\rm I} \tag{18.10}$$

If the displacements at the load points are desired, Eq. 18.10 is substituted into Eq. 18.7 to obtain

$$\{\Delta\}_{\rm I} = ([f]_{\rm I,I} - [f]_{\rm I,II}[f]_{\rm II,II}^{-1}[f]_{\rm II,I})\{P\}_{\rm I} \tag{18.11}$$

With the redundant forces, $\{P\}_{\rm II}$, as determined from Eq. 18.10, and the given loads, $\{P\}_{\rm I}$, the reaction forces $\{P\}_1$ can be determined from statics.

If the support displacements are not zero, then $\{\Delta\}_{\rm II}$ must reflect the presence of these displacements, and Eq. 18.10 will include a term on the right-hand side that is a direct function of $\{\Delta\}_{\rm II}$.

Before the analysis can proceed according to the sequence outlined in this section, it is necessary to generate the structure flexibility matrix $[f]$. The next sections of this chapter address this topic.

18.3

In the previous section, attention was focused on an externally statically indeterminate structure. This orientation will continue for now with the treatment of internal redundancies being deferred until Section 18.6.

Consider again the statically determinate primary structure that results when the redundant reaction forces are removed from the original structure. The structure of Fig. 18.1b serves as an example. Such a structure can be analyzed through the systematic application of the equations of equilibrium. These equations, which evolve from joint equilibrium considerations, take the form

$$\{P\} = [C]\{F\} \tag{18.12}$$

In this equation, $\{P\}$ embraces all of the forces that act on the primary structure, which are divided into two subvectors: $\{P\}_I$ includes the applied forces on the original structure, and $\{P\}_{II}$ includes the redundant reaction components. The $\{F\}$ vector includes the submatrices $\{F\}_{ij}$ for the i end of each member; the forces at the j end need not be included since they can be determined by applying statics to member ij. The $[C]$ matrix is, of course, the global statics matrix.

Since the primary structure is statically determinate and stable, $[C]$ is a nonsingular square matrix. Solving for the member forces, we obtain

$$\{F\} = [C]^{-1}\{P\} = [b]\{P\} \tag{18.13}$$

where $[b]$ is the equilibrium matrix that relates the member forces to the applied forces. In an expanded form, Eq. 18.13 can be written as

$$\{F\} = [[b]_I[b]_{II}] \begin{Bmatrix} \{P\}_I \\ \{P\}_{II} \end{Bmatrix} = [b]_I\{P\}_I + [b]_{II}\{P\}_{II} \tag{18.14}$$

In this form, it is clear that the member forces result from two separate contributions: $[b]_I\{P\}_I$ gives the member forces that result from the applied forces. Whereas $[b]_{II}\{P\}_{II}$ gives the member forces that are induced from the redundant reaction forces.

18.4

For each individual member in the structure, the strain energy can be expressed as

$$U_{ij} = \tfrac{1}{2}\{F\}_{ij}^T\{e\}_{ij} \tag{18.15}$$

where $\{e\}_{ij}$ contains the displacements at the i end relative to the j end of the member ij. Since the member in Fig. 16.4 is fixed at the j end, $\{e\}_{ij} = \{\delta\}_{ij}$, and thus Eq. 16.29 can be written as

$$\{e\}_{ij} = [f]_{ii}^j\{F\}_{ij} \tag{18.16}$$

Substituting Eq. 18.16 into Eq. 18.15, we find

$$U_{ij} = \tfrac{1}{2}\{F\}_{ij}^T[f]_{ii}^j\{F\}_{ij} \tag{18.17}$$

For the entire structure

$$U = \tfrac{1}{2}\{F\}^T[f]_u\{F\} \tag{18.18}$$

where

$$\{F\}^T = \{\{F\}_{12}\{F\}_{23} \cdot \cdot \cdot \{F\}_{ij} \cdot \cdot \cdot\} \tag{18.19}$$

and $[f]_u$ is a matrix containing the individual member flexibility matrices along the diagonal. This is called the *unassembled flexibility matrix* and has the form

$$[f]_u = \begin{bmatrix} [f]_{11}^2 & & & & & \\ & [f]_{22}^3 & & & & \\ & & \cdot & & & \\ & & & \cdot & & \\ & & & & [f]_{ii}^j & \\ & & & & & \cdot \\ & & & & & & \cdot \end{bmatrix} \tag{18.20}$$

Substitution of Eq. 18.13 into Eq. 18.18 gives

$$U = \tfrac{1}{2}\{P\}^T[b]^T[f]_u[b]\{P\} \tag{18.21}$$

Equation 18.21 gives the total strain energy as derived from member considerations. In terms of structure forces and displacements, the strain energy can also be expressed as

$$U = \tfrac{1}{2}\{P\}^T\{\Delta\} \tag{18.22}$$

Substituting Eq. 18.3 into Eq. 18.22, we obtain

$$U = \tfrac{1}{2}\{P\}^T[f]\{P\} \tag{18.23}$$

Comparison of Eqs. 18.21 and 18.23 reveals that

$$[f] = [b]^T[f]_u[b] \tag{18.24}$$

In a partitioned form, Eq. 18.24 becomes

$$[f] = \begin{bmatrix} [b]_I^T \\ [b]_{II}^T \end{bmatrix} [f]_u[[b]_I[b]_{II}]$$

$$= \begin{bmatrix} [b]_I^T[f]_u[b]_I & \vdots & [b]_I^T[f]_u[b]_{II} \\ \hdashline [b]_{II}^T[f]_u[b]_I & \vdots & [b]_{II}^T[f]_u[b]_{II} \end{bmatrix} \tag{18.25}$$

where $[b]_I$ and $[b]_{II}$ are defined by Eq. 18.14.

Comparing Eqs. 18.4 and 18.25, we have

$$\begin{aligned} [f]_{I,I} &= [b]_I^T[f]_u[b]_I \\ [f]_{I,II} &= [b]_I^T[f]_u[b]_{II} \\ [f]_{II,I} &= [b]_{II}^T[f]_u[b]_I \\ [f]_{II,II} &= [b]_{II}^T[f]_u[b]_{II} \end{aligned} \tag{18.26}$$

Application of Eqs. 18.26 yields the submatrices that collectively make up the structure flexibility matrix of Eq. 18.3.

In applying the flexibility method, a very orderly procedure is employed, which **Application of** is outlined by the following steps: **the Flexibility Method**

1. Number the nodal points of the structure. This must include all load points, support points, and points where two or more members join together. This procedure implicitly introduces structure force and displacement vectors in the form of Eq. 16.1 and member force and displacement vectors as described by Eq. 16.2.
2. Select the redundant forces, $\{P\}_{II}$. This defines the primary structure and automatically establishes the arrangements of $\{P\}_I$, $\{\Delta\}_I$, and $\{\Delta\}_{II}$ as indicated by Eq. 18.4.
3. Generate the matrices $[b]_I$ and $[b]_{II}$ of Eq. 18.14 from equilibrium considerations for the primary structure. Here, it is emphasized that the $\{F\}$ vector includes the member-end force vectors at only one end of each member.
4. Establish the individual member flexibility matrices based on Eq. 16.28 and form the unassembled flexibility matrix $[f]_u$ in accordance with Eq. 18.20.
5. Develop the submatrices $[f]_{II,I}$ and $[f]_{II,II}$ of the structure flexibility matrix from Eqs. 18.26.
6. Determine the redundant forces $\{P\}_{II}$ from Eq. 18.10. The other reactions on the primary structure can be determined from statics.
7. Calculate the final member-end forces from Eq. 18.14. The forces at the opposite end of each member can be determined by statics.
8. If the nodal displacements are desired, develop the submatrices $[f]_{I,I}$ and $[f]_{I,II}$ from Eq. 18.26 and obtain the desired displacements from Eq. 18.11.

For certain classes of structures, the force and displacement vectors for both the members and the structure are simplified. For instance, planar truss members have a single member-end force and displacement component, and the structure force and displacement vectors at each joint contain only two components. Similarly, for planar beam-type structures that are subjected to transverse load only, the member and structure force and displacement vectors will exclude axial components. In these simplified cases, the complete member flexibility matrix given in Eq. 16.28 must be modified so as to include only the member actions being considered. This is illustrated in the example problems that follow.

Also, care must be taken to ensure that the rows of the individual $[f]_{ii}^j$ matrices are arranged so that when $[f]_u$ is formed, there is a correspondence between the rows of $[f]_u$ and the rows of the member force matrix $\{F\}$.

Determine the bar forces for the truss structure shown. Assume EA is the same **18.5.1 Example** for each member. **problem**

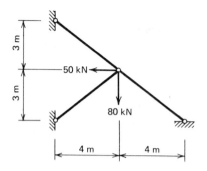

Selection of Node Points:

Structure forces and displacements:

$$\{P\}_1 = \begin{Bmatrix} P_1 \\ P_2 \end{Bmatrix}_1 = \begin{Bmatrix} (P_1)_1 \\ (P_2)_1 \end{Bmatrix} = \begin{Bmatrix} (P_1)_1 \\ 0.75(P_1)_1 \end{Bmatrix}; \qquad (\Delta)_1 = \begin{Bmatrix} \Delta_1 \\ \Delta_2 \end{Bmatrix}_1 = \{0\}$$

$$\{P\}_2 = \begin{Bmatrix} P_1 \\ P_2 \end{Bmatrix}_2 = \begin{Bmatrix} (P_1)_2 \\ (P_2)_2 \end{Bmatrix} = \begin{Bmatrix} -50 \\ 80 \end{Bmatrix}; \qquad \{\Delta\}_2 = \begin{Bmatrix} \Delta_1 \\ \Delta_2 \end{Bmatrix}_2 = \begin{Bmatrix} (\Delta_1)_2 \\ (\Delta_2)_2 \end{Bmatrix}$$

$$\{P\}_4 = \begin{Bmatrix} P_1 \\ P_2 \end{Bmatrix}_4 = \begin{Bmatrix} (P_1)_4 \\ (P_2)_4 \end{Bmatrix} = \begin{Bmatrix} (P_1)_4 \\ 0.75(P_1)_4 \end{Bmatrix}; \qquad (\Delta)_4 = \begin{Bmatrix} \Delta_1 \\ \Delta_2 \end{Bmatrix}_4 = \{0\}$$

$$\{P\}_3 = \begin{Bmatrix} P_1 \\ P_2 \end{Bmatrix}_3 = \begin{Bmatrix} (P_1)_3 \\ (P_2)_3 \end{Bmatrix} = \begin{Bmatrix} (P_1)_3 \\ -0.75(P_1)_3 \end{Bmatrix}; \qquad \{\Delta\}_3 = \begin{Bmatrix} \Delta_1 \\ \Delta_2 \end{Bmatrix}_3 = \{0\}$$

Member forces and displacements:

$$(F_1)_{ij}; (\delta_1)_{ji} \xleftarrow{\quad i \qquad\qquad\qquad\qquad j \quad} (F_1)_{ji} = (F_1)_{ij}; (\delta_1)_{ij}$$

Member 1–2: $\{F\}_{12} = (F_1)_{12};$ $\{\delta\}_{12} = (\delta_1)_{12}$

Member 3–2: $\{F\}_{32} = (F_1)_{32};$ $\{\delta\}_{32} = (\delta_1)_{32}$

Member 4–2: $\{F\}_{42} = (F_1)_{42};$ $\{\delta\}_{42} = (\delta_1)_{42}$

Selection of Redundants:

Select $\{P\}_{II} = \{P\}_4 = \begin{Bmatrix} (P_1)_4 \\ 0.75(P_1)_4 \end{Bmatrix};$ $\{\Delta\}_{II} = \{\Delta\}_4 = \{0\}$

$\{P\}_I = \{P\}_2 = \begin{Bmatrix} (P_1)_2 \\ (P_2)_2 \end{Bmatrix};$ $\{\Delta\}_I = \{\Delta\}_2 = \begin{Bmatrix} (\Delta_1)_2 \\ (\Delta_2)_2 \end{Bmatrix}$

Primary structure

Generation of [b] Matrices:

$$\{F\} = [b]_I\{P\}_I + [b]_{II}\{P\}_{II} \qquad (18.14)$$

Primary structure—loaded with $\{P\}_I$ and $\{P\}_{II}$

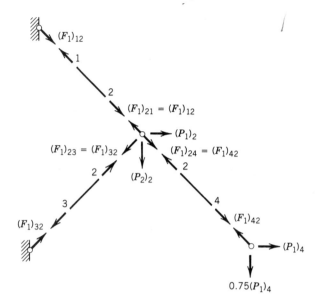

Equilibrium applied to joints 2 and 4 leads to

$$(P_1)_2 - 0.8(F_1)_{12} - 0.8(F_1)_{32} + 0.8(F_1)_{42} = 0$$
$$(P_2)_2 - 0.6(F_1)_{12} + 0.6(F_1)_{32} + 0.6(F_1)_{42} = 0$$
$$(P_1)_4 - 0.8(F_1)_{42} = 0$$

In the form of Eq. 18.12, we have

$$\{P\} = [C]\{F\}; \qquad \begin{Bmatrix} (P_1)_2 \\ (P_2)_2 \\ (P_1)_4 \end{Bmatrix} = \begin{bmatrix} 0.8 & 0.8 & -0.8 \\ 0.6 & -0.6 & -0.6 \\ 0 & 0 & 0.8 \end{bmatrix} \begin{Bmatrix} (F_1)_{12} \\ (F_1)_{32} \\ (F_1)_{42} \end{Bmatrix} \qquad (18.12)$$

Solving for the member forces, we obtain

$$\{F\} = [b]\{P\} = [[b]_{\text{I}} [b]_{\text{II}}] \begin{Bmatrix} \{P\}_{\text{I}} \\ \{P\}_{\text{II}} \end{Bmatrix} \qquad (18.14)$$

$$\begin{Bmatrix} (F_1)_{12} \\ (F_1)_{32} \\ (F_1)_{42} \end{Bmatrix} = \begin{bmatrix} 0.625 & 0.833 & 1.25 \\ 0.625 & -0.833 & 0 \\ 0 & 0 & 1.25 \end{bmatrix} \begin{Bmatrix} (P_1)_2 \\ (P_2)_2 \\ (P_1)_4 \end{Bmatrix}$$

$$\underbrace{\qquad\qquad\qquad}_{[b]_{\text{I}}} \quad \underbrace{\qquad}_{[b]_{\text{II}}}$$

Member Flexibilities:

$$\{\delta\}_{ij} = [f]_{ii}^j \{F\}_{ij} \qquad (16.29)$$

In this case

$$(\delta_1)_{ij} = \left[\frac{l}{EA}\right]_{ij} (F_1)_{ij} \tag{16.28}$$

Member 1–2: $[f]_{11}^2 = \left[\frac{l}{EA}\right] = \left(\frac{5}{EA}\right)$

Member 3–2: $[f]_{33}^2 = \left(\frac{5}{EA}\right)$

Member 4–2: $[f]_{44}^2 = \left(\frac{5}{EA}\right)$

$$[f]_u = \frac{1}{EA}\begin{bmatrix} 5 & 0 & 0 \\ 0 & 5 & 0 \\ 0 & 0 & 5 \end{bmatrix} \tag{18.20}$$

Structure Flexibility Submatrices:

$$[f]_{II,I} = [b]_{II}^T[f]_u[b]_I \tag{18.26}$$

$$[f]_{II,I} = [1.25 \quad 0 \quad 1.25][f]_u \begin{bmatrix} 0.625 & 0.833 \\ 0.625 & -0.833 \\ 0 & 0 \end{bmatrix} = \frac{1}{EA}[3.91 \quad 5.21]$$

$$[f]_{II,II} = [b]_{II}^T[f]_u[b]_{II} \tag{18.26}$$

$$[f]_{II,II} = [1.25 \quad 0 \quad 1.25][f]_u \begin{bmatrix} 1.25 \\ 0 \\ 1.25 \end{bmatrix} = \frac{1}{EA}[15.63]$$

Calculation of Redundants:

$$\{P\}_{II} = -[f]_{II,II}^{-1}[f]_{II,I}\{P\}_I \tag{18.10}$$

$$\{P\}_{II} = -\frac{EA}{15.63} \cdot \frac{1}{EA}[3.91 \quad 5.21]\begin{Bmatrix} -50 \\ 80 \end{Bmatrix} = -14.16 \text{ kN}$$

Final Bar Forces:

$$\{F\} = [b]_I\{P\}_I + [b]_{II}\{P\}_{II} \tag{18.14}$$

$$\begin{Bmatrix} (F_1)_{12} \\ (F_1)_{32} \\ (F_1)_{42} \end{Bmatrix} = \begin{bmatrix} 0.625 & 0.833 \\ 0.625 & -0.833 \\ 0 & 0 \end{bmatrix}\begin{Bmatrix} -50 \\ 80 \end{Bmatrix} + \begin{bmatrix} 1.25 \\ 0 \\ 1.25 \end{bmatrix}\{-14.16\} = \begin{Bmatrix} 17.69 \\ -97.89 \\ -17.70 \end{Bmatrix} \text{ kN}$$

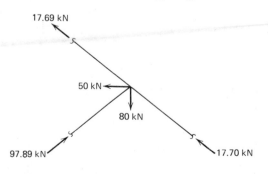

Determine the member-end forces for the structure shown, and construct the shear **18.5.2 Example** and moment diagrams. **problem**

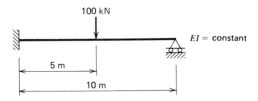

100 kN

EI = constant

5 m

10 m

Selection of Node Points:

Structure forces and displacements:

1   2   3

$$\{P\}_1 = \begin{Bmatrix} P_2 \\ P_6 \end{Bmatrix}_1 = \begin{Bmatrix} (P_2)_1 \\ (P_6)_1 \end{Bmatrix}; \quad \{P\}_2 = \begin{Bmatrix} P_2 \\ P_6 \end{Bmatrix}_2$$

$$= \begin{Bmatrix} (P_2)_2 \\ (P_6)_2 \end{Bmatrix}; \quad \{P\}_3 = \begin{Bmatrix} P_2 \\ P_6 \end{Bmatrix}_3 = \begin{Bmatrix} (P_2)_3 \\ (P_6)_3 \end{Bmatrix}$$

$$\{\Delta\}_1 = \begin{Bmatrix} \Delta_2 \\ \Delta_6 \end{Bmatrix}_1 = \begin{Bmatrix} 0 \\ 0 \end{Bmatrix}; \quad \{\Delta\}_2 = \begin{Bmatrix} \Delta_2 \\ \Delta_6 \end{Bmatrix}_2$$

$$= \begin{Bmatrix} (\Delta_2)_2 \\ (\Delta_6)_2 \end{Bmatrix}; \quad \{\Delta\}_3 = \begin{Bmatrix} \Delta_2 \\ \Delta_6 \end{Bmatrix}_3 = \begin{Bmatrix} 0 \\ (\Delta_6)_3 \end{Bmatrix}$$

Member forces and displacements:

$(F_2)_{ji}$

$(F_6)_{ji}$ i $(F_6)_{ij}$

j

$(F_2)_{ij}$

Member 1–2: $\{F\}_{12} = \begin{Bmatrix} F_2 \\ F_6 \end{Bmatrix}_{12} = \begin{Bmatrix} (F_2)_{12} \\ (F_6)_{12} \end{Bmatrix}$

$$\{\delta\}_{12} = \begin{Bmatrix} \delta_2 \\ \delta_6 \end{Bmatrix}_{12}$$

Member 2–3: $\{F\}_{23} = \begin{Bmatrix} F_2 \\ F_6 \end{Bmatrix}_{23} = \begin{Bmatrix} (F_2)_{23} \\ (F_6)_{23} \end{Bmatrix}$

$$\{\delta\}_{23} = \begin{Bmatrix} \delta_2 \\ \delta_6 \end{Bmatrix}_{23}$$

Selection of Redundants:

Primary structure: 1   2   3

$$\text{Select } \{P\}_{\text{II}} = \{(P_2)_3\}; \quad \{\Delta\}_{\text{II}} = \{(\Delta_2)_3\} = \{0\}$$

then

$$\{P\}_{\text{I}} = \begin{Bmatrix} (P_2)_2 \\ (P_6)_2 \\ (P_6)_3 \end{Bmatrix} = \begin{Bmatrix} 100 \\ 0 \\ 0 \end{Bmatrix}; \quad \{\Delta\}_{\text{I}} = \begin{Bmatrix} (\Delta_2)_2 \\ (\Delta_6)_2 \\ (\Delta_6)_3 \end{Bmatrix}$$

Generation of [b] Matrices:

$$\{F\} = [b]_{\text{I}}\{P\}_{\text{I}} + [b]_{\text{II}}\{P\}_{\text{II}} \tag{18.14}$$

Primary structure loaded with $\{P\}_{\text{I}}$ and $\{P\}_{\text{II}}$:

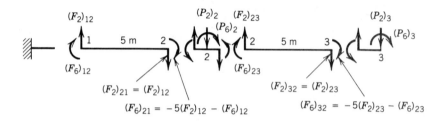

Note:

Member forces at j end are expressed in terms of member forces at i end.

Equilibrium applied at joints 2 and 3 leads to

$$(P_2)_2 - (F_2)_{12} + (F_2)_{23} = 0$$
$$(P_6)_2 + 5(F_2)_{12} + (F_6)_{12} - (F_6)_{23} = 0$$
$$(P_2)_3 - (F_2)_{23} = 0$$
$$(P_6)_3 + 5(F_2)_{23} + (F_6)_{23} = 0$$

In the form of Eq. 18.12, with $\{P\}$ partitioned according to $\{P\}_{\text{I}}$ and $\{P\}_{\text{II}}$, we have

$$\begin{Bmatrix} \{P\}_{\text{I}} \\ \{P\}_{\text{II}} \end{Bmatrix} = [C]\{F\} = \begin{Bmatrix} (P_2)_2 \\ (P_6)_2 \\ (P_6)_3 \\ \hline (P_2)_3 \end{Bmatrix} = \begin{bmatrix} 1 & 0 & -1 & 0 \\ -5 & -1 & 0 & 1 \\ 0 & 0 & -5 & -1 \\ 0 & 0 & 1 & 0 \end{bmatrix} \begin{Bmatrix} \begin{Bmatrix} F_2 \\ F_6 \end{Bmatrix}_{12} \\ \begin{Bmatrix} F_2 \\ F_6 \end{Bmatrix}_{23} \end{Bmatrix} \tag{18.12}$$

Solving for the member forces, we have

$$\{F\} = [b]\{P\} = [[b]_{\text{I}}[b]_{\text{II}}] \begin{Bmatrix} \{P\}_{\text{I}} \\ \{P\}_{\text{II}} \end{Bmatrix} \tag{18.14}$$

$$\begin{Bmatrix} \begin{Bmatrix} F_2 \\ F_6 \end{Bmatrix}_{12} \\ \begin{Bmatrix} F_2 \\ F_6 \end{Bmatrix}_{23} \end{Bmatrix} = \begin{bmatrix} 1 & 0 & 0 & \vdots & 1 \\ -5 & -1 & -1 & \vdots & -10 \\ 0 & 0 & 0 & \vdots & 1 \\ 0 & 0 & -1 & \vdots & -5 \end{bmatrix} \begin{Bmatrix} (P_2)_2 \\ (P_6)_2 \\ (P_6)_3 \\ \hline (P_2)_3 \end{Bmatrix}$$

$$\underbrace{\phantom{\begin{bmatrix} 1 & 0 & 0 \end{bmatrix}}}_{[b]_{\text{I}}} \quad \underbrace{\phantom{\begin{bmatrix} 1 \end{bmatrix}}}_{[b]_{\text{II}}}$$

Member Flexibilities:

$$\{\delta\}_{ij} = [f]_{ii}^j \{F\}_{ij} \tag{16.29}$$

In this case

$$\begin{Bmatrix} \delta_2 \\ \delta_6 \end{Bmatrix}_{ij} = \begin{bmatrix} \dfrac{l^3}{3EI} & \dfrac{l^2}{2EI} \\ \dfrac{l^2}{2EI} & \dfrac{l}{EI} \end{bmatrix}_{ij} \begin{Bmatrix} F_2 \\ F_6 \end{Bmatrix}_{ij} \qquad (16.28)$$

$$\text{Member 1--2: } [f]_{11}^2 = \frac{1}{EI} \begin{bmatrix} 41.7 & 12.5 \\ 12.5 & 5.0 \end{bmatrix}$$

$$\text{Member 2--3: } [f]_{22}^3 = \frac{1}{EI} \begin{bmatrix} 41.7 & 12.5 \\ 12.5 & 5.0 \end{bmatrix}$$

$$[f]_u = \frac{1}{EI} \begin{bmatrix} 41.7 & 12.5 & 0 & 0 \\ 12.5 & 5.0 & 0 & 0 \\ 0 & 0 & 41.7 & 12.5 \\ 0 & 0 & 12.5 & 5.0 \end{bmatrix} \qquad (18.20)$$

Structure Flexibilities:

$$[f]_{\mathrm{II,I}} = [b]_{\mathrm{II}}^T[f]_u[b]_{\mathrm{I}} \qquad (18.26)$$

$$[f]_{\mathrm{II,I}} = [1 \quad -10 \quad 1 \quad -5][f]_u \begin{bmatrix} 1 & 0 & 0 \\ -5 & -1 & -1 \\ 0 & 0 & 0 \\ 0 & 0 & -1 \end{bmatrix} = \frac{1}{EI}[104.2 \quad -37.5 \quad -50]$$

$$[f]_{\mathrm{II,II}} = [b]_{\mathrm{II}}^T[f]_u[b]_{\mathrm{II}} \qquad (18.26)$$

$$[f]_{\mathrm{II,II}} = [1 \quad -10 \quad 1 \quad -5][f]_u \begin{bmatrix} 1 \\ -10 \\ 0 \\ -5 \end{bmatrix} = \frac{1}{EI}[333.4]$$

Determination of Redundants:

$$\{P\}_{\mathrm{II}} = -[f]_{\mathrm{II,II}}^{-1}[f]_{\mathrm{II,I}}\{P\}_{\mathrm{I}} \qquad (18.10)$$

$$\{P\}_{\mathrm{II}} = -\frac{EI}{333.4} \cdot \frac{1}{EI}[104.2 \quad -37.5 \quad -50]\begin{Bmatrix} 100 \\ 0 \\ 0 \end{Bmatrix} = -31.3 \text{ kN}$$

Member-End Forces:

$$\{F\} = [b]_{\mathrm{I}}\{P\}_{\mathrm{I}} + [b]_{\mathrm{II}}\{P\}_{\mathrm{II}} \qquad (18.14)$$

$$\begin{Bmatrix} \begin{Bmatrix} F_2 \\ F_6 \end{Bmatrix}_{12} \\ \begin{Bmatrix} F_2 \\ F_6 \end{Bmatrix}_{23} \end{Bmatrix} = \begin{bmatrix} 1 & 0 & 0 \\ -5 & -1 & -1 \\ 0 & 0 & 0 \\ 0 & 0 & -1 \end{bmatrix}\begin{Bmatrix} 100 \\ 0 \\ 0 \end{Bmatrix} + \begin{bmatrix} 1 \\ -10 \\ 1 \\ -5 \end{bmatrix}\{-31.3\} = \begin{Bmatrix} +68.7 \text{ kN} \\ -187.0 \text{ kN} \cdot \text{m} \\ -31.3 \text{ kN} \\ +156.5 \text{ kN} \cdot \text{m} \end{Bmatrix}$$

Shear and Moment Diagrams:

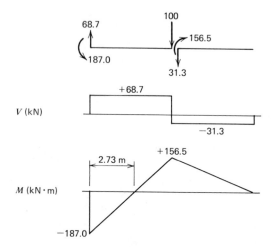

Note:

Start with the forces at the left end of any member and use the static relations for shear and moment to obtain the forces at the right end of the member. These should be consistent with the forces at the left end of the next member.

18.5.3 Example problem   Determine the member-end forces for the frame structure given.

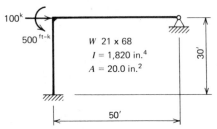

Selection of Node Points:

Structure forces and displacements:

$$\{P\}_2 = \begin{Bmatrix} P_1 \\ P_2 \\ P_6 \end{Bmatrix}_2 = \begin{Bmatrix} (P_1)_2 \\ (P_2)_2 \\ (P_6)_2 \end{Bmatrix} = \begin{Bmatrix} 100 \\ 0 \\ -500 \end{Bmatrix}; \qquad \{\Delta\}_2 = \begin{Bmatrix} \Delta_1 \\ \Delta_2 \\ \Delta_6 \end{Bmatrix}_2 = \begin{Bmatrix} (\Delta_1)_2 \\ (\Delta_2)_2 \\ (\Delta_6)_2 \end{Bmatrix}$$

$$\{P\}_3 = \begin{Bmatrix} P_1 \\ P_2 \\ P_6 \end{Bmatrix}_3 = \begin{Bmatrix} (P_1)_3 \\ (P_2)_3 \\ (P_6)_3 \end{Bmatrix} = \begin{Bmatrix} (P_1)_3 \\ (P_2)_3 \\ 0 \end{Bmatrix}; \qquad \{\Delta\}_3 = \begin{Bmatrix} \Delta_1 \\ \Delta_2 \\ \Delta_6 \end{Bmatrix}_3 = \begin{Bmatrix} (\Delta_1)_3 \\ (\Delta_2)_3 \\ (\Delta_6)_3 \end{Bmatrix} = \begin{Bmatrix} 0 \\ 0 \\ (\Delta_6)_3 \end{Bmatrix}$$

$$\{P\}_1 = \begin{Bmatrix} P_1 \\ P_2 \\ P_6 \end{Bmatrix}_1 = \begin{Bmatrix} (P_1)_1 \\ (P_2)_1 \\ (P_6)_1 \end{Bmatrix}; \qquad \{\Delta\}_1 = \begin{Bmatrix} \Delta_1 \\ \Delta_2 \\ \Delta_6 \end{Bmatrix}_1 = \begin{Bmatrix} (\Delta_1)_1 \\ (\Delta_2)_1 \\ (\Delta_6)_1 \end{Bmatrix} = \begin{Bmatrix} 0 \\ 0 \\ 0 \end{Bmatrix}$$

Member forces and displacements:

Member 1–2: $\{F\}_{12} = \begin{Bmatrix} F_1 \\ F_2 \\ F_6 \end{Bmatrix}_{12} = \begin{Bmatrix} (F_1)_{12} \\ (F_2)_{12} \\ (F_6)_{12} \end{Bmatrix}; \quad \{\delta\}_{12} = \begin{Bmatrix} \delta_1 \\ \delta_2 \\ \delta_6 \end{Bmatrix}_{12} = \begin{Bmatrix} (\delta_1)_{12} \\ (\delta_2)_{12} \\ (\delta_6)_{12} \end{Bmatrix}$

Member 2–3: $\{F\}_{23} = \begin{Bmatrix} F_1 \\ F_2 \\ F_6 \end{Bmatrix}_{23} = \begin{Bmatrix} (F_1)_{23} \\ (F_2)_{23} \\ (F_6)_{23} \end{Bmatrix}; \quad \{\delta\}_{23} = \begin{Bmatrix} \delta_1 \\ \delta_2 \\ \delta_6 \end{Bmatrix}_{23} = \begin{Bmatrix} (\delta_1)_{23} \\ (\delta_2)_{23} \\ (\delta_6)_{23} \end{Bmatrix}$

Selection of Redundants:

$$\text{Select} \quad \{P\}_{\text{II}} = \begin{Bmatrix} (P_1)_3 \\ (P_2)_3 \end{Bmatrix}; \quad \{\Delta\}_{\text{II}} = \begin{Bmatrix} (\Delta_1)_3 \\ (\Delta_2)_3 \end{Bmatrix} = \begin{Bmatrix} 0 \\ 0 \end{Bmatrix}$$

$$\{P\}_{\text{I}} = \begin{Bmatrix} (P_1)_2 \\ (P_2)_2 \\ (P_6)_2 \\ (P_6)_3 \end{Bmatrix} = \begin{Bmatrix} 100 \\ 0 \\ -500 \\ 0 \end{Bmatrix}; \quad \{\Delta\}_{\text{I}} = \begin{Bmatrix} (\Delta_1)_2 \\ (\Delta_2)_2 \\ (\Delta_6)_2 \\ (\Delta_6)_3 \end{Bmatrix}$$

Generation of [b] Matrices:

$$\{F\} = [b]_{\text{I}}\{P\}_{\text{I}} + [b]_{\text{II}}\{P\}_{\text{II}} \tag{18.14}$$

Primary structure loaded with $\{P\}_{\text{I}}$ and $\{P\}_{\text{II}}$:

Note:
Member forces at j end are expressed in terms of member forces at i end.

Equilibrium applied to joints 2 and 3 leads to

$$(P_1)_2 - (F_2)_{12} + (F_1)_{23} = 0$$
$$(P_2)_2 + (F_1)_{12} + (F_2)_{23} = 0$$
$$(P_6)_2 + 30(F_2)_{12} + (F_6)_{12} - (F_6)_{23} = 0$$
$$(P_1)_3 - (F_1)_{23} = 0$$
$$(P_2)_3 - (F_2)_{23} = 0$$
$$(P_6)_3 + 50(F_2)_{23} + (F_6)_{23} = 0$$

In the form of Eq. 18.12, with $\{P\}$ partitioned according to $\{P\}_\mathrm{I}$ and $\{P\}_\mathrm{II}$, we have

$$\begin{Bmatrix} \{P\}_\mathrm{I} \\ \{P\}_\mathrm{II} \end{Bmatrix} = [C]\{F\} = \begin{Bmatrix} (P_1)_2 \\ (P_2)_2 \\ (P_6)_2 \\ (P_6)_3 \\ \hline (P_1)_3 \\ (P_2)_3 \end{Bmatrix} = \begin{bmatrix} 0 & 1 & 0 & -1 & 0 & 0 \\ -1 & 0 & 0 & 0 & -1 & 0 \\ 0 & -30 & -1 & 0 & 0 & 1 \\ 0 & 0 & 0 & 0 & -50 & -1 \\ 0 & 0 & 0 & 1 & 0 & 0 \\ 0 & 0 & 0 & 0 & 1 & 0 \end{bmatrix} \begin{Bmatrix} F_1 \\ F_2 \\ F_6 \end{Bmatrix}_{12} \\ \begin{Bmatrix} F_1 \\ F_2 \\ F_6 \end{Bmatrix}_{23} \qquad (18.12)$$

Solving for the member forces, we have

$$\{F\} = [b]\{P\} = [[b]_\mathrm{I}[b]_\mathrm{II}]\begin{Bmatrix} \{P\}_\mathrm{I} \\ \{P\}_\mathrm{II} \end{Bmatrix} \qquad (18.14)$$

$$\begin{Bmatrix} \begin{Bmatrix} F_1 \\ F_2 \\ F_6 \end{Bmatrix}_{12} \\ \begin{Bmatrix} F_1 \\ F_2 \\ F_6 \end{Bmatrix}_{23} \end{Bmatrix} = \begin{bmatrix} 0 & -1 & 0 & 0 & 0 & -1 \\ 1 & 0 & 0 & 0 & 1 & 0 \\ -30 & 0 & -1 & -1 & -30 & -50 \\ 0 & 0 & 0 & 0 & 1 & 0 \\ 0 & 0 & 0 & 0 & 0 & 1 \\ 0 & 0 & 0 & -1 & 0 & -50 \end{bmatrix} \begin{Bmatrix} (P_1)_2 \\ (P_2)_2 \\ (P_6)_2 \\ (P_6)_3 \\ \hline (P_1)_3 \\ (P_2)_3 \end{Bmatrix}$$

$$\underbrace{\phantom{\begin{matrix} 0 & -1 & 0 & 0 \end{matrix}}}_{[b]_\mathrm{I}} \quad \underbrace{\phantom{\begin{matrix} 0 & -1 \end{matrix}}}_{[b]_\mathrm{II}}$$

Member Flexibilities:

$$\{\delta\}_{ij} = [f]_{ii}^{j}\{F\}_{ij} \qquad (16.29)$$

In this case

$$\begin{Bmatrix} \delta_1 \\ \delta_2 \\ \delta_6 \end{Bmatrix}_{ij} = \begin{bmatrix} \dfrac{l}{EA} & 0 & 0 \\ 0 & \dfrac{l^3}{3EI} & \dfrac{l^2}{2EI} \\ 0 & \dfrac{l^2}{2EI} & \dfrac{l}{EI} \end{bmatrix}_{ij} \begin{Bmatrix} F_1 \\ F_2 \\ F_6 \end{Bmatrix}_{ij} \qquad (16.28)$$

$$[f]_u = \begin{bmatrix} [f]_{11}^2 & \\ & [f]_{22}^3 \end{bmatrix} = \frac{1}{EI}\begin{bmatrix} 18.9 & 0 & 0 & 0 & 0 & 0 \\ 0 & 9{,}000 & 450 & 0 & 0 & 0 \\ 0 & 450 & 30 & 0 & 0 & 0 \\ 0 & 0 & 0 & 31.5 & 0 & 0 \\ 0 & 0 & 0 & 0 & 41{,}667 & 1{,}250 \\ 0 & 0 & 0 & 0 & 1{,}250 & 50 \end{bmatrix} \qquad (18.20)$$

Structure Flexibilities:

$$[f]_{\text{II,I}} = [b]_{\text{II}}^{T}[f]_{u}[b]_{\text{I}} \tag{18.26}$$

$$[f]_{\text{II,I}} = \frac{1}{EI}\begin{bmatrix} 9{,}000 & 0 & 450 & 450 \\ 22{,}500 & 18.9 & 1{,}500 & 2{,}750 \end{bmatrix}$$

$$[f]_{\text{II,II}} = [b]_{\text{II}}^{T}[f]_{u}[b]_{\text{II}} \tag{18.26}$$

$$[f]_{\text{II,II}} = \frac{1}{EI}\begin{bmatrix} 9{,}032 & 22{,}500 \\ 22{,}500 & 116{,}686 \end{bmatrix}$$

Determination of Redundants:

$$\{P\}_{\text{II}} = -[f]_{\text{II,II}}^{-1}[f]_{\text{II,I}}\{P\}_{\text{I}} \tag{18.10}$$

$$\{P\}_{\text{II}} = -\frac{EI}{5.476 \times 10^{8}}\begin{bmatrix} 116{,}686 & -22{,}500 \\ -22{,}500 & 9{,}032 \end{bmatrix}$$

$$\times \frac{1}{EI}\begin{bmatrix} 9{,}000 & 0 & 450 & 450 \\ 22{,}500 & 18.9 & 1{,}500 & 2{,}750 \end{bmatrix}\begin{Bmatrix} 100 \\ 0 \\ -500 \\ 0 \end{Bmatrix} = \begin{Bmatrix} -82.2 \\ +3.0 \end{Bmatrix} \text{ kips}$$

Member-End Forces:

$$\{F\} = [b]_{\text{I}}\{P\}_{\text{I}} + [b]_{\text{II}}\{P\}_{\text{II}} \tag{18.14}$$

$$\begin{Bmatrix} \begin{Bmatrix} F_1 \\ F_2 \\ F_6 \end{Bmatrix}_{12} \\ \begin{Bmatrix} F_1 \\ F_2 \\ F_6 \end{Bmatrix}_{23} \end{Bmatrix} = \begin{bmatrix} 0 & -1 & 0 & 0 \\ 1 & 0 & 0 & 0 \\ -30 & 0 & -1 & -1 \\ 0 & 0 & 0 & 0 \\ 0 & 0 & 0 & 0 \\ 0 & 0 & 0 & -1 \end{bmatrix}\begin{Bmatrix} 100 \\ 0 \\ -500 \\ 0 \end{Bmatrix} + \begin{bmatrix} 0 & -1 \\ 1 & 0 \\ -30 & -50 \\ 1 & 0 \\ 0 & 1 \\ 0 & -50 \end{bmatrix}\begin{Bmatrix} -82.2 \\ +3.0 \end{Bmatrix}$$

$$= \begin{Bmatrix} -3.0^{k} \\ +17.8^{k} \\ -184.0'^{-k} \\ -82.2^{k} \\ +3.0^{k} \\ -150.0'^{-k} \end{Bmatrix}$$

18.6

The procedure developed in Sections 18.2 through 18.4 and illustrated through the examples of Section 18.5 is predicated on the structure's being statically indeterminate externally. For that case, the vector $\{P\}_{\text{II}}$ included the redundant structure forces, and the vector $\{P\}_{\text{I}}$ contained the structure forces corresponding to the free displacements on the original structure. **Internally Redundant Structures**

If the structure is internally redundant, the vector $\{F\}$, which includes the member-end forces at one end of each member, must be partitioned into two parts—vector $\{F\}_{\text{II}}$ contains the redundant member-end forces, and $\{F\}_{\text{I}}$ includes the remaining member-end forces for the primary structure.

When the redundant members are cut at one of their ends, the member-end forces associated with these members become external to the structure. However, compatibility requires that the displacements of each severed member end, with respect to the node from which it is separated, must be zero. These relative displacements are collected in the vector $\{\delta\}_{II}$, as shown for a planar frame member in Fig. 18.2.

In this case, the flexibility equations for the primary structure are expressed in the form

$$\left\{\begin{matrix} \{\Delta\}_I \\ \hline \{\delta\}_{II} \end{matrix}\right\} = \left[\begin{matrix} [/]_{I,I} & [/]_{I,II} \\ \hline [/]_{II,I} & [/]_{II,II} \end{matrix}\right]\left\{\begin{matrix} \{P\}_I \\ \hline \{F\}_{II} \end{matrix}\right\} \qquad (18.27)$$

where $\{\Delta\}_I$ contains the free structure displacements on the primary structure and $\{P\}_I$ includes the corresponding structure forces. Expanding these equations and recalling that $\{\delta\}_{II} = \{0\}$, we obtain

$$\{F\}_{II} = -[/]_{II,II}^{-1}[/]_{II,I}\{P\}_I \qquad (18.28)$$

and

$$\{\Delta\}_I = ([/]_{I,I} - [/]_{I,II}[/]_{II,II}^{-1}[/]_{II,I})\{P\}_I \qquad (18.29)$$

Equilibrium conditions applied to the primary structure give

$$\{F\}_I = [b]_I\{P\}_I + [b]_{II}\{F\}_{II} \qquad (18.30)$$

Following the procedures of Section 18.4, we find

$$[/]_{I,I} = [b]_I^T[f]_{uI}[b]_I$$
$$[/]_{I,II} = [b]_I^T[f]_{uI}[b]_{II}$$
$$[/]_{II,I} = [b]_{II}^T[f]_{uI}[b]_I \qquad (18.31)$$
$$[/]_{II,II} = [b]_{II}^T[f]_{uI}[b]_{II} + [f]_{uII}$$

where $[f]_{uI}$ and $[f]_{uII}$ are unassembled stiffness matrices for the members included in vectors $\{F\}_I$ and $\{F\}_{II}$, respectively.

The application of the flexibility method to internally redundant forces follows the general procedure outlined in Section 18.5, except that the redundants are now

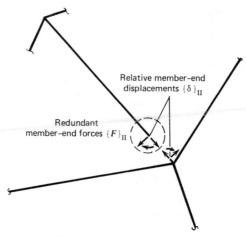

Relative member-end displacements $\{\delta\}_{II}$

Redundant member-end forces $\{F\}_{II}$

Fig. 18.2 *Severed redundant member.*

the member forces $\{F\}_{\mathrm{II}}$ instead of the structure forces $\{P\}_{\mathrm{II}}$ and $[f]_u$ is separated into $[f]_{u\mathrm{I}}$ and $[f]_{u\mathrm{II}}$. The detailed procedure is illustrated by the example problem in the following section.

For a structure that is indeterminate both externally and internally, the procedures developed separately are combined. Here $[b]_{\mathrm{II}}$ must be partitioned to include the effects of the redundant reactions $\{P\}_{\mathrm{II}}$ and the redundant member forces $\{F\}_{\mathrm{II}}$.

Determine the bar forces for the internally statically indeterminate truss shown. **18.6.1 Example**
Assume EA is the same for each member.   **problem**

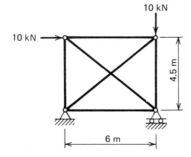

Selection of Node Points:

Structure forces and displacements:

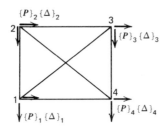

Member forces and displacements:

$$(F_1)_{ij}; \quad (\delta_1)_{ij} \quad i \underline{\hspace{3cm}} j$$

Selection of Redundants:

Select $\{F\}_{\mathrm{II}} = (F_1)_{42};$ $\{\delta\}_{\mathrm{II}} = 0$

Primary structure

$$\{F\}_{\mathrm{I}} = \begin{Bmatrix} (F_1)_{12} \\ (F_1)_{13} \\ (F_1)_{14} \\ (F_1)_{23} \\ (F_1)_{34} \end{Bmatrix}$$

$$\{P\}_{\mathrm{I}} = \begin{Bmatrix} (P_1)_2 \\ (P_2)_2 \\ (P_1)_3 \\ (P_2)_3 \\ (P_1)_4 \end{Bmatrix} = \begin{Bmatrix} 10 \\ 0 \\ 0 \\ 10 \\ 0 \end{Bmatrix}; \quad \{\Delta\}_{\mathrm{I}} = \begin{Bmatrix} (\Delta_1)_2 \\ (\Delta_2)_2 \\ (\Delta_1)_3 \\ (\Delta_2)_3 \\ (\Delta_1)_4 \end{Bmatrix}$$

Generation of [b] Matrices:

$$\{F\}_I = [b]_I\{P\}_I + [b]_{II}\{F\}_{II} \tag{18.30}$$

Primary structure loaded with $\{P\}_I$ and $\{F\}_{II}$:

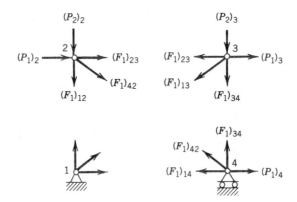

Equilibrium corresponding to kinematic degrees of freedom is applied at joints 2, 3, and 4:

$$(P_1)_2 + 0.8(F_1)_{42} + (F_1)_{23} = 0$$
$$(P_2)_2 + 0.6(F_1)_{42} + (F_1)_{12} = 0$$
$$(P_1)_3 - (F_1)_{23} - 0.8(F_1)_{13} = 0$$
$$(P_2)_3 + 0.6(F_1)_{13} + (F_1)_{34} = 0$$
$$(P_1)_4 - 0.8(F_1)_{42} - (F_1)_{14} = 0$$

In the form of Eq. 18.12, recognizing that $\{P\}$ includes contributions from $\{F\}_{II}$, we have

$$\{P\} = [C]\{F\} = \begin{Bmatrix} (P_1)_2 + 0.8(F_1)_{42} \\ (P_2)_2 + 0.6(F_1)_{42} \\ (P_1)_3 \\ (P_2)_3 \\ (P_1)_4 - 0.8(F_1)_{42} \end{Bmatrix} = \begin{bmatrix} 0 & 0 & 0 & -1 & 0 \\ -1 & 0 & 0 & 0 & 0 \\ 0 & 0.8 & 0 & 1 & 0 \\ 0 & -0.6 & 0 & 0 & -1 \\ 0 & 0 & 1 & 0 & 0 \end{bmatrix} \begin{Bmatrix} (F_1)_{12} \\ (F_1)_{13} \\ (F_1)_{14} \\ (F_1)_{23} \\ (F_1)_{34} \end{Bmatrix}$$

$$(18.12)$$

Solving for the member forces, we obtain

$$\begin{Bmatrix} (F_1)_{12} \\ (F_1)_{13} \\ (F_1)_{14} \\ (F_1)_{23} \\ (F_1)_{34} \end{Bmatrix} = \begin{bmatrix} 0 & -1 & 0 & 0 & 0 \\ 1.25 & 0 & 1.25 & 0 & 0 \\ 0 & 0 & 0 & 0 & 1 \\ -1 & 0 & 0 & 0 & 0 \\ -0.75 & 0 & -0.75 & -1 & 0 \end{bmatrix} \begin{Bmatrix} (P_1)_2 + 0.8(F_1)_{42} \\ (P_2)_2 + 0.6(F_1)_{42} \\ (P_1)_3 \\ (P_2)_3 \\ (P_1)_4 - 0.8(F_1)_{42} \end{Bmatrix}$$

In an expanded form to agree with Eq. 18.30, we have

$$
\begin{bmatrix} (F_1)_{12} \\ (F_1)_{13} \\ (F_1)_{14} \\ (F_1)_{23} \\ (F_1)_{34} \end{bmatrix} = \underbrace{\begin{bmatrix} 0 & -1 & 0 & 0 & 0 \\ 1.25 & 0 & 1.25 & 0 & 0 \\ 0 & 0 & 0 & 0 & 1 \\ -1 & 0 & 0 & 0 & 0 \\ -0.75 & 0 & -0.75 & -1 & 0 \end{bmatrix}}_{[b]_\mathrm{I}} \begin{bmatrix} (P_1)_2 \\ (P_2)_2 \\ (P_1)_3 \\ (P_2)_3 \\ (P_1)_4 \end{bmatrix} + \underbrace{\begin{bmatrix} -0.6 \\ 1.0 \\ -0.8 \\ -0.8 \\ -0.6 \end{bmatrix}}_{[b]_\mathrm{II}} \{(F_1)_{42}\} \qquad (18.30)
$$

Note:

The matrices $[b]_\mathrm{I}$ and $[b]_\mathrm{II}$ may also be established by subjecting the primary structure to the loadings $\{P\}_\mathrm{I}$ and $\{F\}_\mathrm{II}$, respectively, and determining the member forces.

Member Flexibilities:

$$
\{\delta\}_{ij} = [f]_{ii}^j \{F\}_{ij} \qquad (16.29)
$$

In this case

$$
(\delta_1)_{ij} = \left[\frac{l}{EA}\right]_{ij} (F_1)_{ij} \qquad (16.28)
$$

Member 1–2: $[f]_{11}^2 = [4.5/EA]$
Member 1–3: $[f]_{11}^3 = [7.5/EA]$
Member 1–4: $[f]_{11}^4 = [6.0/EA]$
Member 2–3: $[f]_{22}^3 = [6.0/EA]$
Member 3–4: $[f]_{33}^4 = [4.5/EA]$

$$
[f]_{u\mathrm{I}} = \frac{1}{EA}\begin{bmatrix} 4.5 & & & & \\ & 7.5 & & & \\ & & 6.0 & & \\ & & & 6.0 & \\ & & & & 4.5 \end{bmatrix} \qquad (18.20)
$$

Member 4–2: $[f]_{44}^2 = [7.5/EA]$

$$
[f]_{u\mathrm{II}} = \frac{1}{EA}[7.5] \qquad (18.20)
$$

Structure Flexibility Submatrices:

$$
[f]_{\mathrm{II,I}} = [b]_\mathrm{II}^T[f]_{u\mathrm{I}}[b]_\mathrm{I} \qquad (18.31)
$$

$$
[f]_{\mathrm{II,I}} = [-0.6 \quad 1 \quad -0.8 \quad -0.8 \quad -0.6][f]_{u\mathrm{I}} \begin{bmatrix} 0 & -1 & 0 & 0 & 0 \\ 1.25 & 0 & 1.25 & 0 & 0 \\ 0 & 0 & 0 & 0 & 1 \\ -1 & 0 & 0 & 0 & 0 \\ -0.75 & 0 & -0.75 & -1 & 0 \end{bmatrix}
$$

$$= \frac{1}{EA} [16.2 \quad 2.7 \quad 11.4 \quad 4.8 \quad -4.8]$$

$$[/]_{\mathrm{II,II}} = [b]_{\mathrm{II}}^{T}[f]_{u\mathrm{I}}[b]_{\mathrm{II}} + [f]_{u\mathrm{II}} \tag{18.31}$$

$$[/]_{\mathrm{II,II}} = [-0.6 \quad 1 \quad -0.8 \quad -0.8 \quad -0.6][f]_{u\mathrm{I}} \begin{bmatrix} -0.6 \\ 0 \\ -0.8 \\ -0.8 \\ -0.6 \end{bmatrix} + [f]_{u\mathrm{II}}$$

$$= \frac{1}{EA} [18.42] + \frac{1}{EA} [7.5] = \frac{1}{EA} [25.92]$$

Determination of Redundants:

$$\{F\}_{\mathrm{II}} = -[/]_{\mathrm{II,II}}^{-1}[/]_{\mathrm{II,I}}\{P\}_{\mathrm{I}} \tag{18.28}$$

$$\{F\}_{\mathrm{II}} = -\left(\frac{EA}{25.92}\right)\frac{1}{EA} [16.2 \quad 2.7 \quad 11.4 \quad 4.8 \quad -4.8] \begin{bmatrix} 10 \\ 0 \\ 0 \\ 10 \\ 0 \end{bmatrix} = -8.10 \text{ kN}$$

Member-End Forces:

$$\{F\}_{\mathrm{I}} = [b]_{\mathrm{I}}\{P\}_{\mathrm{I}} + [b]_{\mathrm{II}}\{F\}_{\mathrm{II}} \tag{18.30}$$

$$\begin{Bmatrix} (F_1)_{12} \\ (F_1)_{13} \\ (F_1)_{14} \\ (F_1)_{23} \\ (F_1)_{34} \end{Bmatrix} = \begin{bmatrix} 0 & -1 & 0 & 0 & 0 \\ 1.25 & 0 & 1.25 & 0 & 0 \\ 0 & 0 & 0 & 0 & 1 \\ -1 & 0 & 0 & 0 & 0 \\ -0.75 & 0 & -0.75 & -1 & 0 \end{bmatrix} \begin{Bmatrix} 10 \\ 0 \\ 0 \\ 10 \\ 0 \end{Bmatrix}$$

$$+ \begin{bmatrix} -0.6 \\ 1 \\ -0.8 \\ -0.8 \\ -0.6 \end{bmatrix} \{-8.10\} = \begin{Bmatrix} 4.86 \\ 4.40 \\ 6.48 \\ -3.52 \\ -12.64 \end{Bmatrix} \text{ kN}$$

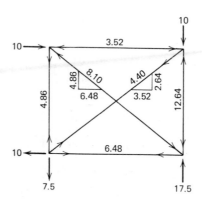

Up until this point in our discussion of the flexibility method, structure forces have been admissible only at the node points. This restriction will now be relaxed.

Two approaches can be employed. In the first approach, the same basic procedure is used that was applied in the stiffness formulation. This method was described in Section 17.6, and it consists of the superposition of two separate solutions: a fixed-jointed analysis, which includes the effects of the member loads acting between the nodes, provides the first solution; the second solution is a standard joint-loaded analysis in which the loads include the applied structure forces and the negative of certain equivalent structure forces. In applying this approach in a flexibility analysis, the fixed-jointed analysis is completed as before, but the joint-loaded analysis is carried out by the flexibility method. For the latter part of the solution, the general approach is the same as that outlined in Section 18.5, with the exceptions that nodal numbering need not include the load points, and the final member-end forces are given by

$$\{F\} = [b]_\text{I}\{P\}_\text{I} + [b]_\text{II}\{P\}_\text{II} + \{F\}^f \qquad (18.32)$$

where $\{F\}^f$ includes the fixed-end forces corresponding to the member ends included in $\{F\}$.

This approach is not a pure flexibility analysis since stiffness considerations are necessary in determining the fixed-end forces of the fixed-jointed analysis.

The second approach for treating loads between the nodes has the elegance of being a pure flexibility formulation. Our discussion will center on a planar structure in flexure, but it could be extended readily to include more complicated structures. We initially consider the fixed-ended member shown in Fig. 18.3a, in which the member is subjected to member-end forces at end i along with the transverse loading along the span. The free-end displacements result from two contributions: there is the contribution from the member-end forces, $\{\delta\}_{ij}$, which is shown in Fig. 18.3b and is given by Eq. 16.29, and there is the contribution from the transverse loading, $\{\delta_0\}_{ij}$, which is shown in Fig. 18.3c; the components of this contribution are given in Table 18.1 for several loading conditions. Therefore, the total member-end displacements at end i, $\{\delta^T\}_{ij}$, are given by

$$\{\delta^T\}_{ij} = [f]_{ii}^j\{F\}_{ij} + \{\delta_0\}_{ij} \qquad (18.33)$$

Associated with the transverse load, there is a set of reaction forces at end j, $\{F\}_{ji}^s$, that are determined from statics. These forces are shown in Fig. 18.3c, and the individual components are given in Table 18.1 for an assortment of loading conditions.

We now return to the arguments that were used in Section 18.4. The strain energy for member ij is again given by Eq. 18.15; however, in this case, the relative displacements, $\{e\}_{ij}$, are given by $\{\delta^T\}_{ij}$ of Eq. 18.33. Therefore the strain energy for member ij becomes

$$U_{ij} = \tfrac{1}{2}\{F\}_{ij}^T[f]_{ii}^j\{F\}_{ij} + \tfrac{1}{2}\{F\}_{ij}^T\{\delta_0\}_{ij} \qquad (18.34)$$

and for the entire structure, the total strain energy is

$$U = \tfrac{1}{2}\{F\}^T[f_u]\{F\} + \tfrac{1}{2}\{F\}^T\{\delta_0\} \qquad (18.35)$$

where $\{F\}$ contains the ordered array of the member-end forces at the i end of all members, $\{\delta_0\}$ includes the displacement components at the i ends that are caused

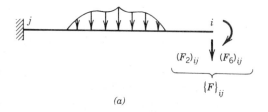

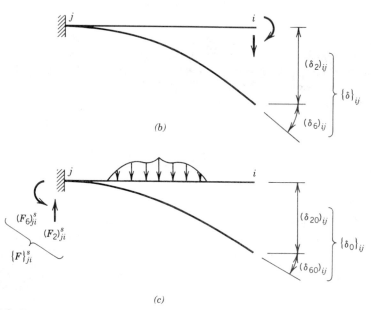

Fig. 18.3 *Forces and displacements on planar member. (a) End-forces and transverse forces between nodes. (b) Displacements from end forces only. (c) Displacements and reactions for transverse load only.*

by transverse loading in the order in which the respective force components are given in $\{F\}$, and $[f_u]$ is as defined in Eq. 18.20.

For an externally redundant structure, we know from Eq. 18.13 that

$$\{F\} = [b]\{P\} = [b]_I\{P\}_I + [b]_{II}\{P\}_{II} \tag{18.36}$$

where $[b]$ is the global statics matrix, $\{P\}_I$ includes the applied forces on the primary structure, and $\{P\}_{II}$ contains the redundant reaction forces.

Substitution of Eq. 18.36 into Eq. 18.35 gives

$$U = \tfrac{1}{2}\{P\}^T[b]^T[f_u][b]\{P\} + \tfrac{1}{2}\{P\}^T[b]^T\{\delta_0\} \tag{18.37}$$

or

$$U = \tfrac{1}{2}\{P\}^T([b]^T[f_u][b]\{P\} + [b]^T\{\delta_0\}) \tag{18.38}$$

However, as was noted in Eq. 18.22, the strain energy can be expressed in terms of structural quantities as

$$U = \tfrac{1}{2}\{P\}^T\{\Delta\} \tag{18.39}$$

TABLE 18.1 Free-End Displacements and Reactions for Transverse Loading

$(F_2)_{ji}^s$	$(F_6)_{ji}^s$	$\{F\}_{ji}^s$ ———EI——— $\{\delta_0\}_{ij}$ (Loading)	$(\delta_{20})_{ij}$	$(\delta_{60})_{ij}$
P	Pa	P at distance a, with b; span l	$\dfrac{Pa^2}{2EI}\left(l-\dfrac{a}{3}\right)$	$\dfrac{Pa^2}{2EI}$
wl	$\dfrac{wl^2}{2}$	uniform load w over span l	$\dfrac{wl^4}{8EI}$	$\dfrac{wl^3}{6EI}$
$\dfrac{wl}{2}$	$\dfrac{wl^2}{3}$	triangular load w over span l	$\dfrac{11wl^4}{60EI}$	$\dfrac{wl^3}{8EI}$
wa	$\dfrac{wa^2}{2}$	partial uniform load w over a, with $l-a$	$\dfrac{wa^3}{24EI}(4l-a)$	$\dfrac{wa^3}{6EI}$
0	M	moment M at distance a, with b	$\dfrac{Ma}{EI}\left(l-\dfrac{a}{2}\right)$	$\dfrac{Ma}{EI}$

Note: Signs are assigned to quantities according to member coordinate system and orientation of member ij in structure.

Comparison of Eqs. 18.38 and 18.39 reveals that

$$\{\Delta\} = [b]^T[f_u][b]\{P\} + [b]^T\{\delta_0\} \tag{18.40}$$

Partitioning Eq. 18.40 according to the redundant and nonredundant components of $\{P\}$, we obtain

$$\begin{Bmatrix} \{\Delta\}_{\text{I}} \\ \{\Delta\}_{\text{II}} \end{Bmatrix} = \begin{bmatrix} [b]_{\text{I}}^T \\ [b]_{\text{II}}^T \end{bmatrix} [f_u][[b]_{\text{I}}][b]_{\text{II}}] \begin{Bmatrix} P_{\text{I}} \\ P_{\text{II}} \end{Bmatrix} + \begin{bmatrix} [b]_{\text{I}}^T \\ [b]_{\text{II}}^T \end{bmatrix} \{\delta_0\} \tag{18.41}$$

Expanding Eq. 18.41 and invoking Eqs. 18.26, we find

$$\{\Delta\}_{\text{I}} = [/]_{\text{I,I}}\{P\}_{\text{I}} + [/]_{\text{I,II}}\{P\}_{\text{II}} + [b]_{\text{I}}^T\{\delta_0\} \tag{18.42}$$

$$\{\Delta\}_{\text{II}} = [/]_{\text{II,I}}\{P\}_{\text{I}} + [/]_{\text{II,II}}\{P\}_{\text{II}} + [b]_{\text{II}}^T\{\delta_0\} \tag{18.43}$$

The individual $[/]$ matrices are defined in Eqs. 18.26.

For nonyielding supports, $\{\Delta\}_{\text{II}} = \{0\}$, and we can then solve for $\{P\}_{\text{II}}$ from Eq. 18.43. This leads to

$$\{P\}_{\text{II}} = -[/]_{\text{II,II}}^{-1}[/]_{\text{II,I}}\{P\}_{\text{I}} - [/]_{\text{II,II}}^{-1}[b]_{\text{II}}^T\{\delta_0\} \tag{18.44}$$

The member forces are then determined from Eq. 18.36, and the structure displacements are given by Eq. 18.42.

In the application of Eq. 18.44, $\{P\}_{\text{II}}$ is given by

$$\{P\}_{\text{I}} = \{P\}_{\text{I}}^a - \{P\}_{\text{I}}^e \qquad (18.45)$$

where $\{P\}_{\text{I}}^a$ are the applied force components associated with the kinematic degrees of freedom of the primary structure, and $\{P\}_{\text{I}}^e$ are the corresponding components of the equivalent structure forces that are needed to sustain the static reactions, $\{F\}_{ji}^s$, of the transversely loaded member. In accordance with the procedure of Section 17.6, the equivalent structure forces at joint j are

$$\{P\}_{j}^e = \sum_k [\beta]_{ji}^T \{F\}_{ji}^s \qquad (18.46)$$

where $[\beta]_{ji}$ is the compatibility matrix for member ij at joint j, and k ranges over all of the members framing into joint j. Only those components associated with the kinematic degrees of freedom of the primary structure are included in $\{P\}_{\text{I}}^e$.

Table 18.1 gives the components of $\{\delta_0\}_{ij}$ and $\{F\}_{ji}^s$ for several transverse load cases. Signs are not assigned to these cases; the signs for forces and displacements must be assigned according to the member coordinate system and the orientation of the member within the structure. This table includes only the information needed for planar flexural members. Axial and torsional effects, as well as flexure in the xz plane, can be added to $\{\delta_0\}_{ij}$ and $\{F\}_{ji}^s$ as needed.

18.7.1 Internal redundants

For internally redundant structures, the same basic procedure is used; however, in this instance, Eq. 18.36 is replaced by Eq. 18.30, and the $[/]$ matrices are defined by Eq. 18.31.

For structures with both internal and external redundancies, the same general procedures are employed. Here, $[b]_{\text{II}}$ is partitioned to include the effects of both the redundant reactions $\{P\}_{\text{II}}$ and the redundant member forces $\{F\}_{\text{II}}$.

18.7.2 Example problem

Determine the member-end forces for the structure shown. Use the pure flexibility approach.

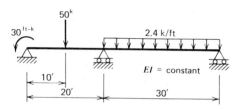

Selection of Node Points:

Structure forces and displacements:

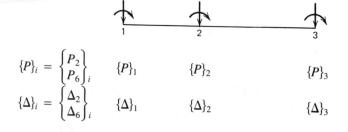

$$\{P\}_i = \begin{Bmatrix} P_2 \\ P_6 \end{Bmatrix}_i, \qquad \{P\}_1 \qquad \{P\}_2 \qquad \{P\}_3$$

$$\{\Delta\}_i = \begin{Bmatrix} \Delta_2 \\ \Delta_6 \end{Bmatrix}_i, \qquad \{\Delta\}_1 \qquad \{\Delta\}_2 \qquad \{\Delta\}_3$$

Member forces and displacements:

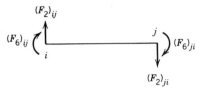

Member 1–2: $\quad \{F\}_{12} = \begin{Bmatrix} F_2 \\ F_6 \end{Bmatrix}_{12}; \quad \{\delta\}_{12} = \begin{Bmatrix} \delta_2 \\ \delta_6 \end{Bmatrix}_{12}$

Member 2–3: $\quad \{F\}_{23} = \begin{Bmatrix} F_2 \\ F_6 \end{Bmatrix}_{23}; \quad \{\delta\}_{23} = \begin{Bmatrix} \delta_2 \\ \delta_6 \end{Bmatrix}_{23}$

Loaded Members—End Displacements and Member-End Reactions:

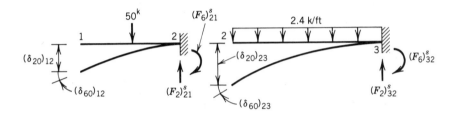

From Table 18.1, with signs established by member coordinates, we have the following in units of kips and feet:

$$(\delta_{20})_{12} = -\frac{50 \times 10^2}{2EI}\left(20 - \frac{10}{3}\right) = -\frac{41,667}{EI} \qquad (\delta_{20})_{23} = -\frac{2.4 \times 30^4}{8EI} = -\frac{243,000}{EI}$$

$$(\delta_{60})_{12} = -\frac{50 \times 10^2}{2EI} = -\frac{2,500}{EI} \qquad (\delta_{60})_{23} = -\frac{2.4 \times 30^3}{6EI} = -\frac{10,800}{EI}$$

$$(F_2)^s_{21} = -50 \qquad\qquad\qquad (F_2)^s_{32} = -2.4 \times 30 = -72$$

$$(F_6)^s_{21} = +50 \times 10 = +500 \qquad\qquad (F_6)^s_{32} = \frac{2.4 \times 30^2}{2} = 1,080$$

In vector form,

$$\{\delta_0\}_{12} = \frac{1}{EI}\begin{Bmatrix} -41,667 \\ -2,500 \end{Bmatrix} \qquad \{\delta_0\}_{23} = \frac{1}{EI}\begin{Bmatrix} -243,000 \\ -10,800 \end{Bmatrix}$$

$$\{F\}^s_{21} = \begin{Bmatrix} -50 \\ 500 \end{Bmatrix} \qquad\qquad \{F\}^s_{32} = \begin{Bmatrix} -72 \\ 1,080 \end{Bmatrix}$$

Equivalent Structure Forces:

$$\{P\}^e_j = \sum_k [\beta]^T_{ji}\{F\}^s_{ji} \qquad\qquad (18.46)$$

$$\{P\}^e_2 = [\beta]^T_{21}\{F\}^s_{21} = \begin{bmatrix} 1 & 0 \\ 0 & 1 \end{bmatrix}\begin{Bmatrix} -50 \\ 500 \end{Bmatrix} = \begin{Bmatrix} -50 \\ 500 \end{Bmatrix}$$

$$\{P\}^e_3 = [\beta]^T_{32}\{F\}^s_{32} = \begin{bmatrix} 1 & 0 \\ 0 & 1 \end{bmatrix}\begin{Bmatrix} -72 \\ 1,080 \end{Bmatrix} = \begin{Bmatrix} -72 \\ 1,080 \end{Bmatrix}$$

Note:

See Example 17.6.1 for $[\beta]$ matrices.

Selection of Redundants:

Primary structure:

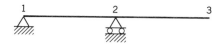

$$\text{Select} \quad \{P\}_{II} = \{(P_2)_3\}; \quad \{\Delta\}_{II} = \{0\}$$

then

$$\{P\}_I = \{P\}_I^a - \{P\}_I^e = \begin{Bmatrix} (P_6)_1 \\ (P_6)_2 \\ (P_6)_3 \end{Bmatrix}_I = \begin{Bmatrix} -30 + 0 \\ 0 - 500 \\ 0 - 1{,}080 \end{Bmatrix} = \begin{Bmatrix} -30 \\ -500 \\ -1{,}080 \end{Bmatrix} \quad (18.45)$$

$$\{\Delta\}_I = \begin{Bmatrix} (\Delta_6)_1 \\ (\Delta_6)_2 \\ (\Delta_6)_3 \end{Bmatrix}$$

Generation of [b] Matrices:

$$\{F\} = [b]_I\{P\}_I + [b]_{II}\{P\}_{II} \quad (18.14)$$

Primary structure loaded with $\{P\}_I$ and $\{P\}_{II}$:

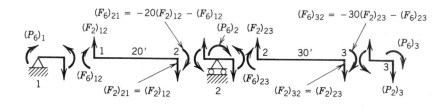

Equilibrium corresponding to kinematic degrees of freedom applied at joints 1, 2, and 3:

$$(P_6)_1 - (F_6)_{12} = 0$$
$$(P_6)_2 + 20(F_2)_{12} + (F_6)_{12} - (F_6)_{23} = 0$$
$$(P_6)_3 + 30(F_2)_{23} + (F_6)_{23} = 0$$
$$(P_2)_3 - (F_2)_{23} = 0$$

In the form of Eq. 18.12, with $\{P\}$ partitioned according to $\{P\}_I$ and $\{P\}_{II}$, we have

$$\begin{Bmatrix} \{P\}_I \\ \{P\}_{II} \end{Bmatrix} = [C]\{F\} = \begin{Bmatrix} (P_6)_1 \\ (P_6)_2 \\ (P_6)_3 \\ (P_2)_3 \end{Bmatrix} = \begin{bmatrix} 0 & 1 & 0 & 0 \\ -20 & -1 & 0 & 1 \\ 0 & 0 & -30 & -1 \\ 0 & 0 & 1 & 0 \end{bmatrix} \begin{Bmatrix} \begin{Bmatrix} F_2 \\ F_6 \end{Bmatrix}_{12} \\ \begin{Bmatrix} F_2 \\ F_6 \end{Bmatrix}_{23} \end{Bmatrix} \quad (18.12)$$

Solving for the member forces, we have

$$\{F\} = [b]\{P\} = [[b]_\text{I}[b]_\text{II}]\begin{Bmatrix} \{P\}_\text{I} \\ \{P\}_\text{II} \end{Bmatrix} \tag{18.14}$$

$$\begin{Bmatrix} \begin{Bmatrix} F_2 \\ F_6 \end{Bmatrix}_{12} \\ \begin{Bmatrix} F_2 \\ F_6 \end{Bmatrix}_{23} \end{Bmatrix} = \begin{bmatrix} -0.05 & -0.05 & -0.05 & \vdots & -1.5 \\ 1 & 0 & 0 & \vdots & 0 \\ 0 & 0 & 0 & \vdots & 1 \\ 0 & 0 & -1 & \vdots & -30 \end{bmatrix} \begin{Bmatrix} (P_6)_1 \\ (P_6)_2 \\ (P_6)_3 \\ \hline (P_2)_3 \end{Bmatrix}$$

$$\underbrace{\hspace{3cm}}_{[b]_\text{I}} \quad \underbrace{\hspace{2cm}}_{[b]_\text{II}}$$

Member Flexibilities:

$$\{\delta\}_{ij} = [f]_{ii}^f\{F\}_{ij} \tag{16.29}$$

In this case

$$\begin{Bmatrix} \delta_2 \\ \delta_6 \end{Bmatrix}_{ij} = \begin{bmatrix} \dfrac{l_3}{3EI} & \dfrac{l^2}{2EI} \\ \dfrac{l^2}{2EI} & \dfrac{l}{EI} \end{bmatrix}_{ij} \begin{Bmatrix} F_2 \\ F_6 \end{Bmatrix}_{ij} \tag{16.28}$$

Member 1–2: $\quad [f]_{11}^2 = \dfrac{1}{EI}\begin{bmatrix} 2{,}667 & 200 \\ 200 & 20 \end{bmatrix}$

Member 2–3: $\quad [f]_{22}^3 = \dfrac{1}{EI}\begin{bmatrix} 9{,}000 & 450 \\ 450 & 30 \end{bmatrix}$

$$[f]_u = \dfrac{1}{EI}\begin{bmatrix} 2{,}667 & 200 & 0 & 0 \\ 200 & 20 & 0 & 0 \\ 0 & 0 & 9{,}000 & 450 \\ 0 & 0 & 450 & 30 \end{bmatrix} \tag{18.20}$$

Structure Flexibilities:

$$[f]_\text{II,I} = [b]_\text{II}^T[f]_u[b]_\text{I} \tag{18.26}$$

$$= \dfrac{1}{EI}[-100 \quad 200 \quad 650]$$

$$[f]_\text{II,II} = [b]_\text{II}^T[f]_u[b]_\text{II} \tag{18.26}$$

$$= \dfrac{1}{EI}[15{,}000]$$

Determination of Redundants:

$$\{P\}_\text{II} = -[f]_\text{II,II}^{-1}[f]_\text{II,I}\{P\}_\text{I} - [f]_\text{II,II}^{-1}[b]_\text{II}^T\{\delta_0\} \tag{18.44}$$

$$\{P\}_\text{II} = -\dfrac{EI}{15{,}000}\cdot\dfrac{1}{EI}[-100 \quad 200 \quad 650]\begin{Bmatrix} -30 \\ -500 \\ -1{,}080 \end{Bmatrix}$$

$$-\dfrac{EI}{15{,}000}[-1.5 \quad 0 \quad 1 \quad -30]\cdot\dfrac{1}{EI}\begin{Bmatrix} -41{,}667 \\ -2{,}500 \\ -243{,}000 \\ -10{,}800 \end{Bmatrix}$$

$$\{P\}_\text{II} = +53.27 - 9.57 = +43.70^k$$

Member-End Forces:

$$\{F\} = [b]_{\mathrm{I}}\{P\}_{\mathrm{I}} + [b]_{\mathrm{II}}\{P\}_{\mathrm{II}} \qquad (18.36)$$

$$
\left\{ \begin{array}{c} \left\{ \begin{array}{c} F_2 \\ F_6 \end{array} \right\}_{12} \\ \left\{ \begin{array}{c} F_2 \\ F_6 \end{array} \right\}_{23} \end{array} \right\} =
\begin{bmatrix} -0.05 & -0.05 & -0.05 \\ 1 & 0 & 0 \\ 0 & 0 & 0 \\ 0 & 0 & -1 \end{bmatrix}
\left\{ \begin{array}{c} -30 \\ -500 \\ -1,080 \end{array} \right\} +
\begin{bmatrix} -1.5 \\ 0 \\ 1 \\ -30 \end{bmatrix} \{43.70\} =
\left\{ \begin{array}{c} 14.95^{k} \\ -30.0'^{-k} \\ 43.70^{k} \\ -231.0'^{-k} \end{array} \right\}
$$

Shear and Moment Diagrams:

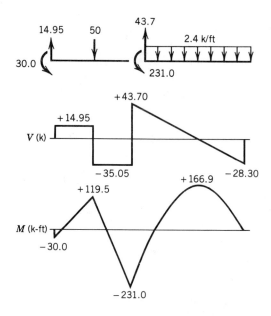

18.8

Self-Straining Problems

The general nature of self-straining problems is discussed in detail in Section 17.7, and the reader is encouraged to review the introductory portion of that section.

It is recalled that member ij was served at one end, and the self-straining action was allowed to accumulate displacements at the i end with respect to the j end. These displacements are shown as $\{\delta_s\}_{ij}$ in Fig. 18.4 and play the same role as $\{\delta_0\}_{ij}$ in Section 18.7 for the member subjected to transverse loads.

Therefore, the procedure of Section 18.7 is applicable with the $\{\delta_0\}$ being replaced by $\{\delta_s\}$. There is one exception to the procedure. Since in the self-straining problem, there are no reactions developed at the j end of the severed member, there are no equivalent structure forces. Thus, the load vector $\{P\}_{\mathrm{I}}$ is composed of applied load components only.

The precise makeup of the $\{\delta_s\}$ vector for each of the self-straining problems is treated in the sections that follow.

18.8.1 Fabrication errors

For fabrication errors, or the lack-of-fit problem, the vector $\{\delta_s\}_{ij}$ includes the displacements at the i end relative to the j end that reflect the components of improper length or alignment.

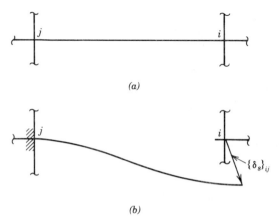

(a)

(b)

Fig. 18.4 *Self-straining displacements.* (a) *Span ij.* (b) *Accumulated displacements for severed member.*

For temperature effects, the vector $\{\delta_s\}_{ij}$ includes the displacements that accrue at the i end of the severed member relative to the j end when temperature variations occur. For a flexural element (including axial effects) these were determined in Section 17.7.3. With the aid of Fig. 17.8, it is clear that

18.8.2 Temperature variation

$$\{\delta_s\}_{ij} = \begin{Bmatrix} \delta_{s1} \\ \delta_{s2} \\ \delta_{s6} \end{Bmatrix} = \begin{Bmatrix} T_{avg}\alpha l \\ -\dfrac{\alpha\Delta T l^2}{2h} \\ -\dfrac{\alpha\Delta T l}{h} \end{Bmatrix} \qquad (18.47)$$

This equation must be used with care. The quantities T_b and T_t, which are needed to establish ΔT, must be interpreted with respect to the assigned locations of i and j.

In the settlement problem, support displacements are imposed. Accordingly, the imposed support displacements are included in $\{\Delta\}_{II}$ of Eq. 18.8. As a result, Eq. 18.10 for $\{P\}_{II}$ is augmented by the term $[/]_{II,II}^{-1}\{\Delta\}_{II}$.

18.8.3 Settlement

Determine the member-end forces for the structure of Section 18.5.3 when the column is subjected to the temperature gradient shown.

18.8.4 Example problem

This is a repeat of the problem that was solved in Section 17.7.4 by the stiffness method.

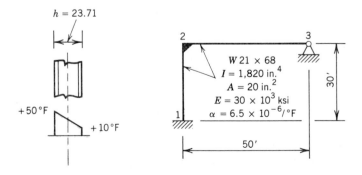

The solution proceeds according to the example of Section 18.5.3. The flexibility matrices and the equilibrium matrices determined there will be used in this example.

Self-Straining Displacements:
Member 1–2:

$$\frac{j = 2 \qquad T_t = +10°F \qquad i = 1}{T_b = +50°F} \qquad \text{(See Fig. 17.8)}$$

$$T_{avg} = (50 + 10)/2 = 30°F; \qquad \Delta T = 50 - 10 = 40°F$$

$$\{\delta_s\}_{12} = \left\{ \begin{array}{c} T_{avg}\alpha l \\ \dfrac{\alpha \Delta T l^2}{2h} \\ -\dfrac{\alpha \Delta T l}{h} \end{array} \right\} = \left\{ \begin{array}{c} 30 \times 6.5 \times 10^{-6} \times 30 \\ \dfrac{6.5 \times 10^{-6} \times 40 \times 30^2}{(2 \times 23.71)/12} \\ -\dfrac{6.5 \times 10^{-6} \times 40 \times 30}{23.71/12} \end{array} \right\} = \left\{ \begin{array}{c} 0.005,85 \text{ ft} \\ -0.059,21 \text{ ft} \\ -0.003,947 \text{ rad} \end{array} \right\} \quad (18.47)$$

Member 2–3:
There is no temperature variation for this member. Therefore,

$$\{\delta_s\}_{23} = [0]$$

Redundant Forces:

$$\{P\}_{II} = -[f]_{II,II}^{-1}[b]_{II}^T\{\delta_s\} \qquad (18.44)$$

Note:

In applying Eq. 18.44, since there are no applied forces, $\{P\}_I = \{0\}$, and $\{\delta_s\}$ replaces $\{\delta_0\}$.

$$\{P\}_{II} = -\frac{EI}{5.476 \times 10^8} \begin{bmatrix} 116,686 & -22,500 \\ -22,500 & 9,032 \end{bmatrix} \begin{bmatrix} 0 & 1 & -30 & 1 & 0 & 0 \\ -1 & 0 & -50 & 0 & 1 & -50 \end{bmatrix} \left\{ \begin{array}{c} 0.005,85 \\ -0.059,21 \\ -0.003,947 \\ 0 \\ 0 \\ 0 \end{array} \right\}$$

For $EI = \dfrac{30 \times 10^3 \times 1,820}{144} = 379.17 \times 10^3 \text{ k} \cdot \text{ft}^2$, we obtain

$$\{P\}_{II} = -\frac{379.17 \times 10^3}{5.476 \times 10^8} \begin{bmatrix} 116,686 & -22,500 \\ -22,500 & 9,032 \end{bmatrix} \left\{ \begin{array}{c} 0.059,2 \\ 0.191,5 \end{array} \right\} = \left\{ \begin{array}{c} -1.80^k \\ -0.27^k \end{array} \right\}$$

Thus, according to the selection of $\{P\}_{II}$ in Section 18.5.3, we have

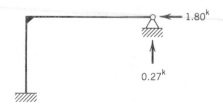

Member Forces:

$$\{F\} = [b]_I\{P\}_I + [b]_{II}\{P\}_{II} + \{F\}^f \qquad (18.32)$$

In the present application, $\{P\}_I = \{0\}$ and $\{F\}^f = 0$. Thus,

$$\{F\} = [b]_{II}\{P\}_{II}$$

$$\begin{Bmatrix} \begin{Bmatrix} F_1 \\ F_2 \\ F_6 \end{Bmatrix}_{12} \\ \begin{Bmatrix} F_1 \\ F_2 \\ F_6 \end{Bmatrix}_{23} \end{Bmatrix} = \begin{bmatrix} 0 & -1 \\ 1 & 0 \\ -30 & -50 \\ 1 & 0 \\ 0 & 1 \\ 0 & -50 \end{bmatrix} \begin{Bmatrix} -1.80 \\ -0.27 \end{Bmatrix} = \begin{Bmatrix} 0.27^k \\ -1.80^k \\ 67.5'^{-k} \\ -1.80^k \\ -0.27^k \\ 13.5'^{-k} \end{Bmatrix}$$

These results agree with those determined in Section 17.5.3.

18.9 Generation of [b] Matrices

The equilibrium matrix, $[b]$, was first introduced in Eq. 18.13. It relates the member forces of the primary structure to the applied structure forces and the redundant reactions, as is evident from the partitioned form of Eq. 18.14. When the structure is internally statically indeterminate, the $[b]$ matrix also serves to relate the primary structure member forces to the redundant member forces, as expressed in Eq. 18.30.

In the example problems that have been considered, equations of equilibrium are written at each joint. These equations relate the structure forces to the member forces and take the form of Eq. 18.12, which involves the global statics matrix, $[C]$. The solution of these equations yields the member forces, as indicated in Eq. 18.13, and the $[b]$ matrix emerges as the inverse of $[C]$.

The procedure for generating $[b]$ can be formalized, and this procedure will be presented here in the context of an example problem. More general developments are available in formal textbooks on matrix methods.

Consider the frame analyzed in Section 18.5.3. This structure is broken down into its members and its structure nodes in Fig. 18.5, with the member-end force vectors and structure force vectors identified in their positive senses.

Equilibrium at each joint can be expressed by the relationship

$$\{P\}_i = \sum_j [\gamma]_{ij}\{F\}_{ij} \qquad (18.48)$$

where $[\gamma]_{ij}$ is as previously defined, and j ranges over all members framing into joint i. For the structure shown in Fig. 18.5, Eq. 18.48 leads to

$$\{P\}_1 = [\gamma]_{12}\{F\}_{12}$$
$$\{P\}_2 = [\gamma]_{21}\{F\}_{21} + [\gamma]_{23}\{F\}_{23} \qquad (18.49)$$
$$\{P\}_3 = [\gamma]_{32}\{F\}_{32}$$

For each member of the structure, the member-end forces at one end can be expressed in terms of the member-end forces at the other end. This can be stated as

$$\{F\}_{ji} = [T]_{ji}\{F\}_{ij} \qquad (18.50)$$

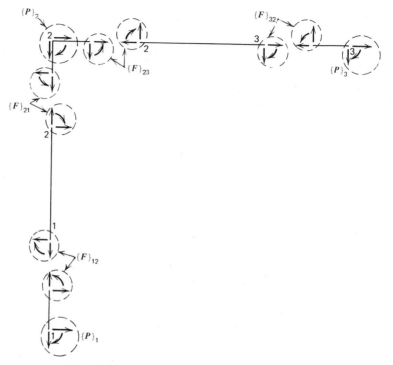

Fig. 18.5 *Member-end forces and structure forces.*

where $[T]_{ji}$ relates the forces at the j end of the member to those at the i end of the member, and it is determined from static equilibrium considerations. For the structure of Fig. 18.5, we have

$$\{F\}_{21} = [T]_{21}\{F\}_{12} \tag{18.51}$$
$$\{F\}_{32} = [T]_{32}\{F\}_{23}$$

Substitution of Eq. 18.51 into 18.49 leads to

$$\begin{Bmatrix} \{P\}_1 \\ \{P\}_2 \\ \{P\}_3 \end{Bmatrix} = \begin{bmatrix} [\gamma]_{12} & 0 \\ [\gamma]_{21}[T]_{21} & [\gamma]_{23} \\ 0 & [\gamma]_{32}[T]_{32} \end{bmatrix} \begin{Bmatrix} \{F\}_{12} \\ \{F\}_{23} \end{Bmatrix} \tag{18.52}$$

However, $\{P\}_1$ can be determined from $\{P\}_2$ and $\{P\}_3$ by applying statics to the primary structure. Thus, the equations for $\{P\}_1$ can be excluded from Eq. 18.52. This leads to

$$\begin{Bmatrix} \{P\}_2 \\ \{P\}_3 \end{Bmatrix} = \begin{bmatrix} [\gamma]_{21}[T]_{21} & [\gamma]_{23} \\ 0 & [\gamma]_{32}[T]_{32} \end{bmatrix} \begin{Bmatrix} \{F\}_{12} \\ \{F\}_{23} \end{Bmatrix} \tag{18.53}$$

This can be expressed in the form of Eq. 18.12 as

$$\{P\} = [C]\{F\} \tag{18.54}$$

Solving for $\{F\}$, we obtain

$$\{F\} = [C]^{-1}\{P\} \tag{18.55}$$

A comparison of Eqs. 18.55 and 18.13 shows that

$$[b] = [C]^{-1} = \begin{bmatrix} [\gamma]_{21}[T]_{21} & [\gamma]_{23} \\ 0 & [\gamma]_{32}[T]_{32} \end{bmatrix}^{-1} \qquad (18.56)$$

It should be recalled that Eq. 18.56 applies only to the structure shown in Fig. 18.5; however, the same procedure would be employed for other structures.

Generate the $[b]$ matrix for the example problem of Section 18.5.3. The structure is shown in Fig. 18.5, and the governing equations are developed in Section 18.9.

18.9.1 Example problem

Development of Eq. 18.53:

Generation of $[\gamma]$ matrices:

$$[\gamma]_{21} \qquad \begin{Bmatrix} P_1 \\ P_2 \\ P_6 \end{Bmatrix}_2 = \begin{bmatrix} 0 & 1 & 0 \\ -1 & 0 & 0 \\ 0 & 0 & 1 \end{bmatrix} \begin{Bmatrix} F_1 \\ F_2 \\ F_6 \end{Bmatrix}_{21}$$

$$\underbrace{\qquad}_{[\gamma]_{21}}$$

$$[\gamma]_{23} \qquad \begin{Bmatrix} P_1 \\ P_2 \\ P_6 \end{Bmatrix}_2 = \begin{bmatrix} -1 & 0 & 0 \\ 0 & -1 & 0 \\ 0 & 0 & 1 \end{bmatrix} \begin{Bmatrix} F_1 \\ F_2 \\ F_6 \end{Bmatrix}_{23}$$

$$\underbrace{\qquad}_{[\gamma]_{23}}$$

$$[\gamma]_{32} \qquad \begin{Bmatrix} P_1 \\ P_2 \\ P_6 \end{Bmatrix}_3 = \begin{bmatrix} 1 & 0 & 0 \\ 0 & 1 & 0 \\ 0 & 0 & 1 \end{bmatrix} \begin{Bmatrix} F_1 \\ F_2 \\ F_6 \end{Bmatrix}_{32}$$

$$\underbrace{\qquad}_{[\gamma]_{32}}$$

Generation of $[T]$ matrices:

$$\{F\}_{21} = [T]_{21}\{F\}_{12} \qquad (18.51)$$

$$\begin{Bmatrix} F_1 \\ F_2 \\ F_6 \end{Bmatrix}_{21} = \begin{bmatrix} 1 & 0 & 0 \\ 0 & 1 & 0 \\ 0 & -30 & -1 \end{bmatrix} \begin{Bmatrix} F_1 \\ F_2 \\ F_6 \end{Bmatrix}_{12}$$

$$\underbrace{\qquad}_{[T]_{21}}$$

$$\{F\}_{32} = [T]_{32}\{F\}_{23} \qquad (18.51)$$

$$\begin{Bmatrix} F_1 \\ F_2 \\ F_6 \end{Bmatrix}_{32} = \begin{bmatrix} 1 & 0 & 0 \\ 0 & 1 & 0 \\ 0 & -50 & -1 \end{bmatrix} \begin{Bmatrix} F_1 \\ F_2 \\ F_6 \end{Bmatrix}_{23}$$

$$\underbrace{\qquad}_{[T]_{32}}$$

Final equations:

$$\begin{Bmatrix} \{P\}_2 \\ \{P\}_3 \end{Bmatrix} = \begin{bmatrix} [\gamma]_{21}[T]_{21} & [\gamma]_{23} \\ 0 & [\gamma]_{32}[T]_{32} \end{bmatrix} \begin{Bmatrix} \{F\}_{12} \\ \{F\}_{23} \end{Bmatrix} \tag{18.53}$$

$$\begin{Bmatrix} \begin{Bmatrix} P_1 \\ P_2 \\ P_6 \end{Bmatrix}_2 \\ \begin{Bmatrix} P_1 \\ P_2 \\ P_6 \end{Bmatrix}_3 \end{Bmatrix} = \begin{bmatrix} 0 & 1 & 0 & -1 & 0 & 0 \\ -1 & 0 & 0 & 0 & -1 & 0 \\ 0 & -30 & -1 & 0 & 0 & 1 \\ 0 & 0 & 0 & 1 & 0 & 0 \\ 0 & 0 & 0 & 0 & 1 & 0 \\ 0 & 0 & 0 & 0 & -50 & -1 \end{bmatrix} \begin{Bmatrix} \begin{Bmatrix} F_1 \\ F_2 \\ F_6 \end{Bmatrix}_{12} \\ \begin{Bmatrix} F_1 \\ F_2 \\ F_6 \end{Bmatrix}_{23} \end{Bmatrix}$$

The above is in the form

$$\{P\} = [C]\{F\} \tag{18.54}$$

from which

$$\{F\} = [C]^{-1}\{P\} = [b]\{P\} \tag{18.55}$$

Or, we finally obtain

$$\begin{Bmatrix} \begin{Bmatrix} F_1 \\ F_2 \\ F_6 \end{Bmatrix}_{12} \\ \begin{Bmatrix} F_1 \\ F_2 \\ F_6 \end{Bmatrix}_{23} \end{Bmatrix} = \begin{bmatrix} 0 & -1 & 0 & 0 & -1 & 0 \\ 1 & 0 & 0 & 1 & 0 & 0 \\ -30 & 0 & -1 & -30 & -50 & -1 \\ 0 & 0 & 0 & 1 & 0 & 0 \\ 0 & 0 & 0 & 0 & 1 & 0 \\ 0 & 0 & 0 & 0 & -50 & -1 \end{bmatrix} \begin{Bmatrix} \begin{Bmatrix} P_1 \\ P_2 \\ P_6 \end{Bmatrix}_2 \\ \begin{Bmatrix} P_1 \\ P_2 \\ P_6 \end{Bmatrix}_3 \end{Bmatrix}$$

With appropriate adjustments in the rows and columns, this agrees with

$$\{F\} = [b]_{\mathrm{I}}\{P\}_{\mathrm{I}} + [b]_{\mathrm{II}}\{P\}_{\mathrm{II}} = [b]\{P\}$$

as developed in section 18.5.3

18.10

Choice of Redundants

Application of the flexibility method, as was also the case with the compatibility methods of Chapters 10 and 11, requires that the analyst select the reactions and member forces that are to be considered as the redundants. There are two considerations: first, the redundants must be selected so that the primary structure is stable; and second, their selection must be such that the accuracy of the solution is ensured.

The question of stability can be resolved in a straightforward fashion. One must merely apply the tests of stability to the primary structure to make certain that it is stable. Mathematically, stability can be tested by checking to see that $[C]$ of Eq. 18.12 possesses an inverse. Only if the inverse exists is it possible to determine the member forces from the structure forces in accordance with Eq. 18.13.

The problem of solution accuracy is more subtle. Regardless of the choice of redundants, the order of $[/]_{\mathrm{II,II}}$ in Eqs. 18.10 or 18.28 is the same. Also, the final results, theoretically, should be the same regardless of the selection of redundants, as long as the primary structure is stable. However, it happens that the arrangement of the elements, as well as their relative magnitudes, will have a bearing on the accuracy of $[/]_{\mathrm{II,II}}^{-1}$ because of limitations in the numerical procedures that are employed. To obtain an accurate inverse, it is best that $[f]_{\text{II,II}}$ be *sparsely banded* and *well-conditioned or diagonally strong*. That is, the matrix should possess terms that are grouped along the diagonal in as narrow a band as possible rather than being fully populated, and the terms along the diagonal should be larger numerically than those that are off the diagonal.

These factors can be illustrated by the example continuous beam given in Fig. 18.6a. If the internal reactions are taken as the redundants, then

$$[f]_{\text{II,II}} = \begin{bmatrix} f_{11} & f_{12} & f_{13} & f_{14} \\ f_{21} & f_{22} & f_{23} & f_{24} \\ f_{31} & f_{32} & f_{33} & f_{34} \\ f_{41} & f_{42} & f_{43} & f_{44} \end{bmatrix} \tag{18.57}$$

This array is fully populated, and it is clear from Fig. 18.6b that $f_{21} > f_{11}$ and $f_{31} > f_{11}$. Thus, the matrix is neither banded nor well conditioned. However, if the support moments are taken as the redundants, we have

$$[f]_{\text{II,II}} = \begin{bmatrix} f_{11} & f_{12} & 0 & 0 \\ f_{21} & f_{22} & f_{23} & 0 \\ 0 & f_{32} & f_{33} & f_{34} \\ 0 & 0 & f_{43} & f_{44} \end{bmatrix} \tag{18.58}$$

Here, the matrix is banded and, as shown in Fig. 18.6c, $f_{11} > f_{21}$, which indicates a well-conditioned matrix.

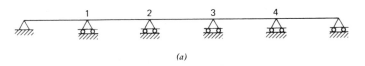

(a)

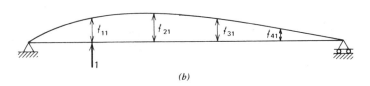

(b)

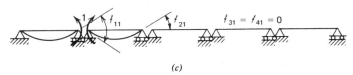

(c)

Fig. 18.6 *Choice of redundants.* (a) *Continuous beam.* (b) *Selected flexibility coefficients with internal reactions as redundants.* (c) *Selected flexibility coefficients with support moments as redundants.*

The contrast between the two cases would be more sharply defined if there were more spans. The case where the internal reactions are taken as the redundants would lead to a fully populated and poorly conditioned matrix, regardless of the order of the array. However, the case with the support moments as redundants would have a well conditioned matrix with a bandwidth of three, regardless of the number of redundants.

Similar considerations would have to be employed for other structures to determine which forces should be taken as the redundants, and herein lies one of the major disadvantages of compatibility or flexibility methods. For complex structures, the question of stability can be complicated, and the optimum selection of the redundants for solution accuracy can be elusive.

18.11

Comparisons between Flexibility and Stiffness Methods

The flexibility and stiffness methods have been presented here as formalizations of the compatibility and equilibrium methods, respectively, that formed the basis of Part II of the book. They are equivalent in that each provides a complete analysis of a structure in which equilibrium and compatibility are fully satisfied. However, the sequence in which these requirements are met is different; the flexibility method considers compatibility explicitly, with equilibrium secondarily woven into the solution, whereas the stiffness method gives primary attention to equilibrium, with compatibility being implicitly satisfied. Of course, both methods reflect the inclusion of the appropriate material characteristics and the governing boundary conditions.

18.12

Additional Reading

Gere, J. M., and Weaver, W., Jr., *Analysis of Framed Structures,* Chapter 3, Van Nostrand, Princeton, N.J., 1965.

Kardestuncer, H., *Elementary Matrix Analysis of Structures,* Chapter 6 and 11, McGraw–Hill, New York, 1974.

Laursen, H. I., *Matrix Analysis of Structures,* Chapter 4, McGraw–Hill, New York, 1966.

Livesley, R. K., *Matrix Methods of Structural Analysis,* Chapter 7, Pergamon Press, Macmillan Co., New York, 1964.

McGuire, W., and Gallagher, R. H., *Matrix Structural Analysis,* Chapters 6 and 7, Wiley, New York, 1979.

Rubinstein, M. F., *Matrix Computer Analysis of Structures,* Chapter 8, Prentice–Hall, Englewood Cliffs, N.J., 1966.

18.13

Suggested Problems

1 through 13. Use the generalized flexibility method to analyze each of the structures described below for the prescribed loading conditions. In each case, determine the member forces and the nodal displacements for the given nodal numbering system.

1. The structure and loading of Problem 3 of Section 17.13.

2. The structure and loading of Problem 4 of Section 17.13.

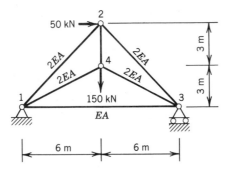

3.

Select member 1–3 as the redundant.

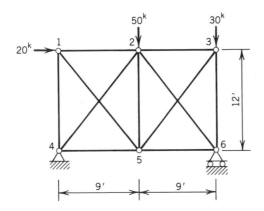

4.

EA is the same for all members.
Select members 2–4 and 2–6 as the redundants.

5. The structure and loading of Problem 11 in Section 17.13.

6. The structure and loading of Problem 12 in Section 17.13.

7. The structure and loading of Problem 13 in Section 17.13.

8. The structure and loading of Problem 14 in Section 17.13.

9. The structure and loading of Problem 16 in Section 17.13.

10. The structure and loading of Problem 22 in Section 17.13.

11. The structure and loading of Problem 23 in Section 17.13.

12. The structure and loading of Problem 24 in Section 17.13.

13. The structure and loading of Problem 25 in Section 17.13.

14 through 19. Use the generalized flexibility method to analyze each of the structures designated below for the given self-straining conditions. In each case, determine the member forces and nodal displacements.

14. The structure and settlement pattern of Problem 27 of Section 17.13.

15. The structure and settlement pattern of Problem 29 of Section 17.13.

16. The structure and temperature changes specified in Problem 31 of Section 17.13.

17. The structure and temperature variation specified in Problem 34 of Section 17.13.

18. The structure and lack-of-fit condition given in Problem 36 of Section 17.13.

19. The structure and lack-of-fit condition given in Problem 38 of Section 17.13.

20 through 22. Use the formal matrix method of Section 18.9 to generate the [b] matrix for each of the structures described below. In each case, the force components that are to be taken as redundants are indicated.

20. The structure and loading of Problem 2. Select members 2–3 and 2–5 as redundants.

21. The structure and loading of Problem 7. Select the reactions at points 2 and 4 as redundants.

22. The structure and loading of Problem 8. Select the reaction components at point 3 as redundants.

The Brooklyn Bridge, New York (courtesy Steinman, Boynton, Gronquist and Birdsall, Consulting Engineers).

A.1

Systems of Units

The U.S. system of weights and measures, which is referred to as the U.S. Customary System (USCS), was officially established by action of the U.S. Treasury Department in the 1830s. Since this system is at variance with a host of other systems throughout the world, there has been much confusion in international exchanges. To resolve the problems that have resulted from this confusion, the International System of Units (Système International d'Unités), commonly referred to as SI, is being adopted throughout the world as a uniform measurement system. Although this is a metric system, it is not the metric system that has been used in various forms throughout Europe and other parts of the world, but is instead a completely new system.

The U.S. Congress passed the Metric Conversion Act in 1975. The changeover is voluntary, and the complete transition from USCS to SI will undoubtedly take many years; however, usage in the engineering and scientific communities is proceeding. Nevertheless, the construction industry might have a tendency to move more slowly toward metrication because, unlike the multinational industries, which have had to change because of economic pressures, the construction industry does not face a pressing need to act immediately. Thus, it seems clear that there will be a time when structural engineers will have to be conversant with both systems of units and be able to convert from one to the other. For these reasons, this textbook presents, in both systems of units, complete problem solutions that are designed to develop students' confidence in their ability to use either system. In certain cases, conversion between systems is employed to aid the student in developing a sense of equivalency; however, the emphasis is on developing a capacity to work in either system and not a mere conversion ability.

The fundamentals concerning SI are briefly outlined in this section.

A.1.1 Base units SI is founded upon seven *base units* on which the entire system is built. These base units represent quantities of length, mass, time, thermodynamic temperature, amount of substance, electric current, and luminous intensity. Each of these units, their definitions, and their symbols have complete international agreement. Only the first four of these base units are of specific interest to the structural engineer. These base units are listed in Table A.1 in both USCS and SI along with the conversion factor relating one to the other.

A.1.2 Supplementary units Two *supplementary units* represent the quantities of plane angle and solid angle. Plane angle is the only one of interest to the structural engineer, and it is entered in Table A.1 along with the units of both USCS and SI.

TABLE A.1 Relationships between USCS and SI Units for Selected Quantities

Quantity	USCS Units		SI Units		Conversion
	Name	Symbol	Name	Symbol	
Base Units					
Length	foot	ft	meter	m	1 ft = 0.304 8 m
	inch	in.			1 in. = 0.025 4 m
Mass	pound	lb	kilogram	kg	1 lb = 0.453 6 kg
	kilopound	kip			1 kip = 453.6 kg
Time	second	sec	second	s	1 sec = 1 s
Temperature	degree Fahrenheit	°F	degree Celsius	°C	$t_F^\circ = 1.8 t_C^\circ + 32$
			kelvin	K	$t_K = t_C^\circ + 273.15$
Supp. Units					
Plane angle	degree	°	radian	rad	1° = 0.017 45 rad
Derived Units					
Area	square foot	ft^2	square meter	m^2	1 ft^2 = 0.092 9 m^2
Volume	cubic foot	ft^3	cubic meter	m^3	1 ft^3 = 0.028 3 m^3
Velocity	foot-second	ft/sec	meter/second	m/s	1 f/sec = 0.304 8 m/s
Acceleration	(foot/second)/second	ft/sec^2	(meter/second)/second	m/s^2	1 ft/sec^2 = 0.304 8 m/s^2
Force	pound	lb	newton, kg · m/sec^2	N	1 lb = 4.448 N
	kilopound	kip			1 kip = 444 8 N
Pressure	pound/foot^2	psf	pascal, N/M^2	Pa	1 psf = 47.88 Pa
	pound/inch^2	psi			1 psi = 689 5 Pa
Work, energy	foot-pound	ft-lb	joule, N · m	J	1 ft-lb = 1.356 J
	foot-kip	ft-kip			1 ft/kip = 135 6 J
Frequency	cycle/second	cps	hertz, l/s	Hz	1 cps = 1 Hz

The base units and supplementary units are combined to produce the required **A.1.3 Derived** derived units. There are 17 of these derived units that have been given special **units** names in SI. The derived units that are of most interest to the structural engineer are given in Table A.1 for both systems. Of the eight derived units that are listed, four represent quantities that have special names for their SI units.

Several features of SI deserve mention. First, SI is a *coherent* system in which all **A.1.4 Special** units are related to each other by unity. For instance, a force of one newton (N) **features of SI** results from a mass of one kilogram (kg) being given an acceleration of one meter per second squared (m/s^2). Or, a force of one newton (N) exerted over a distance of one meter (m) produces energy of one joule (J). Second, SI is an *absolute* system in which there is no dual meaning associated with any unit. The USCS uses the unit of pound (lb) to represent both mass and force. However, SI uses the kilogram (kg) as the unit of mass (not to be confused with the kilogram force of the old metric system) and the newton (N) as the derived unit of force. Weight is not directly employed in SI, but this is understood to be the force caused as gravity acts on a mass. Third, SI is a *unique* system in that each quantity is represented by a single unit. This is not evident for the units given for structural engineers in Table A.1, but it is an important feature in other disciplines. For instance, power in SI is given in the unit of watt (J/s) regardless of whether it is related to mechanical, thermal, or electrical systems.

There is a set of 16 prefixes for use with SI units in forming multiples and sub- **A.1.5 SI prefixes** multiples of these units. These prefixes are given in Table A.2, and they are used to advantage in eliminating insignificant digits and in providing a convenient substitute for writing powers of 10 in computations.

TABLE A.2 SI Prefixes

Factor	Prefix Name	Symbol
10^{18}	exa	E
10^{15}	peta	P
10^{12}	tera	T
10^{9}	giga	G
10^{6}	mega	M
10^{3}	kilo	k
10^{2}	hecto	h
10^{1}	deca	da
10^{-1}	deci	d
10^{-2}	centi	c
10^{-3}	milli	m
10^{-6}	micro	μ
10^{-9}	nano	n
10^{-12}	pico	p
10^{-15}	femto	f
10^{-18}	atto	a

As an example of conversion from USCS to SI and of the use of prefixes, consider the modulus of elasticity of steel. Here,

$$E = 29 \times 10^6 \text{ psi}$$
$$= 29 \times 10^6 \times 6895 \approx 200\ 000 \times 10^6 \text{ Pa (N/m}^2)$$
$$= 200\ 000 \text{ MPa (MN/m}^2) = 200 \text{ GPa (GN/m}^2)$$

Prefixes should not be used in the denominator of compound units. Thus, although $E = 200\ 000$ (MN/m^2) is equivalent to $E = 200\ 000$ (N/mm^2), the former is preferred to the latter.

The kilogram is a special case. Here, since the prefix kilo is part of the base unit itself, it may appear in the denominator. Also, since double prefixes are not to be used, the prefixes of Table A.2 are to be used with gram (symbol g) and not with kilogram. Thus, 8×10^3 kg $= 8 \times 10^6$ g $= 8$ Mg.

It should also be noted that the prefixes centi, deci, deca, and hecto are to be avoided where possible in favor of the prefixes that represent multiples of 1000. This is done to encourage consistency in engineering practice and to avoid errors in calculations. Thus, the conventional units for length are meter (m) and millimeter (mm), and although centimeter (cm) was customarily used in the old metric system, it is discouraged in SI.

TABLE A.3 Convenient Approximate Equivalences

		Units	
Quantity	*USCS*	*SI Exact*	*SI Approx.*
Length	1 in.	25.4 mm	25 mm
	1 ft	0.304 8 m	0.3 m
Mass	1 lb (mass)	0.453 6 kg	0.5 kg
Area	1 ft^2	0.092 90 m^2	0.1 m^2
Volume	1 yd^3	0.765 m^3	0.75 m^3
Force	1 lb	4.448 N	4.5 N
	1 kip	4.448 kN	4.5 kN
	1 lb/ft	14.59 N/m	15 N/m
	1 kip/ft	14.59 kN/m	15 kN/m
Moment	1 ft-lb	1.356 N · m	1.4 N · m
	1 ft-kip	1.356 kN · m	1.4 kN · m
Pressure	1 psi	6.895 kPa	7 kPa
	1 psf	47.88 Pa	48 Pa
Some typical values:			
Water density	62.4 lb/ft^3	1 000 kg/m^3	1 000 kg/m^3
Concrete density	150 lb/ft^3	2 402.81 kg/m^3	2 400 kg/m^3
Water (weight/volume)	62.4 lb/ft^3	9.81 kN/m^3	10 kN/m^3
Concrete (weight/volume)	150 lb/ft^3	23 576 N/m^3	24 kN/m^3
Concrete stress	3,000 psi	20.67 MPa	21 MPa
Steel stress	25,000 psi	172.38 MPa	170 MPa
E steel	29 $\times$ 10^6 psi	199 9 GPa	200 GPa
E concrete	3.5 $\times$ 10^6 psi	24.13 GPa	24 GPa
Coeff. of thermal expansion for steel	6.5 $\times$ 10^{-6}/°F	1.2 $\times$ 10^{-5}/°C	1.2 $\times$ 10^{-5}/°C

Care must be taken to avoid confusion between the use of a prefix and a derived **A.1.6 Additional** compound unit. Thus, a moment in newton meters is represented by N · m, with **notes on the use** the dot representing multiplication, whereas a millinewton is mN without a dot. **of SI** Likewise, a stress in newtons per square meter is given by N/m^2, in which the slash represents division.

Decimal points are used in the same fashion for USCS and SI, but commas are not used in SI to separate multiples of 1000. Instead, a space is used on both sides of the decimal point. Thus, 83,471.321,05 in USCS becomes 83 471.321 05 in SI.

As was mentioned earlier, the emphasis in this book is not on conversion from **A.1.7 Some** one unit system to another. However, in translating the experience that one has in **useful** USCS to SI, it is helpful to note the approximate equivalences given in Table **equivalences** A.3.

A.2

Johnston, N. B., ''Introducing the Modern Metric System into Engineering Education,'' **Additional** *Engineering Education,* **67,** 7, ASEE, Washington, D.C., 1977. **Reading**

Lally, Andrew, ''Metrication in the Construction Trades,'' *Standardization News,* **4,** 2, ASTM, Philadelphia, Pa., 1976.

Metric Editorial Guide, American National Metric Council, Washington, D.C., 1974.

Metric Practice Guide, E380-74, ASTM, Philadelphia, Pa., 1974.

SI Units and Recommendations for the Use of Their Multiples and of Certain Other Units, ISO 1000, American National Standards Institute, New York, 1973.

Students should not cultivate the habit of working for predetermined answers, because such an arrangement is highly artificial. Instead, they should seek independent checks to verify their results and should try to develop an ability to sense whether answers appear to be reasonable. However, in the learning process, the immediate reinforcement of a correct answer has some value. For this reason, selected answers to *most* even-numbered problems are given in the following tabulation. Answers are also included for unique problem types, which appear only as odd-numbered problems.

CHAPTER 1

2. (a) $p_s = 27.7$ lb/ft^2
   (b) $P_r = 8,320$ lb
   (c) Windward, $p = 12.5$ lb/ft^2
       Leeward, $p = -10.4$ lb/ft^2 (suction)
   (d) $P_s = 2,501$ lb
   (e) Windward, $p = 2.2$ lb/ft^2
       Leeward, $p = -4.3$ lb/ft^2
   (f) $P_r = 798$ lb
4. Triangular load with intensity varying from zero at point b to 22.0 kN/m at point a.

CHAPTER 2

1. (b) Determinate, stable
   (d) Geometrically unstable
   (f) Unstable
   (h) Determinate, stable
   (j) Unstable
2. $R_{Ax} = 0$; $R_{Ay} = 163.3$ kN ↑; $R_{By} = 186.7$ kN ↑
4. $R_{Ax} = 60$ kN →; $R_{Ay} = 67.9$ kN ↑; $R_{By} = 167.1$ kN ↑
6. $R_{Ax} = 20$ kN ←; $R_{Ay} = 101.7$ kN ↑; $R_B = 91.7$ kN ↖
8. $R_{Ax} = 42.6$ kN →; $R_{Ay} = 102.3$ kN ↑; $R_{Bx} = 42.6$ kN ←; $R_{By} = 137.8$ kN ↑
10. $R_{Ax} = 20.67^k$ ←; $R_{Ay} = 20.0^k$ ↑; $R_{Bx} = 30.67^k$ →
12. $R_{Ax} = 5.00^k$ ←; $R_{Ay} = 4.96^k$ ↑; $M_A = 50'^{-k}$ ↻; $R_{Bx} = 0$; $R_{By} = 7.04^k$ ↑
14. $R_{Ay} = 0$; $R_{Bx} = 8^k$ ←; $R_{By} = 25^k$ ↑
16. $R_{Ax} = 0$; $R_{Ay} = 160.0^k$ ↑; $M_A = 1,900'^{-k}$ ↻; $R_{Cx} = 45^k$ ←; $R_{Cy} = 30^k$ ↑
18. $R_{Ax} = 30^k$ ←; $R_{Ay} = 10^k$ ↓; $R_{Cy} = 70^k$ ↑; $M_C = 700'^{-k}$ ↻
20. Same answers as for Problem 2.
22. $R_{Ax} = 26.8^k$ →; $R_{Ay} = 30.2^k$ ↑; $R_{By} = 128.5^k$ ↑
24. $R_{Ax} = 0$; $R_{Ay} = 35$ kN ↑; $R_{By} = 80$ kN ↑; $R_{Cy} = 115$ kN ↑
26. Same answers as for Problem 12.
28. Same answers as for Problem 18.
30. Same answers as for Problem 22.
32. Same answers as for Problem 8.
34. $R_{Ax} = 225$ kN →; $M_A = 1\ 200$ kN · m ↻; $R_B = 375$ kN ↖
36. $R_{Ax} = 0$; $R_{Ay} = 40$ kN ↑; $R_{Dy} = 286.7$ kN ↑; $R_{Fx} = 50$ kN →;
    $R_{Fy} = 133.3$ kN ↑; $M_F = 880$ kN · m ↻

CHAPTER 3

1. (b) External: determinate, stable; Internal: indeterminate, stable
   (d) External: unstable; Internal: determinate, stable
   (f) External: determinate, stable; Internal: determinate, stable; Compound
   (h) External: determinate, stable; Internal: unstable

(j) External: determinate, stable; Internal: determinate, stable; Simple

(l) External: determinate, stable; Internal: determinate, stable; Compound

(n) External: determinate, stable; Internal: geometrically unstable

(p) External: determinate, stable; Internal: determinate, stable; Complex

(r) External: determinate, stable; Internal: determinate, stable; Complex

(t) External: determinate, stable; Internal: determinate, stable; Simple

2. $F_{ab} = -326.6$ kN; $F_{cd} = -216.1$ kN; $F_{gd} = +172.9$ kN; $F_{gh} = +108.0$ kN

4. $F_{ab} = -192.69$ kN; $F_{cd} = -133.90$ kN; $F_{fc} = -9.28$ kN; $F_{fg} = +141.94$ kN

6. $F_{ae} = +131.99$ kN; $F_{ef} = +120.94$ kN; $F_{bc} = -99.99$ kN; $F_{cg} = -65.05$ kN

8. $F_{ad} = +40.00^k$; $F_{bc} = -19.56^k$; $F_{ce} = -47.91^k$; $F_{cf} = +20.83^k$

10. $F_{ab} = -10.42^k$; $F_{ad} = -5.02^k$; $F_{be} = -18.42^k$; $F_{ge} = +30.33^k$; $F_{eh} = -50.17^k$; $F_{gh} = +27.50^k$

12. $F_{ab} = -83.09^k$; $F_{cd} = -77.50^k$; $F_{io} = 0$; $F_{pe} = +44.19^k$; $F_{em} = +15.00^k$; $F_{ng} = +61.25^k$

14. $F_{ea} = +301.87$ kN; $F_{bc} = +240.00$ kN; $F_{fg} = -225.00$ kN; $F_{bh} = -67.08$ kN

16. $F_{ab} = -51.5^k$; $F_{bg} = -44.5^k$; $F_{cg} = +23.6^k$; $F_{he} = +37.7^k$

18. $F_{ab} = -65.35^k$; $F_{de} = -45.07^k$; $F_{gh} = -60.84^k$; $F_{lm} = +39.38^k$; $F_{mp} = +12.58^k$; $F_{mp} = +28.13^k$; $F_{ng} = -25.16^k$

20. $F_{ab} = -25.94^k$; $F_{bc} = -38.33^k$; $F_{ed} = +9.17^k$; $F_{dk} = -64.28^k$; $F_{fg} = +47.50^k$; $F_{gl} = +43.75^k$

22. $F_{ea} = -18.75^k$; $F_{ag} = -43.75^k$; $F_{bc} = +15.00^k$; $F_{dh} = -25.00^k$

CHAPTER 4

2. AB: $V(x) = +128.89$ kN; $M(x) = +161.11x$ kN · m

BC: $V(x) = +48.89$ kN; $M(x) = +61.11x + 400$ kN · m

CD: $V(x) = -30x + 301.11$ kN; $M(x) = -15x^2 + 301.11x - 560$ kN · m

4. $0 \le x \le 3$: $V(x) = -\dfrac{25}{9}x^2$; $M(x) = -\dfrac{25}{27}x^3$

$3 \le x \le 9$: $V(x) = 112.5 - \dfrac{25}{9}x^2$; $M(x) = 112.5x - \dfrac{25}{27}x^3 - 337.5$

6. $V(x) = -\dfrac{p_o x^3}{3l^2} + \dfrac{p_o l}{3}$; $M(x) = -\dfrac{p_o x^4}{12l^2} + \dfrac{p_o xl}{3} - \dfrac{p_o l^2}{4}$

8. $V(k)$: 20.72 right of left support; -2.28 left of right support

M(ft-k): 112.8 just left of vertical member

10. $V(k)$: 94.5 right of left support; -45.0 left of right support

M(ft-k): -650 at left support; 244 under 100^k load; -120 at right support

12. $V(k)$: 50.83 right of left support; -9.17 between concentrated loads; -59.17 left of right support

M(ft-k): 483.3 under 20^k load; -100.0 at right support

14. $V(k)$: 75.32 at left support; -4.68 left of joint at crown; -4.13 right of crown; -79.13 at right support (all shears normal to members); M(ft-k): 886.42 maximum on left side; 883.00 at crown; 839.30 maximum on right half

16. Girder shears and moments:

V(kN): 12.5 right of left support; -237.5 left of right support

M(kN · m): 187.5 at interior floor beam; $-1\,000$ at right support

18. $V(k)$: 10 right of left support; 0 between hinges; -18 left of right support

M(ft-k): 60 maximum in left span; -360 at left interior support; 0 at hinges; -72 at right interior support; 40.5 maximum in right span

20. V(kN): 25.56 normal to inclined member; -42.86 normal to horizontal member

M(kN · m): 171.44 at 200 kN load; 150.00 maximum in vertical column

22. V(kN): 40.0 at left end of beam; 53.3 at top of column

M(kN · m): 160.0 maximum in beam; 164.2 maximum in column

24. $V(k)$, average values in 10' panels: 34.16, 4.16, -9.17, -49.17, 10.00

M(ft-k), at 10' panel points: 0, 341.6, 483.2, 391.5, -100.0, 0

26. $V(k)$, average values in 6' panels: 10, -70, 60, 0, 0, -12, 18, -6
    M(ft-k), at 6' panel points: 0, 60, -360, 0, 0, 0, -72, $+36$, 0
28. $V(k)$, average in panels of 10', 10', 10', 2': 18.25, -6.75, -27.58, $+80.00$
    M(ft-k), at corresponding panel points: 0, 182.5, 115.0, -160.8, 0
30. Same answers as for Problem 24.
32. Moments (kN · m) for each span: Left span: 0, 193.88, -235.3; Right span: -336.2,
    228.45, -206.9; Column: -117.6, 130.65, -100.9
33. (a) External: determinate, stable; Overall: determinate, stable
    (b) External: indeterminate (1 degree), stable; Overall: indeterminate (1 degree), stable
    (h) External: indeterminate (1 degree), stable; Overall: indeterminate (1 degree), stable
    (i) External: determinate, stable; Overall: determinate, stable

CHAPTER 5

2. $u_B = 4.58$ mm $\rightarrow$; $v_B = 0.00$ mm; $u_C = 3.50$ mm $\rightarrow$; $v_C = 6.75$ mm $\downarrow$
   $u_D = 1.20$ mm $\rightarrow$; $v_D = 8.60$ mm $\downarrow$; $u_E = 0.50$ mm $\leftarrow$; $v_E = 9.35$ mm $\downarrow$
4. $u_E = 0.053,0$ in. $\rightarrow$; $v_E = 0.309$ in. $\downarrow$
6. $u_C = 15$ mm $\rightarrow$; $v_C = 10$ mm $\downarrow$
8. $u_D = 0.573$ mm $\leftarrow$; $v_D = 15.7$ mm $\downarrow$
10. $u_C = 1.124$ in. $\rightarrow$; $v_C = 1.323$ in. $\downarrow$
12. $u_E = 0.016,1$ in. $\rightarrow$; $v_E = 0.092,2$ in. $\downarrow$
14. $v_D = 46.37$ mm $\downarrow$
16. $\Delta_f = 0.262$ in $\nearrow$ (along sloping surface)
18. $u_D = 0.30$ in. $\leftarrow$; $v_D = 1.66$ in. $\downarrow$
20. $v_E = 0.188$ in. $\uparrow$
22. Shorten BD by 0.044 37 mm
24. $\{\Delta\}^T = \{-0.096,8 \quad 0.089,8 \quad -0.010,4\}$ in.
28. Same answers as for Problem 24.
30. $\{\Delta\}^T = \{-7.5 \quad 15.0 \quad -10.0 \quad 15.0 \quad 17.5\}$ mm

CHAPTER 6

2. $\Delta_B = 26.10$ mm $\downarrow$; $\theta_B = 0.006\ 30$ rad
4. $\Delta_C = 0.701$ in. $\downarrow$; $\Delta_E = 0.855$ in. $\uparrow$; $\theta_B = 0.005,84$ rad $\curvearrowleft$
6. $\Delta_{max} = \Delta_e = 164,014\ P/EI \downarrow$ (k$-$in.^{3})/EI at 9.48 ft left of point d;
   $P_{max} = 212.18$ kips
8. $\Delta_B = 24.16$ mm $\downarrow$; $\theta_B = 0$ rad
10. $\Delta_D = 26.00$ mm $\downarrow$
12. $\Delta = 0.036,6$ in. $\downarrow$; $\theta = 0.000,234$ rad
14. Span AB, midspan $\Delta = 34.46$ mm $\downarrow$; Span CD, midspan $\Delta = 34.72$ mm $\downarrow$
16. $\Delta_f = 0.149$ in. $\downarrow$; $\theta_f = 0.004,811$ rad
18. $\theta_A = 0.012\ 36$ rad $\curvearrowright$; $(\Delta_D)_H = 108.13$ mm $\rightarrow$
20. $(\Delta_b)_H = 0.966$ in. $\rightarrow$; $\theta_b = 0.002,87$ rad $\curvearrowleft$; $(\Delta_c)_V = 0.034,5$ in. $\downarrow$
22. $(\Delta_B)_H = (\Delta_C)_H = 0.049,7$ in. $\rightarrow$; $\theta_D = 0.000,276$ rad. $\curvearrowleft$
24. Same answers as for Problem 2.
26. $\Delta_{max} = 3.727$ mm $\downarrow$ at 4.472 m left of B
28. Same answer as for Problem 10.
30. Same answers as for Problem 12.
32. Check points on diagrams; $\theta_E = 0.002,150$ rad $\curvearrowleft$; $\Delta_E = 0.231$ in. $\downarrow$
34. Check point on deflection diagram: Δ_{max} same as for Problem 26.
36. Check points on diagrams: $\Delta_C = 0.795$ in. $\uparrow$; $\theta_E = 0.002,483$ rad. $\curvearrowright$
38. Defl. (in.) at 5-ft intervals: 0, -0.219, -0.363, -0.382, -0.269, -0.090, 0,
    -0.163, -0.442
40. $\Delta_b = 0.014,83$ in. $\downarrow$; $\Delta_c = 0.042,90$ in. $\downarrow$; $\Delta\theta_b = 0.000,077$ rad; $\Delta\theta_c = 0.000,261$
    rad

42. Same answers as for Problem 38.
46. $\Delta_B = 1.986$ in. $\downarrow$; Angle change at $B = 0.017,4$ rad $\rightarrow$
48. Same answers as for Problem 16.

50. $[f] = \dfrac{1}{EI}\begin{bmatrix} 1,944 & 1,008 \\ 1,008 & 576 \end{bmatrix}; \begin{Bmatrix} \Delta_1 \\ \Delta_2 \end{Bmatrix} = \begin{Bmatrix} 1.306 \\ 0.747 \end{Bmatrix}$ in.

52. $\Delta_C = 0.795$ in. $\uparrow$; $\theta_E = 0.002,483$ rad

54. $[f] = \dfrac{1}{EI}\begin{bmatrix} 173.62 & 215.28 \\ 215.28 & 444.46 \end{bmatrix}; \begin{Bmatrix} \Delta_1 \\ \Delta_2 \end{Bmatrix} = \begin{Bmatrix} 0.395 \\ 0.701 \end{Bmatrix}$ in.

CHAPTER 7

Note:

Influence lines for statically determinate structures are composed of straight-line segments.

2. I.L. R_B (kN/kN): $+1.0$ across entire structure
   I.L. M_B (kN · m/kN): -5.0 at A to 0.0 at B
   I.L. V_C (kN/kN): -1.0 from A to C
   I.L. M_C (kN · m/kN): -2.0 at A to 0.0 at C
4. I.L. R_{Ay} (kN/kN): $+1.0$ from A to C; then to 0.0 at B
   I.L. R_{Ax} = I.L. R_{Bx} (kN/kN): zero from A to C; then to 0.5 at B
   I.L. M_A (kN · m/kN): 0.0 at A to -3.0 at C; then to 0.0 at B
6. I.L. R_{Ay} (kN/kN): $+1.0$ at A to -1.0 at C
   I.L. R_{Dy} (kN/kN): 0.0 at A to $+1.0$ at C
   I.L. M_F (kN · m/kN): 0.0 at A to $+3.0$ at C
8. I.L. R_{Ay} (k/k): 0.0 at A to $+0.643$ at B; then to 0.0 at D
   I.L. R_{Ax} (k/k): 0.0 at A to $+0.179$ at B; then to $+0.500$ at D
   I.L. V_E (k/k): 0.0 at A to $+0.357$ at B; then to -0.714 left of E; $+0.286$ right of E to 0.0 at D
10. I.L. R_{Ay} (k/k): $+1.0$ at A to -0.250 at right end
    I.L. V_C (k/k): 0.0 at A to $+0.625$ left of stringer discontinuity; $+0.375$ right of stringer discontinuity to -0.250 at right end
12. I.L. R_{Ay} (kN/kN): $+1.0$ at B to 0.0 at E
    I.L. R_{Ax} (kN/kN): 0.0 at B to 0.555 at C; then to 0.0 at E
    I.L. M (top of BA) (kN · m/kN): 0.0 at B to -3.33 at C; then to 0.0 at E
14. I.L. R_{ay} (k/k): $+1.0$ at a to 0.0 at e
    I.L. R_{dy} (k/k): 0.0 at a to $+1.50$ at b; then to 0.0 at c
    I.L. M_f (ft-k/k): 0.0 at a to -1.50 at b; then to $+1.50$ at f; then to 0.0 at C
16. I.L. R_{Dy} (k/k): zero from A to C; then to 1.5 at E
    I.L. F_{AB} (k/k): zero from A to B; then to 2.0 at C; then to -1.75 at E
    I.L. F_{FG} (k/k): 0.0 at A to -2.24 at C; then to $+1.12$ at E
18. I.L. R_{L0} (k/k): 1.0 at L_0 to -0.667 at L_5
    I.L. $F_{U_1L_1}$ (k/k): 0.0 at L_0 to 0.667 at L_1; then to -0.667 at L_5
    I.L. $F_{U_2L_3}$ (k/k): 0.0 at L_0 to -0.943 at L_2; then to 0.0 at L_3; then to -0.943 at L_5
20. I.L. R_{fy} (kN/kN): 1.0 at a to $+0.625$ at c; then to 0.0 at e
    I.L. R_{ky} (kN/kN): 0.0 at a to $+0.375$ at c; then to 1.0 at e
    I.L. R_{fx} = I.L. R_{kx} (kN/kN): 0.0 at a to $+0.625$ at c; then to 0.0 at e
22. Same answers as for Problem 4.
24. $R_A = 101.40^k$; positive $M_D = 107.0'^{-k}$; positive $V_D = 21.40^k$
26. $R_{Dx} = 34.28^k$; positive $V_E = 64.02^k$; positive $M_{BD} = 171.4'^{-k}$
28. Positive $F_{U_1L_1} = 59.67^k$; negative $F_{U_1L_1} = -24.67^k$; $F_{U_2U_3} = +194.00^k$;
    $F_{U_2L_3} = -173.19^k$; $F_{U_5L_5} = 90.45^k$
30. Envelope ranges: Shears (kN), -16 to -39 left of point 4, $+25$ to $+65$ right of point 4, -18 to -66 at point 10; Moments (kN · m), -22.5 to -58.5 at point 4, $+13.5$ to -94.5 at point 10

2. $F_{ab} = F_{bd} = +70.71$ kN; $F_{cd} = -66.67$ kN; $F_{ed} = +86.7$ kN; $F_{ec} = -155.5$ kN

4. $F_{ab} = -47.58$ kN; $F_{bc} = +55.08$ kN; $F_{ca} = -47.58$ kN; $F_{ad} = +127.67$ kN;
   $F_{dc} = -72.12$ kN; $F_{db} = -72.12$ kN

6. $F_{ba} = -9.60$ kN; $F_{ca} = 16.80$ kN; $F_{ae} = 12.90$ kN; $F_{ce} = -46.87$ kN;
   $F_{cd} = 33.60$ kN; $F_{de} = -59.39$ kN; $F_{db} = -4.80$ kN; $F_{be} = 16.10$ kN

12. $v_e = 20.84$ mm $\downarrow$

14. $R_{ax} = -10.0^k$; $R_{ay} = 22.5^k$; $R_{az} = 20.0^k$; $M_{ax} = 58.83'^{-k}$; $M_{ay} = 0.0$;
    $M_{az} = 147.5'^{-k}$ (signs based on prescribed axes in problem statement)

16. $R_{ax} = 40.0^k$; $R_{ay} = +12.5^k$; $R_{az} = 5^k$; $M_{ax} = +85'^{-k}$; $R_{fx} = -48^k$; $R_{fy} = +12.5^k$
    (signs based on assumed directions in problem statement)

18. $R_{ax} = 96.65^k$; $R_{ay} = 8.75^k$; $R_{az} = -3.33^k$; $R_{gx} = -116.65^k$; $R_{gy} = 13.75^k$;
    $R_{gz} = 23.33^k$ (signs based on prescribed axes in problem statement)

20. $R_{ay} = 17.92^k$; $M_{ax} = 50.0'^{-k}$; $M_{ay} = -350.0'^{-k}$; $R_{fx} = -15.0^k$; $R_{fy} = 17.08^k$;
    $R_{fz} = 10^k$ (signs based on assumed directions in problem statement)

22. $(\theta_b)_x = -0.016,407$ rad; $v_g = -1.22$ in. (signs based on prescribed axes in problem statement)

2. $F_{AC} = -21.56$ kN; $F_{CB} = -92.27$ kN; $F_{BD} = -77.75$ kN; $F_{DA} = -77.75$ kN;
   $F_{DC} = 80.45$ kN

4. $F_{AB} = -19.73^k$; $F_{CD} = -10.13^k$; $F_{DE} = 10.13^k$; $F_{AC} = 0^k$; $F_{BD} = -15.20^k$;
   $F_{AD} = 25.34^k$; $F_{BE} = -24.66^k$

6. $F_{AB} = -14.06^k$; $F_{HG} = 10.62^k$; $F_{CH} = -49.17^k$; $F_{CJ} = 22.40^k$

10. $F_{AC} = -67.87$ kN; $F_{CB} = -67.87$ kN; $F_{BD} = 107.33$ kN; $F_{DA} = 107.33$ kN;
    $F_{DC} = 96.00$ kN

12. $F_{AC} = -117.26^k$; $F_{CB} = -117.26^k$; $F_{BD} = 82.93^k$; $F_{DA} = 82.93^k$; $F_{DC} = 165.85^k$

16. $F_{AC} = -22.17$ kN; $F_{CB} = -92.88$ kN; $F_{BD} = -76.78$ kN; $F_{DA} = -76.78$ kN;
    $F_{DC} = 81.31$ kN

18. $F_{AC} = -18.30^k$; $F_{CB} = -18.30$ kN; $F_{BD} = 12.94$ kN;
    $F_{DA} = 12.94$ kN; $F_{DC} = 25.88$ kN

20. $F_{AC} = -52.26$ kN; $F_{CB} = -122.97$ kN; $F_{BD} = -29.20$ kN; $F_{DA} = -29.20$ kN;
    $F_{DC} = 123.87$ kN; $F_{AB} = 113.07$ kN

22. $F_{BC} = -328.15$ kN; $F_{GH} = 271.85$ kN; $F_{DH} = 171.85$ kN; $F_{DE} = -353.55$ kN

24. $F_{AC} = -33.27$ kN; $F_{CB} = -33.27$ kN; $F_{AD} = 52.61$ kN; $F_{DB} = 52.61$ kN;
    $F_{CD} = 47.06$ kN; $F_{AB} = -23.53$ kN

26. $F_{BC} = F_{BF} = F_{FG} = -3.45$ EA; $F_{BG} = F_{FC} = +4.88$ EA; $F_{CG} = 0$

28. $F_{CD} = -28.17$ kN; $F_{CE} = -1.64$ kN; $F_{BD} = -30.63$ kN; $F_{AD} = 32.04$ kN;
    $F_{GE} = 1.80$ kN

30. Same answers as for Problem 22.

32. $F_{AC} = F_{BC} = -48.1$ kN; $F_{AD} = F_{BD} = 76.0$ kN; $F_{AB} = -68.1$ kN

34. First equation: $\dfrac{1}{E}[74.54 \quad -31.28 \quad 39.72 \quad -31.28 \quad 45.72]\,\{P\} = \Delta_1$

2. $V_a = 36.14^k$; $V_c = -31.36^k$; $M_a = -296.73'^{-k}$; $M_b = 132.89'^{-k}$

4. $V_a = 145.64$ kN; $V_c = -80.94$ kN; $M_b = -293.6$ kN $\cdot$ m; $M_c = -103.1$ kN $\cdot$ m

6. $V_{b,beam} = 2.0^k$; $V_{c,beam} = -8.0^k$; $M_{bc} = 18.75'^{-k}$; $M_{under\ 10k\ load} = 33.75'^{-k}$;
   $M_{cb} = -26.25'^{-k}$

10. $V_a = 163.29$ kN; $V_c = -205.22$ kN; $M_b = -117.1$ kN $\cdot$ m;
    $M_c = -1\,169.3$ kN $\cdot$ m

12. $R_{ay} = 0.694^k \downarrow$; $M_a = 20.83'^{-k}$; $R_{cy} = 0.694^k \downarrow$

14. $M_b = -276$ kN $\cdot$ m

16. $M_b = -293.64$ kN $\cdot$ m; $M_c = -103.18$ kN $\cdot$ m

18. $M_a = -275.9'^{-k}$
20. $M_b = -6\,750$ kN $\cdot$ m
22. $M_b = -160$ kN $\cdot$ m; $M_c = 80$ kN $\cdot$ m
24. $P_1 = 235.2$ kN; $\{\Delta\}_2^T = \dfrac{1}{EI}\{1\,334.12 \quad 8\,499 \quad 1\,327.47 \quad 80.29 \quad -541.58\}$
26. $\{P\}_I = \{-8 \quad -1.75\}$ kips; $\{\Delta\}_2^T = \dfrac{1}{EI}\{1{,}546.87 \quad 298.82 \quad 84.37 \quad -28.15\}$
28. Same answers as for Problem 22.
30. $\{\Delta\}_2^T = \dfrac{1}{EI}\{11{,}165 \quad 38{,}663\}$ (l in ft)

CHAPTER 12

2. $\Delta_{CH} = 0.017,5$ ft $\rightarrow$; $\Delta_{CV} = 0.021,6$ ft $\downarrow$; $\Delta_{BH} = 0.003$ ft $\leftarrow$; $F_{AC} = 6.70^k$;
$F_{CB} = -10.0^k$; $F_{AB} = -6.00^k$
4. $\Delta_{gH} = 0.001,67$ ft $\rightarrow$; $\Delta_{gV} = 0.001,70$ ft $\downarrow$; $F_{ag} = 16.14^k$;
$F_{dg} = 16.43^k$; $F_{fg} = -2.75^k$
6. $\begin{bmatrix} 47{,}100 & 0 & 0 & 0 \\ 0 & 113{,}800 & 0 & -66{,}700 \\ 0 & 0 & 95{,}400 & 0 \\ 0 & -66{,}700 & 0 & 90{,}507 \end{bmatrix} \begin{Bmatrix} u_c \\ v_c \\ u_d \\ v_d \end{Bmatrix} = \begin{Bmatrix} 50 \\ 0 \\ 0 \\ -150 \end{Bmatrix}$ units of kN and m
10. Same answers as for Problem 2.
12. Same answers as for Problem 4.
14. Same answers as for Problem 6.
18. $\Delta_{gH} = 0.001,29$ ft $\rightarrow$; $\Delta_{gV} = 0.020,51$ ft $\downarrow$; $F_{ag} = 12.5^k$; $F_{dg} = 198.3^k$; $F_{fg} = 98.1^k$
20. $\Delta_{gH} = 0.001,72$ ft $\rightarrow$; $\Delta_{gV} = 0.002,33$ ft $\downarrow$; $\Delta_{dV} = 0.001,62$ ft $\downarrow$; $F_{ag} = 16.43^k$;
$F_{dg} = 5.90^k$; $F_{fg} = -0.33^k$
22. $\Delta_{bH} = 0.003,85$ ft $\rightarrow$; $\Delta_{bV} = 0.019,83$ ft $\uparrow$; $\Delta_{dH} = 0.006,10$ ft $\leftarrow$; $\Delta_{dV} = 0.023,51$
ft $\downarrow$; $\Delta_{cV} = 0.003,20$ ft $\downarrow$; $F_{ab} = 224.75^k$; $F_{bd} = 53.77^k$; $F_{ad} = 64.89^k$
24. $\Delta_{bH} = 0.003,22$ ft $\rightarrow$; $\Delta_{bV} = 0.053,70$ ft $\downarrow$; $\Delta_{dH} = 0.003,68$ ft $\rightarrow$; $\Delta_{dV} = 0.056,24$
ft $\downarrow$; $\Delta_{cV} = 0.001,84$ ft $\downarrow$; $F_{ab} = 93.38^k$; $F_{bd} = 66.93^k$; $F_{ad} = -108.03^k$

CHAPTER 13

2. $M_{ab} = 0$; $M_{ba} = 293.6$ kN $\cdot$ m; $M_{bc} = -293.6$ kN $\cdot$ m; $M_{cb} = 103.2$ kN $\cdot$ m
4. $M_{ab} = -186.8'^{-k}$; $M_{ba} = 1.4'^{-k}$; $M_{bc} = -1.5'^{-k}$; $M_{cb} = 100.0'^{-k}$;
$M_{cd} = -100.0'^{-k}$
6. $M_{ab} = 0$; $M_{ba} = 144.6'^{-k}$; $M_{bc} = -144.6'^{-k}$; $M_{cb} = 0$
8. $M_{ab} = -7.81'^{-k}$; $M_{ba} = -15.6'^{-k}$; $M_{bc} = 15.6'^{-k}$; $M_{cb} = 23.4'^{-k}$
10. $M_{ab} = -55.8'^{-k}$; $M_{ba} = 7.5'^{-k}$; $M_{bc} = -7.5'^{-k}$; $M_{cb} = 42.5'^{-k}$; $M_{cd} = -42.5'^{-k}$;
$M_{dc} = 150.'^{-k}$
12. $M_{ab} = 0$; $M_{ba} = 250.6'^{-k}$; $M_{bc} = -250.6'^{-k}$; $M_{cb} = 0$
14. $M_{ab} = 135.5'^{-k}$; $M_{ba} = 0$
16. $M_{ab} = 0$ kN $\cdot$ m; $M_{ba} = -231.0$ kN $\cdot$ m; $M_{bc} = +231.0$ kN $\cdot$ m; $M_{cb} = -784.5$
kN $\cdot$ m
18. $M_{ab} = -176.0'^{-k}$; $M_{ba} = +109.2'^{-k}$; $M_{bc} = -104.6'^{-k}$; $M_{cb} = -52.3'^{-k}$;
$M_{bd} = -4.5'^{-k}$; $M_{db} = 0'^{-k}$
20. $M_{ab} = 0'^{-k}$; $M_{ba} = -42.6'^{-k}$; $M_{bc} = 42.6'^{-k}$; $M_{cb} = 165.2'^{-k}$; $M_{cd} = -165.2'^{-k}$;
$M_{dc} = -213.1'^{-k}$
22. $M_{ab} = 0$ kN $\cdot$ m; $M_{ba} = 95.8$ kN $\cdot$ m; $M_{bc} = -95.8$ kN $\cdot$ m; $M_{cb} = 77.5$
kN $\cdot$ m; $M_{cd} = -77.5$ kN $\cdot$ m; $M_{dc} = -18.5$ kN $\cdot$ m
26. $M_{ab} = 0.34'^{-k}$; $M_{ba} = 5.84'^{-k}$; $M_{bc} = -5.84'^{-k}$; $M_{cb} = 100.4'^{-k}$;
$M_{cd} = -100.4'^{-k}$; $M_{dc} = 81.7'^{-k}$
28. $M_{ab} = -74.0$ kN $\cdot$ m; $M_{ba} = -55.0$ kN $\cdot$ m; $M_{bc} = 55.0$ kN $\cdot$ m; $M_{cb} = 204.9$
kN $\cdot$ m; $M_{cd} = -204.9$ kN $\cdot$ m; $M_{dc} = -149.0$ kN $\cdot$ m

30. $M_{ab} = 0'^{-k}$; $M_{ba} = -100.4'^{-k}$; $M_{bc} = 100.4'^{-k}$; $M_{cb} = 244.1'^{-k}$;
    $M_{cd} = -244.1'^{-k}$; $M_{dc} = -105.8'^{-k}$

32. $M_{ab} = -60.8'^{-k}$; $M_{ba} = 36.5'^{-k}$; $M_{bc} = -16.2'^{-k}$; $M_{cb} = -8.1'^{-k}$; $M_{bd} = -20.3'^{-k}$; $M_{db} = 0'^{-k}$

34. Same answers as for Problem 4.

36. $M_{ab} = 0$ kN $\cdot$ m; $M_{ba} = 235.3$ kN $\cdot$ m; $M_{bc} = -336.2$ kN $\cdot$ m; $M_{cb} = 206.9$ kN $\cdot$ m; $M_{bd} = 100.9$ kN $\cdot$ m; $M_{db} = -117.6$ kN $\cdot$ m

38. Same answers as for Problem 32.

40. Same answers as for Problem 28.

CHAPTER 14

2. Same answers as for Problem 2 in Section 13.16.

4. Same answers as for Problem 4 in Section 13.16.

6. Same answers as for Problem 6 in Section 13.16.

8. Same answers as for Problem 8 in Section 13.16.

10. Same answers as for Problem 10 in Section 13.16.

12. Same answers as for Problem 12 in Section 13.16.

14. Same answers as for Problem 16 in Section 13.16.

16. Same answers as for Problem 18 in Section 13.16.

18. $M_{ac} = 0$ kN $\cdot$ m; $M_{ca} = -67.5$ kN $\cdot$ m; $M_{cd} = -232.5$ kN $\cdot$ m; $M_{dc} = 0$ kN $\cdot$ m; $M_{bc} = 300.0$ kN $\cdot$ m

20. $M_{ab} = M_{dc} = 0$ kN $\cdot$ m; $M_{ba} = -30.0$ kN $\cdot$ m; $M_{bc} = 30.0$ kN $\cdot$ m; $M_{cb} = 30.0$ kN $\cdot$ m; $M_{cd} = -30.0$ kN $\cdot$ m

22. $M_{ab} = M_{dc} = 0$ kN $\cdot$ m; $M_{ba} = 11.3$ kN $\cdot$ m; $M_{bc} = -11.3$ kN $\cdot$ m; $M_{cb} = 41.3$ kN $\cdot$ m; $M_{cd} = -41.3$ kN $\cdot$ m

24. Same answers as for Problem 22 of Section 13.16.

26. $M_{ab} = -438.1$ kN $\cdot$ m; $M_{ba} = -366.3$ kN $\cdot$ m; $M_{bc} = 366.4$ kN $\cdot$ m; $M_{cb} = 301.9$ kN $\cdot$ m; $M_{ce} = -106.2$ kN $\cdot$ m; $M_{ec} = 0$ kN $\cdot$ m; $M_{cd} = -195.6$ kN $\cdot$ m; $M_{dc} = 0$ kN $\cdot$ m

28. $M_{ab} = 0'^{-k}$; $M_{ba} = -441.4'^{-k}$; $M_{bc} = +441.4'^{-k}$; $M_{cb} = +469.9'^{-k}$

30. Same answers as for Problem 30 of Section 13.16.

32. Same answers as for Problem 32 of Section 13.16.

34. $M_{ab} = M_{dc} = 0'^{-k}$; $M_{ba} = M_{cd} = 153.3'^{-k}$; $M_{bc} = M_{cb} = -153.3'^{-k}$

36. Same answers as for Problem 20.

38. Same answers as for Problem 2.

40. $M_{ab} = 0$ kN $\cdot$ m; $M_{ba} = 235.3$ kN $\cdot$ m; $M_{bc} = -336.3$ kN $\cdot$ m; $M_{cb} = 206.8$ kN $\cdot$ m; $M_{bd} = 100.9$ kN $\cdot$ m; $M_{db} = -117.6$ kN $\cdot$ m

CHAPTER 15

2. I.L. R_B(k/k) at 10' intervals: 0, 0.326, 0.622, 0.857, 1.000, 1.020, 0.939, 0.776, 0.551, 0.286, 0
   I.L. V_D(k/k) at 10' intervals: 0, -0.296, -0.573 (left of D), 0.426 (right of D), 0.186, 0, -0.112, -0.163, -0.165, -0.131, -0.071, 0
   I.L. M_D(ft-k/k) at 10' intervals: 0, 4.08, 8.53, 3.71, 0, -2.24, -3.26, -3.31, -2.61, -1.43, 0

4. I.L. M_E(kN $\cdot$ m/kN) at 2.5 m intervals: 0, 1.109, 2.274, 1.053, 0, -0.846, -1.523, -2.086, -2.594, -1.805, -1.071, -0.451, 0, 0.244, 0.320, 0.188, 0

6. I.L. R_B(k/k) at 10' intervals: 0, 0.512, 0.889, 1.000, 0.771, 0.364, 0, -0.148, -0.119, 0

8.

$$[b]_{MP} = \begin{bmatrix} 0 & 0 & 0 & 0 & 0 & 0 & 0 & 0 & 0 & 0 \\ 0 & 2.36 & 2.95 & 0 & 2.37 & 1.63 & 0 & -0.74 & -0.59 & 0 \\ 0 & -2.36 & -2.95 & 0 & -2.37 & -1.63 & 0 & 0.74 & 0.59 & 0 \\ 0 & -0.59 & -0.75 & 0 & 1.63 & 2.37 & 0 & 2.95 & 2.36 & 0 \\ 0 & 0.59 & 0.75 & 0 & -1.63 & -2.37 & 0 & -2.95 & -2.36 & 0 \\ 0 & 0 & 0 & 0 & 0 & 0 & 0 & 0 & 0 & 0 \end{bmatrix} \frac{\text{ft-k}}{\text{k}}$$

CHAPTER 16

2. Answer given as Eq. 16.6

4. $[k]_r[f] = \begin{bmatrix} \dfrac{EA}{l} & 0 \\ 0 & \dfrac{4EI}{l} \end{bmatrix} \begin{bmatrix} \dfrac{l}{EA} & 0 \\ 0 & \dfrac{l}{4EI} \end{bmatrix} = \begin{bmatrix} 1 & 0 \\ 0 & 1 \end{bmatrix}$

$$[k] = \begin{bmatrix} \dfrac{EA}{l} & 0 & 0 & \dfrac{EA}{l} & 0 & 0 \\ 0 & \dfrac{4EI}{l} & -\dfrac{6EI}{l^2} & 0 & \dfrac{2EI}{l} & -\dfrac{6EI}{l^2} \\ 0 & -\dfrac{6EI}{l^2} & \dfrac{12EI}{l^3} & 0 & -\dfrac{6EI}{l^2} & \dfrac{12EI}{l^3} \\ \dfrac{EA}{l} & 0 & 0 & \dfrac{EA}{l} & 0 & 0 \\ 0 & \dfrac{2EI}{l} & -\dfrac{6EI}{l^2} & 0 & \dfrac{4EI}{l} & -\dfrac{6EI}{l^2} \\ 0 & -\dfrac{6EI}{l^2} & \dfrac{12EI}{l^3} & 0 & -\dfrac{6EI}{l^2} & \dfrac{12EI}{l^3} \end{bmatrix}$$

8. $[f]$ given by Eq. 16.10

$[b] = [0\ 0\ 1]$

$[k] = $ given by Eq. 16.63; result agrees with Eq. 16.6

CHAPTER 17

2. $\{\Delta\}_I^T = \{(\Delta_1)_2(\Delta_2)_2\} = \{-0.001,2 \quad 0.006,8\}$ ft; $\{F\}_{12} = 24.10^k$; $\{F\}_{32} = -29.17^k$; $\{F\}_{42} = -30.06^k$

4. $\{\Delta\}_2^T = \{(\Delta_1)_2(\Delta_2)_2\} = \dfrac{1}{EA}\{-524.93\ 325.54\}$ (l in ft); $\{F\}_{12} = -7.77^k$;
$\{F\}_{42} = -40.32^k$; $\{F\}_{32} = 7.65^k$; $\{F\}_{52} = -26.99^k$

6. $\{\Delta\}_I^T = \{(\Delta_1)_1(\Delta_2)_1\ (\Delta_1)_3(\Delta_1)_4\} = \{0.006,8\ 0.004,0\ 0.002,2\ 0.004,4\}$ ft;
$\{F\}_{12} = 11.04^k$; $\{F\}_{13} = -30.60^k$; $\{F\}_{14} = -24.40^k$; $\{F\}_{23} = 17.21^k$; $\{F\}_{34} = 17.21^k$

8. $\{\Delta\}_I^T = \{(\Delta_1)_2(\Delta_2)_2(\Delta_2)_3(\Delta_1)_4(\Delta_2)_4\} = \{-0.002,17\ 0.008,57\ 0\ -0.001,12\ 0.007,32\}$ ft
$\{F\}_{12} = -19.73^k$; $\{F\}_{13} = 0$; $\{F\}_{14} = 25.34^k$; $\{F\}_{24} = -15.20^k$; $\{F\}_{25} = -24.66^k$;
$\{F\}_{34} = -10.14^k$; $\{F\}_{45} = 10.14^k$

10. $\{\Delta\}_I^T = \{(\Delta_1)_2(\Delta_2)\ (\Delta_2)_3(\Delta_1)_4(\Delta_2)_4\} = \{0.082\ 0.239\ 0.032\ -0.083\ 0.381\}$ in.; $\{F\}_{12} = 197.4^k$; $\{F\}_{13} = 76.5^k$; $\{F\}_{14} = 98.5^k$; $\{F\}_{23} = -108.1^k$; $\{F\}_{24} = 34.6^k$; $\{F\}_{34} = -100.7^k$

12. $\{\Delta\}_I^T = \{(\Delta_2)_2(\Delta_6)_2\ (\Delta_6)_3(\Delta_6)_4\} = \dfrac{1}{EI}\{4,762.32 \quad 142.88 \quad -571.44 \quad 285.68\}$ (l in ft);
$\{F\}_{12}^T = \{48.58^k \quad -257.16'^{-k}\}$; $\{F\}_{23}^T = \{-31.43^k \quad 228.60'^{-k}\}$; $\{F\}_{34}^T = \{4.28^k -85.72'^{-k}\}$

14. $\{\Delta\}_I^T = \{(\Delta_1)_2(\Delta_2)_2\ (\Delta_6)_2(\Delta_6)_3\} = \dfrac{1}{EA}\{204.91 \quad 692.42 \quad -4.29 \quad -49.79\}$ (l in ft);
$\{F\}_{12}^T = \{-21.55^k\ 93.34^k \quad -911.97'^{-k}\}$; $\{F\}_{23}^T = \{-10.25^k \quad -22.75^k\ 454.93'^{-k}\}$

18. $\{\Delta\}_I^T = \{(\Delta_3)_2(\Delta_4)_2\ (\Delta_5)_2\} = \dfrac{M}{EI}\{-4.429 \quad 3.103 \quad -0.213\}$ (l in m)

22. $\{\Delta\}_I^T = \{(\Delta_6)_1(\Delta_6)_2\ (\Delta_6)_3\} = \dfrac{1}{EI}\{1,500 \quad 1,000 \quad 500\}$ (l in ft); $\{F\}_{12}^T = \{32.5^k \quad 0'^{-k}\}$;
$\{F\}_{23}^T = \{+7.5^k \quad -150'^{-k}\}$

24. $\{\Delta\}_I^T = \{(\Delta_6)_2(\Delta_6)_3\} = \dfrac{1}{EI}\{-514.53 \quad 1,007.7\}$ (l in ft); $\{F\}_{12}^T = \{37.41^k \quad -186.8'^{-k}\}$; $\{F\}_{23}^T = \{-3.28^k \quad -1.47'^{-k}\}$

26. $\{\Delta\}_I^T = \{(\Delta_2)_3(\Delta_6)_3\ (\Delta_6)_4\} = \dfrac{1}{EI}\{19,632.7\ 239.1\ -2,093.6\}$ (l in ft); $\{F\}_{13}^T = \{50.86^k \quad -395.6'^{-k}\}$; $\{F\}_{23}^T = \{+9.80^k \quad 131.7'^{-k}\}$; $\{F\}_{34}^T = \{8.84^k \quad 234.6'^{-k}\}$

28. $\{\Delta\}_I^T = \{(\Delta_1)_1(\Delta_2)_1\ (\Delta_1)_3(\Delta_1)_4\} = \{0.053\quad 0.041\quad 0.011\quad 0.022\}$ ft; $\{F\}_{12} = -116^k$; $\{F\}_{13} = 174^k$; $\{F\}_{23} = 81.2^k$; $(P_2)_3 = 170.5^k$

30. $\{\Delta\}_I^T = \{(\Delta_6)_1(\Delta_6)_2\} = \{-0.002\ 64\quad 0.002\ 27\}$ rad; $\{F\}_{12}^T = EI\{-0.000\ 1\quad 0\}$; $\{F\}_{23}^T = EI\{0.000\ 27\quad 0.000\ 98\}$ (l in m)

32. $\{\Delta\}_I^T = \{(\Delta_1)_1(\Delta_2)_1\ (\Delta_1)_3(\Delta_2)_3(\Delta_1)_4\} = -10^{-5}\{69.2\quad 403.2\quad 69.2\quad 334.0\quad 139.5\}$ m; $\{F\}_{12} = -32.7$ kN; $\{F\}_{13} = 46.1$ kN; $\{F\}_{14} = -33.0$ kN; $\{F\}_{23} = 52.2$ kN; $\{F\}_{24} = -23.3$ kN; $\{F\}_{34} = 51.6$ kN

34. Case a: $\{\Delta\}_I = (\Delta_6)_2 = 0.000,585$ rad; $\{F\}_{12}^T = \{1.22^k\quad -60.9'^{-k}\}$; $\{F\}_{23}^T = \{1.22^k\quad -24.3'^{-k}\}$; Case b: $\{\Delta\}_I = (\Delta_6)_2 = 0$; $\{F\}_{12}^T = \{F\}_{23}^T = \{0\quad -48.8'^{-k}\}$

36. $\{\Delta\}_I^T = \{(\Delta_1)_1(\Delta_2)_1\ (\Delta_1)_3(\Delta_2)_3(\Delta_1)_4\} = -10^5\{533.4\quad 677.3\quad 533.4\quad 778.9\quad 1\ 064.7\}$ m; $\{F\}_{12} = -48.0$ kN; $\{F\}_{13} = 67.7$ kN; $\{F\}_{14} = -48.7$ kN; $\{F\}_{23} = 76.8$ kN; $\{F\}_{24} = -34.2$ kN; $\{F\}_{34} = 75.7$ kN

38. $\{\Delta\}_I^T = \{(\Delta_6)_1(\Delta_6)_2\} = \{0.000\ 137\quad -0.000\ 237\}$ rad; $\{F\}_{12}^T = \{-8.76$ kN 0 kN $\cdot$ m$\}$; $\{F\}_{23}^T = \{13.08$ kN -87.36 kN $\cdot$ m$\}$

CHAPTER 18

2. Same answers as for Problem 4 in Section 17.13.

4. $(F_1)_{12} = -22.12^k$; $(F_1)_{15} = 3.53^k$; $(F_1)_{25} = -11.45^k$; $(F_1)_{36} = -38.62^k$; $(F_1)_{56} = 22.28^k$

6. Same answers as for Problem 12 in Section 17.13.

8. Same answers as for Problem 14 in Section 17.13.

10. Same answers as for Problem 22 in Section 17.13.

12. Same answers as for Problem 24 in Section 17.13.

14. $(\Delta_1)_2 = 0.02$ in.; $(\Delta_2)_2 = -0.02$ in.; $\{F\}_{12} = -64.6^k$; $\{F\}_{32} = 0$; $\{F\}_{42} = -64.4^k$

16. $(\Delta_1)_2 = -0.003,2$ in.; $(\Delta_2)_2 = -0.004,3$ in; $\{F\}_{12} = -33.50^k$; $\{F\}_{32} = 0^k$; $\{F\}_{42} = -33.53^k$

18. Same answers as for Problem 36 in Section 17.13.

20.
$$[b]_I = \begin{bmatrix} 1.000 & 1.000 \\ 0 & -1.414 \\ 0 & 0 \\ 0 & 0 \end{bmatrix} \qquad [b]_{II} = \begin{bmatrix} 1.000 & 1.750 \\ 0 & -1.061 \\ 1.000 & 0 \\ 0 & 1.250 \end{bmatrix}$$